Generalized Inverses of Linear Transformations

by

S. L. Campbell
North Carolina State University

and

C. D. Meyer, Jr.
North Carolina State University

Dover Publications, Inc., *New York*

Published in Canada by General Publishing Company, Ltd., 30 Lesmill Road, Don Mills, Toronto, Ontario.

Published in the United Kingdom by Constable and Company, Ltd., 3 The Lanchesters, 162–164 Fulham Palace Road, London W6 9ER.

This Dover edition, first published in 1991, is an unabridged, corrected republication of the work originally published by Pitman Publishing Limited, London, 1979, in the series "Surveys and Reference Works in Mathematics."

Manufactured in the United States of America
Dover Publications, Inc., 31 East 2nd Street, Mineola, N.Y. 11501

Library of Congress Cataloging-in-Publication Data

Campbell, S. L. (Stephen La Vern)
Generalized inverses of linear transformations / by S. L. Campbell and C. D. Meyer.
p. cm.
Reprint. Originally published: London ; San Francisco : Pitman, 1979. Originally published in series: Surveys and reference works in mathematics.
Includes bibliographical references and index.
ISBN 0-486-66693-X (pbk.)
1. Matrix inversion. 2. Transformations (Mathematics) I. Meyer, C. D. (Carl Dean) II. Title.
[QA188.C36 1991]
512.9′434—dc20 90-23852
CIP

To Gail & Becky

Preface

During the last two decades, the study of generalized inversion of linear transformations and related applications has grown to become an important topic of interest to researchers engaged in the study of linear mathematical problems as well as to practitioners concerned with applications of linear mathematics. The purpose of this book is twofold. First, we try to present a unified treatment of the general theory of generalized inversion which includes topics ranging from the most traditional to the most contemporary. Secondly, we emphasize the utility of the concept of generalized inversion by presenting many diverse applications in which generalized inversion plays an integral role. This book is designed to be useful to the researcher and the practitioner, as well as the student. Much of the material is written under the assumption that the reader is unfamiliar with the basic aspects of the theory and applications of generalized inverses. As such, the text is accessible to anyone possessing a knowledge of elementary linear algebra.

This text is not meant to be encyclopedic. We have not tried to touch on all aspects of generalized inversion—nor did we try to include every known application. Due to considerations of length, we have been forced to restrict the theory to finite dimensional spaces and neglect several important topics and interesting applications.

In the development of every area of mathematics there comes a time when there is a commonly accepted body of results, and referencing is limited primarily to more recent and less widely known results. We feel that the theory of generalized inverses has reached that point. Accordingly, we have departed from previous books and not referenced many of the more standard facts about generalized inversion. To the many individuals who have made an original contribution to the theory of generalized inverses we are deeply indebted.

We are especially indebted to Adi Ben-Israel, Thomas Greville, C. R. Rao, and S. K. Mitra whose texts undoubtedly have had an influence on the writing of this book.

In view of the complete (annotated) bibliographies available in other texts, we made no attempt at a complete list of references.

Special thanks are extended to Franklin A. Graybill and Richard J. Painter who introduced author Meyer to the subject of generalized inverses and who provided wisdom and guidance at a time when they were most needed.

S. L. Campbell
C. D. Meyer, Jr
North Carolina State
University at Raleigh

Contents

0
Introduction and other preliminaries

1. Prerequisites and philosophy

The study of generalized inverses has flourished since its rebirth in the early 1950s. Numerous papers have developed both its theory and its applications. The subject has advanced to the point where a unified treatment is possible. It would be desirable to have a book that treated the subject from the viewpoint of linear algebra, and not with regard to a particular application. We do not feel that the world needs another introduction to linear algebra. Accordingly, this book presupposes some familiarity with the basic facts and techniques of linear algebra as found in most introductory courses.

It is our hope that this book would be suitable for self-study by either students or workers in other fields. Needed ideas that a person might well have forgotten or never learned, such as the singular value decomposition, will be stated formally. Unless their proof is illuminating or illustrates an important technique, it will be relegated to the exercises or a reference.

There are three basic kinds of chapters in this book. Chapters 0, 1, 2, 3, 4, 6, 7, 10, and 12 discuss the theory of the generalized inverse and related notions. They are a basic introduction to the mathematical theory. Chapters 5, 8, 9 and 11 discuss applications. These chapters are intended to illustrate the uses of generalized inverses, not necessarily to teach how to use them.

Our goal has been to write a readable, introductory book which will whet the appetite of the reader to learn more. We have tried to bring the reader far enough so that he can proceed into the literature, and yet not bury him under a morass of technical lemmas and concise, abbreviated proofs. This book reflects our rather definite opinions on what an introductory book is and what it should include. In particular, we feel that the numerous applications are necessary for a full appreciation of the theory.

Like most types of mathematics, the introduction of the various generalized inverses is not necessary. One could do mathematics without ever defining a ring or a continuous function. However, the introduction of generalized inverses, as with rings and continuous functions, enables us to more clearly see the underlying structure, to more easily manipulate it, and to more easily express new results.

No attempt has been made to have the bibliography comprehensive. The bibliography of [64] is carefully annotated and contains some 1775 entries. In order to keep this book's size down we have omitted any discussion of the infinite dimensional case. The interested reader is referred to [64].

2. Notation and basic geometry

Theorems, facts, propositions, lemmas, corollaries and examples are numbered consecutively within each section. A reference to Example 3.2.3. refers to the third example in Section 2 of Chapter 3. If the reader were already in Chapter 3, the reference would be just to Example 2.3. Within Section 2, the reference would be to Example 3.

Exercise sections are scattered throughout the chapters. Exercises with an (*) beside them are intended for more advanced readers. In some cases the (*)-exercises require knowledge from an outside area like complex function theory. At other times they involve fairly complicated proofs using basic ideas.

$\mathbb{C}(\mathbb{R})$ is the field of complex (real) numbers. $\mathbb{C}^{m\times n}(\mathbb{R}^{m\times n})$ is the vector-space of $m \times n$ complex (real) matrices over $\mathbb{C}(\mathbb{R})$. $\mathbb{C}^n(\mathbb{R}^n)$ is the vector-space of n-tuples of complex (real) numbers over $\mathbb{C}(\mathbb{R})$. We will frequently not distinguish between $\mathbb{C}^n(\mathbb{R}^n)$ and $\mathbb{C}^{n\times 1}(\mathbb{R}^{n\times 1})$. That is, n-tuples will be written as column vectors. Except where we specifically state otherwise, it is to be assumed that we are working over the complex field.

If $\mathbf{A}\in\mathbb{C}^{m\times n}$, then with respect to the standard basis of $\mathbb{C}^n$ and $\mathbb{C}^m$, $\mathbf{A}$ induces a linear transformation $\underset{\sim}{\mathbf{A}}:\mathbb{C}^n\to\mathbb{C}^m$ by $\underset{\sim}{\mathbf{A}}u=\mathbf{A}\mathbf{u}$ for every $u\in\mathbb{C}^n$. Whenever we go from one of $\mathbf{A}$ or $\underset{\sim}{\mathbf{A}}$ to the other, it is to be understood that it is with respect to the standard basis.

The capital letters, $\mathbf{A}, \mathbf{B}, \mathbf{C}, \mathbf{X}, \mathbf{Y}, \mathbf{Z}$ are reserved for matrices or their corresponding linear transformations. Subspaces are denoted by the capital letters M, N. Subspaces are always linear subspaces. The letters $\mathbf{U}, \mathbf{V}, \mathbf{W}$ are reserved for unitary matrices or partial isometries. $\mathbf{I}$ always denotes the identity matrix. If $\mathbf{I}\in\mathbb{C}^{n\times n}$ we sometimes write $\mathbf{I}_n$. Vectors are denoted by $\mathbf{b}, \mathbf{u}, \mathbf{v}, \mathbf{y}$, etc., scalars by a, b, λ, k, etc. $R(\mathbf{A})$ denotes the range of $\mathbf{A}$, that is, the linear span of the columns of $\mathbf{A}$. The range of $\underset{\sim}{\mathbf{A}}$ is denoted $R(\underset{\sim}{\mathbf{A}})$. Since $\mathbf{A}$ is derived from $\underset{\sim}{\mathbf{A}}$ by way of the standard basis, we have $R(\mathbf{A}) = R(\underset{\sim}{\mathbf{A}})$. The null space of $\mathbf{A}$, $N(\mathbf{A})$, is $\{\mathbf{x}\in\mathbb{C}^n : \mathbf{A}\mathbf{x}=\mathbf{0}\}$.

A matrix $\mathbf{A}$ is *hermitian* if its conjugate transpose, $\mathbf{A}^*$, equals $\mathbf{A}$. If $\mathbf{A}^2=\mathbf{A}$, then $\mathbf{A}$ is called a projector of $\mathbb{C}^n$ onto $R(\mathbf{A})$. Recall that rank $(\mathbf{A}) = \mathrm{Tr}(\mathbf{A})$ if $\mathbf{A}^2=\mathbf{A}$. If $\mathbf{A}^2=\mathbf{A}$ and $\mathbf{A}=\mathbf{A}^*$, then $\mathbf{A}$ is called an orthogonal projector. If $\mathbf{A},\mathbf{B}\in\mathbb{C}^{n\times n}$, then $[\mathbf{A},\mathbf{B}]=\mathbf{AB}-\mathbf{BA}$.

The inner product between two vectors $\mathbf{u},\mathbf{v}\in\mathbb{C}^n$ is denoted by $(\mathbf{u},\mathbf{v})$. If $\mathscr{S}$

is a subset of $\mathbb{C}^n$, then $\mathscr{S}^\perp = \{\mathbf{u}\in\mathbb{C}^n : (u, v) = 0 \text{ for every } \mathbf{v}\in\mathscr{S}\}$. The smallest subspace of $\mathbb{C}^n$ containing $\mathscr{S}$ is denoted $LS(\mathscr{S})$. Notice that $(\mathscr{S}^\perp)^\perp = LS(\mathscr{S})$.

Suppose now that M, N_1, and N_2 are subspaces of $\mathbb{C}^n$. Then $N_1 + N_2 = \{\mathbf{u}+\mathbf{v} : \mathbf{u}\in N_1 \text{ and } \mathbf{v}\in N_2\}$, while $\lambda N_1 = \{\lambda\mathbf{u} : \mathbf{u}\in N_1\}$. If $M = N_1 + N_2$ and $N_1 \cap N_2 = \{\mathbf{0}\}$, then M is called the direct sum of N_1 and N_2. In this case we write $M = N_1 \dot{+} N_2$. If $M = N_1 \dot{+} N_2$ and $N_1 \perp N_2$, that is $(\mathbf{u}, \mathbf{v}) = 0$ for every $\mathbf{u}\in N_1$ and $\mathbf{v}\in N_2$, then M is called the orthogonal sum of N_1 and N_2. This will be written $M = N_1 \oplus N_2$. If two vectors are orthogonal, their sum will frequently be written with a $\oplus$ also. If $\mathbb{C}^n = N_1 \dot{+} N_2$, then N_1 and N_2 are called *complementary subspaces.* Notice that $\mathbb{C}^n = N_1 \oplus N_1^\perp$. The dimension of a subspace M is denoted dim M.

One of the most basic facts used in this book is the next proposition.

Proposition 0.2.1 *Suppose that* $\mathbf{A}\in\mathbb{C}^{m\times n}$. *Then* $R(\mathbf{A}) = N(\mathbf{A}^*)^\perp$.

Proof We will show that $R(\mathbf{A}) \subseteq N(\mathbf{A}^*)^\perp$ and $\dim R(\mathbf{A}) = \dim N(\mathbf{A}^*)^\perp$. Suppose that $\mathbf{u}\in R(\mathbf{A})$. Then there is a $\mathbf{v}\in\mathbb{C}^n$ such that $\mathbf{A}\mathbf{v} = \mathbf{u}$. If $\mathbf{w}\in N(\mathbf{A}^*)$, then $(\mathbf{u}, \mathbf{w}) = (\mathbf{A}\mathbf{v}, \mathbf{w}) = (\mathbf{v}, \mathbf{A}^*\mathbf{w}) = (\mathbf{v}, \mathbf{0}) = 0$. Thus $\mathbf{u}\in N(\mathbf{A}^*)^\perp$ and $R(\mathbf{A}) \subseteq N(\mathbf{A}^*)^\perp$. But $\dim R(\mathbf{A}) = \operatorname{rank} \mathbf{A} = \operatorname{rank} \mathbf{A}^* = m - \dim N(\mathbf{A}^*) = \dim N(\mathbf{A}^*)^\perp$. Thus $R(\mathbf{A}) = N(\mathbf{A}^*)^\perp$. ■

A useful consequence of Proposition 1 is the 'star cancellation law'.

Proposition 0.2.2 *(Star cancellation law) Suppose that* $\mathbf{A}\in\mathbb{C}^{m\times n}$ *and* $\mathbf{B}, \mathbf{C}\in\mathbb{C}^{n\times p}$. *Then* (i) $\mathbf{A}^*\mathbf{A}\mathbf{B} = \mathbf{A}^*\mathbf{A}\mathbf{C}$ *if and only if* $\mathbf{A}\mathbf{B} = \mathbf{A}\mathbf{C}$. *Also* (ii) $N(\mathbf{A}^*\mathbf{A}) = N(\mathbf{A})$, (iii) $R(\mathbf{A}^*\mathbf{A}) = R(\mathbf{A}^*)$.

Proof. (i) may be rewritten as $\mathbf{A}^*\mathbf{A}(\mathbf{B} - \mathbf{C}) = \mathbf{0}$ if and only if $\mathbf{A}(\mathbf{B} - \mathbf{C}) = \mathbf{0}$. Clearly (i) and (ii) are equivalent. To see that (ii) holds notice that by Proposition 1, $N(\mathbf{A}^*) = R(\mathbf{A})^\perp$ and thus $\mathbf{A}^*\mathbf{A}\mathbf{x} = \mathbf{0}$ if and only if $\mathbf{A}\mathbf{x} = \mathbf{0}$. To see (iii), note that using (ii) we have that $R(\mathbf{A}^*\mathbf{A}) = N(\mathbf{A}^*\mathbf{A})^\perp = N(\mathbf{A})^\perp = R(\mathbf{A}^*)$. ■

Propositions 1 and 2 are basic and will be used frequently in what follows without comment.

If M is a subspace of $\mathbb{C}^n$, then we may define the *orthogonal projector*, $\mathbf{P}_M$, *of* $\mathbb{C}^n$ onto M by $\mathbf{P}_M\mathbf{u} = \mathbf{u}$ if $\mathbf{u}\in M$ and $\mathbf{P}_M\mathbf{u} = \mathbf{0}$ if $\mathbf{u}\in M^\perp$. Notice that $\mathbf{I} - \mathbf{P}_M = \mathbf{P}_M^\perp$ and that for any orthogonal projector, $\mathbf{P}$, we have $\mathbf{P} = \mathbf{P}_{R(\mathbf{P})}$.

It is frequently helpful to write a matrix $\mathbf{A}$ in block form, that is, express $\mathbf{A}$ as a matrix made up of matrices, $\mathbf{A} = [\mathbf{A}_{ij}]$. If $\mathbf{B}$ is a second block matrix, then $(\mathbf{A}\mathbf{B})_{ij} = \sum_{r=1}^{n} \mathbf{A}_{ir}\mathbf{B}_{rj}$ and $(\mathbf{A} + \mathbf{B})_{ij} = \mathbf{A}_{ij} + \mathbf{B}_{ij}$ provided the submatrices $\mathbf{A}_{ij}, \mathbf{B}_{ij}$ are of the correct size to permit the indicated multiplications and additions.

Example 0.2.1 Let $\mathbf{A} = \begin{bmatrix} 1 & 2 & 3 & 4 \\ 5 & i & 2 & 0 \\ \pi & 1 & 1 & 1 \end{bmatrix}$ and $\mathbf{B} = \begin{bmatrix} 1 & 0 \\ 0 & 0 \\ 0 & 1 \\ i & 0 \end{bmatrix}$. Then $\mathbf{A}$

can be written as $\mathbf{A} = \begin{bmatrix} \mathbf{A}_{11} & \mathbf{A}_{12} & \mathbf{A}_{13} \\ \mathbf{A}_{21} & \mathbf{A}_{22} & \mathbf{A}_{23} \end{bmatrix}$ where $\mathbf{A}_{11} = [1]$, $\mathbf{A}_{12} = [2, 3]$, $\mathbf{A}_{13} = [4]$, $\mathbf{A}_{21} = \begin{bmatrix} 5 \\ \pi \end{bmatrix}$, $\mathbf{A}_{22} = \begin{bmatrix} i & 2 \\ 1 & 1 \end{bmatrix}$ and $\mathbf{A}_{23} = \begin{bmatrix} 0 \\ 1 \end{bmatrix}$. $\mathbf{B}$ can be written as $\begin{bmatrix} \mathbf{B}_1 \\ \mathbf{B}_2 \\ \mathbf{B}_3 \end{bmatrix}$ where $\mathbf{B}_1 = [1 \quad 0]$, $\mathbf{B}_2 = \begin{bmatrix} 0 & 0 \\ 0 & 1 \end{bmatrix}$, and $\mathbf{B}_3 = [i, 0]$.

Thus $\mathbf{AB} = \begin{bmatrix} \mathbf{A}_{11} & \mathbf{A}_{12} & \mathbf{A}_{13} \\ \mathbf{A}_{21} & \mathbf{A}_{22} & \mathbf{A}_{23} \end{bmatrix} \begin{bmatrix} \mathbf{B}_1 \\ \mathbf{B}_2 \\ \mathbf{B}_3 \end{bmatrix} = \begin{bmatrix} \mathbf{A}_{11}\mathbf{B}_1 + \mathbf{A}_{12}\mathbf{B}_2 + \mathbf{A}_{13}\mathbf{B}_3 \\ \mathbf{A}_{21}\mathbf{B}_1 + \mathbf{A}_{22}\mathbf{B}_2 + \mathbf{A}_{22}\mathbf{B}_3 \end{bmatrix}$.

If all the submatrices are from $\mathbb{C}^{r \times r}$ for a fixed r, then block matrices may be viewed as matrices over a ring. We shall not do so.

This notation is especially useful when dealing with large matrices or matrices which have submatrices of a special type.

There is a close connection between block matrices, invariant subspaces, and projections which we now wish to develop.

Definition 0.2.1 *If M is a subspace of $\mathbb{C}^n$ and $\mathbf{A} \in \mathbb{C}^{n \times n}$, then M is called an* invariant subspace *of $\mathbf{A}$ if and only if $\mathbf{A}M = \{\mathbf{Au} : \mathbf{u} \in M\} \subseteq M$. If M is an invariant subspace for both $\mathbf{A}$ and $\mathbf{A}^*$, then M is called a* reducing subspace *of $\mathbf{A}$.*

Invariant and reducing subspaces have a good characterization in terms of projectors.

Proposition 0.2.3 *Suppose $\mathbf{A} \in \mathbb{C}^{n \times n}$ and M is a subspace of $\mathbb{C}^n$. Let $\mathbf{P}_M$ be the orthogonal projector onto M. Then* (i) *M is an invariant subspace for $\mathbf{A}$ if and only if $\mathbf{P}_M\mathbf{A}\mathbf{P}_M = \mathbf{A}\mathbf{P}_M$.* (ii) *$M$ is a reducing subspace for $\mathbf{A}$ if and only if $\mathbf{A}\mathbf{P}_M = \mathbf{P}_M\mathbf{A}$.*

The proof of Proposition 3 is left to the next set of exercises. Knowledge of invariant subspaces is useful for it enables the introduction of blocks of zeros into a matrix. Invariant subspaces also have obvious geometric importance in studying the effects of $\underset{\sim}{\mathbf{A}}$ on $\mathbb{C}^n$.

Proposition 0.2.4 *Suppose that $\mathbf{A} \in \mathbb{C}^{n \times n}$ and that M is an invariant subspace of $\mathbf{A}$ of dimension r. Then there exists a unitary matrix $\mathbf{U}$ such that*

(i) $\mathbf{A} = \mathbf{U}^* \begin{bmatrix} \mathbf{A}_{11} & \mathbf{A}_{12} \\ \mathbf{0} & \mathbf{A}_{22} \end{bmatrix} \mathbf{U}$ *where* $\mathbf{A}_{11} \in \mathbb{C}^{r \times r}$.

If M is a reducing subspace of $\mathbf{A}$, then (ii) $\mathbf{A} = \mathbf{U}^* \begin{bmatrix} \mathbf{A}_{11} & \mathbf{0} \\ \mathbf{0} & \mathbf{A}_{22} \end{bmatrix} \mathbf{U}$ *where* $\mathbf{A}_{11} \in \mathbb{C}^{r \times r}$, $\mathbf{A}_{22} \in \mathbb{C}^{(n-r) \times (n-r)}$.

Proof Let M be a subspace of $\mathbb{C}^n$. Then $\mathbb{C}^n = M \oplus M^{\perp}$. Let β_1 be an orthonormal basis for M and β_2 be an orthonormal basis for $M^{\perp}$. Then

$\beta = \beta_1 \cup \beta_2$ is an orthonormal basis for $\mathbb{C}^n$. Order the vectors in β so that those in β_1 are listed first.

Let $\mathbf{U}$ be a unitary transformation that maps the standard basis for $\mathbb{C}^n$ onto the ordered basis β. Then

$$\mathbf{U}^*\mathbf{P}_M\mathbf{U} = \begin{bmatrix} \mathbf{I}_r & \mathbf{0} \\ \mathbf{0} & \mathbf{0} \end{bmatrix}, \mathbf{U}^*\mathbf{A}\mathbf{U} = \begin{bmatrix} \mathbf{A}_{11} & \mathbf{A}_{12} \\ \mathbf{A}_{21} & \mathbf{A}_{22} \end{bmatrix} \tag{1}$$

where $\mathbf{A}_{11} \in \mathbb{C}^{r \times r}$, $r = \dim M$. Suppose now that M is an invariant subspace for $\mathbf{A}$. Thus $\mathbf{P}_M\mathbf{A}\mathbf{P}_M = \mathbf{A}\mathbf{P}_M$ by Proposition 3. This is equivalent to $\mathbf{U}^*\mathbf{P}_M\mathbf{U}\mathbf{U}^*\mathbf{A}\mathbf{U}\mathbf{U}^*\mathbf{P}_M\mathbf{U} = \mathbf{U}^*\mathbf{A}\mathbf{U}\mathbf{U}^*\mathbf{P}_M\mathbf{U}$. Substituting (1) into this gives

$$\begin{bmatrix} \mathbf{A}_{11} & \mathbf{0} \\ \mathbf{0} & \mathbf{0} \end{bmatrix} = \begin{bmatrix} \mathbf{A}_{11} & \mathbf{0} \\ \mathbf{A}_{21} & \mathbf{0} \end{bmatrix}.$$

Thus $\mathbf{A}_{21} = \mathbf{0}$ and part (i) of Proposition 4 follows. Part (ii) follows by substituting (1) into $\mathbf{U}^*\mathbf{P}_M\mathbf{U}\mathbf{U}^*\mathbf{A}\mathbf{U} = \mathbf{U}^*\mathbf{A}\mathbf{U}\mathbf{U}^*\mathbf{P}_M\mathbf{U}$. ■

If $\mathbf{A} \in \mathbb{C}^{n \times n}$, then $R(\mathbf{A})$ is always an invariant subspace for $\mathbf{A}$. If $\mathbf{A}$ is hermitian, then every invariant subspace is reducing. In particular, $R(\mathbf{A})$ is reducing.

Proposition 0.2.5 *If* $\mathbf{A} = \mathbf{A}^*$, *then there exists a unitary matrix* $\mathbf{U}$ *and an invertible hermitian matrix* $\mathbf{A}_1$ *such that* $\mathbf{A} = \mathbf{U}^*\begin{bmatrix} \mathbf{A}_1 & \mathbf{0} \\ \mathbf{0} & \mathbf{0} \end{bmatrix}\mathbf{U}$.

Proposition 5 is, of course, a special case of the fact that every hermitian matrix is unitarily equivalent to a diagonal matrix. Viewed in this manner, it is clear that a similar result holds if hermitian is replaced by normal where a matrix is called normal if $\mathbf{A}^*\mathbf{A} = \mathbf{A}\mathbf{A}^*$. We assume that the reader is already familiar with the fact that normal and hermitian matrices are unitarily equivalent to diagonal matrices. Our purpose here is to review some of the 'geometry' of invariant subspaces and to gain a facility with the manipulation of block matrices.

Reducing subspaces are better to work with than invariant subspaces, but reducing subspaces need not always exist. $\mathbf{A} = \begin{bmatrix} 0 & 1 \\ 0 & 0 \end{bmatrix}$ has no reducing subspaces. Every matrix in $\mathbb{C}^{n \times n}$, $n > 1$, does have invariant subspaces since it has eigenvectors. And if $\mathscr{S}$ is a set of eigenvectors corresponding to a particular eigenvalue of $\mathbf{A}$, then $LS(\mathscr{S})$ is an invariant subspace for $\mathbf{A}$.

We shall see later that unitary and hermitian matrices are often easier to work with. Thus it is helpful if a matrix can be written as a product of such factors.

If $\mathbf{A} \in \mathbb{C}^{n \times n}$, there is such a decomposition. It is called the polar form. The name comes from the similarity between it and the polar form of a complex number $z = re^{i\theta}$ where $r \in \mathbb{R}$ and $|e^{i\theta}| = 1$.

Theorem 0.2.1 *If* $\mathbf{A} \in \mathbb{C}^{n \times n}$, *then there exists a unitary matrix* $\mathbf{U}$ *and hermitian matrices* $\mathbf{B}, \mathbf{C}$ *such that* $\mathbf{A} = \mathbf{U}\mathbf{B} = \mathbf{C}\mathbf{U}$.

If $\mathbf{A} \in \mathbb{C}^{m \times n}$ where $m \neq n$, then one cannot hope to get quite as good an expression for $\mathbf{A}$ since $\mathbf{A}$ is not square. There are two ways that Theorem 1 can be extended. One is to replace $\mathbf{U}$ by what is called a partial isometry. This will be discussed in a later section. The other possibility is to replace $\mathbf{B}$ by a matrix like a hermitian matrix. By 'like a hermitian matrix' we are thinking of the block form given in Proposition 5.

Theorem 0.2.2 *(Singular Value Decomposition) Suppose that* $\mathbf{A} \in \mathbb{C}^{m \times n}$. *Then there exist unitary matrices* $\mathbf{U} \in \mathbb{C}^{m \times m}$ *and* $\mathbf{V} \in \mathbb{C}^{n \times n}$, *and an invertible hermitian diagonal matrix* $\mathbf{D} = \mathrm{Diag}\{\sigma_1, \ldots, \sigma_r\}$, *whose diagonal entries are the positive square roots of the eigenvalues of* $(\mathbf{A}^*\mathbf{A})$ *repeated according to multiplicity, such that* $\mathbf{A} = \mathbf{U}\begin{bmatrix} \mathbf{D} & \mathbf{0} \\ \mathbf{0} & \mathbf{0} \end{bmatrix}\mathbf{V}$.

The proofs of Theorems 1 and 2 are somewhat easier if done with the notation of generalized inverses. The proofs will be developed in the exercises following the section on partial isometries.

Two comments are in order. First, the matrix $\begin{bmatrix} \mathbf{D} & \mathbf{0} \\ \mathbf{0} & \mathbf{0} \end{bmatrix}$ is not a square matrix, although $\mathbf{D}$ is square. Secondly, the name 'Singular Value Decomposition' comes from the numbers $\{\sigma_1, \ldots, \sigma_r\}$ which are frequently referred to as the singular values of $\mathbf{A}$.

The notation of the functional calculus is convenient and will be used from time to time. If $\mathbf{C} = \mathbf{U}\,\mathrm{Diag}\{\lambda_1, \ldots, \lambda_n\}\mathbf{U}^*$ for some unitary $\mathbf{U}$ and f is a function defined on $\{\lambda_1, \ldots, \lambda_n\}$, then we define $f(\mathbf{C}) = \mathbf{U}\,\mathrm{Diag}\{f(\lambda_1), \ldots, f(\lambda_n)\}\mathbf{U}^*$. This definition only makes sense if $\mathbf{C}$ is normal. If $p(\lambda) = a_n\lambda^n + \ldots + a_0$, then $p(\mathbf{C})$ as we have defined it here agrees with the standard definition of $p(\mathbf{C}) = a_n\mathbf{C}^n + \ldots + a_0\mathbf{I}$.

For any $\mathbf{A} \in \mathbb{C}^{n \times n}$, $\sigma(\mathbf{A})$ denotes the set of eigenvalues of $\mathbf{A}$.

3. Exercises

1. Prove that if $\mathbf{P}$ is hermitian, then $\mathbf{P}$ is a projector if and only if $\mathbf{P}^3 = \mathbf{P}^2$.
2. Prove that if $M_1 \subseteq M_2 \subseteq \mathbb{C}^n$ are invariant subspaces for $\mathbf{A}$, then $\mathbf{A}$ is unitarily equivalent to a matrix of the form $\begin{bmatrix} \mathbf{X} & \mathbf{X} & \mathbf{X} \\ \mathbf{0} & \mathbf{X} & \mathbf{X} \\ \mathbf{0} & \mathbf{0} & \mathbf{X} \end{bmatrix}$ where $\mathbf{X}$ denotes a non-zero block. Generalize this to t invariant subspaces such that $M_1 \subseteq M_2 \subseteq \ldots \subseteq M_t$.
3. If $M_1 \subseteq M_2 \subseteq \mathbb{C}^n$ are reducing subspaces for $\mathbf{A}$ show that $\mathbf{A}$ is unitarily equivalent to a matrix in the form $\begin{bmatrix} \mathbf{X} & \mathbf{0} & \mathbf{0} \\ \mathbf{0} & \mathbf{X} & \mathbf{0} \\ \mathbf{0} & \mathbf{0} & \mathbf{X} \end{bmatrix}$. Generalize this to t reducing subspaces such that $M_1 \subseteq \ldots \subseteq M_t$.
4. Prove Proposition 2.3.

5. Prove that if $\mathbf{A} \in \mathbb{C}^{n \times n}$ and $M \subseteq \mathbb{C}^n$ is an invariant subspace for $\mathbf{A}$, then $M^{\perp}$ is an invariant subspace for $\mathbf{A}^*$.
6. Suppose $\mathbf{A} = \mathbf{A}^*$. Give necessary and sufficient conditions on $\mathbf{A}$ to guarantee that for every pair of reducing subspaces M_1, M_2 of $\mathbf{A}$ that either $M_1 \perp M_2$ or $M_1 \cap M_2 \neq \{\mathbf{0}\}$.

1
The Moore–Penrose or generalized inverse

1. Basic definitions

Equations of the form

$$\mathbf{Ax} = \mathbf{b}, \mathbf{A} \in \mathbb{C}^{m \times n}, \mathbf{x} \in \mathbb{C}^n, \mathbf{b} \in \mathbb{C}^m \tag{1}$$

occur in many pure and applied problems. If $\mathbf{A} \in \mathbb{C}^{n \times n}$ and is invertible, then the system of equations (1) is, in principle, easy to solve. The unique solution is $\mathbf{x} = \mathbf{A}^{-1}\mathbf{b}$. If $\mathbf{A}$ is an arbitrary matrix in $\mathbb{C}^{m \times n}$, then it becomes more difficult to solve (1). There may be none, one, or an infinite number of solutions depending on whether $\mathbf{b} \in R(\mathbf{A})$ and whether n-rank $(\mathbf{A}) > 0$.

One would like to be able to find a matrix (or matrices) $\mathbf{C}$, such that solutions of (1) are of the form $\mathbf{Cb}$. But if $\mathbf{b} \notin R(\mathbf{A})$, then (1) has no solution. This will eventually require us to modify our concept of what a solution of (1) is. However, as the applications will illustrate, this is not as unnatural as it sounds. But for now we retain the standard definition of solution.

To motivate our first definition of the generalized inverse, consider the functional equation

$$y = f(x), x \in \mathcal{S} \subseteq \mathbb{R}, \tag{2}$$

where f is a real-valued function with domain $\mathcal{S}$. One procedure for solving (2) is to restrict the domain of f to a smaller set $\mathcal{S}'$ so that $\tilde{f} = f|_{\mathcal{S}'}$ is one to one. Then an inverse function $\tilde{f}^{-1}$ from $R(\tilde{f})$ to $\mathcal{S}'$ is defined by $\tilde{f}^{-1}(y) = x$ if $x \in \mathcal{S}'$ and $f(x) = y$. Thus $\tilde{f}^{-1}(y)$ is a solution of (2) for $y \in R(\tilde{f})$. This is how the arcsec, arcsin, and other inverse functions are normally defined.

The same procedure can be used in trying to solve equation (1). As usual, we let $\underset{\sim}{\mathbf{A}}$ be the linear function from $\mathbb{C}^n$ into $\mathbb{C}^m$ defined by $\underset{\sim}{\mathbf{A}}\mathbf{x} = \mathbf{Ax}$ for $\mathbf{x} \in \mathbb{C}^n$. To make $\underset{\sim}{\mathbf{A}}$ a one to one linear transformation it must be restricted to a subspace complementary to $N(\mathbf{A})$. An obvious one is $N(\mathbf{A})^{\perp} = R(\mathbf{A}^*)$. This suggests the following definition of the generalized inverse.

Definition 1.1.1 Functional definition of the generalized inverse *If*

$\mathbf{A}\in\mathbb{C}^{m\times n}$, *define the linear transformation* $\underset{\sim}{\mathbf{A}}^{\dagger}:\mathbb{C}^m\rightarrow\mathbb{C}^n$ *by* $\underset{\sim}{\mathbf{A}}^{\dagger}\mathbf{x}=\mathbf{0}$ *if* $\mathbf{x}\in R(\mathbf{A})^{\perp}$ *and* $\underset{\sim}{\mathbf{A}}^{\dagger}\mathbf{x}=(\underset{\sim}{\mathbf{A}}|_{R(\mathbf{A}^*)})^{-1}\mathbf{x}$ *if* $\mathbf{x}\in R(\mathbf{A})$. *The matrix of* $\underset{\sim}{\mathbf{A}}^{\dagger}$ *is denoted* $\mathbf{A}^{\dagger}$ *and is called the generalized inverse of* $\mathbf{A}$.

It is easy to check that $\mathbf{AA}^{\dagger}\mathbf{x}=\mathbf{0}$ if $\mathbf{x}\in R(\mathbf{A})^{\perp}$ and $\mathbf{AA}^{\dagger}\mathbf{x}=\mathbf{x}$ if $\mathbf{x}\in R(\mathbf{A})$. Similarly, $\mathbf{A}^{\dagger}\mathbf{Ax}=\mathbf{0}$ if $\mathbf{x}\in N(\mathbf{A})=R(\mathbf{A}^*)^{\perp}$ and $\mathbf{A}^{\dagger}\mathbf{Ax}=\mathbf{x}$ if $\mathbf{x}\in R(\mathbf{A}^*)=R(\mathbf{A}^{\dagger})$. Thus $\mathbf{AA}^{\dagger}$ is the orthogonal projector of $\mathbb{C}^m$ onto $R(\mathbf{A})$ while $\mathbf{A}^{\dagger}\mathbf{A}$ is the orthogonal projector of $\mathbb{C}^n$ onto $R(\mathbf{A}^*)=R(\mathbf{A}^{\dagger})$. This suggests a second definition of the generalized inverse due to E. H. Moore

Definition 1.1.2 Moore definition of the generalized inverse *If* $\mathbf{A}\in\mathbb{C}^{m\times n}$, *then the generalized inverse of* $\mathbf{A}$ *is defined to be the unique matrix* $\mathbf{A}^{\dagger}$ *such that*

(a) $\mathbf{AA}^{\dagger}=\mathbf{P}_{R(\mathbf{A})}$, *and*
(b) $\mathbf{A}^{\dagger}\mathbf{A}=\mathbf{P}_{R(\mathbf{A}^{\dagger})}$.

Moore's definition was given in 1935 and then more or less forgotten. This is possibly due to the fact that it was not expressed in the form of Definition 2 but rather in a more cumbersome (no pun intended) notation. An algebraic form of Moore's definition was given in 1955 by Penrose who was apparently unaware of Moore's work.

Definition 1.1.3 Penrose definition of the generalized inverse *If* $\mathbf{A}\in\mathbb{C}^{m\times n}$, *then* $\mathbf{A}^{\dagger}$ *is the unique matrix in* $\mathbb{C}^{n\times m}$ *such that*

(i) $\mathbf{AA}^{\dagger}\mathbf{A}=\mathbf{A}$,
(ii) $\mathbf{A}^{\dagger}\mathbf{AA}^{\dagger}=\mathbf{A}^{\dagger}$,
(iii) $(\mathbf{AA}^{\dagger})^*=\mathbf{AA}^{\dagger}$,
(iv) $(\mathbf{A}^{\dagger}\mathbf{A})^*=\mathbf{A}^{\dagger}\mathbf{A}$.

The first important fact to be established is the equivalence of the definitions.

Theorem 1.1.1 *The functional, Moore and Penrose definitions of the generalized inverse are equivalent.*

Proof We have already noted that if $\mathbf{A}^{\dagger}$ satisfies Definition 1, then it satisfies equations (a) and (b). If a matrix $\mathbf{A}^{\dagger}$ satisfies (a) and (b) then it immediately satisfies (iii) and (iv). Furthermore (i) follows from (a) by observing that $\mathbf{AA}^{\dagger}\mathbf{A}=\mathbf{P}_{R(\mathbf{A})}\mathbf{A}=\mathbf{A}$. (ii) will follow from (b) in a similar manner. Since Definition 1 was constructive and the $\mathbf{A}^{\dagger}$ it constructs satisfies (a), (b) and (i)–(iv), the question of existence in Definitions 2 and 3 is already taken care of. There are then two things remaining to be proven. One is that a solution of equations (i)–(iv) is a solution of (a) and (b). The second is that a solution of (a) and (b) or (i)–(iv) is unique.

Suppose then that $\mathbf{A}^{\dagger}$ is a matrix satisfying (i)–(iv). Multiplying (ii) on the left by $\mathbf{A}$ gives $(\mathbf{AA}^{\dagger})^2=(\mathbf{AA}^{\dagger})$. This and (iii) show that $\mathbf{AA}^{\dagger}$ is an orthogonal projector. We must show that it has range equal to the range of $\mathbf{A}$. Using (i) and the fact that $R(\mathbf{BC}) \subseteq R(\mathbf{B})$ for matrices $\mathbf{B}$ and $\mathbf{C}$, we get

$$R(\mathbf{A}) = R(\mathbf{AA}^{\dagger}\mathbf{A}) \subseteq R(\mathbf{AA}^{\dagger}) \subseteq R(\mathbf{A}), \text{ so that } R(\mathbf{A}) = R(\mathbf{AA}^{\dagger}).$$

Thus $\mathbf{AA}^{\dagger} = \mathbf{P}_{R(\mathbf{A})}$ as desired. The proof that $\mathbf{A}^{\dagger}\mathbf{A} = \mathbf{P}_{R(\mathbf{A}^{\dagger})}$ is similar and is left to the reader as an exercise. One way to show uniqueness is to show that if $\mathbf{A}^{\dagger}$ satisfies (a) and (b), or (i)–(iv), then it satisfies Definition 1. Suppose then that $\mathbf{A}^{\dagger}$ is a matrix satisfying (i)–(iv), (a), and (b). If $\mathbf{x} \in R(\mathbf{A})^{\perp}$, then by (a), $\mathbf{AA}^{\dagger}\mathbf{x} = \mathbf{0}$. Thus by (ii) $\mathbf{A}^{\dagger}\mathbf{x} = \mathbf{A}^{\dagger}\mathbf{AA}^{\dagger}\mathbf{x} = \mathbf{A}^{\dagger}\mathbf{0} = \mathbf{0}$. If $\mathbf{x} \in R(\mathbf{A})$, then there exist $\mathbf{y} \in R(\mathbf{A}^*)$ such that $\mathbf{Ay} = \mathbf{x}$. But $\mathbf{A}^{\dagger}\mathbf{x} = \mathbf{A}^{\dagger}\mathbf{Ay} = \mathbf{y}$. The last equality follows by observing that taking the adjoint of both sides of (i) gives $\mathbf{P}_{R(\mathbf{A}^{\dagger})}\mathbf{A}^* = \mathbf{A}^*$ so that $R(\mathbf{A}^*) \subseteq R(\mathbf{A}^{\dagger})$. But $\mathbf{y} = (\underset{\sim}{\mathbf{A}}|_{R(\mathbf{A}^*)})^{-1}\mathbf{x}$. Thus $\mathbf{A}^{\dagger}$ satisfies Definition 1. ■

As this proof illustrates, equations (i) and (ii) are, in effect, cancellation laws. While we cannot say that $\mathbf{AB} = \mathbf{AC}$ implies $\mathbf{B} = \mathbf{C}$, we can say that if $\mathbf{A}^{\dagger}\mathbf{AB} = \mathbf{A}^{\dagger}\mathbf{AC}$ then $\mathbf{AB} = \mathbf{AC}$. This type of cancellation will frequently appear in proofs and the exercises.

For obvious reasons, the generalized inverse is often referred to as the Moore–Penrose inverse. Note also that if $\mathbf{A} \in \mathbb{C}^{n \times n}$ and $\mathbf{A}$ is invertible, then $\mathbf{A}^{-1} = \mathbf{A}^{\dagger}$ so that the generalized inverse lives up to its name.

2. Basic properties of the generalized inverse

Before proceeding to establish some of what is true about generalized inverses, the reader should be warned about certain things that are not true.

While it is true that $R(\mathbf{A}^*) = R(\mathbf{A}^{\dagger})$, if $\mathbf{A}^{\dagger}$ is the generalized inverse, condition (b) in Definition 2 cannot be replaced by $\mathbf{A}^{\dagger}\mathbf{A} = \mathbf{P}_{R(\mathbf{A}^*)}$.

Example 1.2.1 Let $\mathbf{A} = \begin{bmatrix} 2 & 0 \\ 0 & 0 \end{bmatrix}$, $\mathbf{X} = \frac{1}{2}\begin{bmatrix} 1 & 0 \\ 0 & 1 \end{bmatrix}$. Since $\mathbf{XA} = \mathbf{AX} = \begin{bmatrix} 1 & 0 \\ 0 & 0 \end{bmatrix} = \mathbf{P}_{R(\mathbf{A})} = \mathbf{P}_{R(\mathbf{A}^*)}$, $\mathbf{X}$ satisfies $\mathbf{AX} = \mathbf{P}_{R(\mathbf{A})}$ and $\mathbf{XA} = \mathbf{P}_{R(\mathbf{A}^*)}$. But $\mathbf{A}^{\dagger} = \begin{bmatrix} 1/2 & 0 \\ 0 & 0 \end{bmatrix}$ and hence $\mathbf{X} \neq \mathbf{A}^{\dagger}$. Note that $\mathbf{XA} \neq \mathbf{P}_{R(\mathbf{X})}$ and thus $\mathbf{XAX} \neq \mathbf{X}$. If $\mathbf{XA} = \mathbf{P}_{R(\mathbf{A}^*)}$, $\mathbf{AX} = \mathbf{P}_{R(\mathbf{A})}$, and in addition $\mathbf{XAX} = \mathbf{X}$, then $\mathbf{X} = \mathbf{A}^{\dagger}$. The proof of this last statement is left to the exercises.

In computations involving inverses one frequently uses $(\mathbf{AB})^{-1} = \mathbf{B}^{-1}\mathbf{A}^{-1}$ if $\mathbf{A}$ and $\mathbf{B}$ are invertible. This fails to hold for generalized inverses even if $\mathbf{AB} = \mathbf{BA}$.

Fact 1.2.1 If $\mathbf{A} \in \mathbb{C}^{m \times n}$, $\mathbf{B} \in \mathbb{C}^{n \times p}$, then $(\mathbf{AB})^{\dagger}$ is not necessarily the same as $\mathbf{B}^{\dagger}\mathbf{A}^{\dagger}$. Furthermore $(\mathbf{A}^{\dagger})^2$ is not necessarily equal to $(\mathbf{A}^2)^{\dagger}$.

Example 1.2.2 Let $\mathbf{A} = \begin{bmatrix} 1 & -1 \\ 0 & 0 \end{bmatrix}$. Then $\mathbf{A}^{\dagger} = \frac{1}{2}\begin{bmatrix} 1 & 0 \\ -1 & 0 \end{bmatrix}$. Now

$\mathbf{A}^2 = \mathbf{A}$ while $\mathbf{A}^{\dagger 2} = \frac{1}{2}\mathbf{A}^{\dagger}$. Thus $(\mathbf{A}^{\dagger})^2\mathbf{A}^2 = \frac{1}{2}\mathbf{A}^{\dagger}\mathbf{A}$ which is not a projection. Thus $(\mathbf{A}^{\dagger})^2 \neq (\mathbf{A}^2)^{\dagger}$.

Ways of calculating $\mathbf{A}^{\dagger}$ will be given shortly. The generalized inverses in Examples 1 and 2 can be found directly from Definition 1 without too much difficulty.

Examples 2 illustrates another way in which the properties of the generalized inverse differ from those of the inverse. If $\mathbf{A}$ is invertible, then $\lambda \in \sigma(\mathbf{A})$ if and only if $\frac{1}{\lambda} \in \sigma(\mathbf{A}^{-1})$. If $\mathbf{A} = \begin{bmatrix} 1 & -1 \\ 0 & 0 \end{bmatrix}$ as in Example 2, then $\sigma(\mathbf{A}) = \{1,0\}$ while $\sigma(\mathbf{A}^{\dagger}) = \{\frac{1}{2},0\}$.

If $\mathbf{A}$ is similar to a matrix $\mathbf{C}$, then $\mathbf{A}$ and $\mathbf{C}$ have the same eigenvalues, the same Jordan form, and the same characteristic polynomial. None of these are preserved by taking of the generalized inverse.

Example 1.2.2 Let $\mathbf{A} = \begin{bmatrix} 1 & 1 & -1 \\ 2 & 0 & -2 \\ -1 & 1 & 1 \end{bmatrix}$. Then $\mathbf{A} = \mathbf{BJB}^{-1}$ where $\mathbf{B} = \begin{bmatrix} 1 & 0 & 1 \\ 0 & 1 & 1 \\ 1 & 0 & 0 \end{bmatrix}$ and $\mathbf{J} = \begin{bmatrix} 0 & 1 & 0 \\ 0 & 0 & 0 \\ 0 & 0 & 2 \end{bmatrix}$. The characteristic polynomial of $\mathbf{A}$ and $\mathbf{J}$ is $\lambda^2(\lambda - 2)$ with elementary divisors λ^2 and $\lambda - 2$. $\mathbf{J}^{\dagger} = \begin{bmatrix} 0 & 0 & 0 \\ 1 & 0 & 0 \\ 0 & 0 & 1/2 \end{bmatrix}$ and the characteristic polynomial of $\mathbf{J}^{\dagger}$ is $\lambda^2(\lambda - 1/2)$ with elementary divisors $\lambda^2, (\lambda - 1/2)$. An easy computation gives $\mathbf{A}^{\dagger} = 1/12 \begin{bmatrix} 1 & 2 & -1 \\ 6 & 0 & 6 \\ -1 & -2 & 1 \end{bmatrix}$. But $\mathbf{A}^{\dagger}$ has characteristic polynomial $\lambda(\lambda - (1 + \sqrt{13})/12)(\lambda - (1 - \sqrt{13})/12)$ and hence a diagonal Jordan form.

Thus, if $\mathbf{A}$ and $\mathbf{C}$ are similar, then about the only thing that one can always say about $\mathbf{A}^{\dagger}$ and $\mathbf{C}^{\dagger}$ is that they have the same rank.

A type of inverse that behaves better with respect to similarity is discussed in Chapter VII. Since the generalized inverse does not have all the properties of the inverse, it becomes important to know what properties it does have and which identities it does satisfy. There are, of course, an arbitrarily large number of true statements about generalized inverses. The next theorem lists some of the more basic properties.

Theorem 1.2.1 *Suppose that* $\mathbf{A} \in \mathbb{C}^{m \times n}$. *Then*

(P1) $(\mathbf{A}^{\dagger})^{\dagger} = \mathbf{A}$
(P2) $(\mathbf{A}^{\dagger})^* = (\mathbf{A}^*)^{\dagger}$
(P3) *If* $\lambda \in \mathbb{C}$, $(\lambda\mathbf{A})^{\dagger} = \lambda^{\dagger}\mathbf{A}^{\dagger}$ *where* $\lambda^{\dagger} = \frac{1}{\lambda}$ *if* $\lambda \neq 0$ *and* $\lambda^{\dagger} = 0$ *if* $\lambda = 0$.
(P4) $\mathbf{A}^* = \mathbf{A}^*\mathbf{A}\mathbf{A}^{\dagger} = \mathbf{A}^{\dagger}\mathbf{A}\mathbf{A}^*$
(P5) $(\mathbf{A}^*\mathbf{A})^{\dagger} = \mathbf{A}^{\dagger}\mathbf{A}^{*\dagger}$

(P6) $\mathbf{A}^\dagger = (\mathbf{A}^*\mathbf{A})^\dagger\mathbf{A}^* = \mathbf{A}^*(\mathbf{A}\mathbf{A}^*)^\dagger$
(P7) $(\mathbf{UAV})^\dagger = \mathbf{V}^*\mathbf{A}^\dagger\mathbf{U}^*$ *where* $\mathbf{U}, \mathbf{V}$ *are unitary matrices.*

Proof We will discuss the properties in the order given. A look at Definition 2 and a moment's thought show that (P1) is true. We leave (P2) and (P3) to the exercises. (P4) follows by taking the adjoints of both $\mathbf{A} = (\mathbf{A}\mathbf{A}^\dagger)\mathbf{A}$ and $\mathbf{A} = \mathbf{A}(\mathbf{A}^\dagger\mathbf{A})$. (P5), since it claims that something is a generalized inverse, can be checked by using one of the definitions. Definition 2 is the quickest. $\mathbf{A}^\dagger\mathbf{A}^{*\dagger}\mathbf{A}^*\mathbf{A} = \mathbf{A}^\dagger(\mathbf{A}^{*\dagger}\mathbf{A}^*)\mathbf{A} = \mathbf{A}^\dagger(\mathbf{A}\mathbf{A}^\dagger)^*\mathbf{A} = \mathbf{A}^\dagger\mathbf{A}\mathbf{A}^\dagger\mathbf{A} = \mathbf{A}^\dagger\mathbf{A} = \mathbf{P}_{R(\mathbf{A}^*)} = \mathbf{P}_{R(\mathbf{A}^*\mathbf{A})}$. Similarly, $\mathbf{A}^*\mathbf{A}\mathbf{A}^\dagger\mathbf{A}^{*\dagger} = \mathbf{A}^*(\mathbf{A}\mathbf{A}^\dagger)\mathbf{A}^{*\dagger} = \mathbf{A}^*(\mathbf{A}\mathbf{A}^\dagger)^*\mathbf{A}^{*\dagger} = \mathbf{A}^*(\mathbf{A}^{*\dagger}\mathbf{A}^*)\mathbf{A}^{*\dagger} = (\mathbf{A}^*\mathbf{A}^{*\dagger})(\mathbf{A}^*\mathbf{A}^{*\dagger}) = \mathbf{A}^*\mathbf{A}^{*\dagger} = \mathbf{P}_{R(\mathbf{A}^*)} = \mathbf{P}_{R(\mathbf{A}^*\mathbf{A})} = \mathbf{P}_{R((\mathbf{A}^*\mathbf{A})^*)} = \mathbf{P}_{R((\mathbf{A}^*\mathbf{A})^\dagger)}$. Thus $(\mathbf{A}^*\mathbf{A})^\dagger = \mathbf{A}^\dagger\mathbf{A}^{*\dagger}$ by Definition 2. (P6) follows from (P5). (P7) is left as an exercise. ∎

Proposition 1.2.1 $\begin{bmatrix} \mathbf{A}_1 & \cdots & \mathbf{0} \\ \vdots & \ddots & \vdots \\ \mathbf{0} & \cdots & \mathbf{A}_m \end{bmatrix}^\dagger = \begin{bmatrix} \mathbf{A}_1^\dagger & \cdots & \mathbf{0} \\ \vdots & \ddots & \vdots \\ \mathbf{0} & \cdots & \mathbf{A}_m^\dagger \end{bmatrix}$.

The proof of Proposition 1 is left as an exercise.

As the proof of Theorem 1 illustrates, it is frequently helpful to know the ranges and null spaces of expressions involving $\mathbf{A}^\dagger$. For ease of reference we now list several of these basic geometric properties. (P8) and (P9) have been done already. The rest are left as an exercise.

Theorem 1.2.2 *If* $\mathbf{A} \in \mathbb{C}^{m \times n}$, *then*
(P8) $R(\mathbf{A}) = R(\mathbf{A}\mathbf{A}^\dagger) = R(\mathbf{A}\mathbf{A}^*)$
(P9) $R(\mathbf{A}^\dagger) = R(\mathbf{A}^*) = R(\mathbf{A}^\dagger\mathbf{A}) = R(\mathbf{A}^*\mathbf{A})$
(P10) $R(\mathbf{I} - \mathbf{A}\mathbf{A}^\dagger) = N(\mathbf{A}\mathbf{A}^\dagger) = N(\mathbf{A}^*) = N(\mathbf{A}^\dagger) = R(\mathbf{A})^\perp$
(P11) $R(\mathbf{I} - \mathbf{A}^\dagger\mathbf{A}) = N(\mathbf{A}^\dagger\mathbf{A}) = N(\mathbf{A}) = R(\mathbf{A}^*)^\perp$

3. Computation of $\mathbf{A}^\dagger$

In learning any type of mathematics, the working out of examples is useful, if not essential. The calculation of $\mathbf{A}^\dagger$ from $\mathbf{A}$ can be difficult, and will be discussed more fully later. For the present we will give two methods which will enable the reader to begin to calculate the generalized inverses of small matrices. The first method is worth knowing because using it should help give a feeling of what $\mathbf{A}^\dagger$ is. The method consists of 'constructing' $\mathbf{A}^\dagger$ according to Definition 1.

Example 1.3.1 Let $\mathbf{A} = \begin{bmatrix} 1 & 1 & 2 \\ 0 & 2 & 2 \\ 1 & 0 & 1 \\ 1 & 0 & 1 \end{bmatrix}$. Then $R(\mathbf{A}^*)$ is spanned by

$\left\{ \begin{bmatrix} 1 \\ 1 \\ 2 \end{bmatrix}, \begin{bmatrix} 0 \\ 2 \\ 2 \end{bmatrix}, \begin{bmatrix} 1 \\ 0 \\ 1 \end{bmatrix}, \begin{bmatrix} 1 \\ 0 \\ 1 \end{bmatrix} \right\}$. A subset forming a basis of $R(\mathbf{A}^*)$ is

$\left\{\begin{bmatrix}1\\0\\1\end{bmatrix}, \begin{bmatrix}0\\1\\1\end{bmatrix}\right\}$. Now $\mathbf{A}\begin{bmatrix}1\\0\\1\end{bmatrix} = \begin{bmatrix}3\\2\\2\\2\end{bmatrix}$ and $\mathbf{A}\begin{bmatrix}0\\1\\1\end{bmatrix} = \begin{bmatrix}3\\4\\1\\1\end{bmatrix}$. Thus $\mathbf{A}^\dagger\begin{bmatrix}3\\2\\2\\2\end{bmatrix} = \begin{bmatrix}1\\0\\1\end{bmatrix}$

and $\mathbf{A}^\dagger\begin{bmatrix}3\\4\\1\\1\end{bmatrix} = \begin{bmatrix}0\\1\\1\end{bmatrix}$. We now must calculate a basis for $R(\mathbf{A})^\perp = N(\mathbf{A}^*)$.

Solving the system $\mathbf{A}^*\mathbf{x} = \mathbf{0}$ we get that $\mathbf{x} = x_3\begin{bmatrix}-1\\1/2\\1\\0\end{bmatrix} + x_4\begin{bmatrix}-1\\1/2\\0\\1\end{bmatrix}$ where

$x_3, x_4 \in \mathbb{C}$. Then $\mathbf{A}^\dagger\begin{bmatrix}-1\\1/2\\1\\0\end{bmatrix} = \mathbf{A}^\dagger\begin{bmatrix}-1\\1/2\\0\\1\end{bmatrix} = \mathbf{0} \in \mathbb{C}^3$. Combining all of this

gives $\mathbf{A}^\dagger\begin{bmatrix}3 & 3 & -1 & -1\\2 & 4 & 1/2 & 1/2\\2 & 1 & 1 & 0\\2 & 1 & 0 & 1\end{bmatrix} = \begin{bmatrix}1 & 0 & 0 & 0\\0 & 1 & 0 & 0\\1 & 1 & 0 & 0\end{bmatrix}$ or

$$\mathbf{A}^\dagger = \begin{bmatrix}1 & 0 & 0 & 0\\0 & 1 & 0 & 0\\1 & 1 & 0 & 0\end{bmatrix}\begin{bmatrix}3 & 3 & -1 & -1\\2 & 4 & 1/2 & 1/2\\2 & 1 & 1 & 0\\2 & 1 & 0 & 1\end{bmatrix}^{-1}.$$

The indicated inverse can always be taken since its columns form a basis for $R(\mathbf{A}) \oplus R(\mathbf{A})^\perp$ and hence are a linearly independent set of four vectors in $\mathbb{C}^4$.

Below is a formal statement of the method described in Example 1.

Theorem 1.3.1 *Let $\mathbf{A} \in \mathbb{C}^{m \times n}$ have rank r. If $\{\mathbf{v}_1, \mathbf{v}_2, \ldots, \mathbf{v}_r\}$ is a basis for $R(\mathbf{A}^*)$ and $\{\mathbf{w}_1, \mathbf{w}_2, \ldots, \mathbf{w}_{n-r}\}$ is a basis for $N(\mathbf{A}^*)$, then*

$$\mathbf{A}^\dagger = [\mathbf{v}_1 \mid \mathbf{v}_2 \mid \ldots \mid \mathbf{v}_r \mid \mathbf{0} \mid \ldots \mid \mathbf{0}][\mathbf{A}\mathbf{v}_1 \mid \mathbf{A}\mathbf{v}_2 \mid \ldots \mid \mathbf{A}\mathbf{v}_r \mid \mathbf{w}_1 \mid \mathbf{w}_2 \mid \ldots \mid \mathbf{w}_{n-r}]^{-1}$$

Proof By using Definition 1.1 it is clear that

$$\mathbf{A}^\dagger[\mathbf{A}\mathbf{v}_1 \mid \ldots \mid \mathbf{A}\mathbf{v}_r \mid \mathbf{w}_1 \mid \ldots \mid \mathbf{w}_{n-r}] = [\mathbf{A}^\dagger\mathbf{A}\mathbf{v}_1 \mid \ldots \mid \mathbf{A}^\dagger\mathbf{A}\mathbf{v}_r \mid \mathbf{A}^\dagger\mathbf{w}_1 \mid \ldots \mid \mathbf{A}^\dagger\mathbf{w}_{n-r}]$$
$$= [\mathbf{v}_1 \mid \ldots \mid \mathbf{v}_r \mid \mathbf{0} \mid \ldots \mid \mathbf{0}].$$

Furthermore, $\{\mathbf{A}\mathbf{v}_1, \mathbf{A}\mathbf{v}_2, \ldots, \mathbf{A}\mathbf{v}_r\}$ must be a basis for $R(\mathbf{A})$. Since $R(\mathbf{A})^\perp = N(\mathbf{A}^*)$, it follows that the matrix $[\mathbf{A}\mathbf{v}_1 \mid \ldots \mid \mathbf{A}\mathbf{v}_r \mid \mathbf{w}_1 \mid \ldots \mid \mathbf{w}_{n-r}]$ must be non-singular. The desired result is now immediate. ∎

The second method involves a formula which is sometimes useful. It depends on the following fact:

Proposition 1.3.1 *If* $\mathbf{A}\in\mathbb{C}^{m\times n}$, *then there exists* $\mathbf{B}\in\mathbb{C}^{m\times r}, \mathbf{C}\in\mathbb{C}^{r\times n}$, *such that* $\mathbf{A}=\mathbf{BC}$ *and* $r = \text{rank}(\mathbf{A}) = \text{rank}(\mathbf{B}) = \text{rank}(\mathbf{C})$.

The proof of Proposition 1 is not difficult and is left as an exercise. It means that the next result can, in theory, always be used to calculate $\mathbf{A}^\dagger$. See Chapter 12 for a caution on taking $(\mathbf{CC}^*)^{-1}$ or $(\mathbf{B}^*\mathbf{B})^{-1}$ if $\mathbf{C}$ or $\mathbf{B}$ have small singular values.

Theorem 1.3.2 *If* $\mathbf{A}=\mathbf{BC}$ *where* $\mathbf{A}\in\mathbb{C}^{m\times n}, \mathbf{B}\in\mathbb{C}^{m\times r}, \mathbf{C}\in\mathbb{C}^{r\times n}$, *and* $r = \text{rank}(\mathbf{A}) = \text{rank}(\mathbf{B}) = \text{rank}(\mathbf{C})$, *then* $\mathbf{A}^\dagger = \mathbf{C}^*(\mathbf{CC}^*)^{-1}(\mathbf{B}^*\mathbf{B})^{-1}\mathbf{B}^*$.

Proof. Notice that $\mathbf{B}^*\mathbf{B}$ and $\mathbf{CC}^*$ are rank r matrices in $\mathbb{C}^{r\times r}$ so that it makes sense to take their inverses. Let $\mathbf{X} = \mathbf{C}^*(\mathbf{CC}^*)^{-1}(\mathbf{B}^*\mathbf{B})^{-1}\mathbf{B}^*$. We will show $\mathbf{X}$ satisfies Definition 3. This choice is made on the grounds that the more complicated an expression is, the more difficult it becomes geometrically to work with it, and Definition 3 is algebraic. Now $\mathbf{AX} = \mathbf{BCC}^*(\mathbf{CC}^*)^{-1}(\mathbf{B}^*\mathbf{B})^{-1}\mathbf{B}^* = \mathbf{B}(\mathbf{B}^*\mathbf{B})^{-1}\mathbf{B}^*$, so $(\mathbf{AX})^* = \mathbf{AX}$. Also $\mathbf{XA} = \mathbf{C}^*(\mathbf{CC}^*)^{-1}(\mathbf{B}^*\mathbf{B})^{-1}\mathbf{B}^*\mathbf{BC} = \mathbf{C}^*(\mathbf{CC}^*)^{-1}\mathbf{C}$, so $(\mathbf{XA})^* = \mathbf{XA}$. Thus (iii) and (iv) hold. To check (i) and (ii) use $\mathbf{XA} = \mathbf{C}^*(\mathbf{CC}^*)^{-1}\mathbf{C}$ to get that $\mathbf{A}(\mathbf{XA}) = \mathbf{BC}(\mathbf{C}^*(\mathbf{CC}^*)^{-1}\mathbf{C}) = \mathbf{BC} = \mathbf{A}$. And $(\mathbf{XA})\mathbf{X} = \mathbf{C}^*(\mathbf{CC}^*)^{-1}\mathbf{CC}^*(\mathbf{CC}^*)^{-1} \times (\mathbf{B}^*\mathbf{B})^{-1}\mathbf{B}^* = \mathbf{C}^*(\mathbf{CC}^*)^{-1}(\mathbf{B}^*\mathbf{B})^{-1}\mathbf{B}^* = \mathbf{X}$. Thus $\mathbf{X} = \mathbf{A}^\dagger$ by Definition 3. ■

Example 1.3.2 Let $\mathbf{A} = \begin{bmatrix} 1 & 1 & 2 \\ 2 & 2 & 4 \end{bmatrix}$. $r = \text{rank}(\mathbf{A}) = 1$. $\mathbf{A} = \mathbf{BC}$ where $\mathbf{B}\in\mathbb{C}^{2\times 1}$ and $\mathbf{C}\in\mathbb{C}^{1\times 3}$. In fact, a little thought shows that $\mathbf{A} = \begin{bmatrix} 1 \\ 2 \end{bmatrix}[1\ 1\ 2]$.

Then $\mathbf{B}^*\mathbf{B} = [5]$, $\mathbf{CC}^* = [6]$. Thus $\mathbf{A}^\dagger = \begin{bmatrix} 1 \\ 1 \\ 2 \end{bmatrix}[1,2]\cdot 1/30 = 1/30\begin{bmatrix} 1 & 2 \\ 1 & 2 \\ 2 & 4 \end{bmatrix}$.

Example 1 is typical as the next result shows.

Theorem 1.3.3 *If* $\mathbf{A}\in\mathbb{C}^{m\times n}$ *and rank*$(\mathbf{A}) = 1$, *then* $\mathbf{A}^\dagger = \dfrac{1}{\alpha}\mathbf{A}^*$ *where* $\alpha = \text{Tr}\,\mathbf{A}^*\mathbf{A} = \sum_{i,j}|a_{ij}|^2$.

The proof is left to the exercises.

The method of computing $\mathbf{A}^\dagger$ described in Example 1.3.1 and the method of Theorem 1.3.2 may both be executed by reducing $\mathbf{A}$ by elementary row operations.

Definition 1.3.1 *A matrix* $\mathbf{E}\in\mathbb{C}^{m\times n}$ *which has rank r is said to be in* row echelon form *if* $\mathbf{E}$ *is of the form*

$$\mathbf{E} = \begin{bmatrix} \mathbf{C}_{r\times n} \\ \hline \mathbf{0}_{(m-r)\times n} \end{bmatrix} \tag{1}$$

where the elements c_{ij} *of* $\mathbf{C}\,(=\mathbf{C}_{r\times n})$ *satisfy the following conditions,*

(i) $c_{ij}=0$ *when* $i>j$.
(ii) *The first non-zero entry in each row of* $\mathbf{C}$ *is* 1.
(iii) *If* $c_{ij}=1$ *is the first non-zero entry of the* ith *row, then the* jth *column of* $\mathbf{C}$ *is the unit vector* $\mathbf{e}_i$ *whose only non-zero entry is in the* ith *position.*

For example, the matrix

$$\mathbf{E}=\begin{bmatrix}1&2&0&3&5&0&1\\0&0&1&-2&4&0&0\\0&0&0&0&0&1&3\\0&0&0&0&0&0&0\\0&0&0&0&0&0&0\end{bmatrix} \tag{2}$$

is in row echelon form. Below we state some facts about the row echelon form, the proofs of which may be found in [65].

For $\mathbf{A}\in\mathbb{C}^{m\times n}$ such that $\operatorname{rank}(\mathbf{A})=r$:

(E1) $\mathbf{A}$ can always be row reduced to row echelon form by elementary row operations (i.e. there always exists a non-singular matrix $\mathbf{P}\in\mathbb{C}^{m\times m}$ such that $\mathbf{PA}=\mathbf{E}_{\mathbf{A}}$ where $\mathbf{E}_{\mathbf{A}}$ is in row echelon form).

(E2) For a given $\mathbf{A}$, the row echelon form $\mathbf{E}_{\mathbf{A}}$ obtained by row reducing $\mathbf{A}$ is unique.

(E3) If $\mathbf{E}_{\mathbf{A}}$ is the row echelon form for $\mathbf{A}$ and the unit vectors in $\mathbf{E}_{\mathbf{A}}$ appear in columns $i_1, i_2, \ldots,$ and i_r, then the corresponding columns of $\mathbf{A}$ are a basis for $R(\mathbf{A})$. This particular basis is called the set of *distinguished columns* of $\mathbf{A}$. The remaining columns are called the *undistinguished columns* of $\mathbf{A}$. (For example, if $\mathbf{A}$ is a matrix such that its row echelon form is given by (2) then the first, third, and sixth columns of $\mathbf{A}$ are the distinguished columns.

(E4) If $\mathbf{E}_{\mathbf{A}}$ is the row echelon form (1) for $\mathbf{A}$, then $N(\mathbf{A})=N(\mathbf{E}_{\mathbf{A}})=N(\mathbf{C})$.

(E5) If (1) is the row echelon form for $\mathbf{A}$, and if $\mathbf{B}\in\mathbb{C}^{m\times r}$ is the matrix made up of the distinguished columns of $\mathbf{A}$ (in the same order as they are in $\mathbf{A}$), then $\mathbf{A}=\mathbf{BC}$ where $\mathbf{C}$ is obtained from the row echelon form. This is a full rank factorization such as was described in Proposition 1.

Very closely related to the row echelon form is the hermite echelon form. However, the hermite echelon form is defined only for square matrices.

Definition 1.3.2 *A matrix* $\mathbf{H}\in\mathbb{C}^{n\times n}$ *is said to be in* hermite echelon form *if its elements* h_{ij} *satisfies the following conditions.*

(i) $\mathbf{H}$ *is upper triangular* (i.e. $h_{ij}=0$ *when* $i>j$).
(ii) h_{ii} *is either* 0 *or* 1.
(iii) *If* $h_{ii}=0$, *then* $h_{ik}=0$ *for every* k, $1\le k\le n$.
(iv) *If* $h_{ii}=1$, *then* $h_{ki}=0$ *for every* $k\ne i$.

For example, the matrix

$$\mathbf{H} = \begin{bmatrix} 1 & 2 & 0 & 3 & 5 & 0 & 1 \\ 0 & 0 & 0 & 0 & 0 & 0 & 0 \\ 0 & 0 & 1 & -2 & 4 & 0 & 0 \\ 0 & 0 & 0 & 0 & 0 & 0 & 0 \\ 0 & 0 & 0 & 0 & 0 & 0 & 0 \\ 0 & 0 & 0 & 0 & 0 & 1 & 3 \\ 0 & 0 & 0 & 0 & 0 & 0 & 0 \end{bmatrix}$$

is in hermite form. Below are some facts about the hermite form, the proofs of which may be found in [65].

For $\mathbf{A} \in \mathbb{C}^{n \times n}$:

(H1) $\mathbf{A}$ can always be row reduced to a hermite form. If $\mathbf{A}$ is reduced to its row echelon form, then a permutation of the rows can always be performed to obtain a hermite form.

(H2) For a given matrix $\mathbf{A}$ the hermite form $\mathbf{H}_{\mathbf{A}}$ obtained by row reducing $\mathbf{A}$ is unique.

(H3) $\mathbf{H}_{\mathbf{A}}^2 = \mathbf{H}_{\mathbf{A}}$ (i.e. $\mathbf{H}_{\mathbf{A}}$ is a projection).

(H4) $N(\mathbf{A}) = N(\mathbf{H}_{\mathbf{A}}) = R(\mathbf{I} - \mathbf{H}_{\mathbf{A}})$ and a basis for $N(\mathbf{A})$ is the set of non-zero columns of $\mathbf{I} - \mathbf{H}_{\mathbf{A}}$.

We can now present the methods of Theorems 1 and 2 as algorithms.

Algorithm 1.3.1 To obtain the generalized inverse of a square matrix $\mathbf{A} \in \mathbb{C}^{n \times n}$.

(I) Row reduce $\mathbf{A}^*$ to its hermite form $\mathbf{H}_{\mathbf{A}^*}$.

(II) Select the distinguished columns of $\mathbf{A}^*$. Label these columns $\mathbf{v}_1, \mathbf{v}_2, \ldots, \mathbf{v}_r$ and place them as columns in a matrix $\mathbf{L}$.

(III) Form the matrix $\mathbf{AL}$.

(IV) Form $\mathbf{I} - \mathbf{H}_{\mathbf{A}^*}$ and select the non-zero columns from this matrix. Label these columns $\mathbf{w}_1, \mathbf{w}_2, \ldots, \mathbf{w}_{n-r}$.

(V) Place the columns of $\mathbf{AL}$ and the $\mathbf{w}_i$'s as columns in a matrix $\mathbf{M} = [\mathbf{AL} \mid \mathbf{w}_1 \mid \mathbf{w}_2 \mid \ldots \mid \mathbf{w}_{n-r}]$ and compute $\mathbf{M}^{-1}$. (Actually only the first r rows of $\mathbf{M}^{-1}$ are needed.)

(VI) Place the first r rows of $\mathbf{M}^{-1}$ (in the same order as they appear in $\mathbf{M}^{-1}$) in a matrix called $\mathbf{R}$.

(VII) Compute $\mathbf{A}^\dagger$ as $\mathbf{A}^\dagger = \mathbf{LR}$.

Although Algorithm 1 is stated for square matrices, it is easy to use it for non-square matrices. Add zero rows or zero columns to construct a square matrix and use the fact that $[\mathbf{A} \mid \mathbf{0}]^\dagger = \left[\begin{array}{c} \mathbf{A}^\dagger \\ \hline \mathbf{0}^* \end{array}\right]$ and $\left[\begin{array}{c} \mathbf{A} \\ \hline \mathbf{0} \end{array}\right]^\dagger = [\mathbf{A}^\dagger \mid \mathbf{0}^*]$.

Algorithm 1.3.2 To obtain the full rank factorization and the generalized inverse for any $\mathbf{A} \in \mathbb{C}^{m \times n}$.

(I) Reduce **A** to row echelon form $\mathbf{E}_\mathbf{A}$.

(II) Select the distinguished columns of **A** and place them as the columns in a matrix **B** in the same order as they appear in **A**.

(III) Select the non-zero rows from $\mathbf{E}_\mathbf{A}$ and place them as rows in a matrix **C** in the same order as they appear in $\mathbf{E}_\mathbf{A}$.

(IV) Compute $(\mathbf{CC}^*)^{-1}$ and $(\mathbf{B}^*\mathbf{B})^{-1}$.

(V) Compute $\mathbf{A}^\dagger$ as $\mathbf{A}^\dagger = \mathbf{C}^*(\mathbf{CC}^*)^{-1}(\mathbf{B}^*\mathbf{B})^{-1}\mathbf{B}^*$.

Example 1.3.3 We will use Algorithm 1 to find $\mathbf{A}^\dagger$ where

$$\mathbf{A} = \begin{bmatrix} 1 & 2 & 1 & 4 \\ 2 & 4 & 0 & 6 \\ 0 & 0 & 1 & 1 \\ 0 & 0 & 0 & 1 \end{bmatrix}.$$

(I) Using elementary row operations on $\mathbf{A}^*$ we get that its hermite echelon form is

$$\mathbf{H}_{\mathbf{A}^*} = \begin{bmatrix} 1 & 0 & 1 & 0 \\ 0 & 1 & -\frac{1}{2} & 0 \\ 0 & 0 & 0 & 0 \\ 0 & 0 & 0 & 1 \end{bmatrix}.$$

(II) The first, second and fourth columns of $\mathbf{A}^*$ are distinguished. Thus

$$\mathbf{L} = \begin{bmatrix} 1 & 2 & 0 \\ 2 & 4 & 0 \\ 1 & 0 & 0 \\ 4 & 6 & 1 \end{bmatrix}.$$

(III) Then

$$\mathbf{AL} = \begin{bmatrix} 22 & 34 & 4 \\ 34 & 56 & 6 \\ 5 & 6 & 1 \\ 4 & 6 & 1 \end{bmatrix}.$$

(IV) The non-zero column of $\mathbf{I} - \mathbf{H}_{\mathbf{A}^*}$ is $\mathbf{w}_1^* = [-1, 1/2, 1, 0]^*$.

(V) Putting **AL** and $\mathbf{w}_1$ into a matrix **M** and then computing $\mathbf{M}^{-1}$ gives

$$\mathbf{M} = \begin{bmatrix} 22 & 34 & 4 & -1 \\ 34 & 56 & 6 & 1/2 \\ 5 & 6 & 1 & 1 \\ 4 & 6 & 1 & 0 \end{bmatrix}, \mathbf{M}^{-1} = \frac{1}{90}\begin{bmatrix} 40 & -20 & 50 & -90 \\ -19 & 14 & -26 & 18 \\ -46 & -4 & -44 & 342 \\ -40 & 20 & 40 & 0 \end{bmatrix}.$$

(VI) The first three rows of $\mathbf{M}^{-1}$ give **R** as

$$\mathbf{R} = \frac{1}{90}\begin{bmatrix} 40 & -20 & 50 & -90 \\ -19 & 14 & -26 & 18 \\ -46 & -4 & -44 & 342 \end{bmatrix}.$$

(VII) Thus

$$\mathbf{A}^\dagger = \mathbf{LR} = \frac{1}{45}\begin{bmatrix} 1 & 4 & -1 & -27 \\ 2 & 8 & -2 & -54 \\ 20 & -10 & 25 & -45 \\ 0 & 0 & 0 & 45 \end{bmatrix}.$$

Example 1.3.4 We will now use Algorithm 2 to find $\mathbf{A}^\dagger$ where

$$\mathbf{A} = \begin{bmatrix} 1 & 2 & 1 & 4 & 1 \\ 2 & 4 & 0 & 6 & 6 \\ 1 & 2 & 0 & 3 & 3 \\ 2 & 4 & 0 & 6 & 6 \end{bmatrix}.$$

(I) Using elementary row operations we reduce $\mathbf{A}$ to its row echelon form

$$\mathbf{E}_\mathbf{A} = \begin{bmatrix} 1 & 2 & 0 & 3 & 3 \\ 0 & 0 & 1 & 1 & -2 \\ 0 & 0 & 0 & 0 & 0 \\ 0 & 0 & 0 & 0 & 0 \end{bmatrix}.$$

(II) The first and third columns are distinguished. Thus

$$\mathbf{B} = \begin{bmatrix} 1 & 1 \\ 2 & 0 \\ 1 & 0 \\ 2 & 0 \end{bmatrix}.$$

(III) The matrix $\mathbf{C}$ is made up of the non-zero rows of $\mathbf{E}_\mathbf{A}$ so that

$$\mathbf{C} = \begin{bmatrix} 1 & 2 & 0 & 3 & 3 \\ 0 & 0 & 1 & 1 & -2 \end{bmatrix}.$$

(IV) Now $\mathbf{CC}^* = \begin{bmatrix} 23 & -3 \\ -3 & 6 \end{bmatrix}$ and $\mathbf{B}^*\mathbf{B} = \begin{bmatrix} 10 & 1 \\ 1 & 1 \end{bmatrix}$. Calculating $(\mathbf{CC}^*)^{-1}$ and $(\mathbf{B}^*\mathbf{B})^{-1}$ we get $(\mathbf{CC}^*)^{-1} = \frac{1}{129}\begin{bmatrix} 6 & 3 \\ 3 & 23 \end{bmatrix}$ and $(\mathbf{B}^*\mathbf{B})^{-1} = \frac{1}{9}\begin{bmatrix} 1 & -1 \\ -1 & 10 \end{bmatrix}$.

(V) Substituting the results of steps (II), (III) and (IV) into the formula for $\mathbf{A}^\dagger$ gives

$$\mathbf{A}^\dagger = \mathbf{C}^*(\mathbf{CC}^*)^{-1}(\mathbf{B}^*\mathbf{B})^{-1}\mathbf{B}^* = \frac{1}{1161}\begin{bmatrix} 27 & 6 & 3 & 6 \\ 54 & 12 & 6 & 12 \\ 207 & -40 & -20 & -40 \\ 288 & -22 & -11 & -22 \\ -333 & 98 & 49 & 98 \end{bmatrix}.$$

Theorem 2 is a good illustration of one difficulty with learning from a text in this area. Often the hard part is to come up with the right formula. To verify it is easy. This is not an uncommon phenomenon. In differential equations it is frequently easy to verify that a given function is a solution. The hard part is to show one exists and then to find it. In the study of generalized inverses, existence is usually taken care of early. There remains then the problem of finding the right formula. For these reasons, we urge the reader to try and derive his own theorems as we go. For example, can you come up with an alternative formula to that of Theorem 2? The resulting formula should, of course, only involve $\mathbf{A}^\dagger$ on one side. Ideally, it would not even involve $\mathbf{B}^\dagger$ and $\mathbf{C}^\dagger$. Then ask yourself, can I do better by imposing special conditions on **B** and **C** or **A**? Under what conditions does the formula simplify? The reader who approaches each problem, theorem and exercise in this manner will not only learn the material better, but will be a better mathematician for it.

4. Generalized inverse of a product

As pointed out in Section 2, one of the major shortcomings of the Moore–Penrose inverse is that the 'reverse order law' does not always hold, that is, $(\mathbf{AB})^\dagger$ is not always $\mathbf{B}^\dagger\mathbf{A}^\dagger$. This immediately suggests two questions. What is $(\mathbf{AB})^\dagger$? When does $(\mathbf{AB})^\dagger = \mathbf{B}^\dagger\mathbf{A}^\dagger$?

The question, 'What is $(\mathbf{AB})^\dagger$?' has a lot of useless, or non-answer, answers. For example, $(\mathbf{AB})^\dagger = (\mathbf{AB})^\dagger\mathbf{AB}(\mathbf{AB})^\dagger$ is a non-answer. It merely restates condition (ii) of the Penrose definition of $(\mathbf{AB})^\dagger$. The decision as to whether or not an answer is an answer is subjective and comes with experience. Even then, professional mathematicians may differ on how good an answer is depending on how they happen to view the problem and mathematics.

The authors feel that a really good answer to the question, 'What is $(\mathbf{AB})^\dagger$?' does not, and probably will not exist. However, an answer should:

(A) have some sort of intuitive justification if possible;

(B) suggest at least a partial answer to the other question, 'When does $(\mathbf{AB})^\dagger = \mathbf{B}^\dagger\mathbf{A}^\dagger$?'

Theorem 4.1 is, to our knowledge, the best answer available.

We shall now attack the problem of determining a formula for $(\mathbf{AB})^\dagger$. The first problem is to come up with a theorem to prove. One way to come up with a conjecture would be to perform algebraic manipulations on $(\mathbf{AB})^\dagger$ using the Penrose conditions. Another, and the one we now follow, is to draw a picture and make an educated guess. If that guess does not work, then make another.

Figure 1.1 is, in a sense, not very realistic. However, the authors find it a convenient way to visualize the actions of linear transformations. The vertical lines stand for $\mathbb{C}^p$, $\mathbb{C}^n$, and $\mathbb{C}^m$. A sub-interval is a subspace. The rest of the interval is a (possibly orthogonal) complementary subspace. A

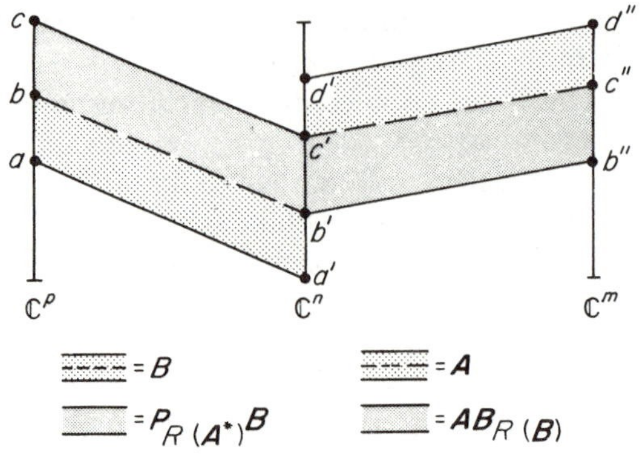

Fig. 1.1

shaded band represents the one to one mapping of one subspace onto another. It is assumed that $\mathbf{A} \in \mathbb{C}^{m \times n}$ and $\mathbf{B} \in \mathbb{C}^{n \times p}$. In the figure: $(a, c) = R(\mathbf{B}^*)$, $(a', c') = R(\mathbf{B})$, $(b', d') = R(\mathbf{A}^*)$, and $(b'', d'') = R(\mathbf{A})$. The total shaded band from $\mathbb{C}^p$ to $\mathbb{C}^n$ represents the action of $\mathbf{B}$ (or $\underset{\sim}{\mathbf{B}}$). The part that is shaded more darkly represents $\mathbf{P}_{R(\mathbf{A}^*)}\mathbf{B} = \mathbf{A}^\dagger\mathbf{A}\mathbf{B}$. The total shaded area from $\mathbb{C}^n$ to $\mathbb{C}^m$ is $\mathbf{A}$. The darker portion is $\mathbf{A}\mathbf{P}_{R(\mathbf{B})} = \mathbf{A}\mathbf{B}\mathbf{B}^\dagger$. The one to one portion of the mapping $\underset{\sim}{\mathbf{AB}}$ from $\mathbb{C}^p$ to $\mathbb{C}^m$ may be viewed as the v-shaped dark band from $\mathbb{C}^p$ to $\mathbb{C}^m$. Thus to 'undo' $\mathbf{AB}$, one would trace the dark band backwards. That is, $(\mathbf{AB})^\dagger = (\mathbf{P}_{R(\mathbf{A}^*)}\mathbf{B})^\dagger(\mathbf{A}\mathbf{P}_{R(\mathbf{B})})^\dagger$.

There are only two things wrong with this conjecture. For one,

$$\mathbf{AB} = \mathbf{A}(\mathbf{A}^\dagger\mathbf{A})(\mathbf{B}\mathbf{B}^\dagger)\mathbf{B} \tag{1}$$

and not $\mathbf{AB} = \mathbf{A}(\mathbf{B}\mathbf{B}^\dagger)(\mathbf{A}^\dagger\mathbf{A})\mathbf{B}$. Secondly, the drawing lies a bit in that the interval (b', c') is actually standing for two slightly skewed subspaces and not just one. However, the factorization (1) does not help and Fig. 1.1 does seem to portray what is happening. We are led then to try and prove the following theorem due to Cline and Greville.

Theorem 1.4.1 *If* $\mathbf{A} \in \mathbb{C}^{m \times n}$ *and* $\mathbf{B} \in \mathbb{C}^{n \times p}$, *then* $(\mathbf{AB})^\dagger = (\mathbf{P}_{R(\mathbf{A}^*)}\mathbf{B})^\dagger(\mathbf{A}\mathbf{P}_{R(\mathbf{B})})^\dagger$.

Proof Assuming that one has stumbled across this formula and is wondering if it is correct, the most reasonable thing to do is see if the formula for $(\mathbf{AB})^\dagger$ satisfies any of the definitions of the generalized inverse. Let $\mathbf{X} = (\mathbf{P}_{R(\mathbf{A}^*)}\mathbf{B})^\dagger(\mathbf{A}\mathbf{P}_{R(\mathbf{B})})^\dagger = (\mathbf{A}^\dagger\mathbf{A}\mathbf{B})^\dagger(\mathbf{A}\mathbf{B}\mathbf{B}^\dagger)^\dagger$. We will proceed as follows. We will assume that $\mathbf{X}$ satisfies condition (i) of the Penrose definition. We will then try and manipulate condition (i) to get a formula that can be verified independently of condition (i). If our manipulations are reversible, we will have shown condition (i) holds. Suppose then that $\mathbf{ABXAB} = \mathbf{AB}$, or equivalently,

$$\mathbf{AB}(\mathbf{A}^\dagger\mathbf{AB})^\dagger(\mathbf{ABB}^\dagger)^\dagger\mathbf{AB} = \mathbf{AB}. \tag{2}$$

Multiply (2) on the left by $\mathbf{A}^\dagger$ and on the right by $\mathbf{B}^\dagger$. Then (2) becomes

$$\mathbf{A}^\dagger\mathbf{AB}(\mathbf{A}^\dagger\mathbf{AB})^\dagger(\mathbf{ABB}^\dagger)^\dagger\mathbf{ABB}^\dagger = \mathbf{A}^\dagger\mathbf{ABB}^\dagger, \tag{3}$$

or, $\mathbf{P}_{R(\mathbf{A}^\dagger\mathbf{AB})}\mathbf{P}_{R(\mathbf{BB}^\dagger\mathbf{A}^*)} = \mathbf{P}_{R(\mathbf{A}^*)}\mathbf{P}_{R(\mathbf{B})}$. Equivalently,

$$\mathbf{P}_{\mathbf{A}^\dagger\mathbf{A}\,R(\mathbf{B})}\mathbf{P}_{\mathbf{BB}^\dagger R(\mathbf{A}^*)} = \mathbf{P}_{R(\mathbf{A}^*)}\mathbf{P}_{R(\mathbf{B})}. \tag{4}$$

To see that (4) is true, we will show that if

$$\mathbf{E}_1 = \mathbf{P}_{\mathbf{A}^\dagger\mathbf{A}\,R(\mathbf{B})}\mathbf{P}_{\mathbf{BB}^\dagger\,R(\mathbf{A}^*)}, \mathbf{E}_2 = \mathbf{P}_{R(\mathbf{A}^*)}\mathbf{P}_{R(\mathbf{B})}, \tag{5}$$

then $\mathbf{E}_1\mathbf{u} = \mathbf{E}_2\mathbf{u}$ for all $\mathbf{u} \in \mathbb{C}^n$. Suppose first that $\mathbf{u} \in R(\mathbf{B})^\perp$. Then $\mathbf{E}_1\mathbf{u} = \mathbf{E}_2\mathbf{u} = \mathbf{0}$. Suppose then that $\mathbf{u} \in R(\mathbf{B})$. Now to find $\mathbf{E}_1\mathbf{u}$ we will need to calculate $\mathbf{P}_{\mathbf{BB}^\dagger R(\mathbf{A}^*)}\mathbf{u}$. But $\mathbf{BB}^\dagger R(\mathbf{A}^*)$ is a subspace of $R(\mathbf{B})$. Let $\mathbf{u} = \mathbf{u}_1 \oplus \mathbf{u}_2$ where $\mathbf{u}_1 \in \mathbf{BB}^\dagger R(\mathbf{A}^*)$ and $\mathbf{u}_2 \in [\mathbf{BB}^\dagger R(\mathbf{A}^*)]^\perp \cap R(\mathbf{B})$. Then

$$\mathbf{E}_1\mathbf{u} = \mathbf{P}_{\mathbf{A}^\dagger\mathbf{A}\,R(\mathbf{B})}\mathbf{P}_{\mathbf{BB}^\dagger\,R(\mathbf{A}^*)}\mathbf{u} \tag{6}$$

$$= \mathbf{P}_{\mathbf{A}^\dagger\mathbf{A}\,R(\mathbf{B})}\mathbf{u}_1 = \mathbf{A}^\dagger\mathbf{A}\mathbf{u}_1. \tag{7}$$

Equality (7) follows since $\mathbf{u}_1 \in R(\mathbf{B})$ and the projection of $R(\mathbf{B})$ onto $\mathbf{A}^\dagger\mathbf{A}R(\mathbf{B})$ is accomplished by $\mathbf{A}^\dagger\mathbf{A}$. Now $\mathbf{E}_2\mathbf{u} = \mathbf{P}_{R(\mathbf{A}^*)}\mathbf{P}_{R(\mathbf{B})}\mathbf{u} = \mathbf{P}_{R(\mathbf{A}^*)}\mathbf{u} = \mathbf{A}^\dagger\mathbf{A}\mathbf{u}$, so (4) will now follow provided that $\mathbf{A}^\dagger\mathbf{A}\mathbf{u} = \mathbf{A}^\dagger\mathbf{A}\mathbf{u}_1$, that is, if $\mathbf{u}_2 \in N(\mathbf{A}) = R(\mathbf{A}^*)^\perp$. Suppose then that $\mathbf{v} \in R(\mathbf{A}^*)$. By definition, $\mathbf{u}_2 \perp \mathbf{BB}^\dagger\mathbf{v}$. Thus $0 = (\mathbf{BB}^\dagger\mathbf{v}, \mathbf{u}_2) = (\mathbf{v}, \mathbf{BB}^\dagger\mathbf{u}_2) = (\mathbf{v}, \mathbf{u}_2)$. Hence $\mathbf{u}_2 \in R(\mathbf{A}^*)^\perp$ as desired. (4) is now established. But (4) is equivalent to (3). Multiply (3) on the left by $\mathbf{A}$ and the right by $\mathbf{B}$. This gives (2). Because of the particular nature of $\mathbf{X}$ it turns out to be easier to use $\mathbf{ABXAB} = \mathbf{AB}$ to show that $\mathbf{X}$ satisfies the Moore definition of $(\mathbf{AB})^\dagger$. Multiply on the right by $(\mathbf{AB})^\dagger$. Then we have $\mathbf{ABXP}_{R(\mathbf{AB})} = \mathbf{P}_{R(\mathbf{AB})}$. But $N(\mathbf{X}) \subseteq N((\mathbf{ABB}^\dagger)^\dagger) = R(\mathbf{ABB}^\dagger)^\perp = R(\mathbf{AB})^\perp$. Thus $\mathbf{XP}_{R(\mathbf{AB})} = \mathbf{P}_{R(\mathbf{AB})}$ and hence

$$\mathbf{ABX} = \mathbf{P}_{R(\mathbf{AB})}. \tag{8}$$

Now multiply $\mathbf{ABXAB} = \mathbf{AB}$ on the left by $(\mathbf{AB})^\dagger$ to get $\mathbf{P}_{R(\mathbf{B}^*\mathbf{A}^*)}\mathbf{XAB} = \mathbf{P}_{R((\mathbf{AB})^\dagger)}$. But $R(\mathbf{X}) \subseteq R((\mathbf{A}^\dagger\mathbf{AB})^\dagger) = R((\mathbf{A}^\dagger\mathbf{AB})^*) = R(\mathbf{B}^*\mathbf{A}^*\mathbf{A}^{*\dagger}) = R(\mathbf{B}^*\mathbf{A}^*)$ and hence $\mathbf{P}_{R(\mathbf{B}^*\mathbf{A}^*)}\mathbf{X} = \mathbf{X}$. Thus

$$\mathbf{XAB} = \mathbf{P}_{R((\mathbf{AB})^\dagger)}. \tag{9}$$

Equations (8) and (9) show that Theorem 1 holds by the Moore definition. ■

Comment It is possible that some readers might have some misgivings about equality (7). An easy way to see that $\mathbf{P}_{\mathbf{A}^\dagger\mathbf{A}\,R(\mathbf{B})}\mathbf{u}_1 = \mathbf{A}^\dagger\mathbf{A}\mathbf{u}_1$ is as follows. Suppose that $\mathbf{u}_1 \in R(\mathbf{B})$. Then $\mathbf{A}^\dagger\mathbf{A}\mathbf{u}_1 \in \mathbf{A}^\dagger\mathbf{A}R(\mathbf{B})$. But $\mathbf{A}(\mathbf{u}_1 - \mathbf{A}^\dagger\mathbf{A}\mathbf{u}_1) = \mathbf{A}\mathbf{u}_1 - \mathbf{A}\mathbf{u}_1 = \mathbf{0}$. Thus $\mathbf{u}_1 - \mathbf{A}^\dagger\mathbf{A}\mathbf{u}_1 \in R(\mathbf{A}^*)^\perp \subseteq (\mathbf{A}^\dagger\mathbf{A}R(\mathbf{B}))^\perp$. Hence $\mathbf{u}_1 = \mathbf{A}^\dagger\mathbf{A}\mathbf{u}_1 \oplus (\mathbf{u}_1 - \mathbf{A}^\dagger\mathbf{A}\mathbf{u}_1)$ where $\mathbf{A}^\dagger\mathbf{A}\mathbf{u}_1 \in \mathbf{A}^\dagger\mathbf{A}R(\mathbf{B})$ and $(\mathbf{u}_1 - \mathbf{A}^\dagger\mathbf{A}\mathbf{u}_1) \in (\mathbf{A}^\dagger\mathbf{A}R(\mathbf{B}))^\perp$. Thus $\mathbf{P}_{R(\mathbf{A}^\dagger\mathbf{AB})}\mathbf{u}_1 = \mathbf{A}^\dagger\mathbf{A}\mathbf{u}_1$.

The formula for $(\mathbf{AB})^\dagger$ simplifies if either $\mathbf{P}_{R(\mathbf{A}^*)} = \mathbf{I}$ or $\mathbf{P}_{R(\mathbf{B})} = \mathbf{I}$.

Corollary 1.4.1 *Suppose that* $\mathbf{A} \in \mathbb{C}^{m \times n}$ *and* $\mathbf{B} \in \mathbb{C}^{n \times p}$.

(i) *If* rank $(\mathbf{A}) = n$, *then* $(\mathbf{AB})^\dagger = \mathbf{B}^\dagger(\mathbf{AP}_{R(\mathbf{B})})^\dagger$.

(ii) *If* rank $(\mathbf{B}) = n$, *then* $(\mathbf{AB})^\dagger = (\mathbf{P}_{R(\mathbf{A}^*)}\mathbf{B})^\dagger\mathbf{A}^\dagger$.

Part (i) (or ii) of Corollary 1 is, of course, valid if **A** (or **B**) is invertible.

Corollary 1.4.2 *Suppose that* $\mathbf{A}\in\mathbb{C}^{m\times n}$, $\mathbf{B}\in\mathbb{C}^{n\times p}$, *and rank* $(\mathbf{A}) =$ *rank* $(\mathbf{B}) = n$. *Then* $(\mathbf{AB})^\dagger = \mathbf{B}^\dagger\mathbf{A}^\dagger$ *and* $\mathbf{A}^\dagger = \mathbf{A}^*(\mathbf{AA}^*)^{-1}$ *while* $\mathbf{B}^\dagger = (\mathbf{B}^*\mathbf{B})^{-1}\mathbf{B}^*$.

The formulas for $\mathbf{A}^\dagger$ and $\mathbf{B}^\dagger$ in Corollary 1 come from Theorem 3.2 and the factoring of $\mathbf{A} = \mathbf{AI}_n$ and $\mathbf{B} = \mathbf{I}_n\mathbf{B}$. In fact Corollary 1 can be derived from Theorem 3.2 also.

The assumptions of Corollary 2 are much more stringent than are necessary to guarantee that $(\mathbf{AB})^\dagger = \mathbf{B}^\dagger\mathbf{A}^\dagger$.

Example 1.4.1 Let $\mathbf{A} = \begin{bmatrix} 0 & 1 & 0 \\ 0 & 0 & 1 \\ 0 & 0 & 0 \end{bmatrix}$ and $\mathbf{B} = \begin{bmatrix} a & 0 & 0 \\ 0 & b & 0 \\ 0 & 0 & c \end{bmatrix}$ where $a, b, c\in\mathbb{C}$.

It is easy to see that $\mathbf{A}^\dagger = \begin{bmatrix} 0 & 0 & 0 \\ 1 & 0 & 0 \\ 0 & 1 & 0 \end{bmatrix}$ and $\mathbf{B}^\dagger = \begin{bmatrix} a^\dagger & 0 & 0 \\ 0 & b^\dagger & 0 \\ 0 & 0 & c^\dagger \end{bmatrix}$. Now

$\mathbf{AB} = \begin{bmatrix} 0 & b & 0 \\ 0 & 0 & c \\ 0 & 0 & 0 \end{bmatrix}$ and $(\mathbf{AB})^\dagger = \begin{bmatrix} 0 & 0 & 0 \\ b^\dagger & 0 & 0 \\ 0 & c^\dagger & 0 \end{bmatrix}$. On the other hand

$\mathbf{B}^\dagger\mathbf{A}^\dagger = \begin{bmatrix} 0 & 0 & 0 \\ b^\dagger & 0 & 0 \\ 0 & c^\dagger & 0 \end{bmatrix}$ so that $(\mathbf{AB})^\dagger = \mathbf{B}^\dagger\mathbf{A}^\dagger$. Notice that $\mathbf{BA} = \begin{bmatrix} 0 & a & 0 \\ 0 & 0 & b \\ 0 & 0 & 0 \end{bmatrix}$.

By varying the values of a, b and c one can see that $(\mathbf{AB})^\dagger = \mathbf{B}^\dagger\mathbf{A}^\dagger$ is possible *without*

(i) $\mathbf{AB} = \mathbf{BA}$ $(a \neq b)$
(ii) rank $(\mathbf{A}) =$ rank $(\mathbf{B})$ $(a = b = 0, c = 1)$
(iii) rank $(\mathbf{AB}) =$ rank $(\mathbf{BA})$ $(a = b \neq 0, c = 0)$.

The list can be continued, but the point has been made. The question remains, when is $(\mathbf{AB})^\dagger = \mathbf{B}^\dagger\mathbf{A}^\dagger$? Consider Example 1 again. What was it about that **A** and **B** that made $(\mathbf{AB})^\dagger = \mathbf{B}^\dagger\mathbf{A}^\dagger$? The only thing that **A** and **B** seem to have in common are some invariant subspaces. The subspaces $R(\mathbf{A}), R(\mathbf{A}^*), N(\mathbf{A})$, and $N(\mathbf{A}^*)$ are all invariant for both **A** and **B**. A statement about invariant subspaces is also a statement about projectors.

A possible method of attack has suggested itself. We will assume that $(\mathbf{AB})^\dagger = \mathbf{B}^\dagger\mathbf{A}^\dagger$. From this we will try to derive statements about projectors. In Example 1, $\mathbf{A}^*\mathbf{A}$ and **B** were simultaneously diagonalizable, so we should see if $\mathbf{A}^*\mathbf{A}$ works in. Finally, we should check to see if our conditions are necessary as well as sufficient.

Assume then that $\mathbf{A}\in\mathbb{C}^{m\times n}$, $\mathbf{B}\in\mathbb{C}^{n\times p}$, and

$$(\mathbf{AB})^\dagger = \mathbf{B}^\dagger\mathbf{A}^\dagger. \tag{10}$$

Theorem 1 gives another formula for $(\mathbf{AB})^\dagger$. Substitute that into (10) to get $(\mathbf{A}^\dagger\mathbf{AB})^\dagger(\mathbf{ABB}^\dagger)^\dagger = \mathbf{B}^\dagger\mathbf{A}^\dagger$. To change this into a projector equation, multiply on the left by $(\mathbf{A}^\dagger\mathbf{A}\mathbf{B})$ and on the right by $(\mathbf{A}\mathbf{B}\mathbf{B}^\dagger)$, to give

$$\mathbf{P}_{R(\mathbf{A}^\dagger\mathbf{AB})}\mathbf{P}_{R(\mathbf{BB}^\dagger\mathbf{A}^*)} = (\mathbf{P}_{R(\mathbf{A}^*)}\mathbf{P}_{R(\mathbf{B})})^2. \tag{11}$$

By equation (4), (11) can be rewritten as $\mathbf{P}_{R(\mathbf{A}^*)}\mathbf{P}_{R(\mathbf{B})} = (\mathbf{P}_{R(\mathbf{A}^*)}\mathbf{P}_{R(\mathbf{B})})^2$ and hence $\mathbf{P}_{R(\mathbf{A}^*)}\mathbf{P}_{R(\mathbf{B})}$ is a projector. But the product of two hermitian projectors is a projector if and only if the two hermitian projectors commute. Thus (recall that $[\mathbf{X}, \mathbf{Y}] = \mathbf{XY} - \mathbf{YX}$)

$$[\mathbf{P}_{R(\mathbf{A}^*)}, \mathbf{P}_{R(\mathbf{B})}] = \mathbf{0}, \text{ or equivalently, } [\mathbf{A}^\dagger\mathbf{A}, \mathbf{BB}^\dagger] = \mathbf{0}, \text{ if } (\mathbf{AB})^\dagger = \mathbf{B}^\dagger\mathbf{A}^\dagger. \tag{12}$$

Is (12) enough to guarantee (10)? We continue to see if we can get any additional conditions. Example 1 suggested that an $\mathbf{A}^*\mathbf{A}$ term might be useful. If (10) holds, then $\mathbf{ABB}^\dagger\mathbf{A}^\dagger$ is hermitian. But then $\mathbf{A}^*(\mathbf{ABB}^\dagger\mathbf{A}^\dagger)\mathbf{A}$ is hermitian. Thus

$$\mathbf{A}^*(\mathbf{ABB}^\dagger)\mathbf{A}^\dagger\mathbf{A} = \mathbf{A}^\dagger\mathbf{ABB}^\dagger\mathbf{A}^*\mathbf{A}. \tag{13}$$

Using (12) and the fact that $\mathbf{A}^\dagger\mathbf{AA}^* = \mathbf{A}^*$, (13) becomes $\mathbf{A}^*\mathbf{ABB}^\dagger = \mathbf{BB}^\dagger\mathbf{A}^*\mathbf{A}$ or

$$[\mathbf{A}^*\mathbf{A}, \mathbf{BB}^\dagger] = \mathbf{0}. \tag{14}$$

Condition (14) is a stronger condition than (12) since it involves $\mathbf{A}^*\mathbf{A}$ and not just $\mathbf{P}_{R(\mathbf{A}^*\mathbf{A})}$.

In Theorem 1 there was a certain symmetry in the formula. It seems unreasonable that in conditions for (10) that either $\mathbf{A}$ or $\mathbf{B}$ should be more important. We return to equation (10) to try and derive a formula like (14) but with the roles of $\mathbf{A}$ and $\mathbf{B}$ 'reversed'. Now $\mathbf{B}^\dagger\mathbf{A}^\dagger\mathbf{AB}$ is hermitian since we are assuming (10) holds. Thus $\mathbf{BB}^\dagger\mathbf{A}^\dagger\mathbf{ABB}^*$ is hermitian. Proceeding as before, we get that

$$[\mathbf{BB}^*, \mathbf{A}^\dagger\mathbf{A}] = \mathbf{0}. \tag{15}$$

Continued manipulation fails to produce any conditions not implied by (14) and (15). We are led then to attempt to prove the following theorem.

Theorem 1.4.2 *Suppose that* $\mathbf{A} \in \mathbb{C}^{m \times n}$ *and* $\mathbf{B} \in \mathbb{C}^{n \times p}$. *Then the following statements are equivalent:*

(i) $(\mathbf{AB})^\dagger = \mathbf{B}^\dagger\mathbf{A}^\dagger$.
(ii) $\mathbf{BB}^*\mathbf{A}^\dagger\mathbf{A}$ *and* $\mathbf{A}^*\mathbf{ABB}^\dagger$ *are hermitian.*
(iii) $R(\mathbf{A}^*)$ *is an invariant subspace for* $\mathbf{BB}^*$ *and* $R(\mathbf{B})$ *is an invariant subspace of* $\mathbf{A}^*\mathbf{A}$.
(iv) $\mathbf{P}_{N(\mathbf{A})}\mathbf{BB}^*\mathbf{P}^\perp_{N(\mathbf{A})} = \mathbf{0}$ *and* $\mathbf{P}_{N(\mathbf{B}^*)}\mathbf{A}^*\mathbf{AP}^\perp_{N(\mathbf{B}^*)} = \mathbf{0}$.
(v) $\mathbf{A}^\dagger\mathbf{ABB}^*\mathbf{A}^* = \mathbf{BB}^*\mathbf{A}^*$ *and* $\mathbf{BB}^\dagger\mathbf{A}^*\mathbf{AB} = \mathbf{A}^*\mathbf{AB}$.

Proof Statement (iii) is equivalent to equations (14) and (15) so that (i) implies (ii). We will first show that (ii)–(v) are all equivalent. Since $\mathbf{BB}^*$ and $\mathbf{A}^*\mathbf{A}$ are hermitian, all of their invariant subspaces are reducing. Thus (ii) and (iii) are equivalent. Now observe that if $\mathbf{C}$ is a matrix and M a subspace, then $\mathbf{CP}_M^\perp = (\mathbf{I} - \mathbf{P}_M)\mathbf{CP}_M^\perp + \mathbf{P}_M\mathbf{CP}_M^\perp = \mathbf{P}_M^\perp\mathbf{CP}_M^\perp + \mathbf{P}_M\mathbf{CP}_M^\perp$. This says that $M^\perp$ is an invariant subspace if and only if $\mathbf{P}_M\mathbf{CP}_M^\perp = 0$. Thus (iii) and (iv) are equivalent. Since (v) is written algebraically, we will show it is equivalent to (ii). Assume then that $\mathbf{BB^*A^\dagger A}$ is hermitian.

Then $\mathbf{BB^*A^\dagger A} = \mathbf{A^\dagger ABB^*}$. Thus $\mathbf{BB^*(A^\dagger A)A^*} = \mathbf{A^\dagger ABB^*A^*}$, or

$$\mathbf{BB^*A^*} = \mathbf{A^\dagger ABB^*A^*}. \tag{16}$$

Similarly if $\mathbf{A^*ABB^\dagger}$ is hermitian, then

$$\mathbf{BB^\dagger A^*AB} = \mathbf{A^*AB} \tag{17}$$

so that (ii) implies (v). Assume now that (16) and (17) hold. Multiply (17) on the right by $\mathbf{B}^\dagger$ and (16) on the right by $\mathbf{A}^{*\dagger}$. The new equations are precisely statement (iii) which is equivalent to (ii). Thus (ii)–(v) are all equivalent. Suppose now that (ii)–(v) hold. We want to prove (i). Observe that $\mathbf{B^\dagger A^\dagger} = \mathbf{B^\dagger(BB^\dagger)(A^\dagger A)A^\dagger} = \mathbf{B^\dagger(A^\dagger A)(BB^\dagger)A^\dagger}$, while by Theorem 1 $\mathbf{(AB)^\dagger} = \mathbf{(A^\dagger AB)^\dagger(ABB^\dagger)^\dagger}$. Theorem 2 will be proven if we can show that

$$\mathbf{(A^\dagger AB)^\dagger} = \mathbf{B^\dagger(A^\dagger A)} \quad \text{and} \tag{18}$$

$$\mathbf{(ABB^\dagger)^\dagger} = \mathbf{BB^\dagger A^\dagger}. \tag{19}$$

To see if (18) holds, check the Penrose equations. Let $\mathbf{X} = \mathbf{B^\dagger(A^\dagger A)}$. Then $\mathbf{A^\dagger ABX} = \mathbf{A^\dagger ABB^\dagger A^\dagger A} = \mathbf{A^\dagger AA^\dagger ABB^\dagger} = \mathbf{A^\dagger ABB^\dagger} = \mathbf{P}_{R(\mathbf{A^\dagger ABB^\dagger})} = \mathbf{P}_{R(\mathbf{A^\dagger AB})}$. Thus Penrose conditions (i) and (iii) are satisfied. Now $\mathbf{X(A^\dagger AB)} = \mathbf{B^\dagger(A^\dagger A)(A^\dagger A)B} = \mathbf{B^\dagger(A^\dagger A)B}$. Thus $\mathbf{X(A^\dagger AB)X} = \mathbf{B^\dagger(A^\dagger A)BB^\dagger(A^\dagger A)} = \mathbf{B^\dagger A^\dagger A} = \mathbf{X}$ and Penrose condition (ii) is satisfied. There remains then only to show that $\mathbf{X(A^\dagger AB)}$ is hermitian. But $\mathbf{A^\dagger ABB^*}$ is hermitian by assumption (ii). Thus $\mathbf{B^\dagger(A^\dagger ABB^*)B^{\dagger *}}$ is hermitian and $\mathbf{B^\dagger(A^\dagger A)BB^*B^{*\dagger}} = \mathbf{B^\dagger(A^\dagger A)B} = \mathbf{X(A^\dagger AB)}$ as desired. The proof of (19) is similar and left to the exercises. ■

It is worth noticing that conditions (ii)–(v) of Theorem 1.4.2 make Fig. 1.1 correct. The interval (b', c') would stand for one subspace $R(\mathbf{A^\dagger ABB^\dagger})$ rather than two skewed subspaces, $R(\mathbf{A^\dagger ABB^\dagger})$ and $R(\mathbf{BB^\dagger A^\dagger A})$. A reader might think that perhaps there is a weaker appearing set of conditions than (ii)–(v) that would imply $\mathbf{(AB^\dagger)} = \mathbf{B^\dagger A^\dagger}$. The next Example shows that even with relatively simple matrices the full statement of the conditions is needed.

Example 1.4.2 Let $\mathbf{A} = \begin{bmatrix} 0 & 0 & 0 \\ 0 & 1 & 1 \\ 0 & 1 & 0 \end{bmatrix}$ and $\mathbf{B} = \begin{bmatrix} 1 & 0 & 0 \\ 0 & 1 & 0 \\ 0 & 0 & 0 \end{bmatrix}$. Then $\mathbf{B} = \mathbf{B}^* = \mathbf{B}^\dagger = \mathbf{B}^2$ and $\mathbf{A}^\dagger = \begin{bmatrix} 0 & 0 \quad 0 \\ 0 & \begin{bmatrix} 1 & 1 \\ 1 & 0 \end{bmatrix}^{-1} \\ 0 & \end{bmatrix} = \begin{bmatrix} 0 & 0 & 0 \\ 0 & 0 & 1 \\ 0 & 1 & -1 \end{bmatrix}$. Now $\mathbf{BB^*A^\dagger A}$ is hermitian so that $[\mathbf{BB^*}, \mathbf{A^\dagger A}] = \mathbf{0}$ and $[\mathbf{BB^\dagger}, \mathbf{A^\dagger A}] = \mathbf{0}$. However,

$\mathbf{A^*ABB^\dagger} = \begin{bmatrix} 0 & 0 & 0 \\ 0 & 2 & 0 \\ 0 & 1 & 0 \end{bmatrix}$ which is not hermitian so that $\mathbf{(AB)^\dagger} \neq \mathbf{B^\dagger A^\dagger}$.

An easy to verify condition that implies $(\mathbf{AB})^{\dagger} = \mathbf{B}^{\dagger}\mathbf{A}^{\dagger}$ is the following.

Corollary 1.4.3 *If* $\mathbf{A}^*\mathbf{ABB}^* = \mathbf{BB}^*\mathbf{A}^*\mathbf{A}$, *then* $(\mathbf{AB})^{\dagger} = \mathbf{B}^{\dagger}\mathbf{A}^{\dagger}$.

The proof of Corollary 2 is left to the exercises.

Corollary 2 has an advantage over conditions (ii)–(iv) of Theorem 2 in that one does not have to calculate $\mathbf{A}^{\dagger}, \mathbf{B}^{\dagger}$, or any projectors to verify it. It has the disadvantage that it is only sufficient and not necessary. Notice that the $\mathbf{A}, \mathbf{B}$ in Example 1 satisfy $[\mathbf{A}^*\mathbf{A}, \mathbf{BB}^*] = \mathbf{0}$ while those in Example 2 do not.

There is another approach to the problem of determining when $(\mathbf{AB})^{\dagger} = \mathbf{B}^{\dagger}\mathbf{A}^{\dagger}$. It is to try and define a different kind of inverse of $\mathbf{A}$, call it $\mathbf{A}^{\sim}$, so that $(\mathbf{AB})^{\sim} = \mathbf{B}^{\sim}\mathbf{A}^{\sim}$. This approach will not be discussed.

5. Exercises

1. Each of the following is an alternative set of equations whose unique solution $\mathbf{X}$ is the generalized inverse of $\mathbf{A}$. For each definition show that it is equivalent to one of the three given in the text.

   (a) $\mathbf{AX} = \mathbf{P}_{R(\mathbf{A})}, N(\mathbf{X}^*) = N(\mathbf{A})$.
   (b) $\mathbf{AX} = \mathbf{P}_{R(\mathbf{A})}, \mathbf{XA} = \mathbf{P}_{R(\mathbf{A}^*)}, \mathbf{XAX} = \mathbf{X}$.
   (c) $\mathbf{XAA}^* = \mathbf{A}^*, \mathbf{XX}^*\mathbf{A}^* = \mathbf{X}$.
   (d) $\mathbf{XAx} = \begin{cases} \mathbf{x} & \text{if } \mathbf{x} \in R(\mathbf{A}^*) \\ \mathbf{0} & \text{if } \mathbf{x} \in N(\mathbf{A}^*). \end{cases}$
   (e) $\mathbf{XA} = \mathbf{P}_{R(\mathbf{A}^*)}, N(\mathbf{X}) = N(\mathbf{A}^*)$.

Comment: Many of these have appeared as definitions or theorems in the literature. Notice the connection between Example 2.1, Exercise 1(b), and Exercise 3 below.

2. Derive a set of conditions equivalent to those given in Definition 1.2 or Definition 1.3. Show they are equivalent. Can you derive others?
3. Suppose that $\mathbf{A} \in \mathbb{C}^{m \times n}$. Prove that a matrix $\mathbf{X}$ satisfies $\mathbf{AX} = \mathbf{P}_{R(\mathbf{A})}$, $\mathbf{XA} = \mathbf{P}_{R(\mathbf{A}^*)}$ if and only if $\mathbf{X}$ satisfies (i), (iii), and (iv) of Definition 1.3. Such an $\mathbf{X}$ is called a (1, 3, 4)-inverse of $\mathbf{A}$ and will be discussed later. Observe that it cannot be unique if it is not $\mathbf{A}^{\dagger}$ since trivially $\mathbf{A}^{\dagger}$ is also a (1, 3, 4)-inverse.
4. Calculate $\mathbf{A}^{\dagger}$ from Theorem 3.1 when $\mathbf{A} = \begin{bmatrix} 1 & 2 & 1 \\ 1 & 2 & 1 \end{bmatrix}$, and when

   $$\mathbf{A} = \begin{bmatrix} 1 & 3 \\ 1 & 1 \\ 1 & 1 \end{bmatrix}.$$

   Hint: For the second matrix see Example 6.
5. Show that if rank $(\mathbf{A}) = 1$, then $\mathbf{A}^{\dagger} = \dfrac{1}{k}\mathbf{A}^*$ where $k = \sum_{i=1}^{m} \sum_{j=1}^{n} |a_{ij}|^2$.
6. If $\mathbf{A} \in \mathbb{C}^{m \times n}$ and rank $(\mathbf{A}) = n$, notice that in Theorem 3.2, $\mathbf{C}$ may be chosen as an especially simple invertible matrix. Derive the formula

for $\mathbf{A}^\dagger$ under the assumption that rank $\mathbf{A} = n$. Do the same for the case when rank $(\mathbf{A}) = m$.

7. Let $\mathbf{A} = \begin{bmatrix} 1 & 0 & 1 & 2 \\ 0 & 0 & 0 & 1 \\ -1 & 0 & -1 & 0 \end{bmatrix}$. Calculate $\mathbf{A}^\dagger$ from Definition 1.1.
8. Verify that $(\mathbf{A}^\dagger)^* = (\mathbf{A}^*)^\dagger$.
9. Verify that $(\lambda\mathbf{A})^\dagger = \lambda^\dagger\mathbf{A}^\dagger, \lambda\in\mathbb{C}$, where $\lambda^\dagger = \frac{1}{\lambda}$ if $\lambda \neq 0$ and $0^\dagger = 0$.
10. If $\mathbf{A}\in\mathbb{C}^{m\times n}$, and $\mathbf{U}, \mathbf{V}$ are unitary matrices, verify that $(\mathbf{UAV})^\dagger = \mathbf{V}^*\mathbf{A}^\dagger\mathbf{U}^*$.
11. Derive an explicit formula for $\mathbf{A}^\dagger$ that involves no $(^\dagger)$'s by using the singular value decomposition.

*12. Verify that $\mathbf{A}^\dagger = \int_0^\infty \mathrm{e}^{-(\mathbf{A}^*\mathbf{A})t}\mathbf{A}^*\mathrm{d}t$.

*13. Verify the $\mathbf{A}^\dagger = \frac{1}{2\pi\mathrm{i}}\int_C \frac{1}{z}(z\mathbf{I} - \mathbf{A}^*\mathbf{A})^{-1}\mathbf{A}^*\mathrm{d}z$ where C is a closed contour containing the non-zero eigenvalues of $\mathbf{A}^*\mathbf{A}$, but not containing the zero eigenvalue of $\mathbf{A}^*\mathbf{A}$ in or on it.

14. Prove Proposition 1.1.

Exercises 15–21 are all drawn from the literature. They were originally done by Schwerdtfeger, Baskett and Katz, Greville, and Erdelyi. Some follow almost immediately from Theorem 4.1. Others require more work.

15. Prove that if $\mathbf{A}\in\mathbb{C}^{m\times n}$ and $\mathbf{B}\in\mathbb{C}^{n\times m}$ such that rank $\mathbf{A}$ = rank $\mathbf{B}$ and if the eigenvectors corresponding to non-zero eigenvalues of the two matrices $\mathbf{A}^*\mathbf{A}$ and $\mathbf{BB}^*$ span the same space, then $(\mathbf{AB})^\dagger = \mathbf{B}^\dagger\mathbf{A}^\dagger$.
16. Prove that if $\mathbf{A}\in\mathbb{C}^{n\times n}, \mathbf{B}\in\mathbb{C}^{n\times n}$ and $\mathbf{AA}^\dagger = \mathbf{A}^\dagger\mathbf{A}, \mathbf{BB}^\dagger = \mathbf{B}^\dagger\mathbf{B}, (\mathbf{AB})^\dagger\mathbf{AB} = \mathbf{AB}(\mathbf{AB})^\dagger$, and rank $\mathbf{A}$ = rank $\mathbf{B}$ = rank$(\mathbf{AB})$, then $(\mathbf{AB})^\dagger = \mathbf{B}^\dagger\mathbf{A}^\dagger$.
17. Assume that $\mathbf{A}\in\mathbb{C}^{m\times n}, \mathbf{B}\in\mathbb{C}^{n\times p}$. Show that the following statements are equivalent.

    (i) $(\mathbf{AB})^\dagger = \mathbf{B}^\dagger\mathbf{A}^\dagger$.
    (ii) $\mathbf{A}^\dagger\mathbf{ABB}^*\mathbf{A}^*\mathbf{ABB}^\dagger = \mathbf{BB}^*\mathbf{A}^*\mathbf{A}$.
    (iii) $\mathbf{A}^\dagger\mathbf{AB} = \mathbf{B}(\mathbf{AB})^\dagger\mathbf{AB}$ and $\mathbf{BB}^\dagger\mathbf{A}^* = \mathbf{A}^*\mathbf{AB}(\mathbf{AB})^\dagger$.
    (iv) $(\mathbf{A}^\dagger\mathbf{ABB}^\dagger)^* = (\mathbf{A}^\dagger\mathbf{ABB}^\dagger)^\dagger$ and the two matrices $\mathbf{ABB}^\dagger\mathbf{A}^\dagger$ and $\mathbf{B}^\dagger\mathbf{A}^\dagger\mathbf{AB}$ are both hermitian.

18. Show that if $[\mathbf{A}, \mathbf{P}_{R(\mathbf{B})}] = \mathbf{0}, [\mathbf{A}^\dagger, \mathbf{P}_{R(\mathbf{B})}] = \mathbf{0}, [\mathbf{B}, \mathbf{P}_{R(\mathbf{A}^*)}] = \mathbf{0}$, and $[\mathbf{B}^\dagger, \mathbf{P}_{R(\mathbf{A}^*)}] = \mathbf{0}$, then $(\mathbf{AB})^\dagger = \mathbf{B}^\dagger\mathbf{A}^\dagger$.
19. Prove that if $\mathbf{A}^* = \mathbf{A}^\dagger$ and $\mathbf{B}^* = \mathbf{B}^\dagger$ and if *any* of the conditions of Exercise 18 hold, then $(\mathbf{AB})^\dagger = \mathbf{B}^\dagger\mathbf{A}^\dagger = \mathbf{B}^*\mathbf{A}^*$.
20. Prove that if $\mathbf{A}^* = \mathbf{A}^\dagger$ and the third and fourth conditions of Exercise 18 hold, then $(\mathbf{AB})^\dagger = \mathbf{B}^\dagger\mathbf{A}^\dagger = \mathbf{B}^\dagger\mathbf{A}^*$.
21. Prove that if $\mathbf{B}^* = \mathbf{B}^\dagger$ and the first and second conditions of Exercise 18 hold, then $(\mathbf{AB})^\dagger = \mathbf{B}^\dagger\mathbf{A}^\dagger$.
22. Prove that if $[\mathbf{A}^*\mathbf{A}, \mathbf{BB}^*] = \mathbf{0}$, then $(\mathbf{AB})^\dagger = \mathbf{B}^\dagger\mathbf{A}^\dagger$.

23. Verify that the product of two hermitian matrices $\mathbf{A}$ and $\mathbf{B}$ is hermitian if and only if $[\mathbf{A}, \mathbf{B}] = \mathbf{0}$.
24. Suppose that $\mathbf{P}, \mathbf{Q}$ are hermitian projectors. Show that $\mathbf{PQ}$ is a projector if and only if $[\mathbf{P}, \mathbf{Q}] = \mathbf{0}$.
25. Assuming (ii)–(v) of Theorem 8, show that $(\mathbf{AP}_{R(\mathbf{B})})^\dagger = \mathbf{P}^\dagger_{R(\mathbf{B})}\mathbf{A}^\dagger = \mathbf{P}_{R(\mathbf{B})}\mathbf{A}^\dagger$ without directly using the fact that $(\mathbf{AB})^\dagger = \mathbf{B}^\dagger\mathbf{A}^\dagger$.
26. Write an expression for $(\mathbf{PAQ})^\dagger$ when $\mathbf{P}$ and $\mathbf{Q}$ are non-singular.
27. Derive necessary and sufficient conditions on $\mathbf{P}$ and $\mathbf{A}$, $\mathbf{P}$ non-singular, for $(\mathbf{P}^{-1}\mathbf{AP})^\dagger$ to equal $\mathbf{P}^{-1}\mathbf{A}^\dagger\mathbf{P}$.
28. Prove that if $\mathbf{A} \in \mathbb{R}^{m \times n}$ and the entries of $\mathbf{A}$ are rational numbers, then the entries of $\mathbf{A}^\dagger$ are rational.

2
Least squares solutions

1. What kind of answer is $\mathbf{A}^\dagger\mathbf{b}$?

At this point the reader should have gained a certain facility in working with generalized inverses, and it is time to find out what kind of solutions they give. Before proceeding we need a simple geometric lemma. Recall that if $\mathbf{w} = [w_1, \ldots, w_p]^* \in \mathbb{C}^p$, then $\|\mathbf{w}\| = \left(\sum_{i=1}^{p} |w_i|^2\right)^{1/2} = (\mathbf{w}^*\mathbf{w})^{1/2}$ denotes the Euclidean norm of $\mathbf{w}$.

Lemma 2.1.1 *If* $\mathbf{u}, \mathbf{v} \in \mathbb{C}^p$ *and* $(\mathbf{u}, \mathbf{v}) = 0$, *then* $\|\mathbf{u} + \mathbf{v}\|^2 = \|\mathbf{u}\|^2 + \|\mathbf{v}\|^2$.

Proof Suppose that $\mathbf{u}, \mathbf{v} \in \mathbb{C}^p$ and $(\mathbf{u}, \mathbf{v}) = 0$. Then

$$\|\mathbf{u} + \mathbf{v}\|^2 = (\mathbf{u} + \mathbf{v}, \mathbf{u} + \mathbf{v}) = (\mathbf{u}, \mathbf{u}) + (\mathbf{v}, \mathbf{u}) + (\mathbf{u}, \mathbf{v}) + (\mathbf{v}, \mathbf{v}) = \|\mathbf{u}\|^2 + \|\mathbf{v}\|^2.$$ ■

Now consider again the problem of finding solutions $\mathbf{u}$ to

$$\mathbf{Ax} = \mathbf{b}, \mathbf{A} \in \mathbb{C}^{m \times n}, \mathbf{b} \in \mathbb{C}^m. \tag{1}$$

If (1) is inconsistent, one could still look for $\mathbf{u}$ that makes $\mathbf{Au} - \mathbf{b}$ as small as possible.

Definition 2.1.1 *Suppose that* $\mathbf{A} \in \mathbb{C}^{m \times n}$ *and* $\mathbf{b} \in \mathbb{C}^m$. *Then a vector* $\mathbf{u} \in \mathbb{C}^n$ *is called a* least squares solution *to* $\mathbf{Ax} = \mathbf{b}$ *if* $\|\mathbf{Au} - \mathbf{b}\| \leq \|\mathbf{Av} - \mathbf{b}\|$ *for all* $\mathbf{v} \in \mathbb{C}^n$. *A vector* $\mathbf{u}$ *is called a* minimal least squares solution *to* $\mathbf{Ax} = \mathbf{b}$ *if* $\mathbf{u}$ *is a least squares solution to* $\mathbf{Ax} = \mathbf{b}$ *and* $\|\mathbf{u}\| < \|\mathbf{w}\|$ *for all other least squares solutions* $\mathbf{w}$.

The name 'least squares' comes from the definition of the Euclidean norm as the square root of a sum of squares. If $\mathbf{b} \in R(\mathbf{A})$, then the notions of solution and least squares solution obviously coincide.

The next theorem speaks for itself.

Theorem 2.1.1 *Suppose that* $\mathbf{A} \in \mathbb{C}^{m \times n}$ *and* $\mathbf{b} \in \mathbb{C}^m$. *Then* $\mathbf{A}^\dagger\mathbf{b}$ *is the* *minimal least squares solution to* $\mathbf{Ax} = \mathbf{b}$.

Proof Notice that $\|\mathbf{Ax}-\mathbf{b}\|^2 = \|(\mathbf{Ax}-\mathbf{AA}^\dagger\mathbf{b}) \oplus -(\mathbf{I}-\mathbf{AA}^\dagger)\mathbf{b}\|^2$

$$= \|\mathbf{Ax}-\mathbf{AA}^\dagger\mathbf{b}\|^2 + \|(\mathbf{I}-\mathbf{AA}^\dagger)\mathbf{b}\|^2.$$

Thus $\mathbf{x}$ will be a least squares solution if and only if $\mathbf{x}$ is a solution of the consistent system $\mathbf{Ax} = \mathbf{AA}^\dagger\mathbf{b}$. But solutions of $\mathbf{Ax} = \mathbf{AA}^\dagger\mathbf{b}$ are of the form $\mathbf{x} = \mathbf{A}^\dagger(\mathbf{AA}^\dagger\mathbf{b}) \oplus (\mathbf{I}-\mathbf{A}^\dagger\mathbf{A})\mathbf{h} = \mathbf{A}^\dagger\mathbf{b} \oplus (\mathbf{I}-\mathbf{A}^\dagger\mathbf{A})\mathbf{h}$. Since $\|\mathbf{x}\|^2 = \|\mathbf{A}^\dagger\mathbf{b}\|^2$ we see that there is exactly one minimal least squares solution $\mathbf{x} = \mathbf{A}^\dagger\mathbf{b}$. ■

As a special case of Theorem 2.1.1, we have the usual description of an orthogonal projection.

Corollary 2.1.1 *Suppose that M is a subspace of $\mathbb{C}^n$ and $\mathbf{P}_M$ is the orthogonal projector of $\mathbb{C}^n$ onto M. If $\mathbf{b}\in\mathbb{C}^n$, then $\mathbf{P}_M\mathbf{b}$ is the unique* closest vector *in M to $\mathbf{b}$ with respect to the Euclidean norm.*

In some applications, the minimality of the norm of a least squares solution is important, in others it is not. If the minimality is not important, then the next theorem can be very useful.

Theorem 2.1.2 *Suppose that $\mathbf{A}\in\mathbb{C}^{m\times n}$ and $\mathbf{b}\in\mathbb{C}^m$. Then the following statements are equivalent*

(i) $\mathbf{u}$ *is a least squares solution of* $\mathbf{Ax} = \mathbf{b}$,
(ii) $\mathbf{u}$ *is a solution of* $\mathbf{Ax} = \mathbf{AA}^\dagger\mathbf{b}$,
(iii) $\mathbf{u}$ *is a solution of* $\mathbf{A}^*\mathbf{Ax} = \mathbf{A}^*\mathbf{b}$.
(iv) $\mathbf{u}$ *is of the form* $\mathbf{A}^\dagger\mathbf{b} + \mathbf{h}$ *where* $\mathbf{h}\in N(\mathbf{A})$.

Proof We know from the proof of Theorem 1 that (i), (ii) and (iv) are equivalent. If (i) holds, then multiplying $\mathbf{Au} = \mathbf{b}$ on the left by $\mathbf{A}^*$ gives (iii). On the other hand, multiplying $\mathbf{A}^*\mathbf{Au} = \mathbf{A}^*\mathbf{b}$ on the left by $\mathbf{A}^{*\dagger}$ gives $\mathbf{Au} = \mathbf{AA}^\dagger\mathbf{b}$. Thus (iii) implies (ii). ■

Notice that the system of equations in statement (iii) of Theorem 2 does not involve $\mathbf{A}^\dagger$ and is a consistent system of equations. They are called the *normal equations* and play an important role in certain areas of statistics.

It was pointed out during the introduction to this section that if $\mathbf{X}$ satisfies $\mathbf{AXA} = \mathbf{A}$, and $\mathbf{b}\in R(\mathbf{A})$, then $\mathbf{Xb}$ is a solution to (1). Thus, for consistent systems a weaker type of inverse than the Moore–Penrose would suffice. However, if $\mathbf{b}\notin R(\mathbf{A})$, then the condition $\mathbf{AXA} = \mathbf{A}$ is not enough to guarantee that $\mathbf{Xb}$ is a least squares solution.

Fact There exist matrices $\mathbf{X}$, $\mathbf{A}$ and vector $\mathbf{b}$, $\mathbf{b}\notin R(\mathbf{A})$, such that $\mathbf{AXA} = \mathbf{A}$ but $\mathbf{Xb}$ is not a least squares solution of $\mathbf{Ax} = \mathbf{b}$.

Example 2.1.1 Let $\mathbf{A} = \begin{bmatrix} 2 & 0 \\ 0 & 0 \end{bmatrix}$. If $\mathbf{X}$ satisfies $\mathbf{AXA} = \mathbf{A}$, then $\mathbf{X}$ is of the form $\begin{bmatrix} 1/2 & x_{12} \\ x_{21} & x_{22} \end{bmatrix}$. Let $\mathbf{b} = \begin{bmatrix} 1 \\ 1 \end{bmatrix}$. Then by Theorem 2 a vector $\mathbf{u}$ is a least squares solution to $\mathbf{Ax} = \mathbf{b}$ if and only if $\mathbf{Ax} = \mathbf{b}_1$ where $\mathbf{b}_1 = \begin{bmatrix} 1 \\ 0 \end{bmatrix}$. If $\mathbf{u}$ is a least squares solution, then $\|\mathbf{Au}-\mathbf{b}\| = \|\mathbf{b}_1 - \mathbf{b}\| = 1$. But $\mathbf{A}(\mathbf{Xb}) = \begin{bmatrix} 1+2x_{12} \\ 0 \end{bmatrix}$.

Thus $\| \mathbf{A}(\mathbf{Xb}) - \mathbf{b} \| = (1 + 4|x_{12}|^2)^{1/2}$. If $x_{12} \neq 0$ then $\| \mathbf{A}(\mathbf{Xb}) - \mathbf{b} \| > 1$ and $\mathbf{Xb}$ will not be a least squares solution $\mathbf{Ax} = \mathbf{b}$.

Example 2 also points out that one can get least squares solutions of the form $\mathbf{Xb}$ where $\mathbf{X}$ is not $\mathbf{A}^\dagger$. Exactly what conditions need to be put on $\mathbf{X}$ to guarantee $\mathbf{Xb}$ is a least squares solution to $\mathbf{Ax} = \mathbf{b}$ will be discussed in Chapter 6.

2. Fitting a linear hypothesis

Consider the law of gravitation which says that the force of attraction y between two unit-mass points is inversely proportional to the square of the distance d between the points. If $x = 1/d^2$, then the mathematical formulation of the relationship between y and x is $y = f(x) = \beta x$ where β is an unknown constant. Because the function f is a linear function, we consider this to be a *linear functional relationship between x and y*. Many such relationships exist. The physical sciences abound with them.

Suppose that an experiment is conducted in which a distance d_o between two unit masses is set and the force of attraction y_o between them is measured. A value for the constant β is then obtained as $\beta = y_o/x_o = y_o d_o^2$. However, if the experiment is conducted a second time, one should not be greatly surprised if a slightly different value of β is obtained. Thus, for the purposes of estimating the value of β, it is more realistic to say that for each fixed value of x, we expect the observed values y_i of y to satisfy an equation of the form $y_i = \beta x + e_i$ where e_i is a measurement error which occurs more or less at random. Furthermore, if continued observations of y were made at a fixed value for x, it is natural to expect that the errors would average out to zero in the long run.

Aside from *measurement errors*, there may be another reason why different observations of y might give rise to different values of β. The force of attraction may vary with unknown quantities other than distance (e.g. the speed of the frame of reference with respect to the speed of light). That is, the *true* functional relationship may be $y = \beta x + g(u_1, u_2, \ldots, u_n)$ where the function g is unknown. Here again, it may not be unreasonable to expect that at each fixed value of x, the function g assumes values more or less at random and which average out to zero in the long run. This second type of error will be called *functional error*.

Many times, especially in the physical sciences, the functional relationship between the quantities in question is beyond reproach so that measurement error is the major consideration. However, in areas such as economics, agriculture, and the social sciences the relationships which exist are much more subtle and one must deal with both types of error.

The above remarks lead us to the following definition.

Definition 2.2.1 *When we hypothesize that y is* related linearly *to* $x_1, x_2, \ldots, x_n$, *we are hypothesizing that for each set of values* $\mathbf{p}_i = (x_{i1}, x_{i2}, \ldots, x_{in})$ *for* $x_1, x_2, \ldots, x_n$, *the observations* y_i *for y at* $\mathbf{p}_i$ *can be expressed as* $y_i = \beta_0 + \beta_1 x_{i1} + \beta_2 x_{i2} + \ldots + \beta_n x_{in} + e_{y_i}$ *where* (i) $\beta_0, \beta_1, \ldots, \beta_n$ *are unknown constants (called* parameters). (ii) e_{y_i} *is a value assumed by an unknown real valued function* $e_{\mathbf{p}_i}$ *such that* $e_{\mathbf{p}_i}$ *has the property that the values which it assumes will 'average out' to zero over all possible observations* y_i *at* $\mathbf{p}_i$.

That is, when we hypothesize that y is related linearly to $x_1, x_2, \ldots, x_n$, we are hypothesizing that for each point $\mathbf{p}_i = (x_{i1}, x_{i2}, \ldots, x_{in})$, the 'expected value', $\mathrm{E}(y_i)$, of the observation y_i at $\mathbf{p}_i$ (that is, the 'average observation' at $\mathbf{p}_i$) satisfies the equation

$$\mathrm{E}(y_i) = \beta_0 + \beta_1 x_{i1} + \beta_2 x_{i2} + \ldots + \beta_n x_{in},$$

and not that y_i does. This can be easily pictured in the case when only two variables are involved. Suppose we hypothesize that y is related linearly to the single variable x. This means that we are assuming the existence of a line $f(x) = \beta_0 + \beta_1 x$ such that each point $(x_i, \mathrm{E}(y_i))$ lies on this straight line. See Fig. 21.

In the case when there are n independent variables, we would be hypothesizing the existence of a surface in $\mathbb{C}^{n+1}$ (which is the translate of a subspace) which passes through the points $(\mathbf{p}_i, \mathrm{E}(y_i))$. We shall refer to such a surface as a *flat*.

In actual practice, the values $\mathrm{E}(y_i)$ are virtually impossible to obtain exactly. Nevertheless, we will see in the next section that it is often possible to obtain good estimates for the unknown parameters, and therefore produce good estimates for the $\mathrm{E}(y_i)$'s while also producing a reasonable facsimile of the hypothesized line of flat.

The statistically knowledgeable reader will by now have observed that we have avoided, as much as possible, introducing the statistical concepts which usually accompany this type of problem. Instead, we have introduced vague terms such as 'average out'. Admittedly, these terms being incorporated in a definition would (and should) make a good mathematician uncomfortable. However, our purpose in this section is to examine just the basic aspects of fitting a linear hypothesis without introducing statistical

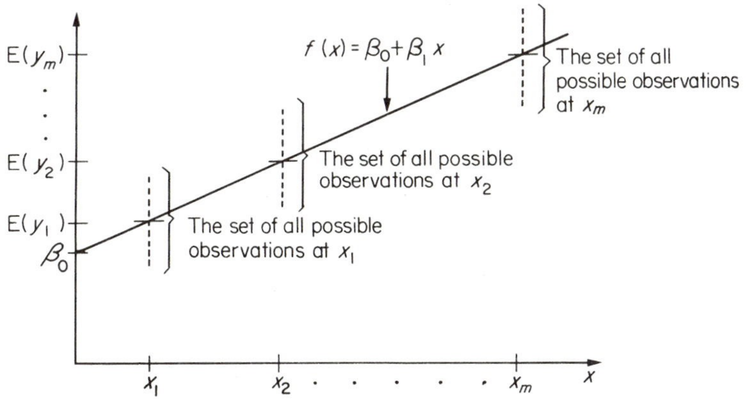

Fig. 2.1

concepts. For some applications, the methods of this section may be sufficient. A rigorous treatment of linear estimation appears in Chapter 6.

In the next two sections we will be concerned with the following two basic problems.

(I) Having hypothesized a linear relationship, $y_i = \beta_0 + \beta_1 x_{i1} + \ldots + \beta_n x_{in} + e_{y_i}$, find estimates $\hat{\beta}_i$ for the unknown parameters $\beta_i, i = 0, 1, \ldots, n$.

(II) Having obtained estimates $\hat{\beta}_i$ for the β_i's, develop a criterion to help decide to what degree the function $f(x_1, x_2, \ldots, x_n) = \hat{\beta}_0 + \hat{\beta}_1 x_1 + \ldots + \hat{\beta}_n x_n$ 'models' the situation under question.

3. Estimating the unknown parameters

We will be interested in two different types of hypotheses.

Definition 2.3.1 *When the term β_0 is present in the expression*

$$y_i = \beta_0 + \beta_1 x_{i1} + \ldots + \beta_n x_{in} + e_{y_i} \tag{1}$$

we shall refer to (1) *as an* intercept hypothesis. *When β_0 does not appear, we will call* (1) *a* no intercept hypothesis.

Suppose that we have hypothesized that y is related linearly to $x_1, x_2, \ldots, x_n$ by the no intercept hypothesis

$$y_i = \beta_1 x_{i1} + \beta_2 x_{i2} + \ldots + \beta_n x_{in} + e_{y_i}. \tag{2}$$

To estimate the parameters $\beta_1, \beta_2, \ldots, \beta_n$, select (either at random or by design) a set of values for the x's. Call them $\mathbf{p}_1 = [x_{11}, x_{12}, \ldots, x_{1n}]$. Then observe a value for y at $\mathbf{p}_1$ and call this observation y_1. Next select a second set of values for the x's and call them $\mathbf{p}_2 = [x_{21}, x_{22}, \ldots, x_{2n}]$(they need not be distinct from the first set) and observe a corresponding value for y. Call it y_2. Continue the process until m sets of values for the x's and m observations for y have been obtained. One usually tries to have $m > n$.

If the observations for the x's are placed as rows in a matrix

$$\mathbf{X} = \begin{bmatrix} x_{11} & x_{12} \cdots x_{1n} \\ x_{21} & x_{22} \cdots x_{2n} \\ \vdots & \\ x_{m1} & x_{m2} \cdots x_{mn} \end{bmatrix} = \begin{bmatrix} \mathbf{p}_1 \\ \mathbf{p}_2 \\ \vdots \\ \mathbf{p}_m \end{bmatrix},$$

which we will call the *design matrix*, and the observed values for y are placed in a vector $\mathbf{y} = [y_1, \ldots, y_m]^{\mathrm{T}}$, we may write our hypothesis (2) as

$$\mathbf{y} = \mathbf{X}\mathbf{b} + \mathbf{e}_y \tag{3}$$

where $\mathbf{b}$ is the vector of unknown parameters $\mathbf{b} = [\beta_1, \ldots, \beta_n]^{\mathrm{T}}$ and $\mathbf{e}_y$ is the unknown $\mathbf{e}_y = [e_{y_1}, \ldots, e_{y_m}]^{\mathrm{T}}$.

In the case of an intercept hypothesis (1) the design matrix $\mathbf{X}_{\mathrm{I}}$ in the equation

$$\mathbf{y} = \mathbf{X}_{\mathrm{I}}\mathbf{b}_{\mathrm{I}} + \mathbf{e}_y \tag{4}$$

takes on a slightly different appearance from the design matrix $\mathbf{X}$ which arose in (3). For an intercept hypothesis, $\mathbf{X}_\mathrm{I}$ is of the form

$$\mathbf{X}_\mathrm{I} = \begin{bmatrix} 1 & x_{11} & x_{12} \cdots x_{1n} \\ 1 & x_{21} & x_{22} \cdots x_{2n} \\ \vdots & \vdots & \vdots & \vdots \\ 1 & x_{m1} & x_{m2} \cdots x_{mn} \end{bmatrix}_{m \times (n+1)} = [\mathbf{j} \mid \mathbf{X}], \qquad \mathbf{j} = \begin{bmatrix} 1 \\ 1 \\ \vdots \\ \vdots \\ 1 \end{bmatrix}.$$

and $\mathbf{b}_\mathrm{I}$ is of the form $\mathbf{b}_\mathrm{I} = [\beta_0, \ldots, \beta_n]^\mathrm{T} = [\beta_0 \mid \mathbf{b}^\mathrm{T}]^\mathrm{T}$, $\mathbf{b} = [\beta_1, \ldots, \beta_n]^\mathrm{T}$.

Consider a no intercept hypothesis and the associated matrix equation (3). One of the most useful ways to obtain estimates $\hat{\mathbf{b}}$ of $\mathbf{b}$ is to use the information contained in $\mathbf{X}$ and $\mathbf{y}$, and impose the demand that $\hat{\mathbf{b}}$ be a vector such that '$\mathbf{X}\hat{\mathbf{b}}$ is as close to $\mathbf{y}$ as possible', or equivalently, '$\mathbf{e}_y$ is as close to $\mathbf{0}$ as possible. That is, we require $\hat{\mathbf{b}}$ to be a least squares solution of $\mathbf{X}\mathbf{b} = \mathbf{y}$. Therefore, from Theorem 1.2, any vector of the form $\hat{\mathbf{b}} = \mathbf{X}^\dagger\mathbf{y} + \mathbf{h}, \mathbf{h} \in N(\mathbf{X})$, could serve as an estimate for $\mathbf{b}$. If $\mathbf{X}$ is not of full column rank, to select a particular estimate one must impose further restrictions on $\hat{\mathbf{b}}$. In passing, we remark that one may always impose the restriction that $\|\hat{\mathbf{b}}\|$ be minimal among all least squares estimates so that the desired estimate is $\hat{\mathbf{b}} = \mathbf{X}^\dagger\mathbf{y}$. Depending on the application, this may or may not be the estimate to choose.

For each least squares estimate $\hat{\mathbf{b}}$ of $\mathbf{b}$, the vector $\hat{\mathbf{y}}$ defined by $\mathbf{X}\hat{\mathbf{b}} = \hat{\mathbf{y}}$ is an estimate for the vector of 'expected values', $\widehat{\mathrm{E}(\mathbf{y})}$, where

$$\mathrm{E}(\mathbf{y}) = [\mathrm{E}(y_1), \ldots, \mathrm{E}(y_n)]^\mathrm{T}.$$

Although it is a trivial consequence of Theorem 1.2, it is useful to observe the following.

Theorem 2.3.1 *The vector* $\hat{\mathbf{y}} = \mathbf{X}\hat{\mathbf{b}}$ *is the same for all least squares solutions* $\hat{\mathbf{b}}$, *of* $\mathbf{X}\hat{\mathbf{b}} = \hat{\mathbf{y}}$. *Moreover, for all least squares solutions* $\hat{\mathbf{b}}$, $\hat{\mathbf{y}} = \mathbf{X}\hat{\mathbf{b}} = \mathbf{P}_{R(\mathbf{X})}\mathbf{y} = \mathbf{X}\mathbf{X}^\dagger\mathbf{y}$ *and* $\mathbf{r} = \mathbf{y} - \hat{\mathbf{y}} = \mathbf{P}_{R(\mathbf{X})}^\perp\mathbf{y} = (\mathbf{I} - \mathbf{X}\mathbf{X}^\dagger)\mathbf{y}$.

A closely related situation is the following. Suppose (as is the case in many applications) one wishes to estimate or predict a value for a particular linear combination

$$y(\mathbf{c}^*) = c_1\beta_1 + c_2\beta_2 + \ldots + c_n\beta_n = \mathbf{c}^*\mathbf{b} \tag{5}$$

of the β_i's on the basis of previous observations. Here, $\mathbf{c}^* = [c_1, \ldots, c_n]$. That is, we want to predict a value, $\widehat{y(\mathbf{c}^*)}$, for y at the point $\mathbf{c}^* = [c_1, c_2, \ldots, c_n]$ on the basis of observations made at the points $\mathbf{p}_1, \mathbf{p}_2, \ldots, \mathbf{p}_m$. If we use $\widehat{y(\mathbf{c}^*)} = \widehat{\mathbf{c}^*\mathbf{b}} = \mathbf{c}^*\hat{\mathbf{b}}$, we know that it may be possible to have infinitely many estimates $\hat{\mathbf{b}}$. Hence $\widehat{y(\mathbf{c}^*)}$ could vary over an infinite set of values in which there may be a large variation. However, there are cases when $\mathbf{c}^*\hat{\mathbf{b}}$ is invariant among all least squares estimates $\hat{\mathbf{b}}$, so that $\widehat{y(\mathbf{c}^*)}$ has a unique value.

Theorem 2.3.2 *Let* $\mathbf{c} \in \mathbb{C}^n$. *The linear form* $\mathbf{c}^*\hat{\mathbf{b}}$ *is invariant among all least squares solutions of* $\mathbf{X}\hat{\mathbf{b}} = \mathbf{y}$ *if and only if* $\mathbf{c} \in R(\mathbf{X}^*)$; *in which case,* $\mathbf{c}^*\hat{\mathbf{b}} = \mathbf{c}^*\mathbf{X}^\dagger\mathbf{y}$.

Proof If $\hat{\mathbf{b}}$ is a least squares solution of $\mathbf{X}\hat{\mathbf{b}} = \mathbf{y}$, then $\hat{\mathbf{b}}$ is of the form $\hat{\mathbf{b}} = \mathbf{X}^\dagger\mathbf{y} + \mathbf{h}$ where $\mathbf{h} \in N(\mathbf{X})$. Thus $\mathbf{c}^*\hat{\mathbf{b}} = \mathbf{c}^*\mathbf{X}^\dagger\mathbf{y} + \mathbf{c}^*\mathbf{h}$ is invariant if and only if $\mathbf{c}^*\mathbf{h} = 0$ for all $\mathbf{h} \in N(\mathbf{X})$. That is, $\mathbf{c} \in R(\mathbf{X}^*) = N(\mathbf{X})^\perp$. ■

Note that in the important special case when $\mathbf{X}$ has full column rank, $\mathbf{c}^*\mathbf{b}$ is trivially invariant for any $\mathbf{c}^*$.

Most of the discussion in this section concerned a no intercept hypothesis and the matrix equation (3). By replacing $\mathbf{X}$ by $\mathbf{X}_\mathrm{I}$ and $\mathbf{b}$ by $\mathbf{b}_\mathrm{I}$, similar remarks can be made about an intercept hypothesis via the matrix equation (4).

4. Goodness of fit

Consider first the case of a no intercept hypothesis. Suppose that for various sets of values $\mathbf{p}_i$ of the x's, we have observed corresponding values y_i for y and set up the equation

$$\mathbf{y} = \mathbf{X}\mathbf{b} + \mathbf{e}_y. \tag{1}$$

From (1) we obtain a set of least squares estimates $\hat{\beta}_1, \hat{\beta}_2, \ldots, \hat{\beta}_n$ for the parameters $\beta_1, \beta_2, \ldots, \beta_n$.

As we have seen, one important application is to use the $\hat{\beta}_i$'s to estimate or predict a value $\widehat{y(\mathbf{c}^*)}$ for y for a given point $\mathbf{c}^* = [c_1, c_2, \ldots, c_n]$ by means of what we shall refer to as the *estimating equation*

$$f(c_1, c_2, \ldots, c_n) = \hat{\beta}_1 c_1 + \hat{\beta}_2 c_2 + \ldots + \hat{\beta}_n c_n. \tag{2}$$

How good is the estimating equation? One way to measure its effectiveness is to use the set of observation points which gave rise to $\mathbf{X}$ and measure how close the vector $\mathbf{y}$ of observed values is to the vector $\hat{\mathbf{y}}$ of estimated values. That is how close does the flat defined by (2) come to passing through the points $(\mathbf{p}_i, y_i)$ in $\mathbb{C}^{n+1}$. One such measure is $\|\mathbf{y} - \hat{\mathbf{y}}\|$. From Theorem 3.1 we know that $\hat{\mathbf{y}}$ is invariant among all least squares estimates $\hat{\mathbf{b}}$ and that $\|\mathbf{y} - \hat{\mathbf{y}}\| = \|\mathbf{r}\| = \|(\mathbf{I} - \mathbf{X}\mathbf{X}^\dagger)\mathbf{y}\|$.

One could be tempted to say that if $\|\mathbf{r}\|$ is small, then our estimating equation provides a good fit for our data. But the term 'small' is relative. If we are dealing with a problem concerning distances between celestial objects, the value $\|\mathbf{r}\| = 10$ ft might be considered small whereas the same value might be considered quite large if we are dealing with distances between electrons in an atomic particle. Therefore we need a measure of relative error rather than absolute error.

Consider Fig. 2.2. This diagram suggests another way to measure how close $\mathbf{y}$ is to $\hat{\mathbf{y}}$. The magnitude of the angle θ between the two vectors is such a measure. In $\mathbb{C}^m$, it is more convenient to measure $|\cos\theta|$, rather than $|\theta|$, by means of the equation

$$|\cos\theta| = \frac{\|\hat{\mathbf{y}}\|}{\|\mathbf{y}\|} = \frac{\|\mathbf{X}\mathbf{X}^\dagger\mathbf{y}\|}{\|\mathbf{y}\|}.$$

(Throughout, we assume $\mathbf{y} \neq \mathbf{0}$, otherwise there is no problem.) Likewise, $|\sin\theta|$ or $|\tan\theta|$ might act as measures of relative error. Since $\mathbf{y}$ can be

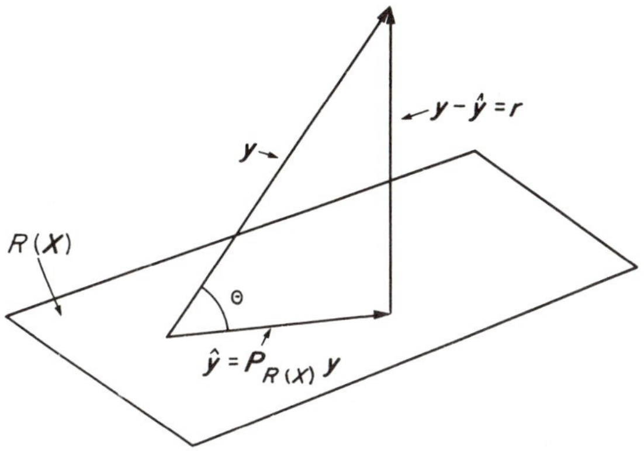

Fig. 2.2

decomposed into two components, one in $R(\mathbf{X})$ and the other in $R(\mathbf{X})^{\perp}$, $\mathbf{y} = \hat{\mathbf{y}} + \mathbf{r}$, one might say, in rough terms, that $|\cos\theta|$ represents the percentage of $\mathbf{y}$ which lies in $R(\mathbf{X})$. Let $\mathrm{R} = \cos\theta$ so that $0 \leq |\mathrm{R}| \leq 1$. Notice that if $|\mathrm{R}| = 1$, then all of $\mathbf{y}$ is in $R(\mathbf{X})$, $\mathbf{y} = \hat{\mathbf{y}}$, and $\mathbf{r} = \mathbf{0}$. If $\mathrm{R} = 0$, then $\mathbf{y} \perp R(\mathbf{X})$, $\hat{\mathbf{y}} = \mathbf{0}$, and $\mathbf{r} = \mathbf{y}$. Thus when $|\mathrm{R}| = 1$, the flat defined by the equation $f(x_1, x_2, \ldots, x_n) = \hat{\beta}_1 x_1 + \hat{\beta}_2 x_2 + \ldots + \hat{\beta}_n x_n$ actually passes through each of the data points $(\mathbf{p}_i, y_i)$ so that we have an 'exact' or 'perfect' fit. When $\mathrm{R} = 0$, $\mathbf{y}$ is as far away from $\mathbf{P}_{R(\mathbf{X})}\mathbf{y} = \hat{\mathbf{y}}$ as possible and we say that we have no fit at all.

In practice, it is common to use the term $\mathrm{R}^2 = \cos^2\theta = \|\hat{\mathbf{y}}\|^2/\|\mathbf{y}\|^2$ rather than $|\mathrm{R}|$ to measure the goodness of fit. Note that $\mathrm{R}^2 < |\mathrm{R}|$ since $|\mathrm{R}| \leq 1$. Thus R^2 is a more conservative measure. For example, if $\mathrm{R}^2 = 0.96$, one would probably consider the fit to be fairly good since this indicates that, in fact, about 98% of $\mathbf{y}$ lies in $R(\mathbf{X})$.

A familiar form for R^2, *when all numbers are real*, is

$$\mathrm{R}^2 = \frac{\left(\sum_{i=1}^{m} y_i \hat{y}_i\right)^2}{\left(\sum_{i=1}^{m} y_i^2\right)\left(\sum_{i=1}^{m} \hat{y}_i^2\right)},$$

where $\hat{y}_i$ denotes the ith entry of $\mathbf{X}\mathbf{X}^{\dagger}\mathbf{y} = \hat{\mathbf{y}}$ and y_i is the ith entry of $\mathbf{y}$.

This follows because $\|\hat{\mathbf{y}}\|^2 = [\mathbf{y}^*\mathbf{X}\mathbf{X}^{\dagger}\mathbf{y}] = [\mathbf{y}^*\hat{\mathbf{y}}] = \sum_{i=1}^{m} y_i \hat{y}_i$.

Hence $\mathrm{R}^2 = \dfrac{\|\hat{\mathbf{y}}\|^2}{\|\mathbf{y}\|^2} = \dfrac{\|\hat{\mathbf{y}}\|^4}{\|\mathbf{y}\|^2\|\hat{\mathbf{y}}\|^2} = \dfrac{\left(\sum_{i=1}^{m} y_i \hat{y}_i\right)^2}{\left(\sum_{i=1}^{m} y_i^2\right)\left(\sum_{i=1}^{m} \hat{y}_i^2\right)}$. Notice that R and R^2 are unit free measures whereas $\|\mathbf{r}\| = \|\mathbf{y} - \hat{\mathbf{y}}\|$ is not.

In statistical circles, R goes by the name of the *product moment correlation between the observed y_i's and the predicted $\hat{y}_i$'s*, and R^2 is known as *the coefficient of determination.*

Consider now the case of the intercept hypothesis

$$y_i = \beta_0 + \beta_1 x_{i1} + \ldots + \beta_n x_{in} + e_{y_i}.$$

As mentioned earlier, this gives rise to the matrix equation $\mathbf{y} = \mathbf{X}_{\mathrm{I}}\mathbf{b} + \mathbf{e}_y$ where $\mathbf{X}_{\mathrm{I}} = [\mathbf{j} \,\vdots\, \mathbf{X}]$.

If one wishes to measure the goodness of fit in this case, one may be tempted to copy the no intercept case and use

$$\frac{\|\mathbf{X}_{\mathrm{I}}\mathbf{X}_{\mathrm{I}}^{\dagger}\mathbf{y}\|}{\|\mathbf{y}\|}. \tag{3}$$

This would be a mistake. In using the matrix $\mathbf{X}_{\mathrm{I}}$, more information would be used than the data provided because of the first column, $\mathbf{j}$. The expression (3) does *not* provide a measure of how well the flat $f(x_1, x_2, \ldots, x_n) = \hat{\beta}_0 + \hat{\beta}_1 x_1 + \ldots + \hat{\beta}_n x_n$ in $\mathbb{C}^{n+1}$ fits the data points $(\mathbf{p}_i, y_i)$. Instead, the expression (3) measures how well the flat $f(x_0, x_1, x_2, \ldots, x_n) = \hat{\beta}_0 x_0 + \hat{\beta}_1 x_1 + \ldots + \hat{\beta}_n x_n$ in $\mathbb{C}^{n+2}$ fits the points $(1, \mathbf{p}_i, y_i)$. In order to decide on a measure of goodness of fit, we need the following fact.

Theorem 2.4.1 *Let* $\mathbf{X}_{\mathrm{I}} = [\mathbf{j} \,\vdots\, \mathbf{X}] \in \mathbb{C}^{m \times (n+1)}$. *The vector* $\hat{\mathbf{b}}_{\mathrm{I}} = [\hat{\beta}_0, \ldots, \hat{\beta}_n]^{\mathrm{T}} = [\hat{\beta}_0 \,\vdots\, \hat{\mathbf{b}}^{\mathrm{T}}]$, $\mathbf{b}^{\mathrm{T}} = [\hat{\beta}_1, \ldots, \hat{\beta}_n]$, *is a least squares solution of* $\mathbf{X}_{\mathrm{I}}\mathbf{b}_{\mathrm{I}} = \mathbf{y}$ *if and only if*

$$\hat{\beta}_0 = \frac{1}{m}\mathbf{j}^*(\mathbf{y} - \mathbf{X}\hat{\mathbf{b}}) \tag{4}$$

and $\hat{\mathbf{b}}$ *is a least squares solution of*

$$\left(\mathbf{I} - \frac{1}{m}\mathbf{J}\right)\mathbf{X}\mathbf{b} = \left(\mathbf{I} - \frac{1}{m}\mathbf{J}\right)\mathbf{y}. \tag{5}$$

Here $\mathbf{J} = \mathbf{j}\mathbf{j}^*$ *is a matrix of ones.*

Proof Suppose first that $\hat{\beta}_0$ satisfies (4) and that $\hat{\mathbf{b}}$ is a least squares solution of (5). To show that $\hat{\mathbf{b}}_{\mathrm{I}} = [\hat{\beta}_0, \hat{\mathbf{b}}^*]^*$ is a least squares solution of $\mathbf{X}_{\mathrm{I}}\mathbf{b}_{\mathrm{I}} = \mathbf{y}$, we shall use Theorem 1.2 and show that $\hat{\mathbf{b}}_{\mathrm{I}}$ satisfies the normal equations, $\mathbf{X}_{\mathrm{I}}^*\mathbf{X}_{\mathrm{I}}\hat{\mathbf{b}}_{\mathrm{I}} = \mathbf{X}_{\mathrm{I}}^*\mathbf{y}$. Note first that

$$\mathbf{X}_{\mathrm{I}}^*\mathbf{X}_{\mathrm{I}} = \left[\begin{array}{c|c} m & \mathbf{j}^*\mathbf{X} \\ \hline \mathbf{X}^*\mathbf{j} & \mathbf{X}^*\mathbf{X} \end{array}\right].$$

$$\begin{aligned} \text{Therefore, } \mathbf{X}_{\mathrm{I}}^*\mathbf{X}_{\mathrm{I}}\hat{\mathbf{b}}_{\mathrm{I}} &= \left[\begin{array}{c} m\hat{\beta}_0 + \mathbf{j}^*\mathbf{X}\hat{\mathbf{b}} \\ \hline \mathbf{X}^*\mathbf{j}\hat{\beta}_0 + \mathbf{X}^*\mathbf{X}\hat{\mathbf{b}} \end{array}\right] \\ &= \left[\begin{array}{c} \mathbf{j}^*(\mathbf{y} - \mathbf{X}\hat{\mathbf{b}}) + \mathbf{j}^*\mathbf{X}\hat{\mathbf{b}} \\ \hline \frac{1}{m}\mathbf{X}^*\mathbf{j}\mathbf{j}^*(\mathbf{y} - \mathbf{X}\hat{\mathbf{b}}) + \mathbf{X}^*\mathbf{X}\hat{\mathbf{b}} \end{array}\right] \\ &= \left[\begin{array}{c} \mathbf{j}^*\mathbf{y} \\ \hline \frac{1}{m}\mathbf{X}^*\mathbf{J}\mathbf{y} - \frac{1}{m}\mathbf{X}^*\mathbf{J}\mathbf{X}\hat{\mathbf{b}} + \mathbf{X}^*\mathbf{X}\hat{\mathbf{b}} \end{array}\right]. \end{aligned} \tag{6}$$

Since $\hat{\mathbf{b}}$ is a least squares solution of (5), we know from Theorem 2.1 that $\hat{\mathbf{b}}$ is a solution of

$$\mathbf{X}^*\left(\mathbf{I}-\frac{1}{m}\mathbf{J}\right)^*\left(\mathbf{I}-\frac{1}{m}\mathbf{J}\right)\mathbf{X}\hat{\mathbf{b}}=\mathbf{X}^*\left(\mathbf{I}-\frac{1}{m}\mathbf{J}\right)^*\left(\mathbf{I}-\frac{1}{m}\mathbf{J}\right)\mathbf{y}. \tag{7}$$

Since $\left(\mathbf{I}-\frac{1}{m}\mathbf{J}\right)=\left(\mathbf{I}-\frac{1}{m}\mathbf{J}\right)^2=\left(\mathbf{I}-\frac{1}{m}\mathbf{J}\right)^*$ $\left(\text{i.e. } \left(\mathbf{I}-\frac{1}{m}\mathbf{J}\right) \text{ is an orthogonal projector onto } N(\mathbf{J})\right)$, the equation (7) is equivalent to

$\mathbf{X}^*\mathbf{X}\hat{\mathbf{b}}-\frac{1}{m}\mathbf{X}^*\mathbf{J}\mathbf{X}\hat{\mathbf{b}}=\mathbf{X}^*\mathbf{y}-\frac{1}{m}\mathbf{X}^*\mathbf{J}\mathbf{y}$, or

$$\frac{1}{m}\mathbf{X}^*\mathbf{J}\mathbf{y}-\frac{1}{m}\mathbf{X}^*\mathbf{J}\mathbf{X}\hat{\mathbf{b}}+\mathbf{X}^*\mathbf{X}\hat{\mathbf{b}}=\mathbf{X}^*\mathbf{y}. \tag{8}$$

Therefore, (6) becomes $\mathbf{X}_I^*\mathbf{X}_I\hat{\mathbf{b}}_I=\left[\begin{array}{c}\mathbf{j}^*\mathbf{y}\\ \hline \mathbf{X}^*\mathbf{y}\end{array}\right]=\mathbf{X}_I^*\mathbf{y}$, which proves that $\hat{\mathbf{b}}_I$ is a least squares solution of $\mathbf{X}_I\hat{\mathbf{b}}_I=\mathbf{y}$. Conversely, assume now that $\hat{\mathbf{b}}_I$ is a least squares solution of $\mathbf{X}_I\hat{\mathbf{b}}_I=\mathbf{y}$. Then $\hat{\mathbf{b}}_I$ satisfies the normal equations $\mathbf{X}_I^*\mathbf{X}_I\hat{\mathbf{b}}_I=\mathbf{X}_I^*\mathbf{y}$. That is, $\hat{\beta}_0$ and $\hat{\mathbf{b}}$ must satisfy

$$\left[\begin{array}{c|c} m & \mathbf{j}^*\mathbf{X}\\ \hline \mathbf{X}^*\mathbf{j} & \mathbf{X}^*\mathbf{X}\end{array}\right]\left[\begin{array}{c}\hat{\beta}_0\\ \hline \hat{\mathbf{b}}\end{array}\right]=\left[\begin{array}{c}\mathbf{j}^*\mathbf{y}\\ \hline \mathbf{X}^*\mathbf{y}\end{array}\right].$$

Direct multiplication yields

$$m\hat{\beta}_0+\mathbf{j}^*\mathbf{X}\hat{\mathbf{b}}=\mathbf{j}^*\mathbf{y}, \tag{9}$$

$$\mathbf{X}^*\mathbf{j}\hat{\beta}_0+\mathbf{X}^*\mathbf{X}\hat{\mathbf{b}}=\mathbf{X}^*\mathbf{y}. \tag{10}$$

Equation (9) implies that $\hat{\beta}_0=\frac{1}{m}\mathbf{j}^*(\mathbf{y}-\mathbf{X}\hat{\mathbf{b}})$, which is (4). Substituting this value of $\hat{\beta}_0$ into (10) yields $\frac{1}{m}\mathbf{X}^*\mathbf{j}\mathbf{j}^*(\mathbf{y}-\mathbf{X}\hat{\mathbf{b}})+\mathbf{X}^*\mathbf{X}\hat{\mathbf{b}}=\mathbf{X}^*\mathbf{y}$ or equivalently, $\frac{1}{m}\mathbf{X}^*\mathbf{J}\mathbf{y}-\frac{1}{m}\mathbf{X}^*\mathbf{J}\mathbf{X}\hat{\mathbf{b}}+\mathbf{X}^*\mathbf{X}\hat{\mathbf{b}}=\mathbf{X}^*\mathbf{y}$, which is equation (9). Hence, $\hat{\mathbf{b}}$ satisfies (7) so that $\hat{\mathbf{b}}$ is a least squares solution of $\left(\mathbf{I}-\frac{1}{m}\mathbf{J}\right)\mathbf{X}\hat{\mathbf{b}}=\left(\mathbf{I}-\frac{1}{m}\mathbf{J}\right)\mathbf{y}$, and the theorem is proven. ■

Let $\mathbf{X}_M$ and $\mathbf{y}_M$ denote the matrices $\mathbf{X}_M=\left(\mathbf{I}-\frac{1}{m}\mathbf{J}\right)\mathbf{X}$ and $\mathbf{y}_M=\left(\mathbf{I}-\frac{1}{m}\mathbf{J}\right)\mathbf{y}$ and let $\bar{x}_j=\frac{1}{m}\sum_{i=1}^{m}x_{ij}$ and $\bar{y}=\frac{1}{m}\sum_{i=1}^{m}y_i$. That is, $\bar{x}_j$, the mean of the jth column of $\mathbf{X}$, is the mean of the values assumed by the jth independent variable x_j. Likewise, $\bar{y}$ is just the mean of all of the observations y_i for y.

Therefore, $\mathbf{X_M}$ and $\mathbf{y_M}$ are the matrices

$$\mathbf{X_M} = \begin{bmatrix} x_{11} - \bar{x}_1 & x_{12} - \bar{x}_2 \ldots & x_{1n} - \bar{x}_n \\ x_{21} - \bar{x}_1 & x_{22} - \bar{x}_2 & x_{2n} - \bar{x}_n \\ \vdots & \vdots & \vdots \\ x_{m1} - \bar{x}_1 & x_{m2} - \bar{x}_2 & x_{mn} - \bar{x}_n \end{bmatrix}, \mathbf{y_M} = \begin{bmatrix} y_1 - \bar{y} \\ y_2 - \bar{y} \\ \vdots \\ y_m - \bar{y} \end{bmatrix}.$$

Theorem 1 says that in obtaining least squares solutions $\hat{\mathbf{b}}_{\mathrm{I}}$ of $\mathbf{X_I}\hat{\mathbf{b}}_{\mathrm{I}} = \mathbf{y}$, we are really only obtaining least squares solutions $\hat{\mathbf{b}}$ of $\mathbf{X_M}\hat{\mathbf{b}} = \mathbf{y_M}$. In effect, this says that fitting the intercept hypothesis

$$y_i = \beta_0 + \beta_1 x_{i1} + \beta_2 x_{i2} + \ldots + \beta_n x_{in} + e_{y_i}$$

by the theory of least squares is equivalent to fitting the no intercept hypothesis

$$y_i - \bar{y} = \beta_1(x_{i1} - \bar{x}_1) + \beta_2(x_{i2} - \bar{x}_2) + \ldots + \beta_n(x_{in} - \bar{x}_n) + e_{(y_i - \bar{y})}.$$

Thus, a measure of goodness of fit for an intercept hypothesis is

$$\mathrm{R}^2 = \frac{\| \mathbf{X_M X_M^\dagger y_M} \|^2}{\| \mathbf{y_M} \|^2} = \frac{\| \hat{\mathbf{y}}_{\mathbf{M}} \|^2}{\| \mathbf{y_M} \|^2}$$

A familiar form for R^2 *when all numbers are real* is

$$\mathrm{R}^2 = \frac{\left(\sum_{i=1}^{m} [y_i - \bar{y}][\hat{y}_i - \bar{y}] \right)^2}{\left(\sum_{i=1}^{m} [y_i - \bar{y}]^2 \right) \left(\sum_{i=1}^{m} [\hat{y}_i - \bar{y}]^2 \right)} \tag{11}$$

where y_i is the ith entry of $\mathbf{y}$ and $\hat{y}_i$ is the ith entry of $\hat{\mathbf{y}} = \mathbf{X_I X_I^\dagger y}$. To prove (11), we must first prove that

$$\mathbf{X_M X_M^\dagger y} = \left(\mathbf{X_I X_I^\dagger} - \frac{1}{m}\mathbf{J} \right) \mathbf{y}. \tag{12}$$

To see that (12) is true, note that $\mathbf{X_M^\dagger y}$ is a least squares solution of $\mathbf{X_M}\hat{\mathbf{b}} = \mathbf{y}$ so that by Theorem 1,

$$\mathbf{s} = \begin{bmatrix} \frac{1}{m}\mathbf{j}^*(\mathbf{y} - \mathbf{X X_M^\dagger y}) \\ \hline \mathbf{X_M^\dagger y} \end{bmatrix}$$

is a least squares solution of $\mathbf{X_I}\hat{\mathbf{b}}_{\mathrm{I}} = \mathbf{y}$. Thus all least squares solutions of $\mathbf{X_I}\hat{\mathbf{b}}_{\mathrm{I}} = \mathbf{y}$ are of the form $\hat{\mathbf{b}}_{\mathrm{I}} = \mathbf{s} + \mathbf{h}$ where $\mathbf{h} \in N(\mathbf{X_I})$. Because $\mathbf{X_I^\dagger y}$ is a least squares solution of $\mathbf{X_I}\hat{\mathbf{b}}_{\mathrm{I}} = \mathbf{y}$, there must be a vector $\mathbf{h}_o \in N(\mathbf{X_I})$ such that $\mathbf{X_I^\dagger y} = \mathbf{s} + \mathbf{h}_o$. Therefore, $\mathbf{X_I X_I^\dagger y} = \mathbf{X_I}(\mathbf{x} + \mathbf{h}_o) = \mathbf{X_I s} = \frac{1}{m}\mathbf{j}\mathbf{j}^*(\mathbf{y} - \mathbf{X X_M^\dagger y}) + \mathbf{X X_M^\dagger y} = \frac{1}{m}\mathbf{J y} - \frac{1}{m}\mathbf{J X X_M^\dagger y} + \mathbf{X X_M^\dagger y} = \frac{1}{m}\mathbf{J y} + \left(\mathbf{I} - \frac{1}{m}\mathbf{J} \right)\mathbf{X X_M^\dagger y} = \frac{1}{m}\mathbf{J y} + \mathbf{X_M X_M^\dagger y} = \left(\frac{1}{m}\mathbf{J} + \mathbf{X_M X_M^\dagger} \right)\mathbf{y}$, from which (12) follows. Now observe that (12) implies that the ith entry $(\hat{\mathbf{y}}_{\mathrm{M}})_i$ of $\hat{\mathbf{y}}_{\mathrm{M}} = \mathbf{X}_{\mathrm{M}}\mathbf{X}_{\mathrm{M}}^{\dagger}\mathbf{y}$ is given by

$$(\hat{\mathbf{y}}_{\mathrm{M}})_i = \hat{y}_i - \bar{y}. \tag{13}$$

We can now obtain (11) as

$$\begin{aligned} \mathrm{R}^2 &= \frac{\|\hat{\mathbf{y}}_{\mathrm{M}}\|^2}{\|\mathbf{y}_{\mathrm{M}}\|^2} = \frac{\|\hat{\mathbf{y}}_{\mathrm{M}}\|^4}{\|\mathbf{y}_{\mathrm{M}}\|^2\|\hat{\mathbf{y}}_{\mathrm{M}}\|^2} \\ &= \frac{(\mathbf{y}_{\mathrm{M}}^*\mathbf{X}_{\mathrm{M}}\mathbf{X}_{\mathrm{M}}^{\dagger}\mathbf{y}_{\mathrm{M}})^2}{\|\mathbf{y}_{\mathrm{M}}\|^2\|\hat{\mathbf{y}}_{\mathrm{M}}\|^2} = \frac{(\mathbf{y}_{\mathrm{M}}, \hat{\mathbf{y}}_{\mathrm{M}})^2}{\|\mathbf{y}_{\mathrm{M}}\|^2\|\hat{\mathbf{y}}_{\mathrm{M}}\|^2}. \end{aligned} \tag{14}$$

By using (13) along with the definition of $\mathbf{y}_{\mathrm{M}}$ we see that (14) reduces to (11).

We summarize the preceding discussion in the following theorem.

Theorem 2.4.2 *For the no intercept hypothesis* $y_i = \beta_1 x_{i1} + \beta_2 x_{i2} + \ldots + \beta_n x_{in} + e_{y_i}$ *the number,*

$$\mathrm{R}^2 = \frac{\|\mathbf{X}\mathbf{X}^{\dagger}\mathbf{y}\|^2}{\|\mathbf{y}\|^2} = \frac{\|\hat{\mathbf{y}}\|^2}{\|\mathbf{y}\|^2} = \frac{\left(\sum_{i=1}^{m} y_i\hat{y}_i\right)^2}{\left(\sum_{i=1}^{m} y_i^2\right)\left(\sum_{i=1}^{m} \hat{y}_i^2\right)},$$

is a measure of goodness of fit. For the intercept hypothesis, $y_i = \beta_o + \beta_1 x_{i1} + \ldots + \beta_n x_{in} + e_{y_i}$, *a measure of goodness of fit is given by*

$$\mathrm{R}^2 = \frac{\|\mathbf{X}_{\mathrm{M}}\mathbf{X}_{\mathrm{M}}^{\dagger}\mathbf{y}_{\mathrm{M}}\|^2}{\|\mathbf{y}_{\mathrm{M}}\|^2} = \frac{\|\hat{\mathbf{y}}_{\mathrm{M}}\|^2}{\|\mathbf{y}_{\mathrm{M}}\|^2} = \frac{\left(\sum_{i=1}^{m} [y_i - \bar{y}][\hat{y}_i - \bar{y}]\right)^2}{\left(\sum_{i=1}^{m} [y_i - \bar{y}]^2\right)\left(\sum_{i=1}^{m} [\hat{y}_i - \bar{y}]^2\right)}$$

where $\mathbf{X}_{\mathrm{M}} = \left(\mathbf{I} - \frac{1}{m}\mathbf{J}\right)\mathbf{X}$, $\mathbf{y}_{\mathrm{M}} = \left(\mathbf{I} - \frac{1}{m}\mathbf{J}\right)\mathbf{y}$, *and* $\hat{y}_i$ *is the ith entry of* $\hat{\mathbf{y}} = \mathbf{X}_{\mathrm{I}}\mathbf{X}_{\mathrm{I}}^{\dagger}\mathbf{y} = \left(\mathbf{X}_{\mathrm{M}}\mathbf{X}_{\mathrm{M}}^{\dagger} + \frac{1}{m}\mathbf{J}\right)\mathbf{y}$. *Here* $\mathbf{X}_{\mathrm{I}} = [\mathbf{j} \,|\, \mathbf{X}]$ *and* $\mathbf{J} = \mathbf{j}\mathbf{j}^*, \mathbf{j} = [1, \ldots, 1]^*$. *In each case* $0 \leq \mathrm{R}^2 \leq 1$ *and is free of units. When* $\mathrm{R}^2 = 1$, *the fit is exact and when* $\mathrm{R} = 0$, *there is no fit at all.*

5. An application to curve fitting

Carl Friedrich Gauss was a famous and extremely gifted scientist who lived from 1777 to 1855. In January of 1801 an astronomer named G. Piazzi briefly observed and then lost a 'new planet' (actually this 'new planet' was the asteroid now known as Ceres). During the rest of 1801 astronomers and other scientists tried in vain to relocate this 'new planet' of Piazzi. The task of finding this 'new planet' on the basis of a few observations seemed hopeless.

Astronomy was one of the many areas in which Gauss took an active interest. In September of 1801, Gauss decided to take up the challenge of finding the lost planet. Gauss hypothesized an elliptical orbit rather than the circular approximation which previously was the assumption of the astronomers of that time. Gauss then proceeded to develop the method of least squares. By December, the task was completed and Gauss informed the scientific community not only where to look, but also predicted the position of the lost planet at any time in the future. They looked and it was where Gauss had predicted it would be.

This extraordinary feat of locating a tiny, distant heavenly body from apparently insufficient data astounded the scientific community. Furthermore, Gauss refused to reveal his methods. These events directly lead to Gauss' fame throughout the entire scientific community (and perhaps most of Europe) and helped to establish his reputation as a mathematical and scientific genius of the highest order. Because of Gauss' refusal to reveal his methods, there were those who even accused Gauss of sorcery.

Gauss waited until 1809 when he published his *Theoria Motus Corporum Coelestium In Sectionibus Conicis Dolem Ambientium* to systematically develop the theory of least squares and his methods of orbit calculation. This was in keeping with Gauss' philosophy to publish nothing but well polished work of lasting significance.

Gauss lived before linear algebra as such existed and he solved the problem of finding Ceres by techniques of calculus. However, it can be done fairly simply without calculus.

For the sake of exposition, we will treat a somewhat simplified version of the problem Gauss faced. To begin with, assume that the planet travels an elliptical orbit centred about a known point and that m observations were made. Our version of Gauss' problem is this.

Problem A Suppose that $(x_1, y_1), (x_2, y_2), \ldots, (x_m, y_m)$ represent the m coordinates in the plane where the planet was observed. Find the ellipse in standard position $x^2/a^2 + y^2/b^2 = 1$, which comes as close to the data points as possible.

If there exists an ellipse which actually passes through the m data points, then there exist parameters $\beta_1 = 1/a^2$, $\beta_2 = 1/b^2$, which satisfy each of the m equations $\beta_1(x_i)^2 + \beta_2(y_i)^2 = 1$ for $i = 1, 2, \ldots, m$. However, due to measurement error, or functional error, or both, it is reasonable to expect that no such ellipse exists. In order to find the ellipse $\beta_1 x^2 + \beta_2 y^2 = 1$ which is 'closest' to our m data points, let

$$e_i = \beta_1(x_i)^2 + \beta_2(y_i)^2 - 1 \text{ for } i = 1, 2, \ldots, m. \tag{1}$$

Then, in matrix notation, (1) is written as

$$\begin{bmatrix} e_1 \\ e_2 \\ \vdots \\ e_m \end{bmatrix} = \begin{bmatrix} x_1^2 & y_1^2 \\ x_2^2 & y_2^2 \\ \vdots & \vdots \\ x_m^2 & y_m^2 \end{bmatrix} \begin{bmatrix} \beta_1 \\ \beta_2 \end{bmatrix} - \begin{bmatrix} 1 \\ 1 \\ \vdots \\ 1 \end{bmatrix}$$

or $\mathbf{e} = \mathbf{Xb} - \mathbf{j}$. There are many ways to minimize the e_i. For example, we could require that $\sum_{i=1}^{m} |e_i|$ be minimal or that $\max\{|e_1|, |e_2|, \ldots, |e_m|\}$ be minimal. However, Gauss himself gave an argument to support the claim that the restriction that

$$\sum_{i=1}^{m} e_i^2 = \|\mathbf{e}\|^2 \text{ be minimal} \tag{2}$$

gives rise to the 'best closest ellipse', a term which we will not define here. (See Chapter 6.) Intuitively, the restriction (2) is perhaps the most reasonable if for no other reason than that it agrees with the usual concept of euclidean length or distance. Thus Problem A can be reformulated as follows.

Problem B Find a vector $\hat{\mathbf{b}} = [\hat{\beta}_1, \hat{\beta}_2]^{\mathrm{T}}$ that is a least squares solution of $\mathbf{Xb} = \mathbf{j}$.

From Theorem 2.4, we know that all possible least squares solutions are of the form $\hat{\mathbf{b}} = \mathbf{X}^{\dagger}\mathbf{j} + \mathbf{h}$ where $\mathbf{h} \in N(\mathbf{X})$. In our example, the rank of $\mathbf{X} \in \mathbb{C}^{m \times 2}$ will be two unless $x_i = \alpha y_i$ for each $i = 1, 2, \ldots, m$; in which case the data points line on a straight line. Assuming non-colinearity, it follows that $N(\mathbf{X}) = \{\mathbf{0}\}$ and there is a unique least squares solution $\hat{\mathbf{b}} = \mathbf{X}^{\dagger}\mathbf{j} = (\mathbf{X}^*\mathbf{X})^{-1}\mathbf{X}^*\mathbf{j}$. That the matrix $\mathbf{X}$ is of full column rank is characteristic of curve fitting by least squares techniques.

Example 2.5.1 We will find the ellipse in standard position which comes as close as possible to the four data points (1, 1), (0, 2), (− 1, 1), and (− 1, 2).

Then $\mathbf{X} = \begin{bmatrix} 1 & 1 \\ 0 & 4 \\ 1 & 1 \\ 1 & 4 \end{bmatrix}$ and $\mathbf{X}^{\dagger} = \frac{1}{66}\begin{bmatrix} 28 & -24 & 28 & 10 \\ -3 & 12 & -3 & 6 \end{bmatrix}$. $\mathbf{X}^{\dagger}\mathbf{j} = \begin{bmatrix} \frac{7}{11} \\ \frac{2}{11} \end{bmatrix}$ is the least squares solution to $\mathbf{X}\hat{\mathbf{b}} = \mathbf{j}$ and $\|\mathbf{e}\| = \|\mathbf{X}\hat{\mathbf{b}} - \mathbf{j}\|$ is approximately 0.5. Thus $\frac{7}{11}x^2 + \frac{2}{11}y^2 = 1$ is the ellipse that fits 'best' (Fig. 2.3). A measure of goodness of fit is $R^2 = \|\mathbf{XX}^{\dagger}\mathbf{j}\|^2 / \|\mathbf{j}\|^2 \doteq 0.932$ ($\doteq$ means 'approximately equal to') which is a decent fit.

Notice that there is nothing in the working of Problem B that forced $\mathbf{X}^{\dagger}\mathbf{j}$ to have positive coefficients. If instead of an ellipse, we had tried to fit a hyperbola in standard position to the data, we would have wound up with the same least squares problem which has only an ellipse as a least squares solutions. To actually get a least squares problem *equivalent* to Problem A it would have to look something like this:

Problem C Find a vector $\mathbf{u}$ with *positive coefficients* such that $\|\mathbf{Au} - \mathbf{b}\| \le \|\mathbf{Av} - \mathbf{b}\|$ for all $\mathbf{v}$ with *positive coefficients.*

The idea of a constrained least squares problem will not be discussed here. It is probably unreasonable to expect to know ahead of time the

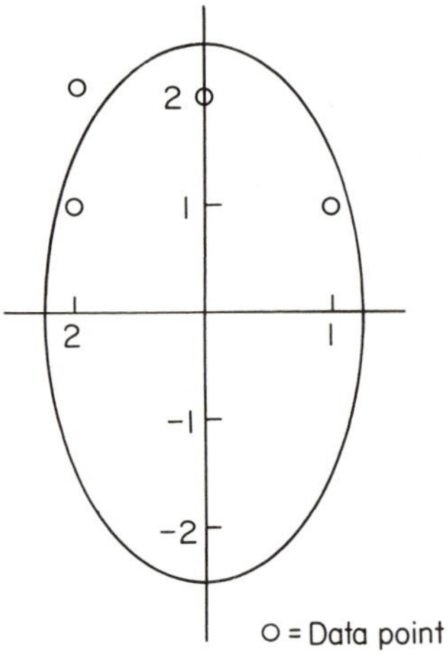

Fig. 2.3 The ellipse $\frac{7}{11}x^2 + \frac{2}{11}y^2 = 1$.

orientation of a trajectory. Perhaps a more reasonable problem than Problem A would be:

Problem D Given some data points find the conic section that provides the 'closest fit.'

For convenience assume that the conic section does not go through the origin. Then it may be written as $ax^2 + by^2 + cx + dy + fxy = 1$. If we use the same four data points of the previous example there are many least squares solutions all of which pass through the four data points. The minimal least squares solution is a hyperbola.

6. Polynomial and more general fittings

In the previous section we were concerned with fitting a conic to a set of data points. A variation which occurs quite frequently is that of trying to find the nth degree polynomial

$$y = \hat{\beta}_0 + \hat{\beta}_1 x + \hat{\beta}_2 x^2 + \ldots + \hat{\beta}_n x^n \tag{1}$$

which best fits m data points $(x_1, y_1), (x_2, y_2), \ldots, (x_m, y_m)$. Usually one has $m > n + 1$; otherwise there is no problem because if $m \leq n + 1$, then the interpolation polynomial of degree $m - 1 \leq n$ provides an exact fit.

We proceed as before by setting

$$e_i = \sum_{j=0}^{n} \beta_j x_i^j - y_i \text{ for each } i = 1, 2, \ldots, m. \tag{2}$$

Thus

$$\begin{bmatrix} e_1 \\ \vdots \\ \vdots \\ e_m \end{bmatrix} = \begin{bmatrix} 1 & x_1 & x_1^2 \ldots x_1^n \\ 1 & x_2 & x_2^2 \ldots x_2^n \\ \vdots & \vdots & \vdots \quad\;\; \vdots \\ 1 & x_m & x_m^2 \cdots x_m^n \end{bmatrix} \begin{bmatrix} \beta_0 \\ \beta_1 \\ \vdots \\ \beta_n \end{bmatrix} - \begin{bmatrix} y_1 \\ y_2 \\ \vdots \\ y_m \end{bmatrix},$$

or $\mathbf{e} = \mathbf{Xb} - \mathbf{y}$. If the restriction that $\|\mathbf{e}\|$ be minimal is imposed, then a closest nth degree polynomial to our data points has as its coefficients $\hat{\beta}_0, \hat{\beta}_1, \ldots, \hat{\beta}_n$. Where the $\hat{\beta}_i$'s are the components of a least squares solution $\hat{\mathbf{b}} = \mathbf{X}^\dagger\mathbf{y} + \mathbf{h}, \mathbf{h} \in N(\mathbf{X})$ of $\mathbf{Xb} = \mathbf{y}$. Notice that if the x_i's are all distinct then $\mathbf{X}$ has full column rank. (It is an example of what is known as a Vandermonde segment.) Hence $N(\mathbf{X}) = \{\mathbf{0}\}$ and there is a unique least squares solution, $\hat{\mathbf{b}} = \mathbf{X}^\dagger\mathbf{y} = (\mathbf{X}^*\mathbf{X})^{-1}\mathbf{X}^*\mathbf{y}$.

To measure goodness of fit, observe that (2) was basically an intercept hypothesis so that from Theorem 4.2, one would use the coefficient of determination $R^2 = \|\mathbf{X}_M\mathbf{X}_M^\dagger\mathbf{y}_M\|^2/\|\mathbf{y}_M\|^2$.

A slightly more general situation than polynomial fitting is the following. Suppose that you are given n functions $g_i(x)$ and n linear functions l_i of k unknown parameters β_i. Now suppose that you are given m data points (x_i, y_i). The problem is to find values $\hat{\beta}_i$ so that

$$y = l_1(\hat{\beta}_1, \ldots, \hat{\beta}_k)g_1(x) + \ldots + l_n(\hat{\beta}_1, \ldots, \hat{\beta}_k)g_n(x)$$

is as close to the data points as possible. Let $l_i(\beta_1, \ldots, \beta_k) = w_{i1}\beta_1 + \ldots + w_{ik}\beta_k$; and define $e_i = \sum_{j=1}^{n} l_j g_j(x_i) - y_i$. Then the corresponding matrix equation is $\mathbf{e} = \mathbf{XWb} - \mathbf{y}$ where $\mathbf{e} = [e_1, \ldots, e_m]^T$, $\mathbf{b} = [\beta_1, \ldots, \beta_k]^T$, $\mathbf{y} = [y_1, \ldots, y_m]^T$,

$$\mathbf{X} = \begin{bmatrix} g_1(x_1) & g_2(x_1) \ldots g_n(x_1) \\ g_1(x_2) & g_2(x_2) \ldots g_n(x_2) \\ \vdots & \vdots \qquad \vdots \\ g_1(x_m) & g_2(x_m) \ldots g_n(x_m) \end{bmatrix}, \text{ and } \mathbf{W} = \begin{bmatrix} w_{11} \ldots w_{1k} \\ \\ w_{n1} \ldots w_{nk} \end{bmatrix}$$

Note that this problem is equivalent to finding values $\hat{\beta}_i$ so that $y = \hat{\beta}_1 L_1(x) + \hat{\beta}_2 L_2(x) + \ldots + \hat{\beta}_k L_k(x)$ is as close to the data points as possible where $L_i(x)$ is a linear combination of the functions $g_1(x), \ldots, g_n(x)$.

To insure that $\|\mathbf{e}\|$ is minimal, the parameters $\hat{\beta}_i$ must be the components of a least squares solution $\hat{\mathbf{b}}_w$ of $\mathbf{XW}\hat{\mathbf{b}}_w = \mathbf{y}$. By Theorem 1.7 we have

$$\hat{\mathbf{b}}_w = (\mathbf{P}_{R(\mathbf{X}^*)}\mathbf{W})^\dagger(\mathbf{XP}_{R(\mathbf{W})})^\dagger\mathbf{y} + \mathbf{h}, \mathbf{h} \in N(\mathbf{XW}).$$

In many situations $\mathbf{W}$ is invertible. It is also frequently the case that $\mathbf{X}$ has full column rank. If $\mathbf{W}$ is invertible and $N(\mathbf{X}) = \{\mathbf{0}\}$, then $N(\mathbf{XW}) = \{\mathbf{0}\}$ so that (2) gives a unique least squares solution

$$\hat{\mathbf{b}}_w = \mathbf{W}^{-1}\mathbf{X}^\dagger\mathbf{y} = \mathbf{W}^{-1}(\mathbf{X}^*\mathbf{X})^{-1}\mathbf{X}^*\mathbf{y} = (\mathbf{X}^*\mathbf{XW})^{-1}\mathbf{X}^*\mathbf{y}. \tag{3}$$

Example 2.6.1 We will find parameters $\hat{\beta}_1$, $\hat{\beta}_2$, and $\hat{\beta}_3$ so that the function

$$f(x) = \hat{\beta}_1 x + \hat{\beta}_2 x^2 + \hat{\beta}_3 (\sin x)$$

best fits the four data points $(0, 1)$, $\left(\frac{\pi}{4}, 2\right)$, $\left(\frac{\pi}{2}, 3\right)$, and $(\pi, 4)$. Then we will find the $\hat{\beta}_i$'s so that

$$\begin{aligned} f(x) &= (\hat{\beta}_1 + \hat{\beta}_2)x + (\hat{\beta}_2 + \hat{\beta}_3)x^2 + \hat{\beta}_3 \sin x \\ &= \hat{\beta}_1 x + \hat{\beta}_2(x + x^2) + \hat{\beta}_3(x^2 + \sin x) \end{aligned}$$

best fits the same four data points. In each case we will also compute R^2, the coefficient of determination.

In the first case, we seek least squares solutions of $\mathbf{X}\hat{\mathbf{b}} = \mathbf{y}$ where

$$\mathbf{X} = \begin{bmatrix} 0 & 0 & 0 \\ \dfrac{\pi}{4} & \dfrac{\pi^2}{16} & \dfrac{1}{\sqrt{2}} \\ \dfrac{\pi}{2} & \dfrac{\pi^2}{4} & 1 \\ \pi & \pi^2 & 0 \end{bmatrix} \text{ and } \mathbf{y} = \begin{bmatrix} 1 \\ 2 \\ 3 \\ 4 \end{bmatrix}$$

In this case, $\mathbf{X}$ has full column rank so that the least squares solution is unique and is given by $\hat{\mathbf{b}} = \mathbf{X}^\dagger \mathbf{y} \doteqdot [9.96749, -2.76747, -5.82845]^{\mathrm{T}}$. Also

$$\mathbf{XX}^\dagger = \begin{bmatrix} 0 & 0 & 0 & 0 \\ 0 & 1 & 0 & 0 \\ 0 & 0 & 1 & 0 \\ 0 & 0 & 0 & 1 \end{bmatrix}, \quad \mathrm{R}^2 = \frac{\|\mathbf{XX}^\dagger \mathbf{y}\|^2}{\|\mathbf{y}\|^2} \doteqdot 0.9667$$

so that we have a fairly good fit.

In the second case, $\mathbf{W} = \begin{bmatrix} 1 & 1 & 0 \\ 0 & 1 & 1 \\ 0 & 0 & 1 \end{bmatrix}$ so that from (3), the least squares solution is given by

$$\hat{\mathbf{b}}_w = \mathbf{W}^{-1}\mathbf{X}^\dagger \mathbf{y} = \begin{bmatrix} 1 & -1 & 1 \\ 0 & 1 & 1 \\ 0 & 0 & 1 \end{bmatrix} \begin{bmatrix} 9.96749 \\ -2.76747 \\ -5.82845 \end{bmatrix} = \begin{bmatrix} 6.90651 \\ 3.06098 \\ -5.82845 \end{bmatrix}.$$

Since $N(\mathbf{X}) = \{\mathbf{0}\}$ and $\mathbf{W}$ is invertible we have $\mathbf{XW}(\mathbf{XW})^\dagger = \mathbf{XWW}^{-1}\mathbf{X}^\dagger = \mathbf{XX}^\dagger$ and

$$R_w^2 = \frac{\|\mathbf{XW}(\mathbf{XW})^\dagger \mathbf{y}\|^2}{\|\mathbf{y}\|^2} = \frac{\|\mathbf{XX}^\dagger \mathbf{y}\|^2}{\|\mathbf{y}\|^2} = \mathrm{R}^2 \doteqdot 0.9667.$$

In closing this section we would like to note that while in curve fitting problems it is usually possible to get $\mathbf{X}$ of full column rank. There are times when this is not the case. In the study of Linear Statistical Models, in say Economics, one sometimes has several x_i that tend to move together (are highly correlated). As a result some columns of $\mathbf{X}$, if not linearly dependent, will be nearly so. In these cases $\mathbf{X}$ is either not of full rank or is ill-conditioned (see Chapter 12) [71].

The reader interested in a more statistically rigorous treatment of least squares problems is referred to Section 6.4.

7. Why $\mathbf{A}^\dagger$?

It may come as a surprise to the reader who is new to the ideas in this book that not everyone bestows upon the Moore–Penrose inverse the same central role that we have bestowed upon it so far.

The reason for this disfavour has to do with computability. There is another type of inverse, which we denote for now by $\mathbf{A}^l$ such that $\mathbf{A}^l\mathbf{b}$ is a least squares solution to $\mathbf{Ax} = \mathbf{b}$. ($\mathbf{A}^l$ is any matrix such that $\mathbf{AA}^l = \mathbf{P}_{R(\mathbf{A})}$.) The computation of $\mathbf{A}^l$ or $\mathbf{A}^l\mathbf{b}$ frequently requires fewer arithmetic operations than the computation of $\mathbf{A}^\dagger$. Thus, if one is only interested in finding a least squares solution, then $\mathbf{A}^l\mathbf{b}$ is fine and there would appear to be no need for $\mathbf{A}^\dagger$. Since this is the case in certain areas, such as parts of statistics, they are usually happy with an $\mathbf{A}^l\mathbf{b}$ and are not too concerned with $\mathbf{A}^\dagger$. Because they are useful we will discuss the $\mathbf{A}^l$ in the chapter on other types of inverses (Chapter 6).

We feel, however, that the generalized inverse deserves the central role it has played so far.

The first and primary reason is pedagogical. $\mathbf{A}^\dagger$ stands for a particular matrix while $\mathbf{A}^l$ is not unique. Two different formulas for an $\mathbf{A}^l$ of a given matrix $\mathbf{A}$ might lead to two different matrices. The authors believe very strongly that for the readers with only an introductory knowledge of linear algebra and limited 'mathematical maturity' it is much better to first learn thoroughly the theory of the Moore–Penrose generalized inverse. Then with a firm foundation they can easily learn about the other types of inverses, some of which are not unique and some of which need not always exist.

Secondly, a standard way to check an answer is to calculate it again by a different means. This may not work if one calculates an $\mathbf{A}^l$ by two different techniques, for it is quite possible that the two different correct approaches will produce very different appearing answers. But no matter how one calculates $\mathbf{A}^\dagger$ for a given matrix $\mathbf{A}$, the answer should be the same.

3
Sums, partitioned matrices and the constrained generalized inverse

1. The generalized inverse of a sum

For non-singular matrices $\mathbf{A}$, $\mathbf{B}$, and $\mathbf{A}+\mathbf{B}$, the inverse of the sum is rarely the sum of the inverses. In fact, most would agree that a worthwhile expression for $(\mathbf{A}+\mathbf{B})^{-1}$ is not known in the general case. This would tend to make one believe that there is not much that can be said, in general, about $(\mathbf{A}+\mathbf{B})^{\dagger}$. Although this may be true, there are some special cases which may prove to be useful. In the first section we will state two results and prove a third. The next sections apply the ideas of the first to develop computational algorithms for $\mathbf{A}^{\dagger}$ and prove some results on the generalized inverse of special kinds of partitioned matrices. Our first result is easily verified by checking the four Penrose conditions of Definition 1.1.3.

Theorem 3.1.1 *If* $\mathbf{A}, \mathbf{B} \in \mathbb{C}^{m \times n}$, *and if* $\mathbf{AB}^* = \mathbf{0}$ *and* $\mathbf{B}^*\mathbf{A} = \mathbf{0}$, *then* $(\mathbf{A}+\mathbf{B})^{\dagger} = \mathbf{A}^{\dagger} + \mathbf{B}^{\dagger}$.

The hypothesis of Theorem 1 is equivalent to requiring that $R(\mathbf{A}^*) \perp R(\mathbf{B}^*)$ and $R(\mathbf{A}) \perp R(\mathbf{B})$. Clearly this is very restrictive. If the hypothesis of Theorem 1 is relaxed to require only that $R(\mathbf{A}^*) \perp R(\mathbf{B}^*)$, which is still a very restrictive condition, or if we limit our attention to special sums which have the form $\mathbf{AA}^* + \mathbf{BB}^*$, it is then possible to prove that the following rather complicated formulas hold.

Theorem 3.1.2 *If* $\mathbf{A} \in \mathbb{C}^{m \times n}$, $\mathbf{B} \in \mathbb{C}^{m \times p}$, *then* $(\mathbf{AA}^* + \mathbf{BB}^*)^{\dagger} = (\mathbf{I} - \mathbf{C}^{\dagger *}\mathbf{B}^*)\mathbf{A}^{\dagger}[\mathbf{I} - \mathbf{A}^{\dagger}\mathbf{B}(\mathbf{I} - \mathbf{C}^{\dagger}\mathbf{C})\mathbf{KB}^*\mathbf{A}^{\dagger *}]\mathbf{A}^{\dagger}(\mathbf{I} - \mathbf{BC}^{\dagger}) + \mathbf{C}^{\dagger *}\mathbf{C}^{\dagger}$ *where* $\mathbf{C} = (\mathbf{I} - \mathbf{AA}^{\dagger})\mathbf{B}$, $\mathbf{K} = [\mathbf{I} + (\mathbf{I} - \mathbf{C}^{\dagger}\mathbf{C})\mathbf{B}^*\mathbf{A}^{\dagger *}\mathbf{A}^{\dagger}\mathbf{B}(\mathbf{I} - \mathbf{C}^{\dagger}\mathbf{C})]^{-1}$. *If* $\mathbf{A}$, $\mathbf{B} \in \mathbb{C}^{m \times n}$ *and* $\mathbf{AB}^* = \mathbf{0}$, *then* $(\mathbf{A}+\mathbf{B})^{\dagger} = \mathbf{A}^{\dagger} + (\mathbf{I} - \mathbf{A}^{\dagger}\mathbf{B})[\mathbf{C}^{\dagger} + (\mathbf{I} - \mathbf{C}^{\dagger}\mathbf{C}) \times \mathbf{KB}^*\mathbf{A}^{\dagger *}\mathbf{A}^{\dagger}(\mathbf{I} - \mathbf{BC}^{\dagger})]$ *where* $\mathbf{C}$ *and* $\mathbf{K}$ *are defined above.*

Since Theorem 2 is stated only to give the reader an idea of what types of statements about sums are possible and will not be used in the sequel, its proof is omitted. The interested reader may find the proof in Cline's paper [30].

We will now develop a useful formulation for the generalized inverse of a particular kind of sum. For any matrix $\mathbf{B}\in\mathbb{C}^{m\times n}$, the matrix $\mathbf{B}$ can be written as a sum of matrices of rank one. For example, let $\mathbf{E}_{ij}$ denote the matrix in $\mathbb{C}^{m\times n}$ which contains a 1 in the (i,j)th position and 0's elsewhere. If $\mathbf{B}=[\beta_{ij}]$. then

$$\mathbf{B}=\sum_{i,j}\beta_{ij}\mathbf{E}_{ij}, \tag{1}$$

is the sum of mn rank one matrices. It can be easily shown that if rank$(\mathbf{B})=r$, then $\mathbf{B}$ can be written as the sum of just r matrices of rank one. Furthermore, if $\mathbf{F}\in\mathbb{C}^{m\times n}$ is any matrix of rank one, then $\mathbf{F}$ can be written as the product of two vectors, $\mathbf{F}=\mathbf{cd}^*$, where $\mathbf{c}\in\mathbb{C}^m, \mathbf{d}\in\mathbb{C}^n$. Thus $\mathbf{B}\in\mathbb{C}^{m\times n}$ can always be written as

$$\mathbf{B}=\mathbf{c}_1\mathbf{d}_1^*+\mathbf{c}_2\mathbf{d}_2^*+\ldots+\mathbf{c}_k\mathbf{d}_k^*. \tag{2}$$

Throughout this chapter $\mathbf{e}_i$ will denote a vector with a 1 in the ith place and zeros elsewhere. Thus if $\{\mathbf{e}_1,\ldots,\mathbf{e}_m\}\subseteq\mathbb{C}^m$, then $\{\mathbf{e}_1,\ldots,\mathbf{e}_m\}$ would be the *standard basis* for $\mathbb{C}^m$. If $\mathbf{B}$ has the decomposition given in (1), let $\mathbf{b}_{ij}=\beta_{ij}\mathbf{e}_i$ where $\mathbf{e}_i\in\mathbb{C}^m$. Then the representation (2) assumes the form $\mathbf{B}=\sum_{i,j}\mathbf{b}_{ij}\mathbf{e}_j^*$. It should be clear that a representation such as (2) is not unique.

Now if one had at one's disposal a formula by which one could *g-invert* (invert in the generalized sense) a sum of the form $\mathbf{A}+\mathbf{cd}^*$ where $\mathbf{c}\in\mathbb{C}^m$ and $\mathbf{d}\in\mathbb{C}^n$, then $\mathbf{B}$ could be written as in (2) and $(\mathbf{A}+\mathbf{B})^\dagger$ could be obtained by recursively using this formula. In order for a formula for $(\mathbf{A}+\mathbf{cd}^*)^\dagger$ to be useful, it is desirable that it be of the form $(\mathbf{A}+\mathbf{cd}^*)^\dagger=\mathbf{A}^\dagger+\mathbf{G}$ where $\mathbf{G}$ is a matrix made up of sums and products of only the matrices $\mathbf{A},\mathbf{A}^\dagger,\mathbf{c},\mathbf{d}$, and their conjugate transposes. The reason for this requirement will become clearer in Sections 2 and 3.

Rather than present one long complicated expression for $(\mathbf{A}+\mathbf{cd}^*)^\dagger$ (see Exercise 3.7.18.), it is more convenient to consider the following six logical possibilities which are clearly exhaustive.

(i) $\mathbf{c}\notin R(\mathbf{A})$ and $\mathbf{d}\notin R(\mathbf{A}^*)$ and $1+\mathbf{d}^*\mathbf{A}^\dagger\mathbf{c}$ arbitrary;
(ii) $\mathbf{c}\in R(\mathbf{A})$ and $\mathbf{d}\notin R(\mathbf{A}^*)$ and $1+\mathbf{d}^*\mathbf{A}^\dagger\mathbf{c}=0$;
(iii) $\mathbf{c}\in R(\mathbf{A})$ and $\mathbf{d}$ arbitrary and $1+\mathbf{d}^*\mathbf{A}^\dagger\mathbf{c}\neq 0$;
(iv) $\mathbf{c}\notin R(\mathbf{A})$ and $\mathbf{d}\in R(\mathbf{A}^*)$ and $1+\mathbf{d}^*\mathbf{A}^\dagger\mathbf{c}=0$;
(v) $\mathbf{c}$ arbitrary and $\mathbf{d}\in R(\mathbf{A}^*)$ and $1+\mathbf{d}^*\mathbf{A}^\dagger\mathbf{c}\neq 0$;
(vi) $\mathbf{c}\in R(\mathbf{A})$ and $\mathbf{d}\in R(\mathbf{A}^*)$ and $1+\mathbf{d}^*\mathbf{A}^\dagger\mathbf{c}=0$.

Throughout the following discussion, we will make frequent use of the fact that the generalized inverse of a non-zero vector $\mathbf{x}$ is given by $\mathbf{x}^\dagger=\mathbf{x}^*/\|\mathbf{x}\|^2$ where $\|\mathbf{x}\|^2=(\mathbf{x},\mathbf{x})$.

Theorem 3.1.3 *For* $\mathbf{A}\notin\mathbb{C}^{m\times n}$, $\mathbf{c}\in\mathbb{C}^m$ *and* $\mathbf{d}\in\mathbb{C}^n$ *let* $\mathbf{k}=$ *the column* $\mathbf{A}^\dagger\mathbf{c}$, $\mathbf{h}=$ *the* row $\mathbf{d}^*\mathbf{A}^\dagger$, $\mathbf{u}=$ *the column* $(\mathbf{I}-\mathbf{AA}^\dagger)\mathbf{c}$, $\mathbf{v}=$ *the* row $\mathbf{d}^*(\mathbf{I}-\mathbf{A}^\dagger\mathbf{A})$, *and* $\beta=$ *the scalar* $1+\mathbf{d}^*\mathbf{A}^\dagger\mathbf{c}$. (*Notice that* $\mathbf{c}\in R(\mathbf{A})$ *if and only if* $\mathbf{u}=\mathbf{0}$ *and*

$\mathbf{d} \in R(\mathbf{A}^*)$ *if and only* if $\mathbf{v} = \mathbf{0}$.) *Then the generalized inverse of* $(\mathbf{A} + \mathbf{cd}^*)$ *is as follows.*

(i) *If* $\mathbf{u} \neq \mathbf{0}$ *and* $\mathbf{v} \neq \mathbf{0}$, *then* $(\mathbf{A} + \mathbf{cd}^*)^\dagger = \mathbf{A}^\dagger - \mathbf{k}\mathbf{u}^\dagger - \mathbf{v}^\dagger\mathbf{h} + \beta\mathbf{v}^\dagger\mathbf{u}^\dagger$.

(ii) *If* $\mathbf{u} = \mathbf{0}$, $\mathbf{v} \neq \mathbf{0}$, *and* $\beta = 0$, *then* $(\mathbf{A} + \mathbf{cd}^*)^\dagger = \mathbf{A}^\dagger - \mathbf{k}\mathbf{k}^\dagger\mathbf{A}^\dagger - \mathbf{v}^\dagger\mathbf{h}$.

(iii) *If* $\mathbf{u} = \mathbf{0}$ *and* $\beta \neq 0$, *then* $(\mathbf{A} + \mathbf{cd}^*)^\dagger = \mathbf{A}^\dagger + \frac{1}{\bar{\beta}}\mathbf{v}^*\mathbf{k}^*\mathbf{A}^\dagger - \frac{\bar{\beta}}{\sigma_1}\mathbf{p}_1\mathbf{q}_1^*$,

where $\mathbf{p}_1 = -\left(\frac{\|\mathbf{k}\|^2}{\bar{\beta}}\mathbf{v}^* + \mathbf{k}\right)$, $\mathbf{q}_1^* = -\left(\frac{\|\mathbf{v}\|^2}{\bar{\beta}}\mathbf{k}^*\mathbf{A}^\dagger + \mathbf{h}\right)$,

and $\sigma_1 = \|\mathbf{k}\|^2\|\mathbf{v}\|^2 + |\beta|^2$.

(iv) *If* $\mathbf{u} \neq \mathbf{0}$, $\mathbf{v} = \mathbf{0}$, *and* $\beta = 0$, *then* $(\mathbf{A} + \mathbf{cd}^*)^\dagger = \mathbf{A}^\dagger - \mathbf{A}^\dagger\mathbf{h}^\dagger\mathbf{h} - \mathbf{k}\mathbf{u}^\dagger$.

(v) *If* $\mathbf{v} = \mathbf{0}$ *and* $\beta \neq 0$, *then* $(\mathbf{A} + \mathbf{cd}^*)^\dagger = \mathbf{A}^\dagger + \frac{1}{\bar{\beta}}\mathbf{A}^\dagger\mathbf{h}^*\mathbf{u}^* - \frac{\bar{\beta}}{\sigma_2}\mathbf{p}_2\mathbf{q}_2^*$,

where $\mathbf{p}_2 = -\left(\frac{\|\mathbf{u}\|^2}{\bar{\beta}}\mathbf{A}^\dagger\mathbf{h}^* + \mathbf{k}\right)$, $\mathbf{q}_2^* = -\left(\frac{\|\mathbf{h}\|^2}{\bar{\beta}}\mathbf{u}^* + \mathbf{h}\right)$, *and*

$\sigma_2 = \|\mathbf{h}\|^2\|\mathbf{u}\|^2 + |\beta|^2$.

(vi) *If* $\mathbf{u} = \mathbf{0}$, $\mathbf{v} = \mathbf{0}$, *and* $\beta = 0$, *then* $(\mathbf{A} + \mathbf{cd}^*)^\dagger = \mathbf{A}^\dagger - \mathbf{k}\mathbf{k}^\dagger\mathbf{A}^\dagger - \mathbf{A}^\dagger\mathbf{h}^\dagger\mathbf{h} + (\mathbf{k}^\dagger\mathbf{A}^\dagger\mathbf{h}^\dagger)\mathbf{k}\mathbf{h}$.

Before proving Theorem 1, two preliminary facts are needed. We state them as lemmas.

Lemma 3.1.1 $\;$ *rank*$(\mathbf{A} + \mathbf{cd}^*) = $ *rank* $\begin{bmatrix} \mathbf{A} & \mathbf{u} \\ \mathbf{v} & -\beta \end{bmatrix} - 1$.

Proof $\;$ This follows immediately from the factorization

$$\begin{bmatrix} \mathbf{A} + \mathbf{cd}^* & \mathbf{c} \\ \mathbf{0}^* & -1 \end{bmatrix} = \begin{bmatrix} \mathbf{I} & \mathbf{0} \\ \mathbf{h} & 1 \end{bmatrix}\begin{bmatrix} \mathbf{A} & \mathbf{u} \\ \mathbf{v} & -\beta \end{bmatrix}\begin{bmatrix} \mathbf{I} & \mathbf{k} \\ \mathbf{0}^* & 1 \end{bmatrix}\begin{bmatrix} \mathbf{I} & \mathbf{0} \\ \mathbf{d}^* & 1 \end{bmatrix}. \quad \blacksquare$$

Lemma 3.1.2 $\;$ *If* $\mathbf{M}$ *and* $\mathbf{X}$ *are matrices such that* $\mathbf{X}\mathbf{M}\mathbf{M}^\dagger = \mathbf{X}$ *and* $\mathbf{M}^\dagger\mathbf{M} = \mathbf{X}\mathbf{M}$, *then* $\mathbf{X} = \mathbf{M}^\dagger$.

Proof $\;$ $\mathbf{M}^\dagger = (\mathbf{M}^\dagger\mathbf{M})\mathbf{M}^\dagger = \mathbf{X}\mathbf{M}\mathbf{M}^\dagger = \mathbf{X}$. $\blacksquare$

We now proceed with the proof of Theorem 3. Throughout, we assume $\mathbf{c} \neq \mathbf{0}$ and $\mathbf{d} \neq \mathbf{0}$.

Proof of (i). Let $\mathbf{X}_1$ denote the right-hand side of the equation in (i) and let $\mathbf{M} = \mathbf{A} + \mathbf{cd}^*$. The proof consists of showing that $\mathbf{X}_1$ satisfies the four Penrose conditions. Using $\mathbf{A}\mathbf{v}^\dagger = \mathbf{0}$, $\mathbf{d}^*\mathbf{v}^\dagger = 1$, $\mathbf{d}^*\mathbf{k} = \beta - 1$, and $\mathbf{c} - \mathbf{A}\mathbf{k} = \mathbf{u}$, it is easy to see that $\mathbf{M}\mathbf{X}_1 = \mathbf{A}\mathbf{A}^\dagger + \mathbf{u}\mathbf{u}^\dagger$ so that the third Penrose condition holds. Using $\mathbf{u}^\dagger\mathbf{A} = \mathbf{0}$, $\mathbf{u}^\dagger\mathbf{c} = 1$, $\mathbf{h}\mathbf{c} = \beta - 1$, and $\mathbf{d}^* - \mathbf{h}\mathbf{A} = \mathbf{v}$, one obtains $\mathbf{X}_1\mathbf{M} = \mathbf{A}^\dagger\mathbf{A} = \mathbf{v}^\dagger\mathbf{v}$ and hence the fourth condition holds. The first and second conditions follow easily.

Proof of (ii). Let $\mathbf{X}_2$ denote the right-hand side of the equality (ii). By using $\mathbf{A}\mathbf{k} = \mathbf{c}$, $\mathbf{A}\mathbf{v}^\dagger = \mathbf{0}$, $\mathbf{d}^*\mathbf{v}^\dagger = 1$, and $\mathbf{d}^*\mathbf{k} = -1$, it is seen that $(\mathbf{A} + \mathbf{cd}^*)\mathbf{X}_2 = \mathbf{A}\mathbf{A}^\dagger$, which is hermitian. From the facts that $\mathbf{k}^\dagger \mathbf{A}^\dagger \mathbf{A} = \mathbf{k}^\dagger$, $\mathbf{hc} = -1$, and $\mathbf{d}^* - \mathbf{hA} = \mathbf{v}$, it follows that $\mathbf{X}_2(\mathbf{A} + \mathbf{cd}^*) = \mathbf{A}^\dagger\mathbf{A} - \mathbf{kk}^\dagger + \mathbf{v}^\dagger\mathbf{v}$, which is also hermitian. The first and second Penrose conditions are now easily verified.

Proof of (iii) This case is the most difficult. Here $\mathbf{u} = \mathbf{0}$ so that $\mathbf{c} \in R(\mathbf{A})$ and hence it follows that $R(\mathbf{A} + \mathbf{cd}^*) \subseteq R(\mathbf{A})$. Since $\beta \neq 0$ it is clear from Lemma 1 that $\operatorname{rank}(\mathbf{A} + \mathbf{cd}^*) = \operatorname{rank}(\mathbf{A})$ so that $R(\mathbf{A} + \mathbf{cd}^*) = R(\mathbf{A})$. Therefore

$$(\mathbf{A} + \mathbf{cd}^*)(\mathbf{A} + \mathbf{cd}^*)^\dagger = \mathbf{AA}^\dagger \tag{3}$$

because $\mathbf{AA}^\dagger$ is the unique orthogonal projector onto $R(\mathbf{A})$. Let $\mathbf{X}_3$ denote the right-hand side of the equation in (iii). Because $\mathbf{q}_1^*\mathbf{AA}^\dagger = \mathbf{q}_1^*$ it follows immediately from (3) that $\mathbf{X}_3(\mathbf{A} + \mathbf{cd}^*)(\mathbf{A} + \mathbf{cd}^*)^\dagger = \mathbf{X}_3$. Hence the first condition of Lemma 2 is satisfied.

To show that the second condition of Lemma 2 is also satisfied, we first show that $(\mathbf{A} + \mathbf{cd}^*)^\dagger(\mathbf{A} + \mathbf{cd}^*) = \mathbf{A}^\dagger\mathbf{A} - \mathbf{kk}^\dagger + \mathbf{p}_1\mathbf{p}_1^\dagger$. The matrix $\mathbf{A}^\dagger\mathbf{A} - \mathbf{kk}^\dagger + \mathbf{p}_1\mathbf{p}_1^\dagger$ is hermitian and idempotent. The fact that it is hermitian is clear and the fact that it is idempotent follows by direct computation using $\mathbf{A}^\dagger\mathbf{Ak} = \mathbf{k}$, $\mathbf{A}^\dagger\mathbf{Ap}_1 = -\mathbf{k}$, and $\mathbf{kk}^\dagger\mathbf{p}_1 = -\mathbf{k}$. Since the rank of an idempotent matrix is equal to its trace and since trace is a linear function, it follows that $\operatorname{rank}(\mathbf{A}^\dagger\mathbf{A} - \mathbf{kk}^\dagger + \mathbf{p}_1\mathbf{p}_1^\dagger) = \operatorname{Tr}(\mathbf{A}^\dagger\mathbf{A} - \mathbf{kk}^\dagger + \mathbf{p}_1\mathbf{p}_1^\dagger) = \operatorname{Tr}(\mathbf{A}^\dagger\mathbf{A}) - \operatorname{Tr}(\mathbf{kk}^\dagger) + \operatorname{Tr}(\mathbf{p}_1\mathbf{p}_1^\dagger)$. Now, $\mathbf{kk}^\dagger$ and $\mathbf{p}_1\mathbf{p}_1^\dagger$ are idempotent matrices of rank = trace = 1 and $\mathbf{A}^\dagger\mathbf{A}$ is an idempotent matrix whose rank is equal to rank(**A**), so that

$$\operatorname{rank}(\mathbf{A}^\dagger\mathbf{A} - \mathbf{kk}^\dagger + \mathbf{p}_1\mathbf{p}_1^\dagger) = \operatorname{rank}(\mathbf{A} + \mathbf{cd}^*). \tag{4}$$

Using the facts $\mathbf{Ak} = \mathbf{c}$, $\mathbf{Ap}_1 = -\mathbf{c}$, $\mathbf{d}^*\mathbf{k} = \beta - 1$, $\mathbf{d}^*\mathbf{p}_1 = 1 - \sigma_1\bar{\beta}^{-1}$, and $\mathbf{d}^*\mathbf{A}^\dagger\mathbf{A} = \mathbf{d}^* - \mathbf{v}$, one obtains $(\mathbf{A} + \mathbf{cd}^*)(\mathbf{A}^\dagger\mathbf{A} - \mathbf{kk}^\dagger + \mathbf{p}_1\mathbf{p}_1^\dagger) = \mathbf{A} + \mathbf{cd}^* - \mathbf{c}(\mathbf{v} + \beta\mathbf{k}^\dagger + \sigma_1\bar{\beta}^{-1}\mathbf{p}_1^\dagger)$. Now, $\|\mathbf{p}_1\|^2 = \|\mathbf{k}\|^2\sigma_1|\beta|^{-2}$, so that $\sigma_1\bar{\beta}^{-1}\|\mathbf{p}_1\|^{-2} = \beta\|\mathbf{k}\|^{-2}$ and hence $\sigma_1\bar{\beta}^{-1}\mathbf{p}_1^\dagger + \beta\|\mathbf{k}\|^{-2}\mathbf{p}_1^* = -\mathbf{v} - \beta\mathbf{k}^\dagger$. Thus, $(\mathbf{A} + \mathbf{cd}^*)(\mathbf{A}^\dagger\mathbf{A} - \mathbf{kk}^\dagger + \mathbf{p}_1\mathbf{p}_1^\dagger) = \mathbf{A} + \mathbf{cd}^*$. Because $\mathbf{A}^\dagger\mathbf{A} - \mathbf{kk}^\dagger + \mathbf{p}_1\mathbf{p}_1^\dagger$ is an orthogonal projector, it follows that $R(\mathbf{A}^* + \mathbf{dc}^*) \subseteq R(\mathbf{A}^\dagger\mathbf{A} - \mathbf{kk}^\dagger + \mathbf{p}_1\mathbf{p}_1^\dagger)$. By virtue of (4), we conclude that $R(\mathbf{A}^* + \mathbf{dc}^*) = R(\mathbf{A}^\dagger\mathbf{A} - \mathbf{kk}^\dagger + \mathbf{p}_1\mathbf{p}_1^\dagger)$, and hence $(\mathbf{A}^* + \mathbf{dc}^*)(\mathbf{A}^* + \mathbf{dc}^*)^\dagger = \mathbf{A}^\dagger\mathbf{A} - \mathbf{kk}^\dagger + \mathbf{p}_1\mathbf{p}_1^\dagger$, or equivalently, $(\mathbf{A} + \mathbf{cd}^*)^\dagger(\mathbf{A} + \mathbf{cd}^*) = \mathbf{A}^\dagger\mathbf{A} - \mathbf{kk}^\dagger + \mathbf{p}_1\mathbf{p}_1^\dagger$.

To show that $\mathbf{X}_3(\mathbf{A} + \mathbf{cd}^*) = \mathbf{A}^\dagger\mathbf{A} + \mathbf{p}_1\mathbf{p}_1^\dagger - \mathbf{kk}^\dagger$, we compute $\mathbf{X}_3(\mathbf{A} + \mathbf{cd}^*)$. Observe that $\mathbf{k}^*\mathbf{A}^\dagger\mathbf{A} = \mathbf{k}^*$, $\mathbf{q}_1^*\mathbf{c} = 1 - \sigma_1\bar{\beta}^{-1}$, and $\mathbf{q}_1^*\mathbf{A} + \mathbf{d}^* = -\|\mathbf{v}\|^2\bar{\beta}^{-1}\mathbf{k}^* + \mathbf{v}$.

Now, $\mathbf{X}_3(\mathbf{A} + \mathbf{cd}^*)$

$$= \mathbf{A}^\dagger\mathbf{A} + \frac{1}{\bar{\beta}}\mathbf{v}^*\mathbf{k}^* - \frac{\bar{\beta}}{\sigma_1}\mathbf{p}_1\mathbf{q}_1^*\mathbf{A} + \left(\mathbf{k} + \frac{\|\mathbf{k}\|^2}{\bar{\beta}}\mathbf{v}^*\right)\mathbf{d}^* - \frac{\bar{\beta}}{\sigma_1}\mathbf{p}_1\mathbf{q}_1^*\mathbf{cd}^*$$

$$= \mathbf{A}^\dagger\mathbf{A} + \frac{1}{\bar{\beta}}\mathbf{v}^*\mathbf{k}^* - \frac{\bar{\beta}}{\sigma_1}\mathbf{p}_1\mathbf{q}_1^*\mathbf{A} - \mathbf{p}_1\mathbf{d}^* - \frac{\bar{\beta}}{\sigma_1}\mathbf{p}_1\mathbf{d}^* + \mathbf{p}_1\mathbf{d}^*$$

$$= \mathbf{A}^\dagger\mathbf{A} + \frac{1}{\bar{\beta}}\mathbf{v}^*\mathbf{k}^* - \frac{\bar{\beta}}{\sigma_1}\mathbf{p}_1(\mathbf{q}_1^*\mathbf{A} + \mathbf{d}^*)$$

$$= \mathbf{A}^\dagger\mathbf{A} + \frac{1}{\bar\beta}\mathbf{v}^*\mathbf{k}^* - \frac{\bar\beta}{\sigma_1}\mathbf{p}_1(\mathbf{v} - \|\mathbf{v}\|^2\bar\beta^{-1}\mathbf{k}^*).$$

Write $\mathbf{v}$ as $\mathbf{v} = -\beta\|\mathbf{k}\|^{-2}(\mathbf{p}_1^* + \mathbf{k}^*)$ and substitute this in the expression in parentheses and use the fact that $\|\mathbf{p}_1\|^{-2} = |\beta|^2\sigma_1^{-1}\|\mathbf{k}\|^{-2}$ to obtain

$$\mathbf{X}_3(\mathbf{A} + \mathbf{cd}^*) = \mathbf{A}^\dagger\mathbf{A} + \frac{1}{\bar\beta}\mathbf{v}^*\mathbf{k}^* + \mathbf{p}_1\mathbf{p}_1^\dagger + \frac{1}{\|\mathbf{k}\|^2}\mathbf{p}_1\mathbf{k}^*.$$

Since $\frac{1}{\bar\beta}\mathbf{v}^* + \frac{1}{\|\mathbf{k}\|^2}\mathbf{p}_1 = \frac{1}{\bar\beta}\mathbf{v}^* - \frac{1}{\bar\beta}\mathbf{v}^* - \frac{1}{\|\mathbf{k}\|^2}\mathbf{k} = -\frac{1}{\|\mathbf{k}\|^2}\mathbf{k}$,
we arrive at $\mathbf{X}_3(\mathbf{A} + \mathbf{cd}^*) = \mathbf{A}^\dagger\mathbf{A} + \mathbf{p}_1\mathbf{p}_1^\dagger - \mathbf{kk}^\dagger$.
Thus $(\mathbf{A} + \mathbf{cd}^*)^\dagger(\mathbf{A} + \mathbf{cd}^*) = \mathbf{X}_3(\mathbf{A} + \mathbf{cd}^*)$ so that $\mathbf{X}_3 = (\mathbf{A} + \mathbf{cd}^*)^\dagger$ by Lemma 2.

Proof of (iv) and (v). (iv) follows from (ii) and (v) follows from (iii) by taking conjugate transposes and using the fact that for any matrix $\mathbf{M}, (\mathbf{M}^\dagger)^* = (\mathbf{M}^*)^\dagger$.

Proof of (vi). Each of the matrices $\mathbf{AA}^\dagger - \mathbf{h}^\dagger\mathbf{h}$ and $\mathbf{A}^\dagger\mathbf{A} - \mathbf{kk}^\dagger$ is an orthogonal projector. The fact that they are idempotent follows from $\mathbf{AA}^\dagger\mathbf{h}^\dagger = \mathbf{h}^\dagger$, $\mathbf{hAA}^\dagger = \mathbf{h}$, $\mathbf{A}^\dagger\mathbf{Ak} = \mathbf{k}$ and $\mathbf{k}^\dagger\mathbf{A}^\dagger\mathbf{A} = \mathbf{k}^\dagger$. It is clear that each is hermitian. Moreover, the rank of each is equal to its trace and hence each has rank equal to rank$(\mathbf{A}) - 1$. Also, since $\mathbf{u} = \mathbf{0}$, $\mathbf{v} = \mathbf{0}$, and $\beta = 0$, it follows from Lemma 1.1 that rank$(\mathbf{A} + \mathbf{cd}^*) = $ rank$(\mathbf{A}) - 1$. Hence,

$$\text{rank}(\mathbf{A} + \mathbf{cd}^*) = \text{rank}(\mathbf{AA}^\dagger - \mathbf{h}^\dagger\mathbf{h}) = \text{rank}(\mathbf{AA}^\dagger - \mathbf{k}^\dagger\mathbf{k}). \tag{5}$$

With the facts $\mathbf{AA}^\dagger\mathbf{c} = \mathbf{c}$, $\mathbf{hc} = -1$, and $\mathbf{hA} = \mathbf{d}^*$, it is easy to see that $(\mathbf{AA}^\dagger - \mathbf{h}^\dagger\mathbf{h})(\mathbf{A} + \mathbf{cd}^*) = (\mathbf{A} + \mathbf{cd}^*)$, so that $R(\mathbf{A} + \mathbf{cd}^*) \subseteq R(\mathbf{AA}^\dagger - \mathbf{h}^\dagger\mathbf{h})$. Likewise, using $\mathbf{d}^*\mathbf{A}^\dagger\mathbf{A} = \mathbf{d}^*$, $\mathbf{d}^*\mathbf{k} = -1$, and $\mathbf{Ak} = \mathbf{c}$, one sees that $(\mathbf{A} + \mathbf{cd}^*)(\mathbf{A}^\dagger\mathbf{A} - \mathbf{kk}^\dagger) = \mathbf{A} + \mathbf{cd}^*$. Hence $R(\mathbf{A}^* + \mathbf{dc}^*) \subseteq R(\mathbf{A}^\dagger\mathbf{A} - \mathbf{kk}^\dagger)$. By virtue of (5), it now follows that

$$(\mathbf{A} + \mathbf{cd}^*)(\mathbf{A} + \mathbf{cd}^*)^\dagger = \mathbf{AA}^\dagger - \mathbf{h}^\dagger\mathbf{h}, \text{ and} \tag{6}$$

$$(\mathbf{A} + \mathbf{cd}^*)^\dagger(\mathbf{A} + \mathbf{cd}^*) = \mathbf{A}^\dagger\mathbf{A} - \mathbf{kk}^\dagger. \tag{7}$$

If $\mathbf{X}_4$ denotes the right-hand side of (vi), use (6) and the fact that $\mathbf{hAA}^\dagger = \mathbf{h}$ to obtain $\mathbf{X}_4(\mathbf{A} + \mathbf{cd}^*)(\mathbf{A} + \mathbf{cd}^*)^\dagger = \mathbf{X}_4$ which is the first condition of Lemma 2. Use $\mathbf{k}^\dagger\mathbf{A}^\dagger\mathbf{A} = \mathbf{k}^\dagger$, $\mathbf{hA} = \mathbf{d}^*$, and $\mathbf{hc} = -1$ to obtain $\mathbf{X}_4(\mathbf{A} + \mathbf{cd}^*) = \mathbf{A}^\dagger\mathbf{A} - \mathbf{kk}^\dagger$. Then by (7), we have that the second condition of Lemma 2 is satisfied. Hence $\mathbf{X}_4 = (\mathbf{A} + \mathbf{cd}^*)^\dagger$. ■

Corollary 3.1.1 *When* $\mathbf{c} \in R(\mathbf{A})$, $\mathbf{d} \in R(\mathbf{A}^*)$, *and* $\beta \neq 0$, *the generalized inverse of* $\mathbf{A} + \mathbf{cd}^*$ *is given by* $(\mathbf{A} + \mathbf{cd}^*)^\dagger = \mathbf{A}^\dagger - \frac{1}{\beta}\mathbf{A}^\dagger\mathbf{cd}^*\mathbf{A}^\dagger = \mathbf{A}^\dagger - \frac{1}{\beta}\mathbf{kh}$.

Proof Set $\mathbf{v} = \mathbf{0}$ in (iii), $\mathbf{u} = \mathbf{0}$ in (v). ■

Corollary 1 is the analogue of the well known formula which states that if both $\mathbf{A}$ and $\mathbf{A} + \mathbf{cd}^*$ are non-singular, then $(\mathbf{A} + \mathbf{cd}^*)^{-1} = \mathbf{A}^{-1} - \frac{1}{\beta}\mathbf{A}^{-1}\mathbf{cd}^*\mathbf{A}^{-1}$

2. Modified matrices

At first glance, the results of Theorem 1.3 may appear too complicated to be any practical value. However, a closer examination of Theorem 1.3 will reveal that it may be very useful and it is not difficult to apply to a large class of problems.

Suppose that one is trying to model a particular situation with a mathematical expression which involves a matrix $\mathbf{A} \in \mathbb{C}^{m \times n}$ and its generalized inverse $\mathbf{A}^\dagger$. For a variety of reasons, it is frequently the case that one wishes to modify the model by changing one or more entries of $\mathbf{A}$ to produce a 'modified' matrix $\tilde{\mathbf{A}}$, and then to compute $\tilde{\mathbf{A}}^\dagger$. The modified model involving $\tilde{\mathbf{A}}$ and $\tilde{\mathbf{A}}^\dagger$ may then be analysed and compared with the original model to determine what effects the modifications produce.

A similar situation which is frequently encountered is that an error is discovered in a matrix $\mathbf{A}$ of data for which $\mathbf{A}^\dagger$ has been previously calculated. It then becomes necessary to correct or modify $\mathbf{A}$ to produce a matrix $\tilde{\mathbf{A}}$ and then to compute the generalized inverse $\tilde{\mathbf{A}}^\dagger$ of the modified matrix.

In each of the above situations, it is highly desirable to use the already known information; $\mathbf{A}, \mathbf{A}^\dagger$ and the modifications made, in the computation of $\tilde{\mathbf{A}}^\dagger$ rather than starting from scratch. Theorem 1.3 allows us to do this since any matrix modification can always be accomplished by the addition of one or more rank one matrices.

To illustrate these ideas, consider the common situation in which one wishes to add a scalar α to the (i,j)th entry of $\mathbf{A} \in \mathbb{C}^{m \times n}$ to produce the modified matrix $\tilde{\mathbf{A}}$. Write $\tilde{\mathbf{A}}$ as

$$\tilde{\mathbf{A}} = \mathbf{A} + \alpha \mathbf{e}_i \mathbf{e}_j^* \quad \text{where} \quad \mathbf{e}_i \in \mathbb{C}^m, \; \mathbf{e}_j \in \mathbb{C}^n. \tag{1}$$

Write $\mathbf{A}^\dagger$ as $\mathbf{A}^\dagger = [g_{ij}] = [\mathbf{c}_1 \,\vdots\, \cdots \,\vdots\, \mathbf{c}_m] = \begin{bmatrix} \mathbf{r}_1 \\ \vdots \\ \mathbf{r}_n \end{bmatrix}$.

That is, g_{ij} denotes the (i,j)-entry of $\mathbf{A}^\dagger, \mathbf{c}_i$ is the ith column of $\mathbf{A}^\dagger$, and $\mathbf{r}_i$ is the ith row of $\mathbf{A}^\dagger$. The dotted lines which occur in the block matrix of $\mathbf{A}^\dagger$ are included to help the reader distinguish the blocks and their arrangement. They will be especially useful in Section 3 where some blocks have rather complicated expressions.

To use Theorem 1.3 on the modified matrix (1), order the computation as follows.

Algorithm 3.2.1 To g-invert the modified matrix $\mathbf{A} + \alpha \mathbf{e}_i \mathbf{e}_j^*$.

(I) Compute $\mathbf{k}$ and $\mathbf{h}$. This is easy since $\mathbf{k} = \mathbf{A}^\dagger \mathbf{e}_i$ and $\mathbf{h} = \alpha \mathbf{e}_j^* \mathbf{A}^\dagger$ so that $\mathbf{k} = \mathbf{c}_i$ and $\mathbf{h} = \alpha \mathbf{r}_j$.

(II) Compute $\mathbf{u}$ and $\mathbf{v}$ by $\mathbf{u} = \mathbf{e}_i - \mathbf{A}\mathbf{c}_i$ and $\mathbf{v} = \alpha(\mathbf{e}_j^* - \mathbf{r}_j \mathbf{A})$.

(III) Compute β, this is also easy since $\beta = 1 + \mathbf{h}\mathbf{e}_i$, so that $\beta = 1 + \alpha g_{ji}$.

(IV) Decide which of the six cases to use according as $\mathbf{u}$, $\mathbf{v}$, and β are zero or non-zero.

(V) Depending on which case is to be used, carefully arrange the computation of the terms involved so as to minimize the number of multiplications and divisions performed.

To illustrate step (V) of Algorithm 1, consider the term $\mathbf{k}\mathbf{k}^\dagger\mathbf{A}^\dagger$, which has to be computed in cases (ii) and (vi) of Theorem 1.3. It could be computed in several ways. Let us examine two of them. To obtain $\mathbf{k}^\dagger$ (we assume) $\mathbf{k} \neq \mathbf{0}$) we use $\mathbf{k}^\dagger = \mathbf{k}/\|\mathbf{k}\|^2 = \mathbf{c}_i^*/(\mathbf{c}_i, \mathbf{c}_i)$. This requires $2n$ *operations* (an operation is either a multiplication or division). If we perform the calculations by next forming the product $\mathbf{k}\mathbf{k}^\dagger$ and then the product $(\mathbf{k}\mathbf{k}^\dagger)\mathbf{A}^\dagger$, it would be necessary to do an additional $mn^2 + n^2$ operations, making a total of $n^2(m+1) + 2n$ operations. However, if $\mathbf{k}\mathbf{k}^\dagger\mathbf{A}^\dagger$ is computed by first obtaining $\mathbf{k}^\dagger$ and then forming the product $\mathbf{k}^\dagger\mathbf{A}^\dagger$, followed then by forming the product $(\mathbf{k}(\mathbf{k}^\dagger\mathbf{A}^\dagger))$, the number of operations required is reduced to $2n(m+1)$. This could amount to a significant saving in time and effort as compared to the former operational count.

It is important to observe that the products $\mathbf{A}\mathbf{A}^\dagger$ or $\mathbf{A}^\dagger\mathbf{A}$ do not need to be explicitly computed in order to use Theorem 1.3. If one were naive enough to form the products $\mathbf{A}\mathbf{A}^\dagger$ or $\mathbf{A}^\dagger\mathbf{A}$, a large amount of unnecessary effort would be expended.

Example 3.2.1 Suppose that

$$\mathbf{A} = \begin{bmatrix} 1 & 2 & 0 & 1 \\ 0 & 1 & -1 & 0 \\ 0 & 0 & 1 & -1 \end{bmatrix}, \text{ and } \mathbf{A}^\dagger = \frac{1}{12}\begin{bmatrix} 3 & -3 & 0 \\ 3 & 5 & 4 \\ 3 & -7 & 4 \\ 3 & -7 & -8 \end{bmatrix}$$

has been previously computed. Assume that an error has been discovered in $\mathbf{A}$ in that the (3, 3)-entry of $\mathbf{A}$ should have been zero instead of one. Then $\mathbf{A}$ is corrected by adding -1 to the (3, 3)-entry. Thus the modified matrix is $\tilde{\mathbf{A}} = \mathbf{A} + \mathbf{e}_3(-\mathbf{e}_3^*)$. To obtain $\tilde{\mathbf{A}}^\dagger$ we proceed as follows.

(I) The terms $\mathbf{k}$ and $\mathbf{h}$ are first read from $\mathbf{A}^\dagger$ as

$$\mathbf{k} = \mathbf{c}_3 = \frac{1}{12}\begin{bmatrix} 0 \\ 4 \\ 4 \\ -8 \end{bmatrix} \text{ and } \mathbf{h} = -\mathbf{r}_3 = \frac{1}{12}[-3 \quad 7 \quad -4].$$

(II) The terms $\mathbf{u}$ and $\mathbf{v}$ are easily calculated. $\mathbf{A}\mathbf{c}_3 = \mathbf{e}_3$ so that $\mathbf{u} = \mathbf{0}$.

$$\mathbf{v} = -\mathbf{e}_3^* + \mathbf{r}_3\mathbf{A} = \frac{1}{12}[3 \quad -1 \quad -1 \quad -1].$$

(III) The term β is also read from $\mathbf{A}^\dagger$ as $\beta = 1 - g_{33} = \dfrac{2}{3}$.

(IV) Since $\mathbf{u} = \mathbf{0}$, $\mathbf{v} \neq \mathbf{0}$, and $\beta \neq 0$, case (iii) of Theorem 1.3 must be used to obtain $\tilde{\mathbf{A}}^\dagger$.

(V) Computing the terms in case (iii) we get $\|\mathbf{k}\|^2 = \|\mathbf{c}_3\|^2 = \dfrac{2}{3}$ and

$$\|\mathbf{v}\|^2 = \frac{1}{12}, \quad \sigma_1 = \|\mathbf{c}_3\|^2\|\mathbf{v}\|^2 + \frac{4}{9} = \frac{1}{2}, \quad \frac{1}{\bar{\beta}}\mathbf{k}^*\mathbf{A}^\dagger = \frac{1}{\bar{\beta}}\mathbf{c}_3^*\mathbf{A}^\dagger = \frac{1}{2}[0 \quad 1 \quad 2],$$

$$\frac{1}{\bar{\beta}}\mathbf{v}^*\mathbf{k}^*\mathbf{A}^\dagger = \frac{1}{24}\begin{bmatrix} 0 & 3 & 6 \\ 0 & -1 & -2 \\ 0 & -1 & -2 \\ 0 & -1 & -2 \end{bmatrix},$$

$$\frac{\bar{\beta}}{\sigma_1}\mathbf{p}_1 = -\frac{1}{\sigma_1}(\|\mathbf{k}\|^2\mathbf{v}^* + \bar{\beta}\mathbf{k}) = -\frac{1}{3}\begin{bmatrix} 1 \\ 1 \\ 1 \\ -3 \end{bmatrix}$$

$$\mathbf{q}_1^* = -\left(\|\mathbf{v}\|^2\left(\frac{1}{\bar{\beta}}\mathbf{k}^*\mathbf{A}^\dagger\right) + \mathbf{h}\right) = -\frac{1}{8}[-2 \quad 5 \quad -2], \text{ and}$$

$$\frac{\bar{\beta}}{\sigma_1}\mathbf{p}_1\mathbf{q}_1^* = \frac{1}{24}\begin{bmatrix} -2 & 5 & -2 \\ -2 & 5 & -2 \\ -2 & 5 & -2 \\ 6 & -15 & 6 \end{bmatrix}.$$

Then by (iii) of Theorem 1.3., $\tilde{\mathbf{A}}^\dagger = \frac{1}{6}\begin{bmatrix} 2 & -2 & 2 \\ 2 & 1 & 2 \\ 2 & -5 & 2 \\ 0 & 0 & -6 \end{bmatrix}.$

3. Partitioned matrices

Let $\mathbf{E} = \begin{bmatrix} \mathbf{A} & \mathbf{B} \\ \mathbf{C} & \mathbf{D} \end{bmatrix}$ where **A**, **B**, **C** and **D** are four matrices such that **E** is also a matrix. Then **A**, **B**, **C**, **D** are called *conformable*. There are two ways to think of **E**. One is that **E** is made up of blocks **A**, **B**, **C**, **D**. In this case the blocks are, in a sense, considered fixed. If one is trying to 'build' an **E** to have a certain property, then one might experiment with a particular size or kind of blocks. This is especially the case in certain more advanced areas of mathematics such as Operator Theory where specific examples of linear operators on infinite dimensional vector spaces are often defined in terms of block matrices.

One can also view **E** as partitioned into its blocks. In this viewpoint one starts with **E**, views it as a *partitioned matrix* and tries to compute things about **E** from the blocks it is partitioned into. In this viewpoint **E** is fixed and different arrangements of blocks may be considered.

Partitioned matrix and block matrix are equivalent mathematical terms. However, in a given area of mathematics one of the two terms is more likely to be in vogue. We shall try to be in style.

Of course, a partitioned or block matrix may have more or less than four blocks.

This section is concerned with how to compute the generalized inverse of a matrix in terms of various partitions of it. As with Section 2, it is virtually impossible to come up with usable results in the general case. However, various special cases can be handled and, as in Section 2, they are not only theoretically interesting, but lead to useful algorithms. In fact. Theorem 1.3 will be the basis of much of this section, including the first case we consider.

Let $\mathbf{P} \in \mathbb{C}^{m \times (n+1)}$ be partitioned by slicing off its last column, so that $\mathbf{P} = [\mathbf{B} \,|\, \mathbf{c}]$ where $\mathbf{B} \in \mathbb{C}^{m \times n}$ and $\mathbf{c} \in \mathbb{C}^m$. Our objective is to obtain a useful expression for $\mathbf{P}^\dagger$. $\mathbf{P}$ may also be written as $\mathbf{P} = [\mathbf{B} \,|\, \mathbf{0}_1] + \mathbf{c}[\mathbf{0}_2^* \,|\, 1]$ where $\mathbf{0}_1 \in \mathbb{C}^m$, $\mathbf{0}_2 \in \mathbb{C}^n$.

Then $\mathbf{P}$ is in a form for which Theorem 1.3 applies. Let $\mathbf{A} = [\mathbf{B} \,|\, \mathbf{0}]$ and $\mathbf{d}^* = [\mathbf{0}^* \,|\, 1]$. Using the notation of Theorem 1.3 and the fact that $\mathbf{A}^\dagger = \begin{bmatrix} \mathbf{B}^\dagger \\ \mathbf{0}^* \end{bmatrix}$, one easily obtains $\mathbf{h} = \mathbf{d}^*\mathbf{A}^\dagger = \mathbf{0}$ so that $\beta = 1 + \mathbf{d}^*\mathbf{A}^\dagger\mathbf{c} = 1$ and $\mathbf{v} = \mathbf{d}^* \neq \mathbf{0}$. Also, $\mathbf{u} = (\mathbf{I} - \mathbf{A}\mathbf{A}^\dagger)\mathbf{c} = (\mathbf{I} - \mathbf{B}\mathbf{B}^\dagger)\mathbf{c}$. Thus, there are two cases to consider, according as to whether $\mathbf{u} \neq \mathbf{0}$ or $\mathbf{u} = \mathbf{0}$. Consider first the case when $\mathbf{u} \neq \mathbf{0}$ (i.e. $\mathbf{c} \notin R(\mathbf{B})$). Then case (i) in Theorem 1.3 is used to obtain $\mathbf{P}^\dagger$. In this case

$$\mathbf{A}^\dagger\mathbf{c} = \begin{bmatrix} \mathbf{B}^\dagger\mathbf{c} \\ 0 \end{bmatrix}, \text{ and } \mathbf{v}^\dagger = \begin{bmatrix} 0 \\ 1 \end{bmatrix}. \text{ Then } \mathbf{P}^\dagger = \begin{bmatrix} \mathbf{B}^\dagger \\ \hline \mathbf{0}^* \end{bmatrix} - \begin{bmatrix} \mathbf{B}^\dagger\mathbf{c}\mathbf{u}^\dagger \\ \hline \mathbf{0}^* \end{bmatrix} + \begin{bmatrix} \mathbf{0} \\ \hline \mathbf{u}^\dagger \end{bmatrix} = \begin{bmatrix} \mathbf{B}^\dagger - \mathbf{B}^\dagger\mathbf{c}\mathbf{u}^\dagger \\ \hline \mathbf{u}^\dagger \end{bmatrix}.$$

Next, consider the case when $\mathbf{u} = \mathbf{0}$ so that (iii) of Theorem 1.3 must be used. Let $\mathbf{k} = \mathbf{B}^\dagger\mathbf{c}$. Then $\sigma_1 = 1 + \mathbf{c}^*\mathbf{B}^{\dagger *}\mathbf{B}^\dagger\mathbf{c} = 1 + \mathbf{k}^*\mathbf{k}$, $\mathbf{p}_1 = -\begin{bmatrix} \mathbf{k} \\ \mathbf{k}^*\mathbf{k} \end{bmatrix}$, and $\mathbf{q}_1 = -\mathbf{k}^*\mathbf{B}^\dagger$ so that

$$\mathbf{P}^\dagger = \begin{bmatrix} \mathbf{B}^\dagger - \dfrac{\mathbf{k}\mathbf{k}^*\mathbf{B}^\dagger}{1 + \mathbf{k}^*\mathbf{k}} \\ \hline \dfrac{\mathbf{k}^*\mathbf{B}^\dagger}{1 + \mathbf{k}^*\mathbf{k}} \end{bmatrix}.$$ Thus we have the following theorem.

Theorem 3.3.1 *For* $\mathbf{B} \in \mathbb{C}^{m \times n}$, $\mathbf{c} \in \mathbb{C}^m$, *and* $\mathbf{P} = [\mathbf{B} \,|\, \mathbf{c}]$, *let* $\mathbf{k} = \mathbf{B}^\dagger\mathbf{c}$ *and* $\mathbf{u} = (\mathbf{I} - \mathbf{B}\mathbf{B}^\dagger)\mathbf{c} = \mathbf{c} - \mathbf{B}\mathbf{k}$. *The generalized inverse of* $\mathbf{P}$ *is given by*

$$\mathbf{P}^\dagger = \begin{bmatrix} \mathbf{B}^\dagger - \mathbf{k}\mathbf{y} \\ \mathbf{y} \end{bmatrix} \text{ where } \mathbf{y} = \begin{cases} \mathbf{u}^\dagger & \text{if } \mathbf{u} \neq \mathbf{0}, \\ (1 + \mathbf{k}^*\mathbf{k})^{-1}\mathbf{k}^*\mathbf{B}^\dagger & \text{if } \mathbf{u} = \mathbf{0}. \end{cases}$$

Theorem 4 can be applied to matrices of the form $\mathbf{P} = \begin{bmatrix} \mathbf{B} \\ \mathbf{r} \end{bmatrix}$, $\mathbf{r}$ a row vector, by using the conjugate transpose.

By using Theorem 1 together with Theorem 1.3, it is possible to consider the more general partitioned matrix $\mathbf{M} = \begin{bmatrix} \mathbf{A} & \mathbf{c} \\ \mathbf{d}^* & \alpha \end{bmatrix}$ where $\mathbf{A} \in \mathbb{C}^{m \times n}$, $\mathbf{c} \in \mathbb{C}^m$,

$\mathbf{d}\in\mathbb{C}^n$, and $\alpha\in\mathbb{C}$. Let $\mathbf{P}=[\mathbf{A}\,|\,\mathbf{c}]$ so that $\mathbf{M}$ can be written as

$$\mathbf{M}=\left[\begin{array}{c|c}\mathbf{A} & \mathbf{c}\\ \hline \mathbf{0}^* & 0\end{array}\right]+\begin{bmatrix}\mathbf{0}\\ \hline 1\end{bmatrix}[\mathbf{d}^*\,|\,\alpha]=\begin{bmatrix}\mathbf{P}\\ \hline \mathbf{0}^*\end{bmatrix}+\begin{bmatrix}\mathbf{0}\\ \hline 1\end{bmatrix}[\mathbf{d}^*\,|\,\alpha].$$

Theorem 1.3 can be applied to obtain $\mathbf{M}^\dagger$ as $\mathbf{M}^\dagger=[\mathbf{P}^\dagger\,|\,\mathbf{0}]+\mathbf{G}$. $\mathbf{P}^\dagger$ is known from Theorem 1. Clearly, $\begin{bmatrix}\mathbf{0}\\ 1\end{bmatrix}\notin R\left(\begin{bmatrix}\mathbf{P}\\ \mathbf{0}^*\end{bmatrix}\right)$ and $1+[\mathbf{d}^*\,|\,\alpha]\times[\mathbf{P}^\dagger\,|\,\mathbf{0}]\begin{bmatrix}\mathbf{0}\\ 1\end{bmatrix}=1\neq 0$. Thus either case (i) or case (v) of Theorem 1.3 must be used, depending on whether or not $\begin{bmatrix}\mathbf{d}\\ \bar{\alpha}\end{bmatrix}\in R\left(\begin{bmatrix}\mathbf{A}^*\\ \mathbf{c}^*\end{bmatrix}\right)$.

The details, which are somewhat involved, are left to the interested reader, or one may see Meyer's paper [55]. We state the end result.

Theorem 3.3.2 *For* $\mathbf{A}\in\mathbb{C}^{m\times n}$, $\mathbf{c}\in\mathbb{C}^m$, $\mathbf{d}\in\mathbb{C}^n$, $\alpha\in\mathbb{C}$, *let*

$$\mathbf{M}=\begin{bmatrix}\mathbf{A} & \mathbf{c}\\ \mathbf{d}^* & \alpha\end{bmatrix},\ \mathbf{k}=\mathbf{A}^\dagger\mathbf{c},\ \mathbf{h}=\mathbf{d}^*\mathbf{A}^\dagger,\ \mathbf{u}=(\mathbf{I}-\mathbf{A}\mathbf{A}^\dagger)\,\mathbf{c},$$
$$\mathbf{v}=\mathbf{d}^*(\mathbf{I}-\mathbf{A}^\dagger\mathbf{A}),\ \omega_1=1+\|\mathbf{k}\|^2,\ \omega_2=1+\|\mathbf{h}\|^2,\ \textit{and}\ \delta=\alpha-\mathbf{d}^*\mathbf{A}^\dagger\mathbf{c}.$$

The generalized inverse for $\mathbf{M}$ *is as follows.*

(i) *If* $\mathbf{u}\neq\mathbf{0}$ *and* $\mathbf{v}\neq\mathbf{0}$, *then* $\mathbf{M}^\dagger=\left[\begin{array}{c|c}\mathbf{A}^\dagger-\mathbf{k}\mathbf{u}^\dagger-\mathbf{v}^\dagger\mathbf{h}-\delta\mathbf{v}^\dagger\mathbf{u}^\dagger & \mathbf{v}^\dagger\\ \hline \mathbf{u}^\dagger & 0\end{array}\right]$.

(ii) *If* $\mathbf{u}=\mathbf{0}$ *and* $\delta=0$, *then* $\mathbf{M}^\dagger=\left[\begin{array}{c|c}\mathbf{A}^\dagger-\omega_1^{-1}\mathbf{k}\mathbf{k}^*\mathbf{A}^\dagger-\mathbf{v}^\dagger\mathbf{h} & \mathbf{v}^\dagger\\ \hline \omega_1^{-1}\mathbf{k}^*\mathbf{A}^\dagger & 0\end{array}\right]$.

(iii) *If* $\mathbf{u}=\mathbf{0}$ *and* $\delta\neq 0$, *then* $\mathbf{M}^\dagger=\left[\begin{array}{c|c}\mathbf{A}^\dagger-\bar{\delta}^{-1}\mathbf{v}^*\mathbf{k}^*\mathbf{A}^\dagger & \mathbf{0}\\ \hline \mathbf{0}^* & 0\end{array}\right]+\dfrac{\bar{\delta}}{\phi_1}\begin{bmatrix}\mathbf{p}_1\\ 1\end{bmatrix}[\mathbf{q}_1\,|\,1]$.

where $\mathbf{p}_1=\bar{\delta}^{-1}\omega_1\mathbf{v}^*-\mathbf{k}$, $\mathbf{q}_1^*=\bar{\delta}^{-1}\|\mathbf{v}\|^2\mathbf{k}^*\mathbf{A}^\dagger-\mathbf{h}$ *and* $\phi_1=\omega_1\|\mathbf{v}\|^2+|\delta|^2$.

(iv) *If* $\mathbf{v}=\mathbf{0}$ *and* $\delta=0$, *then* $\mathbf{M}^\dagger=\left[\begin{array}{c|c}\mathbf{A}^\dagger-\omega_2^{-1}\mathbf{A}^\dagger\mathbf{h}^*\mathbf{h}-\mathbf{k}\mathbf{u}^\dagger & \omega_2^{-1}\mathbf{A}^\dagger\mathbf{h}^*\\ \hline \mathbf{u}^\dagger & 0\end{array}\right]$.

(v) *If* $\mathbf{v}=\mathbf{0}$ *and* $\delta\neq 0$, *then* $\mathbf{M}^\dagger=\left[\begin{array}{c|c}\mathbf{A}^\dagger-\bar{\delta}^{-1}\mathbf{A}^\dagger\mathbf{h}^*\mathbf{u}^* & \mathbf{0}\\ \hline \mathbf{0}^* & 0\end{array}\right]+\dfrac{\bar{\delta}}{\phi_2}\begin{bmatrix}\mathbf{p}_2\\ 1\end{bmatrix}[\mathbf{q}_2^*\,|\,1]$.

where $\mathbf{p}_2=\bar{\delta}^{-1}\|\mathbf{u}\|^2\mathbf{A}^\dagger\mathbf{h}^*-\mathbf{k}$, $\mathbf{q}_2^*=\delta^{-1}\omega_2\mathbf{u}^*-\mathbf{h}$, *and* $\phi_2=\omega_2\|\mathbf{u}\|^2+|\delta|^2$.

(vi) *If* $\mathbf{u}=\mathbf{0}$, $\mathbf{v}=\mathbf{0}$, *and* $\delta=0$, *then*

$$\mathbf{M}^\dagger=\left[\begin{array}{c|c}\mathbf{A}^\dagger-\omega_1^{-1}\mathbf{k}\mathbf{k}^*\mathbf{A}^\dagger-\omega_2^{-1}\mathbf{A}^\dagger\mathbf{h}^*\mathbf{h} & \omega_2^{-1}\mathbf{A}^\dagger\mathbf{h}^*\\ \hline -\mathbf{k}^*\mathbf{A} & 0\end{array}\right]+\frac{\mathbf{k}^*\mathbf{A}^\dagger\mathbf{h}^*}{\omega_1\omega_2}\begin{bmatrix}\mathbf{k}\\ \hline -1\end{bmatrix}[\mathbf{h}\,|\,-1].$$

Frequently one finds that one is dealing with a hermitian or a real symmetric matrix. The following corollary might then be useful.

Corollary 3.3.1 *For* $\mathbf{A}\in\mathbb{C}^{m\times m}$ *such that* $\mathbf{A}=\mathbf{A}^*$ *and for* $\alpha\in\mathbb{R}$, *the generalized inverse of the hermitian matrix* $\mathbf{H}=\left[\begin{array}{c|c}\mathbf{A} & \mathbf{c}\\ \hline \mathbf{c}^* & \alpha\end{array}\right]$ *is as follows:*

(i) *If* $\mathbf{u}\neq\mathbf{0}$, *then* $\mathbf{H}^\dagger=\left[\begin{array}{c|c}\mathbf{A}^\dagger-\mathbf{k}\mathbf{u}^\dagger-\mathbf{u}^{\dagger}\mathbf{k}^*-\delta\mathbf{u}^{\dagger *}\mathbf{u}^\dagger & \mathbf{u}^{\dagger *}\\ \hline \mathbf{u}^\dagger & 0\end{array}\right]$.

(ii) *If* $\mathbf{u}=\mathbf{0}$ *and* $\delta\neq 0$, *then* $\mathbf{H}^\dagger=\left[\begin{array}{c|c}\mathbf{A}^\dagger & \mathbf{0}\\ \hline \mathbf{0}^* & 0\end{array}\right]+\delta^{-1}\left[\begin{array}{c}\mathbf{k}\\ \hline -1\end{array}\right][\mathbf{k}^* \,\vert\, -1]$.

(iii) *If* $\mathbf{u}=\mathbf{0}$ *and* $\delta=0$, *then*

$$\mathbf{H}^\dagger=\left[\begin{array}{c|c}\mathbf{A}^\dagger-\omega^{-1}\mathbf{k}\mathbf{k}^*\mathbf{A}^\dagger-\omega_1^{-1}\mathbf{A}^\dagger\mathbf{k}\mathbf{k}^* & \omega_1^{-1}\mathbf{A}^\dagger\mathbf{k}\\ \hline \omega_1^{-1}\mathbf{k}^*\mathbf{A}^\dagger & 0\end{array}\right]+\frac{\mathbf{k}^*\mathbf{A}^\dagger\mathbf{k}}{\omega_1^2}\left[\begin{array}{c}\mathbf{k}\\ \hline -1\end{array}\right][\mathbf{k}^* \,\vert\, -1].$$

Theorems 1 and 2 may be used to recursively compute a generalized inverse, and were the basis of some early methods. However, the calculations are sensitive to ill conditioning (see Chapter 12). The next two algorithms, while worded in terms of calculating $\mathbf{A}^\dagger$ should only be used for that purpose on small sized, reasonably well conditioned matrices.

The real value of the next two algorithms, like Algorithm 2.1, is in updating. Only in this case instead of updating $\mathbf{A}$ itself, one is adding rows and/or columns to $\mathbf{A}$. This corresponds, for example in least squares problems, to either making an additional observation or adding a new parameter to the model.

Algorithm 3.3.1 To g-invert any $\mathbf{A}\in\mathbb{C}^{m\times n}$.

(I) Partition $\mathbf{A}$ as $\mathbf{A}=[\mathbf{c}_1 \,\vert\, \dots \,\vert\, \mathbf{c}_n]$ and set $\mathbf{B}_1=\mathbf{c}_1$ and $\mathbf{B}_1^\dagger=\mathbf{c}_1^\dagger$.
(II) For $i\geq 2$, set $\mathbf{B}_i=[\mathbf{B}_{i-1} \,\vert\, \mathbf{c}_i]$,
(III) $\mathbf{k}_i=\mathbf{B}_{i-1}^\dagger\mathbf{c}_i$,
(IV) $\mathbf{u}_i=\mathbf{c}_i-\mathbf{B}_{i-1}\mathbf{k}_i$,

$$\mathbf{y}_i=\begin{cases}\dfrac{\mathbf{u}_i^*}{\mathbf{u}_i^*\mathbf{u}_i} & \text{if } \mathbf{u}_i\neq\mathbf{0},\\[2ex] \dfrac{\mathbf{k}_i^*\mathbf{B}_{i-1}^\dagger}{1+\mathbf{k}_i^*\mathbf{k}_i} & \text{if } \mathbf{u}_i=\mathbf{0},\end{cases}\qquad\text{and}$$

(VI) $\mathbf{B}_i^\dagger=\left[\begin{array}{c}\mathbf{B}_{i-1}^\dagger-\mathbf{k}_i\mathbf{y}_i\\ \hline \mathbf{y}_i\end{array}\right]$.
(VII) Then $\mathbf{A}^\dagger=\mathbf{B}_n^\dagger$.

Example 3.3.1 Suppose that we have the linear system $\mathbf{A}\mathbf{x}=\mathbf{b}$, and have computed $\mathbf{A}^\dagger$ so we know that

$$\mathbf{A} = \begin{bmatrix} 1 & 0 \\ 2 & 1 \\ 3 & 1 \\ 1 & 1 \end{bmatrix}, \mathbf{A}^\dagger = \frac{1}{3}\begin{bmatrix} 1 & 0 & 1 & -1 \\ -2 & 1 & -1 & 3 \end{bmatrix}.$$

But now we wish to add another independent variable and solve $\tilde{\mathbf{A}}\tilde{\mathbf{x}} = \mathbf{b}$ where

$$\tilde{\mathbf{A}} = \begin{bmatrix} 1 & 0 & 1 \\ 2 & 1 & 0 \\ 3 & 1 & 1 \\ 1 & 1 & -1 \end{bmatrix}, \text{ by computing } \tilde{\mathbf{A}}^\dagger.$$

We will use Algorithm 1. In the notation of Algorithm 1, $\tilde{\mathbf{A}} = \mathbf{B}_3$, $\mathbf{A} = \mathbf{B}_2$, $\mathbf{c}^* = [1 \quad 0 \quad 1 \quad -1]$. For $i = 3$, $\mathbf{k}_3 = \begin{bmatrix} 1 \\ -2 \end{bmatrix}$, $\mathbf{u}_3 = \mathbf{0}$, and $\mathbf{y}_3 = \frac{1}{18}[5 \quad -2 \quad 3 \quad -7]$, so that

$$\mathbf{B}_3^\dagger = \tilde{\mathbf{A}}^\dagger = \frac{1}{18}\begin{bmatrix} 1 & 2 & 3 & 1 \\ -2 & 2 & 0 & 4 \\ 5 & -2 & 3 & -7 \end{bmatrix}.$$

If the matrix in our model is always hermitian and we add both an independent variable and a row, the next Algorithm could be useful.

Algorithm 3.3.2 To g-invert $\mathbf{H} = [h_{ij}] \in \mathbb{C}^{n \times n}$ such that $\mathbf{H} = \mathbf{H}^*$

(I) Set $\mathbf{A}_1 = \mathbf{h}_{11}$.

(II) For $i \geq 2$, set $\mathbf{A}_i = \left[\begin{array}{c|c} \mathbf{A}_{i-1} & \mathbf{c}_i \\ \hline \mathbf{c}_i^* & h_{ii} \end{array}\right]$ where $\mathbf{c}_i = \begin{bmatrix} h_{1i} \\ h_{2i} \\ \vdots \\ h_{i-1,i} \end{bmatrix}$

(III) Let $\mathbf{k}_i = \mathbf{A}_{i-1}\mathbf{c}_i$,

(IV) $\delta_i = h_{ii} - \mathbf{c}_i^*\mathbf{k}_i$, and

(V) $\mathbf{u}_i = \mathbf{c}_i - \mathbf{A}_{i-1}\mathbf{k}_i$.

(VI) If $\mathbf{u}_i \neq \mathbf{0}$, then $\mathbf{u}_i^\dagger = \dfrac{\mathbf{u}_i^*}{\mathbf{u}_i^*\mathbf{u}_i}$, $\mathbf{Y}_i = \mathbf{k}_i\mathbf{u}_i^\dagger$, and

$$\mathbf{A}_i^\dagger = \left[\begin{array}{c|c} \mathbf{A}_{i-1}^\dagger - (\mathbf{Y}_i + \mathbf{Y}_i^*) - \delta_i\mathbf{u}_i^{\dagger *}\mathbf{u}_i & \mathbf{u}_i^{\dagger *} \\ \hline \mathbf{u}_i^\dagger & 0 \end{array}\right].$$

(VII) If $\mathbf{u}_i = \mathbf{0}$ and $\delta_i \neq 0$, then $\mathbf{A}_i^\dagger = \left[\begin{array}{c|c} \mathbf{A}_{i-1}^\dagger + \delta_i^{-1}\mathbf{k}_i\mathbf{k}_i^* & -\delta_i^{-1}\mathbf{k}_i \\ \hline -\delta_i^{-1}\mathbf{k}_i^* & \delta_i^{-1} \end{array}\right]$.

(VIII) If $\mathbf{u}_i = \mathbf{0}$ and $\delta_i = 0$, then let $\mathbf{r}_i = \mathbf{A}_{i-1}^\dagger\left(\dfrac{\mathbf{k}_i}{1 + \mathbf{k}_i^*\mathbf{k}_i}\right)$,

$\psi_i = \left(\dfrac{\mathbf{k}_i}{1+\mathbf{k}_i^*\mathbf{k}_i}\right)\mathbf{r}_i$, $\mathbf{Y}_i = \mathbf{r}_i\mathbf{k}_i^*$ and $\mathbf{z}_i = \psi_i\mathbf{k}_i$, so that

$$\mathbf{A}_i^\dagger = \left[\begin{array}{c|c} \mathbf{A}_{i-1}^\dagger - (\mathbf{Y}_i + \mathbf{Y}_i^*) + \mathbf{z}_i\mathbf{k}_i^* & \mathbf{r}_i - \mathbf{z}_i \\ \hline (\mathbf{r}_i - \mathbf{z}_i)^* & \psi_i \end{array}\right].$$

(IX) Then $\mathbf{H}^\dagger = \mathbf{A}_n^\dagger$.

For a general hermitian matrix it is difficult to compare operational counts, however Algorithm 2 is usually more efficient than Algorithm 1 when applicable since it utilizes the symmetry.

There is, of course, no clear cut point at which it is better to recompute $\tilde{\mathbf{A}}^\dagger$ from scratch then use augmentation methods. If $\mathbf{A}$ were 11×4 and 2 rows were added, the authors would go with Algorithm 1. If $\mathbf{A}$ were 11×4 and 7 rows were added, we would recompute $\tilde{\mathbf{A}}^\dagger$ directly from $\tilde{\mathbf{A}}$.

It is logical to wonder what the extensions of Theorems 1 and 2 are. That is, what are $[\mathbf{A}\,|\,\mathbf{C}]^\dagger$ and $\begin{bmatrix} \mathbf{A} & \mathbf{C} \\ \mathbf{D} & \mathbf{B} \end{bmatrix}^\dagger$ when $\mathbf{C}$ and $\mathbf{D}$ no longer are just columns and $\mathbf{B}$ is no longer a scalar? When $\mathbf{A}, \mathbf{B}, \mathbf{C}$ and $\mathbf{D}$ are general conformable matrices, the answer to 'what is $[\mathbf{A}\,|\,\mathbf{C}]$?' is difficult. A *useful* answer to 'what is $\begin{bmatrix} \mathbf{A} & \mathbf{C} \\ \mathbf{D} & \mathbf{B} \end{bmatrix}^\dagger$?' is not yet known though formulas exist.

The previous discussion suggests that in some cases, at least when $\mathbf{C}$ has a 'small' number of columns, such as extensions could be useful.

We will begin by examining matrices of the form $[\mathbf{A}\,|\,\mathbf{C}]$ where $\mathbf{A}$ and $\mathbf{C}$ are general conformable matrices. One representation for $[\mathbf{A}\,|\,\mathbf{C}]^\dagger$ is as follows.

Theorem 3.3.3 *For $\mathbf{A} \in \mathbb{C}^{m \times n}$ and $\mathbf{C} \in \mathbb{C}^{m \times p}$, the generalized inverse of $[\mathbf{A}\,|\,\mathbf{C}]$ can be written as*

$$[\mathbf{A}\,|\,\mathbf{C}]^\dagger = \left[\begin{array}{c} (\mathbf{I} + \mathbf{TT}^*)^{-1}(\mathbf{A}^\dagger - \mathbf{A}^\dagger\mathbf{CB}^\dagger) \\ \hline \mathbf{T}^*(\mathbf{I} + \mathbf{TT}^*)^{-1}(\mathbf{A}^\dagger - \mathbf{A}^\dagger\mathbf{CB}^\dagger) + \mathbf{B}^\dagger \end{array}\right]$$

where $\mathbf{B} = (\mathbf{I} - \mathbf{AA}^\dagger)\mathbf{C}$ and $\mathbf{T} = \mathbf{A}^\dagger\mathbf{C}(\mathbf{I} - \mathbf{B}^\dagger\mathbf{B})$.

Proof One verifies that the four Penrose conditions are satisfied. ∎

A representation similar to that of Theorem 3 is possible for matrices partitioned in the form $\left[\begin{array}{c} \mathbf{A} \\ \hline \mathbf{R} \end{array}\right]$ by taking transposes.

The reader should be aware that there are many other known ways of representing the generalized inverse for matrices partitioned as $[\mathbf{A}\,|\,\mathbf{C}]$ or as $\left[\begin{array}{c} \mathbf{A} \\ \hline \mathbf{R} \end{array}\right]$. The interested reader is urged to consult the following references to obtain several other useful representations. (See R. E. Cline [31], A. Ben-Israel [12], and J. V. Rao [72].)

As previously mentioned, no useful representation for $\begin{bmatrix} \mathbf{A} & \mathbf{C} \\ \mathbf{R} & \mathbf{D} \end{bmatrix}^\dagger$ has, up to this time, been given where $\mathbf{A}, \mathbf{C}, \mathbf{R}$ and $\mathbf{D}$ are general comformable matrices. However, if we place some restrictions on the blocks in question, we can obtain some useful results.

Lemma 3.3.1 *If $\mathbf{A}, \mathbf{C}, \mathbf{R}$ and $\mathbf{D}$ are conformable matrices such that $\mathbf{A}$ is square and non-singular and rank $(\mathbf{A}) =$ rank $\begin{bmatrix} \mathbf{A} & \mathbf{C} \\ \mathbf{R} & \mathbf{D} \end{bmatrix}$, then $\mathbf{D} = \mathbf{R}\mathbf{A}^{-1}\mathbf{C}$. Furthermore, if $\mathbf{P} = \mathbf{R}\mathbf{A}^{-1}$ and $\mathbf{Q} = \mathbf{A}^{-1}\mathbf{C}$ then*

$$\begin{bmatrix} \mathbf{A} & \mathbf{C} \\ \mathbf{R} & \mathbf{D} \end{bmatrix} = \begin{bmatrix} \mathbf{I} \\ \mathbf{P} \end{bmatrix} \mathbf{A} [\mathbf{I} \quad \mathbf{Q}].$$

Proof The factorization

$$\left[\begin{array}{c|c} \mathbf{I} & \mathbf{0} \\ \hline -\mathbf{R}\mathbf{A}^{-1} & \mathbf{I} \end{array}\right] \left[\begin{array}{c|c} \mathbf{A} & \mathbf{C} \\ \hline \mathbf{R} & \mathbf{D} \end{array}\right] \left[\begin{array}{c|c} \mathbf{I} & -\mathbf{A}^{-1}\mathbf{C} \\ \hline \mathbf{0} & \mathbf{I} \end{array}\right] = \left[\begin{array}{c|c} \mathbf{A} & \mathbf{0} \\ \hline \mathbf{0} & \mathbf{D} - \mathbf{R}\mathbf{A}^{-1}\mathbf{C} \end{array}\right]$$

yields rank $\left[\begin{array}{c|c} \mathbf{A} & \mathbf{C} \\ \hline \mathbf{R} & \mathbf{D} \end{array}\right] =$ rank $\left[\begin{array}{c|c} \mathbf{A} & \mathbf{0} \\ \hline \mathbf{0} & \mathbf{D} - \mathbf{R}\mathbf{A}^{-1}\mathbf{C} \end{array}\right] =$ rank$(\mathbf{A})$ + rank$(\mathbf{D} - \mathbf{R}\mathbf{A}^{-1}\mathbf{C})$. Therefore, it can be concluded that rank$(\mathbf{D} - \mathbf{R}\mathbf{A}^{-1}\mathbf{C}) = 0$, or equivalently, $\mathbf{D} = \mathbf{R}\mathbf{A}^{-1}\mathbf{C}$. The factorization (1) follows directly. ■

Matrices of the type discussed in Lemma 1 have generalized inverses which possess a relatively simple form.

Theorem 3.3.4 *Let $\mathbf{A}, \mathbf{C}, \mathbf{R}$ and $\mathbf{D}$ be conformable matrices such that $\mathbf{A}$ is square, non-singular, and rank$(\mathbf{A}) =$ rank $\begin{bmatrix} \mathbf{A} & \mathbf{C} \\ \mathbf{R} & \mathbf{D} \end{bmatrix}$. If $\mathbf{P}$ and $\mathbf{Q}$ are any matrices such that $\begin{bmatrix} \mathbf{A} & \mathbf{C} \\ \mathbf{R} & \mathbf{D} \end{bmatrix} = \begin{bmatrix} \mathbf{I} \\ \mathbf{P} \end{bmatrix} \mathbf{A}[\mathbf{I} \quad \mathbf{Q}]$, then*

$$\begin{bmatrix} \mathbf{A} & \mathbf{C} \\ \mathbf{R} & \mathbf{D} \end{bmatrix}^\dagger = \begin{bmatrix} \mathbf{I} \\ \mathbf{Q}^* \end{bmatrix} ([\mathbf{I} + \mathbf{P}^*\mathbf{P}]\mathbf{A}[\mathbf{I} + \mathbf{Q}\mathbf{Q}^*])^{-1} [\mathbf{I} \quad \mathbf{P}^*].$$

Proof Let $\mathbf{A} \in \mathbb{C}^{r \times r}$, $\mathbf{M} = \begin{bmatrix} \mathbf{A} & \mathbf{C} \\ \mathbf{R} & \mathbf{D} \end{bmatrix}$, $\mathbf{B} = \begin{bmatrix} \mathbf{I}_r \\ \mathbf{P} \end{bmatrix} \mathbf{A}$, and $\mathbf{G} = [\mathbf{I}_r \quad \mathbf{Q}]$. Notice that rank$(\mathbf{B}) =$ rank$(\mathbf{G}) =$ rank$(\mathbf{A}) =$ rank$(\mathbf{M}) = r =$ (number of columns of $\mathbf{B}$) = number of rows of $\mathbf{G}$). Thus, we may apply Theorem 1.3.2 to obtain $\mathbf{M}^\dagger = (\mathbf{B}\mathbf{G})^\dagger = \mathbf{G}^*(\mathbf{G}\mathbf{G}^*)^{-1}(\mathbf{B}^*\mathbf{B})^{-1}\mathbf{B}^*$. Since $(\mathbf{B}^*\mathbf{B})^{-1}\mathbf{B}^* = [\mathbf{A}^*(\mathbf{I} + \mathbf{P}^*\mathbf{P})\mathbf{A}]^{-1}\mathbf{A}^*[\mathbf{I} \mid \mathbf{P}^*] = \mathbf{A}^{-1}(\mathbf{I} + \mathbf{P}^*\mathbf{P})^{-1}[\mathbf{I} \mid \mathbf{P}^*]$ and $\mathbf{G}^*(\mathbf{G}\mathbf{G}^*)^{-1} = \begin{bmatrix} \mathbf{I} \\ \mathbf{Q}^* \end{bmatrix} (\mathbf{I} + \mathbf{Q}\mathbf{Q}^*)^{-1}$ the desired result is obtained. ■

It is always possible to perform a permutation of rows and columns to any matrix so as to bring a full-rank non-singular block to the upper left hand corner. Theorem 3.3.4 may then be used as illustrated below.

Example 3.3.3 In order to use Theorem 4 to find $\mathbf{M}^\dagger$, the first step is to reduce $\mathbf{M}$ to row echelon form, $\mathbf{E}_\mathbf{M}$. This not only will reveal the rank of $\mathbf{M}$, but will also indicate a full set of linearly independent columns.

$$\text{Let } \mathbf{M} = \begin{bmatrix} 1 & 2 & 1 & 1 & 3 \\ 2 & 4 & 2 & 2 & 6 \\ 1 & 2 & -1 & 0 & 2 \\ 2 & 4 & 0 & 1 & 5 \end{bmatrix}, \text{ so that } \mathbf{E}_\mathbf{M} = \begin{bmatrix} 1 & 2 & 0 & 1/2 & 5/2 \\ 0 & 0 & 1 & 1/2 & 1/2 \\ 0 & 0 & 0 & 0 & 0 \\ 0 & 0 & 0 & 0 & 0 \end{bmatrix}.$$

Thus, rank$(\mathbf{M}) = 2$ and the first and third columns of $\mathbf{M}$ form a full independent set. Let $\mathbf{F}$ be the 5×5 permutation matrix obtained by exchanging the second and third rows of $\mathbf{I}_5$ so that

$$\mathbf{MF} = \left[\begin{array}{cc|ccc} 1 & 1 & 2 & 1 & 3 \\ 2 & 2 & 4 & 2 & 6 \\ 1 & -1 & 2 & 0 & 2 \\ 2 & 0 & 4 & 1 & 5 \end{array}\right] = [\mathbf{X}_1 \mid \mathbf{X}_2].$$ The next step is to select two independent rows from the matrix $\mathbf{X}_1$. This may be accomplished in several ways. One could have obtained this information by noting which rows were interchanged during the reduction to echelon form, or one might just look at $\mathbf{X}_1$ and select the appropriate rows, or one might reduce $\mathbf{X}_1^T$ to echelon form. In our example, it is easy to see that the first and third rows of $\mathbf{X}_1$ are independent. Let $\mathbf{E}$ be the 4×4 permutation matrix obtained by exchanging the second and third rows of $\mathbf{I}_4$ so that

$$\mathbf{EMF} = \left[\begin{array}{cc|ccc} 1 & 1 & 2 & 1 & 3 \\ 1 & -1 & 2 & 0 & 2 \\ \hline 2 & 2 & 4 & 2 & 6 \\ 2 & 0 & 4 & 1 & 5 \end{array}\right] = \left[\begin{array}{c|c} \mathbf{A} & \mathbf{C} \\ \hline \mathbf{R} & \mathbf{D} \end{array}\right]$$

Since permutation matrices are unitary, Theorem 1.2.1 allows us to write $(\mathbf{EMF})^\dagger = \mathbf{F}^*\mathbf{M}^\dagger\mathbf{E}^*$ so that $\mathbf{M}^\dagger = \mathbf{F}(\mathbf{EMF})^\dagger\mathbf{E}$. Now apply Theorem 4 to obtain $(\mathbf{EMF})^\dagger$. In our example,

$$\mathbf{A}^{-1} = 1/2\begin{bmatrix} 1 & 1 \\ 1 & -1 \end{bmatrix}, \mathbf{P} = \begin{bmatrix} 2 & 0 \\ 1 & 1 \end{bmatrix}, \mathbf{Q} = 1/2\begin{bmatrix} 4 & 1 & 5 \\ 0 & 1 & 1 \end{bmatrix}$$

$$\text{so that } (\mathbf{EMF})^\dagger = \frac{1}{330}\begin{bmatrix} -3 & 18 & 6 & 15 \\ 33 & -88 & 66 & -55 \\ -6 & 36 & -12 & 30 \\ 15 & -35 & 30 & -20 \\ 9 & 1 & 18 & 10 \end{bmatrix},$$

$$\mathbf{M}^\dagger = \frac{1}{330}\begin{bmatrix} -3 & -6 & 18 & 15 \\ -6 & -12 & 36 & 30 \\ 33 & 66 & -88 & -55 \\ 15 & 30 & -35 & -20 \\ 9 & 18 & 1 & 10 \end{bmatrix}.$$

Theorem 3.3.5. *Let* **A**. **C**, **R** *and* **D** *be conformable matrices such that* **A** *is square, non-singular, and* $\text{rank}(\mathbf{A}) = \text{rank}\begin{bmatrix} \mathbf{A} & \mathbf{C} \\ \mathbf{R} & \mathbf{D} \end{bmatrix}$. *Let* **M** *denote the matrix* $\mathbf{M} = \begin{bmatrix} \mathbf{A} & \mathbf{C} \\ \mathbf{R} & \mathbf{D} \end{bmatrix}$ *and let* $\mathbf{W} = [\mathbf{A}^* \mid \mathbf{R}^*]\mathbf{M}\begin{bmatrix} \mathbf{A}^* \\ \mathbf{C}^* \end{bmatrix}$. *The matrix* **W** *is non-singular and* $\mathbf{M}^\dagger = \begin{bmatrix} \mathbf{A}^* \\ \mathbf{C}^* \end{bmatrix}\mathbf{W}^{-1}[\mathbf{A}^* \mid \mathbf{R}^*]$.

Proof We first prove the matrix **W** is non-singular. Write **W** as $\mathbf{W} = \mathbf{A}^*\mathbf{A}\mathbf{A}^* + \mathbf{R}^*\mathbf{R}\mathbf{A}^* + \mathbf{A}^*\mathbf{C}\mathbf{C}^* + \mathbf{R}^*\mathbf{D}\mathbf{C}^*$. From Lemma 1 we know that $\mathbf{D} = \mathbf{R}\mathbf{A}^{-1}\mathbf{C}$ so that $\mathbf{W} = \mathbf{A}^*\mathbf{A}\mathbf{A} + \mathbf{R}^*\mathbf{R}\mathbf{A}^* + \mathbf{A}^*\mathbf{C}\mathbf{C}^* + \mathbf{R}^*\mathbf{R}\mathbf{A}^{-1}\mathbf{C}\mathbf{C}^* = (\mathbf{A}^*\mathbf{A} + \mathbf{R}^*\mathbf{R})\mathbf{A}^{-1}(\mathbf{A}\mathbf{A}^* + \mathbf{C}\mathbf{C}^*)$. Because **A** is non-singular, the matrices $(\mathbf{A}^*\mathbf{A} + \mathbf{R}^*\mathbf{R})$ and $(\mathbf{A}\mathbf{A}^* + \mathbf{C}\mathbf{C}^*)$, are both positive definite, and thus non-singular. Therefore, **W** must be non-singular. Furthermore, $\mathbf{W}^{-1} = (\mathbf{A}\mathbf{A}^* + \mathbf{C}\mathbf{C}^*)^{-1}\mathbf{A}(\mathbf{A}^*\mathbf{A} + \mathbf{R}^*\mathbf{R})^{-1}$. Using this, one can now verify the four Penrose conditions are satisfied. ■

In both Theorems 4 and 5, it is necessary to invert only one matrix whose dimensions are equal to those of **A**. In Theorem 5, it is not necessary to obtain the matrices **P** and **Q** as in Theorem 4. However, where problems of ill-conditioning are encountered (see Chapter 12), Theorem 4 might be preferred over Theorem 5.

4. Block triangular matrices

Definition 3.4.1 *For conformable matrices* $\mathbf{T}_{11}, \mathbf{T}_{12}, \mathbf{T}_{21}$, *and* $\mathbf{T}_{22}$, *matrices of the form*

$$\begin{bmatrix} \mathbf{T}_{11} & \mathbf{T}_{12} \\ \mathbf{0} & \mathbf{T}_{22} \end{bmatrix} \text{ and } \begin{bmatrix} \mathbf{T}_{11} & \mathbf{0} \\ \mathbf{T}_{21} & \mathbf{T}_{22} \end{bmatrix}$$

are called upper block triangular *and* lower block triangular, *respectively.*

It is important to note that neither $\mathbf{T}_{11}$ nor $\mathbf{T}_{22}$ are required to be square in the above definition.

Throughout this section, we will discuss only upper block triangular matrices. For each statement we make about upper block triangular matrices, there is a corresponding statement possible for lower block triangular matrices.

Definition 3.4.2 *For an upper block triangular matrix,*

$$\mathbf{T} = \begin{bmatrix} \mathbf{T}_{11} & \mathbf{T}_{12} \\ \mathbf{0} & \mathbf{T}_{22} \end{bmatrix}, \tag{1}$$

we shall say **T** *is a* properly partitioned upper block triangular *matrix if* $\mathbf{T}^\dagger$ *is upper block triangular of the form* $\mathbf{T}^\dagger = \begin{bmatrix} \mathbf{G}_{11} & \mathbf{G}_{12} \\ \mathbf{0} & \mathbf{G}_{22} \end{bmatrix}$ *where the* *dimensions of* $\mathbf{G}_{11}$ *and* $\mathbf{G}_{22}$ *are the same as the dimensions of the transposes of* $\mathbf{T}_{11}$ *and* $\mathbf{T}_{22}$, *respectively. Any partition of* $\mathbf{T}$ *which makes* $\mathbf{T}$ *into a properly partitioned matrix is called a* proper partition for $\mathbf{T}$.

Example 3.4.1 Let $\mathbf{T} = \begin{bmatrix} 1 & 1 & 1 & 0 \\ 0 & 0 & 0 & 1 \\ 0 & 0 & 0 & 1 \end{bmatrix}$ so that $\mathbf{T}^{\dagger} = \begin{bmatrix} 1/3 & 0 & 0 \\ 1/3 & 0 & 0 \\ 1/3 & 0 & 0 \\ 0 & 1/2 & 1/2 \end{bmatrix}$

There are several ways to partition $\mathbf{T}$ so that $\mathbf{T}$ will have an upper block triangular form. For example,

$$\mathbf{T}_1 = \left[\begin{array}{ccc|c} 1 & 1 & 1 & 0 \\ \hline 0 & 0 & 0 & 1 \\ 0 & 0 & 0 & 1 \end{array}\right] \text{ and } \mathbf{T}_2 = \left[\begin{array}{cc|cc} 1 & 1 & 1 & 0 \\ \hline 0 & 0 & 0 & 1 \\ 0 & 0 & 0 & 1 \end{array}\right]$$

are two different partitions of $\mathbf{T}$ which both give rise to upper block triangular forms. Clearly, $\mathbf{T}_1$ is a proper partition of $\mathbf{T}$ while $\mathbf{T}_2$ is not. In fact, $\mathbf{T}_1$ is the only proper partition of $\mathbf{T}$.

Example 3.4.2 If $\mathbf{T}$ is an upper block triangular matrix which is partitioned as (1) whether $\mathbf{T}_{11}$ and $\mathbf{T}_{22}$ are both non-singular, then $\mathbf{T}$ is properly partitioned because

$$\mathbf{T}^{-1} = \left[\begin{array}{c|c} \mathbf{T}_{11}^{-1} & -\mathbf{T}_{11}^{-1}\mathbf{T}_{12}\mathbf{T}_{22}^{-1} \\ \hline \mathbf{0} & \mathbf{T}_{22}^{-1} \end{array}\right].$$

Not all upper block triangular matrices can be properly partitioned.

Example 3.4.3 Let $\mathbf{T} = \begin{bmatrix} 1 & 2 & 1 \\ 2 & 4 & 0 \\ 0 & 0 & 1 \end{bmatrix}$. Since there are no zeros in

$\mathbf{T}^{\dagger} = \frac{1}{45}\begin{bmatrix} 1 & 4 & -1 \\ 2 & 8 & -2 \\ 20 & -10 & 25 \end{bmatrix}$, there can be no proper partition for $\mathbf{T}$.

The next theorem characterizes properly partitioned matrices.

Theorem 3.4.1 *Let* $\mathbf{T}$ *be an upper block triangular matrix partitioned as* (1). $\mathbf{T}$ *is properly partitioned if and only if* $R(\mathbf{T}_{12}) \subseteq R(\mathbf{T}_{11})$ *and* $R(\mathbf{T}_{12}^*) \subseteq R(\mathbf{T}_{22}^*)$. *Furthermore, when* $\mathbf{T}$ *is properly partitioned,* $\mathbf{T}^{\dagger}$ *is given by*

$$\mathbf{T}^{\dagger} = \left[\begin{array}{c|c} \mathbf{T}_{11}^{\dagger} & -\mathbf{T}_{11}^{\dagger}\mathbf{T}_{12}\mathbf{T}_{22}^{\dagger} \\ \hline \mathbf{0} & \mathbf{T}_{22}^{\dagger} \end{array}\right]. \tag{2}$$

(*Note the resemblance between this expression and that of Example* 2.)

Proof Suppose first that $\mathbf{T}$ is properly partitioned so that $\mathbf{T}^{\dagger}$ is upper block triangular. It follows that $\mathbf{T}\mathbf{T}^{\dagger}$ and $\mathbf{T}^{\dagger}\mathbf{T}$ must also be upper block triangular. Since $\mathbf{T}\mathbf{T}^{\dagger}$ and $\mathbf{T}^{\dagger}\mathbf{T}$ are hermitian, it must be the case that they are of the form

$$\mathbf{TT}^{\dagger} = \begin{bmatrix} \mathbf{R}_1 & \mathbf{0} \\ \mathbf{0} & \mathbf{R}_2 \end{bmatrix} \text{ and } \mathbf{T}^{\dagger}\mathbf{T} = \begin{bmatrix} \mathbf{L}_1 & \mathbf{0} \\ \mathbf{0} & \mathbf{L}_2 \end{bmatrix}.$$

By using the fact that $\mathbf{TT}^{\dagger}\mathbf{T} = \mathbf{T}$, one obtains

$$\begin{aligned} &\mathbf{R}_1\mathbf{T}_{11} = \mathbf{T}_{11} \text{ and } \mathbf{L}_2\mathbf{T}_{22} = \mathbf{T}_{22}, \\ &\mathbf{R}_1\mathbf{T}_{12} = \mathbf{T}_{12} \text{ and } \mathbf{T}_{12}\mathbf{L}_2 = \mathbf{T}_{12} \end{aligned} \tag{3}$$

Also, $\mathbf{R}_1, \mathbf{R}_2, \mathbf{L}_1$ and $\mathbf{L}_2$ must be orthogonal projectors because they are hermitian and idempotent. Since $\mathbf{R}_1 = \mathbf{T}_{11}\mathbf{X}$ for some $\mathbf{X}$ and $\mathbf{R}_1\mathbf{T}_{11} = \mathbf{T}_{11}$, we can conclude $R(\mathbf{R}_1) = R(\mathbf{T}_{11})$, and hence $\mathbf{R}_1 = \mathbf{P}_{R(\mathbf{T}_{11})}$. Likewise, $\mathbf{L}_2 = \mathbf{YT}_{22}$ for some $\mathbf{Y}$ and $\mathbf{T}_{22} = \mathbf{T}_{22}\mathbf{L}_2$ implies $\mathbf{L}_2 = \mathbf{P}_{R(\mathbf{T}_{22}^*)}$. From (3), we now have

$$\mathbf{P}_{R(\mathbf{T}_{11})}\mathbf{T}_{12} = \mathbf{T}_{12} \text{ and } \mathbf{P}_{R(\mathbf{T}_{22}^*)}\mathbf{T}_{12}^* = \mathbf{T}_{12}^*, \tag{4}$$

and therefore

$$R(\mathbf{T}_{12}) \subseteq R(\mathbf{T}_{11}) \text{ and } R(\mathbf{T}_{12}^*) \subseteq R(\mathbf{T}_{22}^*). \tag{5}$$

To prove the converse, one first notes that (5) implies (4) and then uses this to show that the four Penrose conditions of Definition 1.1.3 are satisfied by the matrix (2). ■

A necessary condition for an upper block triangular to be properly partitioned is easily obtained from Theorem 1.

Corollary 3.4.1 *Let* $\mathbf{T}$ *be partitioned as in* (1). *If* $\mathbf{T}$ *is properly partitioned, then* $rank(\mathbf{T}) = rank(\mathbf{T}_{11}) + rank(\mathbf{T}_{22})$.

Proof If $\mathbf{T}$ is properly partitioned, then $\mathbf{T}^{\dagger}$ is given by (2) and $\mathbf{T}_{11}\mathbf{T}_{11}^{\dagger}\mathbf{T}_{12} = \mathbf{T}_{12}$ so that $\mathbf{TT}^{\dagger} = \left[\begin{array}{c|c} \mathbf{T}_{11}\mathbf{T}_{11}^{\dagger} & \mathbf{0} \\ \hline \mathbf{0} & \mathbf{T}_{22}\mathbf{T}_{22}^{\dagger} \end{array}\right]$. Thus, $\text{rank}(\mathbf{T}) = \text{rank}(\mathbf{TT}^{\dagger}) = \text{rank}(\mathbf{T}_{11}\mathbf{T}_{11}^{\dagger}) + \text{rank}(\mathbf{T}_{22}\mathbf{T}_{22}^{\dagger}) = \text{rank}(\mathbf{T}_{11}) + \text{rank}(\mathbf{T}_{22})$. ■

5. The fundamental matrix of constrained minimization

Definition 3.5.1 *Let* $\mathbf{V} \in \mathbb{C}^{n \times n}$ *be a positive semi-definite matrix and let* $\mathbf{C}^*$ *be any matrix in* $\mathbb{C}^{n \times r}$. *The block matrix* $\mathbf{B} = \begin{bmatrix} \mathbf{V} & \mathbf{C}^* \\ \mathbf{C} & \mathbf{0} \end{bmatrix}$ *is called the* fundamental matrix of constrained minimization.

This matrix is so named because of its importance in the theory of constrained minimization, as is demonstrated in the next section. It also plays a fundamental role in the theory of linear estimation. (See Section 4 of Chapter 6.) Throughout this section, the letter $\mathbf{B}$ will always denote the matrix of Definition 1. Our purpose here is to obtain a form for $\mathbf{B}^{\dagger}$.

Note that if $\mathbf{S}$ is the permutation matrix $\mathbf{S} = \begin{bmatrix} \mathbf{0} & \mathbf{I}_n \\ \mathbf{I}_r & \mathbf{0} \end{bmatrix}$, then

$\mathbf{B} = \begin{bmatrix} \mathbf{C}^* & \mathbf{V} \\ \mathbf{0} & \mathbf{C} \end{bmatrix} \mathbf{S}^*$ so that $\mathbf{B}^\dagger = \mathbf{S} \begin{bmatrix} \mathbf{C}^* & \mathbf{V} \\ \mathbf{0} & \mathbf{C} \end{bmatrix}^\dagger$.

Thus, we may use Theorem 3.4.1 to obtain the following result.

Theorem 3.5.1 $\mathbf{B}^\dagger = \left[\begin{array}{c|c} \mathbf{0} & \mathbf{C}^\dagger \\ \hline \mathbf{C}^{\dagger *} & -\mathbf{C}^{\dagger *}\mathbf{V}\mathbf{C}^\dagger \end{array}\right]$ *if and only if* $R(\mathbf{V}) \subseteq R(\mathbf{C}^*)$.

In the case when $R(\mathbf{V}) \not\subseteq R(\mathbf{C}^*)$, it is possible to add an appropriate term to the expression in Theorem 1 to get $\mathbf{B}^\dagger$.

Theorem 3.5.2 *For any positive semi-definite* $\mathbf{V} \in \mathbb{C}^{n \times n}$ *and any* $\mathbf{C}^* \in \mathbb{C}^{n \times r}$ *let* $\mathbf{E} = \mathbf{I} - \mathbf{C}^\dagger\mathbf{C}$ *and let* $\mathbf{Q} = (\mathbf{EVE})^\dagger$. *Then*

$$\mathbf{B}^\dagger = \left[\begin{array}{c|c} \mathbf{0} & \mathbf{C}^\dagger \\ \hline \mathbf{C}^{\dagger *} & -\mathbf{C}^{\dagger *}\mathbf{V}\mathbf{C}^\dagger \end{array}\right] + \begin{bmatrix} \mathbf{I} \\ -\mathbf{C}^{\dagger *}\mathbf{V} \end{bmatrix} \mathbf{Q}[\mathbf{I} \mid -\mathbf{V}\mathbf{C}^\dagger]. \tag{1}$$

Proof Since $\mathbf{V}$ is positive semi-definite, there exists $\mathbf{A} \in \mathbb{C}^{n \times n}$ such that $\mathbf{V} = \mathbf{A}^*\mathbf{A}$. Now, $\mathbf{E} = \mathbf{E}^* = \mathbf{E}^2$ so that

$$\mathbf{Q} = (\mathbf{E}^*\mathbf{A}^*\mathbf{A}\mathbf{E})^\dagger = ([\mathbf{AE}]^*\mathbf{AE})^\dagger \tag{2}$$

and hence $R(\mathbf{Q}) = R([\mathbf{AE}]^*) = R(\mathbf{E}\mathbf{A}^*) \subseteq R(\mathbf{E})$. This together with the fact that $\mathbf{Q} = \mathbf{Q}^*$ implies

$$\mathbf{EQ} = \mathbf{Q} \text{ and } \mathbf{QE} = \mathbf{Q}, \tag{3}$$

so that $\mathbf{CQ} = \mathbf{0}$ and $\mathbf{Q}^*\mathbf{C} = \mathbf{0}$. Let $\mathbf{X}$ denote the right-hand side of (1). We shall show that $\mathbf{X}$ satisfies the four Penrose conditions. Using the above information, calculate $\mathbf{BX}$ as

$$\mathbf{BX} = \left[\begin{array}{c|c} \mathbf{C}^\dagger\mathbf{C} + \mathbf{EVQ} & \mathbf{EVC}^\dagger - \mathbf{EVQVC}^\dagger \\ \hline \mathbf{0} & \mathbf{CC}^\dagger \end{array}\right].$$

Use (3) to write $\mathbf{EVQ} = \mathbf{EVEQ} = \mathbf{Q}^\dagger\mathbf{Q}$ and

$$\begin{aligned} \mathbf{EVQV} &= \mathbf{EVEQEV} = \mathbf{EA}^*\mathbf{AE}(\mathbf{AE})^\dagger(\mathbf{AE})^{*\dagger}(\mathbf{AE})^*\mathbf{A} = \mathbf{EA}^*\mathbf{AE}(\mathbf{AE})^\dagger(\mathbf{AE}) \\ &\quad \times (\mathbf{AE})^\dagger\mathbf{A} = \mathbf{EA}^*\mathbf{AE}(\mathbf{AE})^\dagger\mathbf{A} = (\mathbf{AE})^*(\mathbf{AE})(\mathbf{AE})^\dagger\mathbf{A} = (\mathbf{AE})^*\mathbf{A} \\ &= \mathbf{EA}^*\mathbf{A} = \mathbf{EV}. \end{aligned} \tag{4}$$

Thus,

$$\mathbf{BX} = \left[\begin{array}{c|c} \mathbf{C}^\dagger\mathbf{C} + \mathbf{Q}^\dagger\mathbf{Q} & \mathbf{0} \\ \hline \mathbf{0} & \mathbf{CC}^\dagger \end{array}\right], \tag{5}$$

which is hermitian. Using (5), compute $\mathbf{BXB} = \left[\begin{array}{c|c} \mathbf{C}^\dagger\mathbf{CV} + \mathbf{EVEQV} & \mathbf{C}^* \\ \hline \mathbf{C} & \mathbf{0} \end{array}\right]$.

From (3) and (4) it is easy to get that $\mathbf{EVEQV} = \mathbf{EVQV} = \mathbf{EV}$, and hence $\mathbf{BXB} = \mathbf{B}$. It follows by direct computation using (5) that $\mathbf{XBX} = \mathbf{X}$.

Finally, compute $\mathbf{XB} = \left[\begin{array}{c|c} \mathbf{QVE} + \mathbf{C}^\dagger\mathbf{C} & \mathbf{0} \\ \hline -\mathbf{C}^{\dagger *}\mathbf{VQVE} + \mathbf{C}^{\dagger *}\mathbf{VE} & \mathbf{CC}^\dagger \end{array}\right]$. From (3) $\mathbf{QVE} = \mathbf{QEVE} = \mathbf{QQ}^\dagger$. In a manner similar to that used in obtaining (4), one can show that $\mathbf{VQVE} = \mathbf{VE}$, so that $\mathbf{XB} = \left[\begin{array}{c|c} \mathbf{C}^\dagger\mathbf{C} + \mathbf{QQ}^\dagger & \mathbf{0} \\ \hline \mathbf{0} & \mathbf{CC}^\dagger \end{array}\right]$, which is hermitian. We have shown that $\mathbf{X}$ satisfies all four Penrose conditions so that $\mathbf{B}^\dagger = \mathbf{X}$. ■

6. Constrained least squares and constrained generalized inverses

In this section, we deal with two problems in constrained minimization. Let $\mathbf{A} \in \mathbb{C}^{m \times n}$, $\mathbf{b} \in \mathbb{C}^m$, $\mathbf{C} \in \mathbb{C}^{p \times n}$, and $\mathbf{f} \in \mathbb{C}^p$. Let $\mathscr{S}$ denote the set

$$\mathscr{S} = \{\mathbf{x} \,|\, \mathbf{x} = \mathbf{C}^\dagger\mathbf{f} + N(\mathbf{C})\}.$$

That is, $\mathscr{S}$ is the set of solutions (or least squares solutions) of $\mathbf{Cx} = \mathbf{f}$, depending on whether or not $\mathbf{f} \in R(\mathbf{C})$. $\mathscr{S}$ will be the *set of constraints*. It can be argued that the function $\|\mathbf{Ax} - \mathbf{b}\|$ attains a minimum value on $\mathscr{S}$. The two problems which we will consider are as follows.

Problem 1 Find $\min_{\mathbf{x} \in \mathscr{S}} \|\mathbf{Ax} - \mathbf{b}\|$ and describe the points in $\mathscr{S}$ at which the minimum is attained as a function of $\mathbf{A}, \mathbf{b}, \mathbf{C}$, and $\mathbf{f}$.

Problem 2 Among the points in $\mathscr{S}$ for which the minimum of Problem 1 is attained, show there is a unique point of minimal norm and then describe it as a function of $\mathbf{A}, \mathbf{b}, \mathbf{C}$, and $\mathbf{f}$.

The solutions of these two problems rest on the following fundamental theorem. This theorem also indicates why the term 'fundamental matrix' was used in Definition 3.5.1.

Theorem 3.6.1 *Let* $\mathbf{A}, \mathbf{b}, \mathbf{C}, \mathbf{f}$, *and* $\mathscr{S}$ *be as described above, and let* $q(\mathbf{x}) = \|\mathbf{Ax} - \mathbf{b}\|^2$. *A vector* $\mathbf{x}_0$ *satisfies the conditions that* $\mathbf{x}_0 \in \mathscr{S}$ *and* $\|\mathbf{Ax}_0 - \mathbf{b}\| \leq \|\mathbf{Ax} - \mathbf{b}\|$ *for all* $\mathbf{x} \in \mathscr{S}$ *if and only if there is a vector* $\mathbf{y}_0$ *such that* $\mathbf{z}_0 = \begin{bmatrix} \mathbf{x}_0 \\ \mathbf{y}_0 \end{bmatrix}$ *is a least squares solution of the system*

$$\left[\begin{array}{c|c} \mathbf{A}^*\mathbf{A} & \mathbf{C}^* \\ \hline \mathbf{C} & \mathbf{0} \end{array}\right] \begin{bmatrix} \mathbf{x} \\ \mathbf{y} \end{bmatrix} = \begin{bmatrix} \mathbf{A}^*\mathbf{b} \\ \mathbf{f} \end{bmatrix}.$$

Proof Let $\mathbf{B}$ and $\mathbf{v}$ denote the block matrices $\mathbf{B} = \left[\begin{array}{c|c} \mathbf{A}^*\mathbf{A} & \mathbf{C}^* \\ \hline \mathbf{C} & \mathbf{0} \end{array}\right]$, $\mathbf{v} = \begin{bmatrix} \mathbf{A}^*\mathbf{b} \\ \mathbf{f} \end{bmatrix}$. Suppose first that $\mathbf{z}_0$ is a least squares solution of $\mathbf{Bz} = \mathbf{v}$. From Theorem 2.1.2, we have that $\mathbf{Bz}_0 = \mathbf{BB}^\dagger\mathbf{v}$. From equations (2) and (5) of Section 5, we have

$$\mathbf{BB}^\dagger = \left[\begin{array}{c|c} \mathbf{C}^\dagger\mathbf{C} + (\mathbf{AE})^\dagger(\mathbf{AE}) & \mathbf{0} \\ \hline \mathbf{0} & \mathbf{CC}^\dagger \end{array}\right] \text{ where } \mathbf{E} = \mathbf{I} - \mathbf{C}^\dagger\mathbf{C},$$

so that $\mathbf{Bz}_o = \mathbf{BB}^\dagger\mathbf{v}$ implies

$$\begin{aligned} \mathbf{A}^*\mathbf{Ax}_o = \mathbf{C}^*\mathbf{y}_o &= \mathbf{C}^\dagger\mathbf{C}\mathbf{A}^*\mathbf{b} + (\mathbf{AE})^\dagger(\mathbf{AE})\mathbf{A}^*\mathbf{b} \\ &= \mathbf{C}^\dagger\mathbf{C}\mathbf{A}^*\mathbf{b} + (\mathbf{AE})^\dagger(\mathbf{AE})(\mathbf{AE})^*\mathbf{b} \\ &= \mathbf{C}^\dagger\mathbf{C}\mathbf{A}^*\mathbf{b} + (\mathbf{AE})^*\mathbf{b} = \mathbf{A}^*\mathbf{b} \end{aligned} \tag{1}$$

and

$$\mathbf{Cx}_o = \mathbf{CC}^\dagger\mathbf{f}. \tag{2}$$

From (2) we know $\mathbf{x}_o \in \mathscr{S}$. Write $\mathbf{x}_o = \mathbf{C}^\dagger\mathbf{f} + \mathbf{h}_o$ where $\mathbf{h}_o \in N(\mathbf{C})$. For every $\mathbf{x} \in \mathscr{S}$ we have $\mathbf{x} = \mathbf{C}^\dagger\mathbf{f} + \mathbf{h}_x$, $\mathbf{h}_x \in N(\mathbf{C})$ so that

$$\begin{aligned} q(\mathbf{x}) = q(\mathbf{C}^\dagger\mathbf{f} + \mathbf{h}_x) &= \| \mathbf{AC}^\dagger\mathbf{f} + \mathbf{Ah}_x - \mathbf{Ax}_o + \mathbf{Ax}_o - \mathbf{b} \|^2 \\ &= \| \mathbf{AC}^\dagger\mathbf{f} + \mathbf{Ah}_x - \mathbf{AC}^\dagger\mathbf{f} - \mathbf{Ah}_o + \mathbf{Ax}_o - \mathbf{b} \|^2 \\ &= \| \mathbf{A}(\mathbf{h}_x - \mathbf{h}_o) + \mathbf{Ax}_o - \mathbf{b} \|^2 \end{aligned} \tag{3}$$

For all $\mathbf{h} \in N(\mathbf{C})$, we may use (1) to get $(\mathbf{Ah}, \mathbf{Ax}_o - \mathbf{b}) = (\mathbf{h}, \mathbf{A}^*(\mathbf{Ax}_o - \mathbf{b})) = (\mathbf{h}, -\mathbf{C}^*\mathbf{y}_o) = -(\mathbf{Ch}, \mathbf{y}_o) = 0$. Hence (3) becomes $q(x) = \| \mathbf{A}(\mathbf{h}_x - \mathbf{h}_o) \|^2 + q(\mathbf{x}_o)$ so that $q(\mathbf{x}_o) \le q(\mathbf{x})$ for all $\mathbf{x} \in \mathscr{S}$, as desired.

Conversely, suppose $\mathbf{x}_o \in \mathscr{S}$ and $q(\mathbf{x}_o) \le q(\mathbf{x})$. If $\mathbb{C}^m$ is decomposed as $\mathbb{C}^m = \mathbf{A}(N(\mathbf{C})) \dotplus [\mathbf{A}(N(\mathbf{C}))]^\perp$, then

$$\mathbf{Ax}_o - \mathbf{b} = \mathbf{Ah} + \omega, \text{ where } \mathbf{h} \in N(\mathbf{C}), \omega \in [\mathbf{A}(N(\mathbf{C}))]^\perp. \tag{4}$$

We can write

$$q(\mathbf{x}_o) = \| \mathbf{Ah} + \omega \|^2 = \| \mathbf{Ah} \|^2 + \| \omega \|^2. \tag{5}$$

Now observe that $(\mathbf{x}_o - \mathbf{h}) \in \mathscr{S}$ because $\mathbf{x}_o \in \mathscr{S}$ and $\mathbf{h} \in N(\mathbf{C})$ implies $\mathbf{C}(\mathbf{x}_o - \mathbf{h}) = \mathbf{CC}^\dagger\mathbf{f}$. By hypothesis, we have $q(\mathbf{x}_o) \le q(\mathbf{x})$ for all $\mathbf{x} \in \mathscr{S}$ so that $q(\mathbf{x}_o) \le q(\mathbf{x}_o - \mathbf{h}) = \| (\mathbf{Ax}_o - \mathbf{b}) - \mathbf{Ah} \|^2 = \| (\mathbf{Ah} + \omega) - \mathbf{Ah} \|^2 = \| \omega \|^2$ (from (4)) $= q(\mathbf{x}_o) - \| \mathbf{Ah} \|^2$ (from (5)). Thus $\mathbf{Ah} = \mathbf{0}$ and $(\mathbf{Ax}_o - \mathbf{b}) \in [\mathbf{A}(N(\mathbf{C}))]^\perp$ by (4). Hence for any $\mathbf{g} \in N(\mathbf{C})$, $0 = (\mathbf{Ag}, \mathbf{Ax}_o - \mathbf{b}) = (\mathbf{g}, \mathbf{A}^*\mathbf{Ax}_o - \mathbf{A}^*\mathbf{b})$, and $(\mathbf{A}^*\mathbf{Ax}_o - \mathbf{A}^*\mathbf{b}) \in N(\mathbf{C})^\perp = R(\mathbf{C}^*)$. This means there exists a vector $(-\mathbf{y}_o)$ such that $\mathbf{C}^*(-\mathbf{y}_o) = \mathbf{A}^*\mathbf{Ax}_o - \mathbf{A}^*\mathbf{b}$ or

$$\begin{aligned} \mathbf{A}^*\mathbf{Ax}_o + \mathbf{C}^*\mathbf{y}_o = \mathbf{A}^*\mathbf{b} &= \mathbf{A}^*\mathbf{b} - (\mathbf{AE})^*\mathbf{b} + (\mathbf{AE})^*\mathbf{b} \\ &= \mathbf{A}^*\mathbf{b} - (\mathbf{AE})^*\mathbf{b} + (\mathbf{AE})^\dagger(\mathbf{AE})(\mathbf{AE})^*\mathbf{b} \\ &= \mathbf{A}^*\mathbf{b} - \mathbf{EA}^*\mathbf{b} + (\mathbf{AE})^\dagger\mathbf{AEA}^*\mathbf{b} \\ &= [\mathbf{C}^\dagger\mathbf{C} + (\mathbf{AE})^\dagger(\mathbf{AE})]\mathbf{A}^*\mathbf{b} \end{aligned} \tag{6}$$

Now (6) together with the fact that $\mathbf{x}_o \in \mathscr{S}$, gives $\mathbf{B}\begin{bmatrix}\mathbf{x}_o \\ \mathbf{y}_o\end{bmatrix} = \mathbf{BB}^\dagger\begin{bmatrix}\mathbf{A}^*\mathbf{b} \\ \mathbf{f}\end{bmatrix}$ and therefore $\begin{bmatrix}\mathbf{x}_o \\ \mathbf{y}_o\end{bmatrix}$ is a least squares solution of

$$\left[\begin{array}{c|c} \mathbf{A}^*\mathbf{A} & \mathbf{C}^* \\ \hline \mathbf{C} & \mathbf{0} \end{array}\right]\begin{bmatrix}\mathbf{x} \\ \mathbf{y}\end{bmatrix} = \begin{bmatrix}\mathbf{A}^*\mathbf{b} \\ \mathbf{f}\end{bmatrix}. \quad \blacksquare$$

The solution to Problem 1 is obtained directly from Theorem 1.

Theorem 3.6.2 *The set of vectors $M \subseteq \mathcal{S}$ at which $\min_{\mathbf{x}\in\mathcal{S}} \|\mathbf{Ax} - \mathbf{b}\|$ is attained is given by*

$$M = \{(\mathbf{AE})^\dagger(\mathbf{b} - \mathbf{AC}^\dagger\mathbf{f}) + \mathbf{C}^\dagger\mathbf{f} + (\mathbf{I} - (\mathbf{AE})^\dagger\mathbf{A})\xi \,|\, \xi \in N(\mathbf{C})\}$$

where $\mathbf{E} = \mathbf{I} - \mathbf{C}^\dagger\mathbf{C}$. (*$M$ will be called the set of* constrained least squares solutions). *Furthermore*

$$\min_{\mathbf{x}\in\mathcal{S}} \|\mathbf{Ax} - \mathbf{b}\| = \|(\mathbf{I} - \mathbf{A}(\mathbf{AE})^\dagger)(\mathbf{AC}^\dagger\mathbf{f} - \mathbf{b})\|. \tag{7}$$

Proof From Theorem 1, we have

$$M = \left\{\mathbf{x} \,\middle|\, \begin{bmatrix}\mathbf{x}\\ \cdot\end{bmatrix} = \mathbf{B}^\dagger\begin{bmatrix}\mathbf{A}^*\mathbf{b}\\ \mathbf{f}\end{bmatrix} + (\mathbf{I} - \mathbf{B}^\dagger\mathbf{B})\begin{bmatrix}\mathbf{v}_1\\ \mathbf{v}_2\end{bmatrix}, \mathbf{v}_1 \text{ and } \mathbf{v}_2 \text{ arbitrary}\right\}.$$

By Theorem 5.2, $M = \{\mathbf{QA}^*(\mathbf{b} - \mathbf{AC}^\dagger\mathbf{f}) + \mathbf{C}^\dagger\mathbf{f} + (\mathbf{E} - \mathbf{QQ}^\dagger)\mathbf{v} \,|\, \mathbf{v}\in\mathbb{C}^n\}$ where $\mathbf{Q} = (\mathbf{EA}^*\mathbf{AE})^\dagger$. Observe that $\mathbf{QQ}^\dagger = [(\mathbf{AE})^*(\mathbf{AE})]^\dagger(\mathbf{AE})^*(\mathbf{AE}) = (\mathbf{AE})^\dagger(\mathbf{AE})^{*\dagger}(\mathbf{AE})^*(\mathbf{AE}) = (\mathbf{AE})^\dagger(\mathbf{AE})$ so that M becomes $M = \{\mathbf{QA}^*(\mathbf{b} - \mathbf{AC}^\dagger\mathbf{f}) + \mathbf{C}^\dagger\mathbf{f} + (\mathbf{I} - (\mathbf{AE})^\dagger(\mathbf{AE}))\mathbf{Ev} \,|\, \mathbf{v}\in\mathbb{C}^n\}$, Note that, $R((\mathbf{AE})^\dagger) = R((\mathbf{AE})^*) = R(\mathbf{EA}^*) \subseteq R(\mathbf{E})$, so that

$$\mathbf{E}(\mathbf{AE})^\dagger = (\mathbf{AE})^\dagger \tag{8}$$

and (3) of Section 5 yields $\mathbf{QA}^* = (\mathbf{EQE})\mathbf{A}^* = \mathbf{E}(\mathbf{AE})^\dagger(\mathbf{AE})^{*\dagger}(\mathbf{AE})^* = \mathbf{E}(\mathbf{AE})^\dagger = (\mathbf{AE})^\dagger$. Thus M becomes $M = \{(\mathbf{AE})^\dagger(\mathbf{b} - \mathbf{AC}^\dagger\mathbf{f}) + \mathbf{C}^\dagger\mathbf{f} + (\mathbf{I} - (\mathbf{AE})^\dagger(\mathbf{AE}))\xi \,|\, \xi\in N(\mathbf{C})\}$. For each $\mathbf{m}\in M$, we wish to write the expression $\|\mathbf{Am} - \mathbf{b}\|$. In order to do this, observe (8) implies that $\mathbf{A}(\mathbf{AE})^\dagger(\mathbf{AE}) = \mathbf{AE}$ and $\xi = \mathbf{E}\xi$ when $\xi\in N(\mathbf{C})$, so that $\mathbf{A}(\mathbf{I} - (\mathbf{AE})^\dagger(\mathbf{AE}))\xi = \mathbf{0}$ for all $\xi\in N(\mathbf{C})$. Expression (7) now follows. ∎

The solution to Problem 2 also follows quickly.

Theorem 3.6.3 *Let M denote the set of constrained least squares solutions as given in Theorem 6.2. If* $\mathbf{u}$ *denotes the vector* $\mathbf{u} = (\mathbf{AE})^\dagger(\mathbf{b} - \mathbf{AC}^\dagger\,\mathbf{f}) + \mathbf{C}^\dagger\mathbf{f}$, *then* $\mathbf{u}$ *is the unique constrained least squares solution of minimal norm. That is,* $\mathbf{u}\in M$ *and* $\|\mathbf{u}\| < \|\mathbf{x}\|$ *for all* $\mathbf{x}\in M$ *such that* $\mathbf{x} \neq \mathbf{u}$.

Proof The fact that $\mathbf{u}\in M$ is a consequence of Theorem 2 by taking $\xi = \mathbf{0}$. To see $\mathbf{u}$ has minimal norm, suppose $\mathbf{x}\in M$ and use Theorem 2 to write $\mathbf{x} = \mathbf{u} + (\mathbf{I} - (\mathbf{AE})^\dagger(\mathbf{AE}))\xi$, $\xi\in N(\mathbf{C}) = N(\mathbf{C}^{\dagger *})$ Since $R((\mathbf{AE})^\dagger(\mathbf{AE})) = R((\mathbf{AE})^*) = R(\mathbf{EA}^*) \subseteq R(\mathbf{E}) = N(\mathbf{C}^{\dagger *})$, it follows that $\mathbf{C}^{\dagger *}(\mathbf{AE})^\dagger\,\mathbf{AE} = \mathbf{0}$ and it is now a simple matter to verify that $\mathbf{u} \perp (\mathbf{I} - (\mathbf{AE})^\dagger(\mathbf{AE}))\xi$. Therefore $\|\mathbf{x}\|^2 = \|\mathbf{u}\|^2 + \|(\mathbf{I} - (\mathbf{AE})^\dagger(\mathbf{AE}))\xi\|^2 \geq \|\mathbf{u}\|^2$ with equality holding if and only if $(\mathbf{I} - (\mathbf{AE})^\dagger(\mathbf{AE}))\xi = \mathbf{0}$, i.e if and only if $\mathbf{u} = \mathbf{x}$. ∎

From Theorems 2 and 3, one sees that the matrix $(\mathbf{AE})^\dagger$ is the basic quantity which allows the solution of the constrained least squares problem to be written in a fashion analogous to that of the solution of the unconstrained problem.

Suppose one wished to define a 'constrained generalized inverse for

A with respect to **C**' so that it would have the same type of least squares properties in the constrained sense as $\mathbf{A}^\dagger$ has in the unconstrained sense. Suppose you also wanted it to reduce to $\mathbf{A}^\dagger$ when no constraints are present (i.e. $\mathbf{C} = \mathbf{0}$). The logical definition would be the matrix $(\mathbf{AE})^\dagger$.

Definition 3.6.1 *For* $\mathbf{A} \in \mathbb{C}^{m \times n}$ *and* $\mathbf{C} \in \mathbb{C}^{p \times n}$ *the* constrained generalized inverse of **A** with respect to **C**, *denoted by* $\mathbf{A}_\mathbf{C}^\dagger$, *is defined to be* $\mathbf{A}_\mathbf{C}^\dagger = (\mathbf{AP}_{\mathbf{N(C)}})^\dagger = (\mathbf{A}(\mathbf{I} - \mathbf{C}^\dagger\mathbf{C}))^\dagger$. (*Notice that* $\mathbf{A}_\mathbf{C}^\dagger$ *reduces to* $\mathbf{A}^\dagger$ *when* $\mathbf{C} = \mathbf{0}$.)

The definition of $\mathbf{A}_\mathbf{C}^\dagger$ could also have been formulated algebraically, see Exercise 7.18. The solutions of Problem 1 and Problem 2 now take on a familiar form. The constrained least squares solution of $\mathbf{Ax} = \mathbf{b}$ of minimal norm is.

$$\mathbf{x}_m = \mathbf{A}_\mathbf{C}^\dagger\mathbf{b} + (\mathbf{I} - \mathbf{A}_\mathbf{C}^\dagger\mathbf{A})\mathbf{C}^\dagger\mathbf{f}. \tag{9}$$

The set of constrained least squares solutions is given by

$$\mathbf{M} = \{\mathbf{A}_\mathbf{C}^\dagger\mathbf{b} + (\mathbf{I} - \mathbf{A}_\mathbf{C}^\dagger\mathbf{A})(\mathbf{C}^\dagger\mathbf{f} + \xi) \mid \xi \in N(\mathbf{C})\}. \tag{10}$$

Furthermore,

$$\min_{\mathbf{x} \in \mathscr{S}} \|\mathbf{Ax} - \mathbf{b}\| = \|(\mathbf{I} - \mathbf{AA}_\mathbf{C}^\dagger)(\mathbf{AC}^\dagger\mathbf{f} - \mathbf{b})\|. \tag{11}$$

The special case when the set of constraints defines a subspace instead of just a flat deserves mention as a corollary.

Corollary 3.6.1 *Let* V *be a subspace of* $\mathbb{C}^n$, *and* $\mathbf{P} = \mathbf{P}_V^\perp$. *The point* $\mathbf{x}_m \in V$ *of minimal norm at which* $\min_{\mathbf{x} \in V} \|\mathbf{Ax} - \mathbf{b}\|$ *is attained is given by*

$$\mathbf{x}_m = \mathbf{A}_\mathbf{P}^\dagger\mathbf{b} \tag{12}$$

and the set of points $M \subseteq V$ *at which* $\min_{\mathbf{x} \in V} \|\mathbf{Ax} - \mathbf{b}\|$ *is attained is*

$$M = \{\mathbf{A}_\mathbf{P}^\dagger\mathbf{b} + (\mathbf{I} - \mathbf{A}_\mathbf{P}^\dagger\mathbf{A})\xi \mid \xi \in V\} \tag{13}$$

Furthermore,

$$\min_{\mathbf{x} \in V} \|\mathbf{Ax} - \mathbf{b}\| = \|(\mathbf{I} - \mathbf{AA}^\dagger)\mathbf{b}\|. \tag{14}$$

Proof $\mathbf{C} = \mathbf{P}_V^\perp$ and $\mathbf{f} = \mathbf{0}$, in (9), (10), and (11).

Whether or not the constrained problem $\mathbf{Ax} = \mathbf{b}, \mathbf{x} \in V$ is consistent also has an obvious answer.

Corollary 3.6.2 *If* V *is a subspace of* $\mathbb{C}^n$ *and* $\mathbf{P} = \mathbf{P}_V^\perp$ *then the problem* $\mathbf{Ax} = \mathbf{b}$, $\mathbf{x} \in V$ *has a solution if and only if* $\mathbf{AA}_\mathbf{P}^\dagger\mathbf{b} = \mathbf{b}$ (i.e. $\mathbf{b} \in R(\mathbf{AP}_V)$). *If the problem is consistent, then the solution set is given by* $\{\mathbf{A}_\mathbf{P}^\dagger\mathbf{b} + (\mathbf{I} - \mathbf{A}_\mathbf{P}^\dagger\mathbf{A})\xi \mid \xi \in V\}$ *and the minimal norm solution is* $\mathbf{x}_m = \mathbf{A}_\mathbf{P}^\dagger\mathbf{b}$.

Proof The problem is consistent if and only if the quantity in (14) is zero, that is, $\mathbf{AA}_\mathbf{P}^\dagger\mathbf{b} = \mathbf{b}$. That this is equivalent to saying $\mathbf{b} \in R(\mathbf{AP}_V)$

follows from (8). The rest of the proof follows from (13) and (12). ■

In the same fashion one can analyse the consistency of the problem $\mathbf{Ax} = \mathbf{b}$, $\mathbf{x} \in \{\mathbf{f} + V \mid V \text{ is a subspace}\}$ or one can decide when two systems possess a common solution. This topic will be discussed in Chapter 6 from a different point of view.

7. Exercises

1. Use Theorem 3.1.3 to prove Theorem 3.3.2.
2. Prove that if $\operatorname{rank}(\mathbf{A}) = \operatorname{rank}\begin{bmatrix} \mathbf{A} & \mathbf{C} \\ \mathbf{R} & \mathbf{D} \end{bmatrix}$, $R(\mathbf{C}) \subseteq R(\mathbf{A})$, and $R(\mathbf{R}^*) \subseteq R(\mathbf{A}^*)$, then $\mathbf{D} = \mathbf{RA}^\dagger\mathbf{C}$.
3. Let $\mathbf{Q} = \mathbf{D} - \mathbf{RA}^\dagger\mathbf{C}$. If $R(\mathbf{C}) \subseteq R(\mathbf{A})$, $R(\mathbf{R}^*) \subseteq R(\mathbf{A}^*)$, $R(\mathbf{R}) \subseteq R(\mathbf{Q})$, and $R(\mathbf{C}^*) \subseteq R(\mathbf{Q}^*)$, prove that
$$\left[\begin{array}{c|c} \mathbf{A} & \mathbf{C} \\ \hline \mathbf{R} & \mathbf{D} \end{array}\right]^\dagger = \left[\begin{array}{c|c} \mathbf{A}^\dagger + \mathbf{A}^\dagger\mathbf{CQ}^\dagger\mathbf{RA}^\dagger & -\mathbf{A}^\dagger\mathbf{CQ}^\dagger \\ \hline -\mathbf{Q}^\dagger\mathbf{RA}^\dagger & \mathbf{Q}^\dagger \end{array}\right]$$
4. Let $\mathbf{P} = \mathbf{A} - \mathbf{CD}^\dagger\mathbf{R}$. If $R(\mathbf{R}) \subseteq R(\mathbf{D})$, $R(\mathbf{C}^*) \subseteq R(\mathbf{D}^*)$, $R(\mathbf{R}^*) \subseteq R(\mathbf{P}^*)$, and $R(\mathbf{C}) \subseteq R(\mathbf{P})$, write an expression for $\begin{bmatrix} \mathbf{A} & \mathbf{C} \\ \mathbf{R} & \mathbf{D} \end{bmatrix}^\dagger$ in terms of $\mathbf{P}^\dagger, \mathbf{C}, \mathbf{R}$ and $\mathbf{D}$.
5. If $\mathbf{M} = \begin{bmatrix} \mathbf{A} & \mathbf{C} \\ \mathbf{C}^* & \mathbf{D} \end{bmatrix}$ is a positive semi-definite hermitian matrix such that $R(\mathbf{C}^*\mathbf{A}^\dagger) \subseteq R(\mathbf{D} - \mathbf{C}^*\mathbf{A}^\dagger\mathbf{C})$, write an expression for $\mathbf{M}^\dagger$.
6. Suppose $\mathbf{A}$ is non-singular in the matrix $\mathbf{M}$ of Exercise 5. Under this assumption, write an expression for $\mathbf{M}^\dagger$.
7. If $\mathbf{T}_{22}$ is non-singular in (1) of Section 4 prove that $\mathbf{B} = \mathbf{T}_{12}^*(\mathbf{I} - \mathbf{T}_{11}\mathbf{T}_{11}^\dagger)\mathbf{T}_{12} + \mathbf{T}_{22}^*\mathbf{T}_{22}$ is non-singular and then prove that
$$\mathbf{T}^\dagger = \begin{bmatrix} \mathbf{T}_{11}^\dagger & \mathbf{0} \\ \mathbf{0} & \mathbf{0} \end{bmatrix} + \begin{bmatrix} -\mathbf{T}_{11}^\dagger\mathbf{T}_{12} \\ \mathbf{I} \end{bmatrix}\mathbf{B}^{-1}[\mathbf{T}_{12}^*(\mathbf{I} - \mathbf{T}_{11}\mathbf{T}_{11}^\dagger) \mid \mathbf{T}_{22}^*].$$
8. If $\mathbf{T}_{11}$ is non-singular in Exercise 7, write an expression for $\mathbf{T}^\dagger$.
9. Let $\mathbf{T} = \begin{bmatrix} \mathbf{A} & \mathbf{c} \\ \mathbf{0}^* & \alpha \end{bmatrix}$ where $\mathbf{A} \in \mathbb{C}^{m \times n}$, $\mathbf{c} \in \mathbb{C}^m$, and $\alpha \in \mathbb{C}$. Derive an expression for $\mathbf{T}^\dagger$.
10. Prove that the generalized inverse of an upper (lower) triangular matrix $\mathbf{T}$ of rank r is again upper (lower) triangular if and only if there exists a permutation matrix $\mathbf{P}$ such that $\mathbf{P}^*\mathbf{TP} = \begin{bmatrix} \mathbf{T}_1 & \mathbf{0} \\ \mathbf{0} & \mathbf{0} \end{bmatrix}$ where $\mathbf{T}_1 \in \mathbb{C}^{r \times r}$ is a non-singular upper (lower) triangular matrix.
11. For $\mathbf{A} \in \mathbb{C}^{n \times n}$ such that $\operatorname{rank}(\mathbf{A}) = r$, prove that $\mathbf{A}^\dagger\mathbf{A} = \mathbf{AA}^\dagger$ if and only if there exists a unitary matrix $\mathbf{W}$ such that $\mathbf{W}^*\mathbf{AW} = \begin{bmatrix} \mathbf{T}_1 & \mathbf{0} \\ \mathbf{0} & \mathbf{0} \end{bmatrix}$ where

$\mathbf{T}_1 \in \mathbb{C}^{r \times r}$ is a nonsingular triangular matrix. Use Exercise 10 and the fact that any square matrix is unitarily equivalent to an upper (lower) triangular matrix.

12. For $\mathbf{A} \in \mathbb{C}^{m \times n}$ and $\mathbf{R} \in \mathbb{C}^{s \times n}$ write an expression for $\begin{bmatrix} \mathbf{A} \\ \mathbf{R} \end{bmatrix}^{\dagger}$.
13. Prove Theorem 3.4.1 for lower block triangular matrices.
14. Give an example to show that the condition $\operatorname{rank}(\mathbf{T}) = \operatorname{rank}(\mathbf{T}_{11}) + \operatorname{rank}(\mathbf{T}_{22})$ is not sufficient for (1) of Section 4 to be properly partitioned.
15. Complete the proof of Theorem 3.3.3.
16. Complete the proof of Theorem 3.3.5.
17. If $\mathbf{V}$ is a positive definite hermitian matrix and if $\mathbf{C}$ is conformable, let $\mathbf{K} = \mathbf{V} + \mathbf{CC}^*$ and $\mathbf{R} = \mathbf{C}^*\mathbf{K}^{\dagger}\mathbf{C}$. Show that

$$\left[\begin{array}{c|c} \mathbf{V} & \mathbf{C} \\ \hline \mathbf{C}^* & \mathbf{0} \end{array}\right]^{\dagger} = \left[\begin{array}{c|c} \mathbf{K}^{\dagger} - \mathbf{K}^{\dagger}\mathbf{C}\mathbf{R}^{\dagger}\mathbf{C}^*\mathbf{K}^{\dagger} & \mathbf{K}^{\dagger}\mathbf{C}\mathbf{R}^{\dagger} \\ \hline \mathbf{R}^{\dagger}\mathbf{C}^*\mathbf{K}^{\dagger} & \mathbf{R}^{\dagger}\mathbf{R} - \mathbf{R}^{\dagger} \end{array}\right].$$

18. The constrained generalized inverse of $\mathbf{A}$ with respect to $\mathbf{C}$ is the unique solution $\mathbf{X}$ of the five equations (1) $\mathbf{AXA} = \mathbf{A}$ on $N(\mathbf{C})$, (2) $\mathbf{XAX} = \mathbf{X}$, (3) $(\mathbf{AX})^* = \mathbf{AX}$, (4) $\mathbf{P}_{N(\mathbf{C})}(\mathbf{XA})^* = \mathbf{XA}$, on $N(\mathbf{C})$ (5) $\mathbf{CX} = \mathbf{0}$.
19. Complete the proof of Theorem 3.1.1
20. Derive Theorem 3.3.1 from Theorem 3.3.3.

4 Partial isometries and EP matrices

1. Introduction

There are certain special types of matrices which occur frequently and have useful properties. For example, $\mathbf{A}\in\mathbb{C}^{n\times n}$ is called unitary if $\mathbf{A}^* = \mathbf{A}^{-1}$, hermitian if $\mathbf{A} = \mathbf{A}^*$, and normal if $\mathbf{A}^*\mathbf{A} = \mathbf{A}\mathbf{A}^*$. This should suggest to the reader questions like: when is $\mathbf{A}^* = \mathbf{A}^\dagger$?, when is $\mathbf{A} = \mathbf{A}^\dagger$?, and when is $\mathbf{A}^\dagger\mathbf{A} = \mathbf{A}\mathbf{A}^\dagger$? The answering of such questions is useful in understanding the generalized inverse and is probably worth doing for that reason alone. It turns out, however, that the matrices involved are useful.

It should probably be pointed out that one very rarely *has* to use partial isometries or the polar form. The ideas discussed in this short chapter tend to be geometrical in nature and if there is a geometrical way of doing something then there is probably an algebraic way (and conversely). It is the feeling of the authors, however, that to be able to view a problem from more than one viewpoint is advantageous. Accordingly, we have tried to develop both the geometric and algebraic theory as we proceed.

Throughout this chapter $\|\cdot\|$ denotes the Euclidean norm on $\mathbb{C}^p$.

2. Partial isometries

Part of the difficulty with generalizing the polar form in Theorem 0.3.1 form $\mathbf{A}\in\mathbb{C}^{n\times n}$ to $\mathbf{A}\in\mathbb{C}^{m\times n}$, $m \neq n$, was the need for a 'non-square unitary'. We will now develop the appropriate generalization of a unitary matrix.

Definition 4.2.1 *Suppose that* $\mathbf{V}\in\mathbb{C}^{m\times n}$, $n \leq m$. *Then* $\mathbf{V}$ *is called an* isometry *if* $\|\mathbf{V}\mathbf{u}\| = \|\mathbf{u}\|$ *for all* $\mathbf{u}\in\mathbb{C}^n$.

The equation $\|\mathbf{V}\mathbf{u}\| = \|\mathbf{u}\|$ may be rewritten $(\mathbf{V}\mathbf{u}, \mathbf{V}\mathbf{u}) = (\mathbf{u}, \mathbf{u})$ or $(\mathbf{V}^*\mathbf{V}\mathbf{u}, \mathbf{u}) = (\mathbf{u}, \mathbf{u})$. Now if $\mathbf{C}_1, \mathbf{C}_2$ are *hermitian* matrices in $\mathbb{C}^{n\times n}$, then $(\mathbf{C}_1\mathbf{u}, \mathbf{u}) = (\mathbf{C}_2\mathbf{u}, \mathbf{u})$ for all $\mathbf{u}\in\mathbb{C}^n$ if and only if $\mathbf{C}_1 = \mathbf{C}_2$. Thus we have:

Proposition 4.2.1 $\mathbf{V}\in\mathbb{C}^{m\times n}$ *is an isometry if and only if* $\mathbf{V}^*\mathbf{V} = \mathbf{I}_n$.

A more general concept than isometry is that of a partial isometry.

Definition 4.2.2 *Let* $\mathbb{C}^n = M \oplus M^\perp$ *for a subspace* M. *Then* $\mathbf{V} \in \mathbb{C}^{m\times n}$ *is a* partial isometry (*of* M *into* $\mathbb{C}^m$) *if and only if*

(i) $\|\mathbf{Vu}\| = \|\mathbf{u}\|$ *for all* $\mathbf{u} \in M$ *and*
(ii) $\mathbf{Vu} = \mathbf{0}$ *if* $\mathbf{u} \in M^\perp$.

The subspace M *is called the* initial space *of* $\mathbf{V}$ *and* $R(\mathbf{V})$ *is called the* final space.

A partial isometry $\mathbf{V}$ (or $\underset{\sim}{V}$) sends its initial space onto its final space without changing the lengths of vectors in its initial space or the angles between them. In other words, a partial isometry can be viewed as the identification of two subspaces. Orthogonal projections are a special type of partial isometry.

Partial isometries are easy to characterize.

Theorem 4.2.1 *Suppose that* $\mathbf{V} \in \mathbb{C}^{m\times n}$. *Then the following are equivalent.*

(i) $\mathbf{V}$ *is a partial isometry*
(ii) $\mathbf{V}^* = \mathbf{V}^\dagger$.
(iii) $\mathbf{VV}^* = \mathbf{P}_{R(\mathbf{V})}$ *and* $\mathbf{V}^*\mathbf{V} = \mathbf{P}_{R(\mathbf{V}^*)} = \mathbf{P}_{\text{Initial space of } \mathbf{V}}$
(iv) $\mathbf{V} = \mathbf{VV}^*\mathbf{V}$.
(v) $\mathbf{V}^* = \mathbf{V}^*\mathbf{VV}^*$.
(vi) $(\mathbf{V}^*\mathbf{V})^2 = (\mathbf{V}^*\mathbf{V})$.
(vii) $(\mathbf{VV}^*)^2 = \mathbf{VV}^*$.

Proof The equivalence of (i) and (iv)–(vii) is left to the exercises, while the equivalence of (ii) and (iii) is the Moore definition of $\mathbf{V}^\dagger$. Suppose then that $\mathbf{V}$ is a partial isometry and M is its initial space. If $\mathbf{u} \in M$, then $(\mathbf{Vu}, \mathbf{Vu}) = (\mathbf{V}^*\mathbf{Vu}, \mathbf{u}) = (\mathbf{u}, \mathbf{u})$. But also $R(\mathbf{V}^*\mathbf{V}) = R(\mathbf{V}^*) = N(\mathbf{V})^\perp = M$. Thus $\underset{\sim}{\mathbf{V}^*\mathbf{V}}|_M = \underset{\sim}{\mathbf{I}}|_M$ since $\mathbf{V}^*\mathbf{V}$ is hermitian. If $\mathbf{u} \in M^\perp$, then $\mathbf{V}^*\mathbf{Vu} = \mathbf{0}$ since $\mathbf{Vu} = \mathbf{0}$. Thus $\mathbf{V}^*\mathbf{V} = \mathbf{P}_M$. Similar arguments show that $\mathbf{VV}^* = \mathbf{P}_{R(\mathbf{V})}$ and (iii) follows. To show that (iii) implies (i) the above argument can be done in reverse. ■

Corollary 4.2.1 *If* $\mathbf{V}$ *is a partial isometry, then so is* $\mathbf{V}^*$.

For partial isometries the Singular Value Decomposition mentioned in Chapter 0 takes a form that is worth noting. We are not going to prove the Singular Value Decomposition but our proof of this special case and of the general polar form should help the reader do so for himself.

Proposition 4.2.2 *Suppose that* $\mathbf{V} \in \mathbb{C}^{m\times n}$ *is a partial isometry of rank* r. *Then there exist unitary matrices* $\mathbf{U} \in \mathbb{C}^{m\times m}$ *and* $\mathbf{W} \in \mathbb{C}^{n\times n}$ *such that*

$$\mathbf{V} = \mathbf{U}\begin{bmatrix} \mathbf{I}_r & \mathbf{0} \\ \mathbf{0} & \mathbf{0} \end{bmatrix}\mathbf{W}.$$

Proof Suppose that $\mathbf{V} \in \mathbb{C}^{m\times n}$ is a partial isometry. Let $M = R(\mathbf{V}^*)$ be its initial space. Let $\{\mathbf{b}_1, \ldots, \mathbf{b}_r\}$ be an orthonormal basis for M. Extend this to an orthonormal basis $\beta_1 = \{\mathbf{b}_1, \ldots, \mathbf{b}_r, \mathbf{b}_{r+1}, \ldots, \mathbf{b}_n\}$ of $\mathbb{C}^n$. Since $\mathbf{V}$ is isometric on M, $\{\mathbf{Vb}_1, \ldots, \mathbf{Vb}_r\}$ is an orthonormal basis for $R(\mathbf{V})$. Extend $\{\mathbf{Vb}_1, \ldots, \mathbf{Vb}_r\}$ to an orthonormal basis $\beta_2 = \{\mathbf{Vb}_1, \ldots, \mathbf{Vb}_r, \mathbf{c}_{r+1}, \ldots, \mathbf{c}_m\}$ of $\mathbb{C}^m$. Let $\mathbf{W}$ be the unitary transformation which changes a vector to its coordinates with respect to basis β_1. Let $\mathbf{U}$ be the unitary transformation which changes a β_2-coordinate vector into a coordinate vector with respect to the standard basis of $\mathbb{C}^m$. Then (1) follows. ■

We are now in a position to prove the general polar form.

Theorem 4.2.2 (*General Polar Form*). *Suppose that* $\mathbf{A} \in \mathbb{C}^{m \times n}$. *Then*

(i) *There exists a hermitian* $\mathbf{B} \in \mathbb{C}^{n \times n}$ *such that* $N(\mathbf{B}) = N(\mathbf{A})$ *and a partial isometry* $\mathbf{V} \in \mathbb{C}^{m \times n}$ *such that* $R(\mathbf{V}) = R(\mathbf{A})$, $N(\mathbf{V}) = N(\mathbf{B})$, *and* $\mathbf{A} = \mathbf{VB}$.

(ii) *There exists a hermitian* $\mathbf{C} \in \mathbb{C}^{m \times m}$ *such that* $R(\mathbf{C}) = R(\mathbf{A})$ *and a partial isometry* $\mathbf{W}$ *such that* $N(\mathbf{W}) = N(\mathbf{A})$, $R(\mathbf{W}) = R(\mathbf{C})$, *and* $\mathbf{A} = \mathbf{CW}$.

Proof The proof is motivated by the complex number idea it generalizes. If $z = re^{i\theta}$, then $r = (z\bar{z})^{1/2}$ and $e^{i\theta} = z(z\bar{z})^{-1/2}$. We will prove (i) of Theorem 2 and leave (ii), which is similar, to the exercises.

Let $\mathbf{B} = (\mathbf{A}^*\mathbf{A})^{1/2}$. (Recall the notation of page 6.) Then $\mathbf{B} \in \mathbb{C}^{n \times n}$ and $N(\mathbf{B}) = N(\mathbf{A}^*\mathbf{A}) = N(\mathbf{A})$. Let $\mathbf{V} = \mathbf{AB}^\dagger$. We must show that $\mathbf{V}$ is the required partial isometry. Notice that $\mathbf{B}^\dagger$ is hermitian, $N(\mathbf{B}^\dagger) = N(\mathbf{B})$, and $R(\mathbf{B}^\dagger) = R(\mathbf{B})$. Thus $R(\mathbf{V}) = R(\mathbf{AB}^\dagger) = R(\mathbf{AB}) = R(\mathbf{A}(\mathbf{A}^*\mathbf{A})) = R(\mathbf{A})$ and $N(\mathbf{V}) = N(\mathbf{AA}^*\mathbf{A}) = N(\mathbf{A}) = N(\mathbf{B})$ as desired. Suppose then that $\mathbf{u} \in N(\mathbf{V})^\perp = R(\mathbf{B})$. Then $\|\mathbf{Vu}\|^2 = (\mathbf{Vu}, \mathbf{Vu}) = (\mathbf{AB}^\dagger\mathbf{u}, \mathbf{AB}^\dagger\mathbf{u}) = (\mathbf{B}^\dagger\mathbf{A}^*\mathbf{AB}^\dagger\mathbf{u}, \mathbf{u}) = (\mathbf{B}^\dagger\mathbf{B}^2\mathbf{B}^\dagger\mathbf{u}, \mathbf{u}) = (\mathbf{P}_{R(\mathbf{B})}\mathbf{u}, \mathbf{u}) = (\mathbf{u}, \mathbf{u}) = \|\mathbf{u}\|^2$. Thus $\mathbf{V}$ is the required partial isometry. ■

The proof of the singular value decomposition theorem is left to the exercises. Note that if $\mathbf{D}$ is square, then $\begin{bmatrix} \mathbf{D} & \mathbf{0} \\ \mathbf{0} & \mathbf{0} \end{bmatrix}$ can be factored as $\begin{bmatrix} \mathbf{D} & \mathbf{0}' \\ \mathbf{0}' & \mathbf{0}' \end{bmatrix}\begin{bmatrix} \mathbf{I} & \mathbf{0} \\ \mathbf{0} & \mathbf{0} \end{bmatrix}$ where $\begin{bmatrix} \mathbf{D} & \mathbf{0}' \\ \mathbf{0}' & \mathbf{0}' \end{bmatrix}$ is square and $\begin{bmatrix} \mathbf{I} & \mathbf{0} \\ \mathbf{0} & \mathbf{0} \end{bmatrix}$ is a partial isometry. A judicious use of this observation, Theorem 2, and the *proof* of Proposition 2 should lead to a proof of the singular value decomposition.

While partial isometries are a generalization of unitary matrices there are some differences. For example, the columns or rows of a partial isometry need not be an orthonormal set unless a subset of the standard basis for $\mathbb{C}^n$ or $\mathbb{C}^m$ is a basis for $R(\mathbf{V}^*)$ or $R(\mathbf{V})$.

Example 4.2.1 Let $\mathbf{V} = \dfrac{1}{4}\begin{bmatrix} 4 & 0 & 0 \\ 0 & \sqrt{3}+1 & -\sqrt{3}-1 \\ 0 & -\sqrt{3}+1 & \sqrt{3}-1 \end{bmatrix}$. Then $\mathbf{V}$ is a partial isometry but neither the columns nor the rows (or a subset thereof) form an orthonormal basis for $R(\mathbf{V})$ or $R(\mathbf{V}^*)$.

It should also be noted that, in general, the product of a pair of partial isometries need not be a partial isometry. Also, unlike unitary operators, square partial isometries can have eigenvalues of modulus unequal to one or zero.

Example 4.2.2 Let $\mathbf{V} = \frac{1}{2}\begin{bmatrix} \sqrt{3} & 0 \\ 1 & 0 \end{bmatrix}$. Then $\mathbf{V}$ is a partial isometry and $\sigma(\mathbf{V}) = \{0, \sqrt{3}/2\}$.

Example 4.2.3 Let $\mathbf{V} = \begin{bmatrix} 0 & 1 & 0 \\ 0 & 0 & 1 \\ 0 & 0 & 0 \end{bmatrix}$. Then $\mathbf{V}$ is a partial isometry and $\sigma(\mathbf{V}) = \{0\}$.

3. EP matrices

The identities $\mathbf{A}^*\mathbf{A} = \mathbf{A}\mathbf{A}^*$ for normal matrices and $\mathbf{A}^{-1}\mathbf{A} = \mathbf{A}\mathbf{A}^{-1}$ for invertible matrices are sometimes useful. This suggests that it might be helpful to know when $\mathbf{A}^\dagger\mathbf{A} = \mathbf{A}\mathbf{A}^\dagger$.

Definition 4.3.1 *Suppose that* $\mathbf{A} \in \mathbb{C}^{n \times n}$ *and rank* $(\mathbf{A}) = r$. *If* $\mathbf{A}^\dagger\mathbf{A} = \mathbf{A}\mathbf{A}^\dagger$, *then* $\mathbf{A}$ *is called an* EP_r, *or simply* EP, *matrix.*

The basic facts about EP matrices are set forth in the next theorem.

Theorem 4.3.1 *Suppose that* $\mathbf{A} \in \mathbb{C}^{n \times n}$. *Then the following are equivalent.*

(i) $\mathbf{A}$ *is* EP

(ii) $R(\mathbf{A}) = R(\mathbf{A}^*)$

(iii) $\mathbb{C}^n = R(\mathbf{A}) \oplus N(\mathbf{A})$

(iv) *There exists a unitary matrix* $\mathbf{U}$ *and an invertible* $r \times r$ *matrix* $\mathbf{A}_1$, $r =$ *rank* $(\mathbf{A})$, *such that*

$$\mathbf{A} = \mathbf{U}\begin{bmatrix} \mathbf{A}_1 & \mathbf{0} \\ \mathbf{0} & \mathbf{0} \end{bmatrix}\mathbf{U}^*. \tag{1}$$

Proof (i), (ii) and (iii) are clearly equivalent. That (iv) implies (iii) is obvious. To see that (iii) implies (iv) let β be an orthonormal basis for $\mathbb{C}^n$ consisting of first an orthonormal basis for $R(\mathbf{A})$ and then an orthonormal basis for $N(\mathbf{A})$. $\mathbf{U}^*$ is then the coordinate transformation from standard coordinates to β-coordinates. ■

If $\mathbf{A}$ is EP and has the factorization given by (1), then since $\mathbf{U}, \mathbf{U}^*$ are unitary

$$\mathbf{A}^\dagger = \mathbf{U}\begin{bmatrix} \mathbf{A}_1^{-1} & \mathbf{0} \\ \mathbf{0} & \mathbf{0} \end{bmatrix}\mathbf{U}^*. \tag{2}$$

Since EP matrices have a nice form it is helpful if one can tell when a matrix is EP. This problem will be discussed again later. Several conditions implying EP are given in the exercises.

It was pointed out in Chapter 1 that, unlike the taking of an inverse, the taking of a generalized inverse does not have a nice 'spectral mapping property'.

If $\mathbf{A}\in\mathbb{C}^{n\times n}$ is invertible, then

$$\lambda\in\sigma(\mathbf{A}) \text{ if and only if } \frac{1}{\lambda}\in\sigma(\mathbf{A}^{-1}) \tag{3}$$

and

$$\mathbf{A}\mathbf{x}=\lambda\mathbf{x} \text{ if and only if } \mathbf{A}^{-1}\mathbf{x}=\left(\frac{1}{\lambda}\right)\mathbf{x}. \tag{4}$$

While it is difficult to characterize matrices which satisfy condition (3), it is relatively easy to characterize those that satisfy condition (4). Notice that (4) implies (3).

Theorem 4.3.2 *Suppose that* $\mathbf{A}\in\mathbb{C}^{n\times n}$. *Then* $\mathbf{A}$ *is* EP *if and only if*

$$(\mathbf{A}\mathbf{x}=\lambda\mathbf{x} \textit{ if and only if } \mathbf{A}^{\dagger}\mathbf{x}=\lambda^{\dagger}\mathbf{x}). \tag{5}$$

Proof Suppose that $\mathbf{A}$ is EP. By Theorem 1, $\mathbf{A}=\mathbf{U}\begin{bmatrix}\mathbf{A}_1 & \mathbf{0}\\ \mathbf{0} & \mathbf{0}\end{bmatrix}\mathbf{U}^*$ where $\mathbf{U}$ is unitary and $\mathbf{A}_1^{-1}$ exists. Then $\mathbf{A}\mathbf{x}=\lambda\mathbf{x}$ if and only if $\begin{bmatrix}\mathbf{A}_1 & \mathbf{0}\\ \mathbf{0} & \mathbf{0}\end{bmatrix}\mathbf{U}^*\mathbf{x}=\lambda(\mathbf{U}^*\mathbf{x})$. Let $\mathbf{U}^*\mathbf{x}=\begin{bmatrix}\mathbf{u}_1\\ \mathbf{u}_2\end{bmatrix}$.

If $\lambda=0$, then $\mathbf{u}_1=\mathbf{0}$, and $\mathbf{A}^{\dagger}\mathbf{x}=\mathbf{0}$. Thus (5) holds for $\lambda=0$. If $\lambda\neq 0$, then $\mathbf{u}_2=\mathbf{0}$ and $\mathbf{u}_1$ is an eigenvector for $\mathbf{A}_1$. Thus (5) follows from (2) and (4).

Suppose now that (5) holds. Then $N(\mathbf{A})=N(\mathbf{A}^{\dagger})=R(\mathbf{A})^{\perp}$. Thus $\mathbf{A}$ is EP by condition (iii) of Theorem 1. ■

Corollary 4.3.1 *If* $\mathbf{A}$ *is* EP, *then* $\lambda\in\sigma(\mathbf{A})$ *if and only if* $\lambda^{\dagger}\in\sigma(\mathbf{A}^{\dagger})$.

Corollary 1 does not of course, characterize when $\mathbf{A}$ is EP.

Example 4.3.1 Let $\mathbf{A}=\begin{bmatrix}0 & 1\\ 0 & 0\end{bmatrix}$. Notice that $\sigma(\mathbf{A})=\{0\}$ and $\mathbf{A}^{\dagger}=\mathbf{A}^*$. Thus $\lambda\in\sigma(\mathbf{A})$ if and only if $\lambda^{\dagger}\in\sigma(\mathbf{A})$. However, $\mathbf{A}^{\dagger}\mathbf{A}=\begin{bmatrix}0 & 0\\ 0 & 1\end{bmatrix}$ while $\mathbf{A}\mathbf{A}^{\dagger}=\begin{bmatrix}1 & 0\\ 0 & 0\end{bmatrix}$. Thus $\mathbf{A}\mathbf{A}^{\dagger}\neq\mathbf{A}^{\dagger}\mathbf{A}$ and $\mathbf{A}$ is not EP.

4. Exercises

1. If $\mathbf{V}$, $\mathbf{W}$, and $\mathbf{V}\mathbf{W}$ are partial isometries, show that $(\mathbf{V}\mathbf{W})^{\dagger}=\mathbf{W}^{\dagger}\mathbf{V}^{\dagger}$ using only Theorem 1.

2. If **V**, **W** are partial isometries and $[\mathbf{W}, \mathbf{V}\mathbf{V}^\dagger] = \mathbf{0}$ *or* $[\mathbf{V}, \mathbf{W}^\dagger\mathbf{W}] = \mathbf{0}$, then **WV** is a partial isometry.
3. If $\mathbf{V} \in \mathbb{C}^{n \times n}$ is a partial isometry and **U**, **W** are unitary, show that **UVW** is a partial isometry.
4. Show that the following conditions are equivalent.

   (i) **V** is a partial isometry
   (ii) $\mathbf{V}\mathbf{V}^*\mathbf{V} = \mathbf{V}$
   (iii) $\mathbf{V}^*\mathbf{V}\mathbf{V}^* = \mathbf{V}^*$
   (iv) $(\mathbf{V}^*\mathbf{V})^2 = \mathbf{V}^*\mathbf{V}$
   (v) $(\mathbf{V}\mathbf{V}^*)^2 = \mathbf{V}\mathbf{V}^*$

5. Prove part (ii) of Theorem 2.

*6. Prove the Singular Value Decomposition Theorem (Theorem 0.2).

7. Prove that if $\mathbf{A}^*\mathbf{A} = \mathbf{A}\mathbf{A}^*$, then **A** is EP.
8. Prove that the following are equivalent.

   (a) **A** is EP
   (b) $[\mathbf{A}^\dagger\mathbf{A}, \mathbf{A} + \mathbf{A}^\dagger] = \mathbf{0}$
   (c) $[\mathbf{A}\mathbf{A}^\dagger, \mathbf{A} + \mathbf{A}^\dagger] = \mathbf{0}$
   (d) $[\mathbf{A}^\dagger\mathbf{A}, \mathbf{A} + \mathbf{A}^*] = \mathbf{0}$
   (e) $[\mathbf{A}\mathbf{A}^\dagger, \mathbf{A} + \mathbf{A}^*] = \mathbf{0}$
   (f) $[\mathbf{A}, \mathbf{A}^\dagger\mathbf{A}] = \mathbf{0}$
   (g) $[\mathbf{A}, \mathbf{A}\mathbf{A}^\dagger] = \mathbf{0}$

9. Prove that if **A** is EP, then $(\mathbf{A}^\dagger)^2 = (\mathbf{A}^2)^\dagger$. Find an example of a matrix $\mathbf{A} \neq \mathbf{0}$, such that $(\mathbf{A}^2)^\dagger = (\mathbf{A}^\dagger)^2$ but **A** is not EP.

*10. Prove that **A** is EP if and only if both $(\mathbf{A}^\dagger)^2 = (\mathbf{A}^2)^\dagger$ and $R(\mathbf{A}) = R(\mathbf{A}^2)$.

*11. Prove that **A** is EP if and only if $R(\mathbf{A}^2) = R(\mathbf{A})$ and $[\mathbf{A}^\dagger\mathbf{A}, \mathbf{A}\mathbf{A}^\dagger] = \mathbf{0}$. Comment: If $(\mathbf{A}^\dagger)^2 = (\mathbf{A}^2)^\dagger$, then $[\mathbf{A}^\dagger\mathbf{A}, \mathbf{A}\mathbf{A}^\dagger] = \mathbf{0}$ but not conversely. Thus the result of Exercise 11 implies the result of Exercise 10. Exercise 11 has a fairly easy proof if the condition $[\mathbf{A}^\dagger\mathbf{A}, \mathbf{A}\mathbf{A}^\dagger] = \mathbf{0}$ is translated into a decomposition of $\mathbb{C}^n$.

12. Suppose that $\mathbf{X} = \mathbf{X}^\dagger$. What can you say about **X**? Give an example of a **X** such that $\mathbf{X} = \mathbf{X}^\dagger$ and **X** is not a partial isometry. What conditions in addition to $\mathbf{X} = \mathbf{X}^\dagger$ are needed to make **X** a partial isometry?
13. Prove that **V** is an orthogonal projector if and only if $\mathbf{V} = \mathbf{V}^\dagger = \mathbf{V}^*\mathbf{V}$.
14. Prove that if **A**, **B** are EP (not necessarily of the same rank) and $\mathbf{AB} = \mathbf{BA}$, then $(\mathbf{AB})^\dagger = \mathbf{B}^\dagger\mathbf{A}^\dagger$.

5
The generalized inverse in electrical engineering

1. Introduction

In almost any situation where a system of linear equations occurs there is the possibility of applications of the generalized inverse. This chapter will describe a place where the generalized inverse appears in electrical engineering.

To make the exposition easily accessible to those with little knowledge of circuit theory, we have kept the examples and discussion at an elementary level. Technical terms will often be followed by intuitive definitions. No attempt has been made to describe all the uses of generalized inverses in electrical engineering, but rather, one particular use will be discussed in some detail. Additional uses will be mentioned in the closing paragraphs to this chapter. It should be understood that almost everything done here can be done for more complex circuits.

Of course, curve fitting and least squares analysis as discussed in Chapter 2 is useful in electrical engineering. The applications of this chapter are of a different sort.

The Drazin Inverse of Chapter 7 as shown in Chapter 9 can be used to study linear systems of differential equations with singular coefficients. Such equations sometimes occur in electrical circuits if, for example, there are dependent sources.

2. *n*-port network and the impedance matrix

It is sometimes desirable, or necessary, to consider an electrical network in terms of how it appears from the outside. One should visualize a box (the network) from which lead several *terminals* (wires). The idea is to describe the network in terms of measurements made at the terminals. One thus characterizes the network by what it does, rather than what it physically is. This is the so-called 'black box' approach. This approach appears in many other fields such as nuclear engineering where the black box might be a nuclear reactor and the terminals might represent measurements of variables such as neutron flow, temperature, etc.

We will restrict ourselves to the case when the terminals may be treated in pairs. Each pair is called a *port*. It is assumed that the amount of current going into one terminal of a port is the same as that coming out of the other terminal of the same port. This is a restriction on the types of devices that might be attached to the network at the port. It is not a restriction on the network. A network with n ports is called an *n-port network*.

Given an n-port network there are a variety of ways to characterize it depending on what one wants to do. In particular, there are different kinds of readings that can be taken at the ports. Those measurements thought of as independent variables are called *inputs*. Those thought of as dependent variables are called *outputs*. We will assume that our networks have the properties of *homogeneity* and *superposition*. Homogeneity says that if the inputs are multiplied by a factor, then the outputs are multiplied by that same factor. If the network has the property of superposition, then the output for the sum of several inputs is the sum of the outputs for each input. We will use current as our input and voltage as our output.

Kirchhoff's laws are useful in trying to determine if a particular pair of terminals are acting like a port. We will also use them to analyse a particular circuit. A *node* is the place where two or more wires join together. A *loop* is any closed conducting path.

KIRCHHOFF'S CURRENT LAW: *The algebraic sum of all the instantaneous currents leaving a node is zero.*

KIRCHHOFF'S VOLTAGE LAW: *The algebraic sum of all the voltage drops around any loop is zero.*

Kirchhoff's current law may also be applied to the currents entering and leaving a network if there are no current sources inside the network.

Suppose that r denotes a certain amount of *resistance* to the current in a wire. We will assume that our wires have no resistance and that the resistance is located in certain devices called *resistors*. Provided that the resistance of the wires is 'small' compared with that of other devices in the circuit this is not a 'bad' approximation of a real circuit. Let v denote the *voltage* (pressure forcing current) across the resistor. The voltage across the resistor is also sometimes referred to as the 'voltage drop' across the resistor, or the 'change in potential'. Let i denote the current in the resistor. Then

$$v = ir. \tag{1}$$

Suppose that r is constant but v and i vary with time. If the one-sided Laplace transform is taken of both sides of (1), then $v = ir$ where v and i are now functions of a frequency variable rather than of a time variable. When v and i are these transformed functions (for *any* circuit), then the ratio v/i is called the *impedance* of the circuit. Impedance is in the same units (ohms) as resistance but is a function of frequency. If the circuit

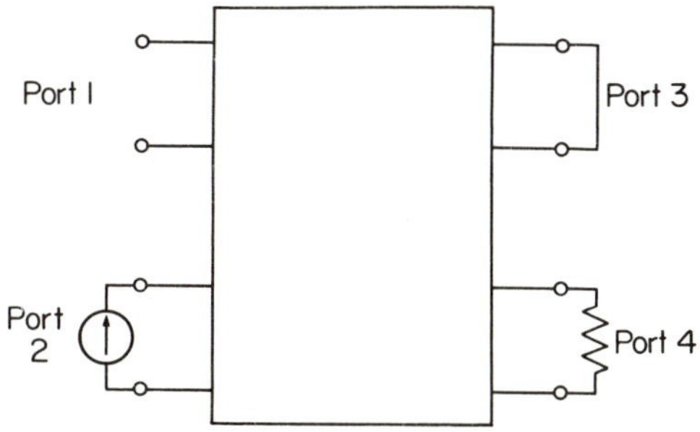

Fig. 5.1 A 4-port network.

consists only of resistors, then the impedance is constant and equals the resistance. Impedance is usually denoted by a z.

In order to visualize what is happening it is helpful to be able to picture a network. We will denote a *current source* by (↑), a resistor by —ww—, a terminal by —o and a node by —o—. The reader should be aware that not all texts distinguish between terminals and nodes as we do. We reserve the word 'terminal' for the ports. Our current sources will be *ideal current sources* in that they are assumed to have zero resistance. Resistors are assumed to have constant resistance.

Before proceeding let us briefly review the definition of an n-port network.

Figure 5.1 is a 4-port network where port 1 is open, port 2 has a current source applied across it, port 3 is *short-circuited*, and port 4 has a resistor across it. Kirchhoff's current law can be applied to show that ports 1, 2 and 3 actually are ports, that is, the current entering one terminal is the same as that leaving the other terminal. Port 1 is a port since it is open and there is no current at all.

Now consider the network in Fig. 5.2. The network in Fig. 5.2 is not a 4-port network. As before, the pairs of terminals 5 and 6, 7 and 8, do form ports. But there is no way to guarantee,

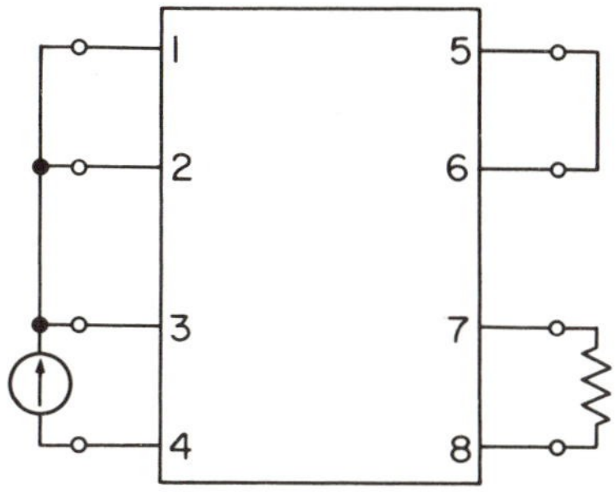

Fig. 5.2 A network which is not an n-port.

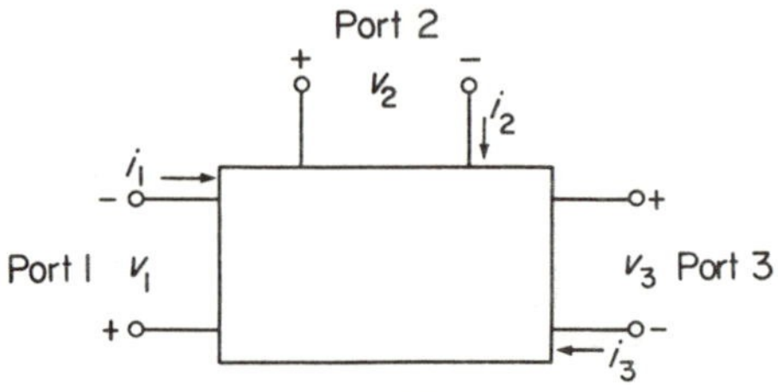

Fig. 5.3

without looking inside the box, that the current coming out of terminal 4 is the same as that flowing into any of terminals 1, 2 or 3. Thus terminal 4 cannot be teamed up with any other terminal to form a port.

There are, of course, ways of working with terminals that cannot be considered as ports, but we will not discuss them here.

It is time to introduce the matrices. Consider a 3-port network, Fig. 5.3, which may, in fact, be hooked up to other networks not shown. Let v_j be the potential (voltage) across the jth port. Let i_j be the current through one of the terminals of the jth port.

Since the v_j, i_j are variables, it really does not matter which way the arrow for i_j points. Given the values of i_1, i_2 and i_3, the voltages v_1, v_2, v_3 are determined. But we have assumed our network was homogeneous and had the property of superposition. Thus the v_j can be written in terms of the i_j by a system of linear equations.

$$\begin{aligned} v_1 &= z_{11}i_1 + z_{12}i_2 + z_{13}i_3 \\ v_2 &= z_{21}i_1 + z_{22}i_2 + z_{23}i_3, \\ v_3 &= z_{31}i_1 + z_{32}i_2 + z_{33}i_3 \end{aligned} \tag{2}$$

or in matrix notation,

$$\mathbf{v} = \mathbf{Z}\mathbf{i}, \text{ where } \mathbf{v}, \mathbf{i} \in \mathbb{C}^3,\ \mathbf{Z} \in \mathbb{C}^{3\times 3}. \tag{3}$$

Z is called the *impedance matrix* of the network since it has the same units as impedance and (3) looks like (1). In the system of equations (2), v_j, i_j, and the z_{jk} are all functions of the frequency variable mentioned earlier. If there are devices other than just resistors, such as capacitors, in the network, then **Z** will vary with the frequency.

The numbers z_{jk} have a fairly elementary physical meaning. Suppose that we take the 3-port of Fig. 5.3 and apply a current of strength i_1 across the terminals forming port 1, leave ports 2 and 3 open, and measure the voltage across port 3. Now an *ideal voltmeter* has infinite resistance, that is, there is no current in it. (In reality a small amount of current goes through it.) Thus $i_3 = 0$. Since port 2 was left open, we have $i_2 = 0$. Then (2) says that $v_3 = z_{31}i_1$ or $z_{31} = v_3/i_1$ when $i_2 = i_3 = 0$. The other z_{kj} have similar interpretations.

Example 1 We shall calculate the impedance matrix of the network in Fig. 5.4. In practice **Z** would be calculated by actual physical

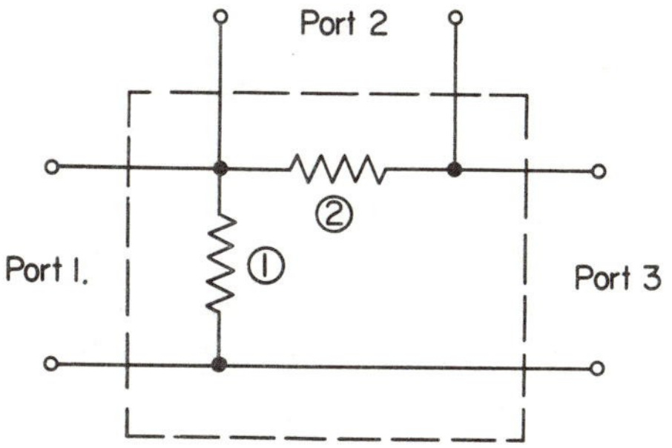

Fig. 5.4 A particular 3-port network. The circled number give the resistance in ohms of the resistor.

measurements of currents and voltages. We shall calculate it by looking 'inside the box'. If a current i_1 is applied across port 1 we have the situation in Fig. 5.5.

The only current is around the indicated loop. There is a resistance of 1 ohm on this loop so that $v_1 = 1 \cdot i_1$. Thus $z_{11} = v_1/i_1 = 1$. Now there is no current in the rest of the network so there can be no changes in potential. This means that $v_2 = 0$ since there is no potential change across the terminals forming port 2. It also means that the potential v_3 across port 3 is the same as the potential between nodes a and b in Fig. 5.5. That is, $v_3 = 1$ also. Hence $z_{21} = v_2/i_1 = 0$ and $z_{31} = v_3/i_1 = 1$. Continuing we get

$$\mathbf{Z} = \begin{bmatrix} 1 & 0 & 1 \\ 0 & 2 & 2 \\ 1 & 2 & 3 \end{bmatrix}. \tag{4}$$

In order to calculate z_{33} recall that if two resistors are connected in series (Fig. 5.6), then the resistance of the two considered as one resistor is the sum of the resistance of each.

Several comments about the matrix (4) are in order. First, the matrix (4) is hermitian. This happened because our network was *reciprocal.* A network is reciprocal if when input and output terminals are interchanged, the relationship between input and output is unchanged. That is, $z_{jk} = z_{kj}$. Second, the matrix (4) had only constant terms since the network in Fig. 5.5

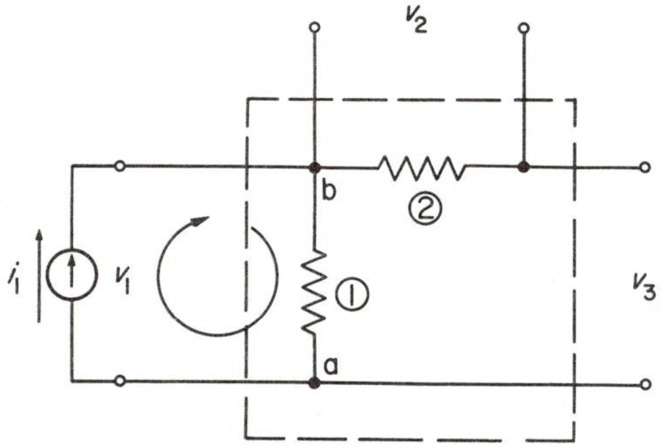

Fig. 5.5 Application of a current to port 1.

Fig. 5.6 Two resistors in series.

was *resistive*, that is, composed of only resistors. Finally, notice that (4) was not invertible. This, of course, was due to the fact that $v_3 = v_1 + v_2$. One might argue that v_3 could thus be eliminated. However, this dependence might not be known *a priori*. Also the three-port might be needed for joining with other networks. We shall also see later that theoretical considerations sometimes lead to singular matrices.

3. Parallel sums

Suppose that R_1 and R_2 are two resistors with resistances r_1 and r_2. Then if R_1 and R_2 are in series (see Fig. 5.6) we have that the total resistance is $r_1 + r_2$. The resistors may also be wired in *parallel* (Fig. 5.7). The total resistance of the circuit elements in Fig. 5.7 is $r_1 r_2/(r_1 + r_2)$ unless $r_1 = r_2 = 0$ in which case it is zero. The number

$$r_1 r_2/(r_1 + r_2) \tag{1}$$

is called the *parallel sum* of r_1 and r_2. It is sometimes denoted $r_1 : r_2$.

This section will discuss to what extent the impedance matrices of two n-ports, in series or in parallel, can be computed from formulas like those of simple resistors.

It will be convenient to alter our notation of an n-port slightly by writing the ‘input’ terminals on the left and the ‘output’ terminals on the right. The numbers j, j' will label the two terminals forming the jth port. Thus the 3-port in Fig. 5.8a would now be written as in Fig. 5.8b. The notation of Fig. 5.8a is probably more intuitive while that of Fig. 5.8b is more convenient for what follows.

The parallel and series connection of two n-ports is done on a port basis.

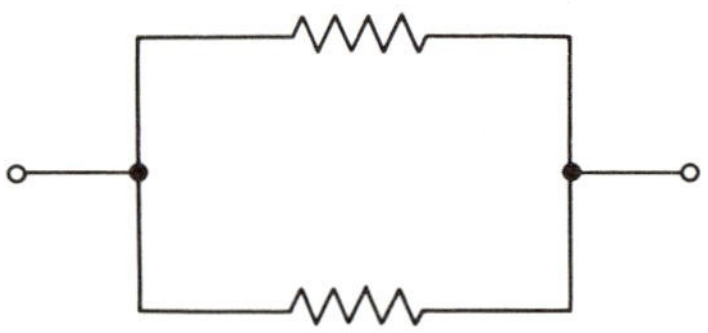

Fig. 5.7 Two resistors wired in parallel.

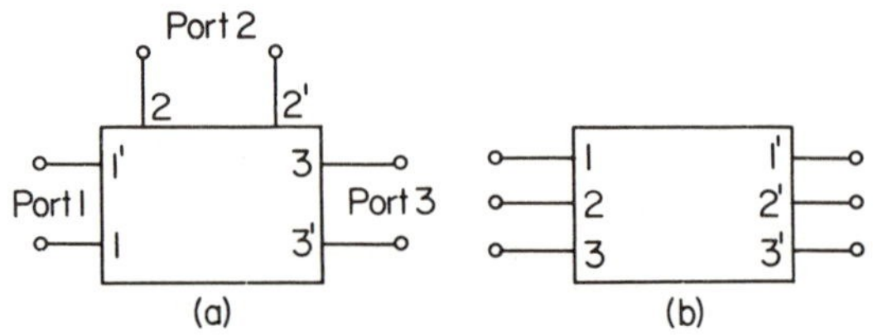

Fig. 5.8 Two ways of writing a 3-port network.

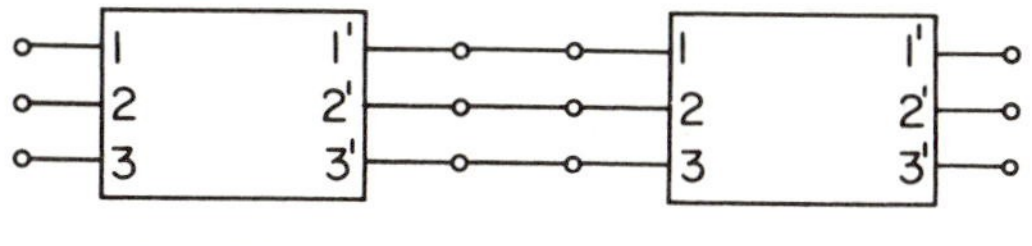

Fig. 5.9 Series connection of two 3-ports.

That is, in the series connection of two n-port networks, the ports labelled 1 are in series, the ports labelled 2 are in series, etc. (Fig. 5.9). Note, though, that the designation of a port as 1, 2, 3, ... is arbitrary. Notice that the parallel or series connection of two n-ports forms what appears to be a new n-port (Fig. 5.10).

Proposition 5.3.1 *Suppose that one has two n-ports* N_1 *and* N_2 *with impedance matrices* $\mathbf{Z}_1$ *and* $\mathbf{Z}_2$. *Then the impedance matrix of the series connection of* N_1 *and* N_2 *will be* $\mathbf{Z}_1 + \mathbf{Z}_2$ *provided that the two n-ports are still functioning as n-ports.*

Basically, the provision says that one cannot expect to use $\mathbf{Z}_1$ and $\mathbf{Z}_2$, if in the series connection, N_1 and N_2 no longer act like they did when $\mathbf{Z}_1$ and $\mathbf{Z}_2$ were computed.

It is not too difficult to see why Proposition 1 is true. Let N be the network formed by the series connection of two n-ports N_1 and N_2. Suppose in the series connection that N_1 and N_2 still function as n-ports. Apply a current of I amps across the ith port of N. Then a current of magnitude I goes into the 'first' terminal of port i of N_1. Since N_1 is an n-port, the same amount of current comes out of the second terminal of port i of N_1 and into the first terminal of port i of N_2. But N_2 is also functioning as an n-port. Thus I amps flow out of terminal 2 of the ith port of N_2. The resulting current is thus equivalent to having applied a current of I amps across the ith ports of N_1 and N_2 separately. But the potential across the jth port of N, denoted v_j, is the sum of the potentials across the jth ports of N_1 and N_2 since the second terminal of port j of N_1 and the first terminal of port j of N_2 are at the same potential. But N_k, $k = 1, 2$ are functioning as n-ports so we have that $v_j^{(1)} = z_{ij}^{(1)} I$, and $v_j^{(2)} = z_{ij}^{(2)} I$ where the superscript refers to the network. Thus $z_{ij} = v_j / I = (v_j^{(1)} + v_j^{(2)})/I = z_{ij}^{(1)} + z_{ij}^{(2)}$ as desired.

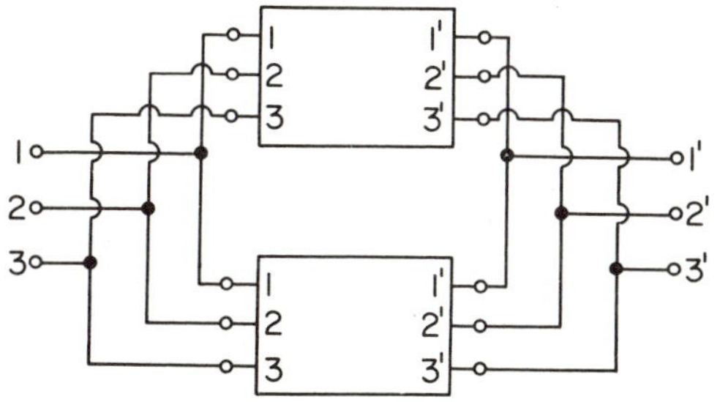

Fig. 5.10 Parallel connection of two 3-ports.

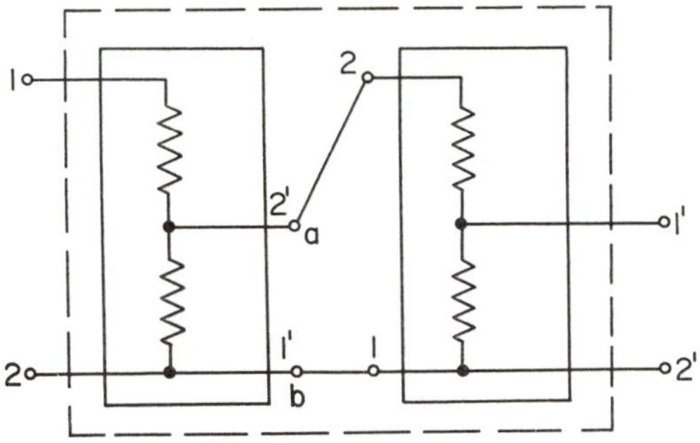

Fig. 5.11 Two 2-ports connected in series.

Example 5.3.1 Consider the series connection of two 2-ports shown in Fig. 5.11. All resistors are assumed to be 1 ohm. The impedance matrix of the first 2-port in Fig. 5.11 is $\mathbf{Z}_1 = \begin{bmatrix} 2 & 1 \\ 1 & 1 \end{bmatrix}$, while that of the second is $\mathbf{Z}_2 = \begin{bmatrix} 1 & 1 \\ 1 & 2 \end{bmatrix}$. Suppose now that a current of magnitude i is applied across port 1 of the combined 2-port of Fig. 5.11. The resistance between nodes a and b is $1:2 = 2/3$. The potential between a and b is thus $2/3\ I$. But what is important, 1/3 of the current goes through branch b and 2/3 through branch a. Thus in the series hookup of Fig. 5.11, there is i amperes of current going in terminal 1 of the first 2-port but only $2i/3$ coming out of terminal 1 of the first 2-port. Thus the first port of the first 2-port is no longer acting like a port. If the impedance matrix $\mathbf{Z}$ of the entire network of Fig. 5.11 is calculated, we get $\mathbf{Z} = \begin{bmatrix} 3 & 0 \\ 0 & 0 \end{bmatrix} \neq \mathbf{Z}_1 + \mathbf{Z}_2$. In many cases, however, the n-ports still act like n-ports when in series and one may add the impedance matrices. In practice, there is a simple procedure that can be used to check if the n-ports are still functioning as n-ports.

We see then that when n-ports are in series, that the impedance matrix of the whole network can frequently be calculated by adding the individual impedance matrices. Likewise, when in parallel a formula similar to (1) can sometimes be used to calculate the impedance matrix.

Suppose that $\mathbf{A}, \mathbf{B} \in \mathbb{C}^{n \times n}$. Then define the *parallel sum* $\mathbf{A}:\mathbf{B}$ of $\mathbf{A}$ and $\mathbf{B}$ by

$$\mathbf{A}:\mathbf{B} = \mathbf{A}(\mathbf{A}+\mathbf{B})^{\dagger}\mathbf{B}.$$

If a reciprocal network is composed solely of resistive elements, then the impedance matrix $\mathbf{Z}$ is not only hermitian but also *positive semi-definite.* That is, $(\mathbf{Zx}, \mathbf{x}) \geq 0$ for all $\mathbf{x} \in \mathbb{C}^n$. If $\mathbf{Z}$ is positive semi-definite, we sometimes write $\mathbf{Z} \geq \mathbf{0}$. If $\mathbf{Z}$ is positive semi-definite, then $\mathbf{Z}$ is hermitian. (This depends on the fact that $\mathbf{x} \in \mathbb{C}^n$ and not just $\mathbb{R}^n$.) If $\mathbf{A} - \mathbf{B} \geq \mathbf{0}$ for $\mathbf{A}, \mathbf{B} \in \mathbb{C}^{n \times n}$, then we write $\mathbf{A} \geq \mathbf{B}$ and say $\mathbf{A}$ is greater than or equal to $\mathbf{B}$.

Proposition 5.3.2 *Suppose that* N_1 *and* N_2 *are two reciprocal n-ports which are resistive networks with impedance matrices* $\mathbf{Z}_1$ *and* $\mathbf{Z}_2$. *Then the impedance matrix of the parallel connection of* N_1 *and* N_2 *is* $\mathbf{Z}_1 : \mathbf{Z}_2$.

Proof In order to prove Proposition 2 we need to use three facts about the parallel sum of hermitian positive semi-definite matrices $\mathbf{Z}_1, \mathbf{Z}_2$. The first is that $(\mathbf{Z}_1 : \mathbf{Z}_2) = (\mathbf{Z}_2 : \mathbf{Z}_1)$. The second is that $R(\mathbf{Z}_1) + R(\mathbf{Z}_2) = R(\mathbf{Z}_1 + \mathbf{Z}_2)$, so that, in particular, $R(\mathbf{Z}_1), R(\mathbf{Z}_2) \subseteq R(\mathbf{Z}_1 + \mathbf{Z}_2)$. The third is that $R(\mathbf{Z}_i)^\perp = N(\mathbf{Z}_i)$, $i = 1, 2$, since $\mathbf{Z}_i$ is hermitian. The proof of these facts is left to the exercises.

Let N_1 and N_2 be two n-ports connected in parallel to form an n-port N. Let $\mathbf{Z}_1$, $\mathbf{Z}_2$ and $\mathbf{Z}$ be the impedance matrices of N_1, N_2 and N respectively. Similarly, let $\mathbf{i}_1$, $\mathbf{i}_2$ and $\mathbf{i}$; $\mathbf{v}_1$, $\mathbf{v}_2$ and $\mathbf{v}$ be the current and voltage vectors for N_1, N_2 and N. To prove Proposition 2 we must show that

$$\mathbf{v} = \mathbf{Z}_1[\mathbf{Z}_1 + \mathbf{Z}_2]^\dagger \mathbf{Z}\mathbf{i} = (\mathbf{Z}_1 : \mathbf{Z}_2)\mathbf{i}. \tag{2}$$

The proof of (2) will follow the derivation of the simple case when N_1, N_2 are two resistors and $\mathbf{Z}_1$, $\mathbf{Z}_2$ are positive real numbers.

The current vector $\mathbf{i}$ may be decomposed as

$$\mathbf{i} = \mathbf{i}_1 + \mathbf{i}_2. \tag{3}$$

But $\mathbf{v} = \mathbf{v}_1 = \mathbf{v}_2$ since N_1 and N_2 are connected in parallel. Thus

$$\mathbf{v} = \mathbf{Z}_1\mathbf{i}_1, \text{ and } \mathbf{v} = \mathbf{Z}_2\mathbf{i}_2. \tag{4}$$

We will now transform (3) into the form of (2). Multiply (3) by $\mathbf{Z}_1$ and $\mathbf{Z}_2$ to get the two equations $\mathbf{Z}_1\mathbf{i} = \mathbf{Z}_1\mathbf{i}_1 + \mathbf{Z}_1\mathbf{i}_2 = \mathbf{v} + \mathbf{Z}_1\mathbf{i}_2$, and $\mathbf{Z}_2\mathbf{i} = \mathbf{Z}_2\mathbf{i}_1 + \mathbf{Z}_2\mathbf{i}_2 = \mathbf{v} + \mathbf{Z}_2\mathbf{i}_1$. Now multiply both of these equations by $(\mathbf{Z}_1 + \mathbf{Z}_2)^\dagger$. This gives

$$(\mathbf{Z}_1 + \mathbf{Z}_2)^\dagger \mathbf{Z}_1\mathbf{i} = (\mathbf{Z}_1 + \mathbf{Z}_2)^\dagger \mathbf{v} + (\mathbf{Z}_1 + \mathbf{Z}_2)^\dagger \mathbf{Z}_1\mathbf{i}_2, \text{ and} \tag{5}$$

$$(\mathbf{Z}_1 + \mathbf{Z}_2)^\dagger \mathbf{Z}_2\mathbf{i} = (\mathbf{Z}_1 + \mathbf{Z}_2)^\dagger \mathbf{v} + (\mathbf{Z}_1 + \mathbf{Z}_2)^\dagger \mathbf{Z}_2\mathbf{i}_1. \tag{6}$$

Multiply (5) on the left by $\mathbf{Z}_2$ and (6) on the left by $\mathbf{Z}_1$. Equations (5) and (6) become

$$\begin{aligned}(\mathbf{Z}_2 : \mathbf{Z}_1)\mathbf{i} &= \mathbf{Z}_2(\mathbf{Z}_1 + \mathbf{Z}_2)^\dagger \mathbf{v} + (\mathbf{Z}_2 : \mathbf{Z}_1)\mathbf{i}_2, \text{ and} \\ (\mathbf{Z}_1 : \mathbf{Z}_2)\mathbf{i} &= \mathbf{Z}_1(\mathbf{Z}_1 + \mathbf{Z}_2)^\dagger \mathbf{v} + (\mathbf{Z}_1 : \mathbf{Z}_2)\mathbf{i}_1.\end{aligned} \tag{7}$$

But $(\mathbf{Z}_1 : \mathbf{Z}_2) = (\mathbf{Z}_2 : \mathbf{Z}_1)$, $\mathbf{i} = \mathbf{i}_1 + \mathbf{i}_2$, and $\mathbf{Z}_1 + \mathbf{Z}_2$ is hermitian. Thus addition of the two equations in (7) gives us that

$$(\mathbf{Z}_1 : \mathbf{Z}_2)\mathbf{i} = (\mathbf{Z}_1 + \mathbf{Z}_2)(\mathbf{Z}_1 + \mathbf{Z}_2)^\dagger \mathbf{v} = \mathbf{P}_{R(\mathbf{Z}_1 + \mathbf{Z}_2)}\mathbf{v}. \tag{8}$$

Now the impedance matrix gives $\mathbf{v}$ from $\mathbf{i}$. Thus $\mathbf{v}$ must be in $R(\mathbf{Z}_1)$ and $R(\mathbf{Z}_2)$ by (4) so that $\mathbf{P}_{R(\mathbf{Z}_1 + \mathbf{Z}_2)}\mathbf{v} = \mathbf{v}$ and (8) becomes $(\mathbf{Z}_1 : \mathbf{Z}_2)\mathbf{i} = \mathbf{v}$ as desired. ■

Example 5.3.2 Consider the parallel connection of two 3-port networks

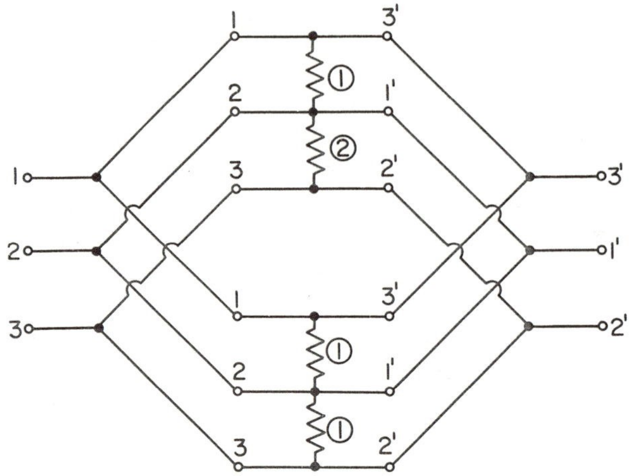

(9)

Fig. 5.12 The parallel connection of N_3 and N_2.

shown in Fig. 5.12. The impedance matrices of N_1 and N_2 are

$$\mathbf{Z}_1 = \begin{bmatrix} 1 & 0 & 1 \\ 0 & 2 & 2 \\ 1 & 2 & 3 \end{bmatrix} \text{ and } \mathbf{Z}_2 = \begin{bmatrix} 1 & 0 & 1 \\ 0 & 1 & 1 \\ 1 & 1 & 2 \end{bmatrix}.$$

By Proposition 2, the impedance matrix of circuit (9) is $\mathbf{Z}_1 : \mathbf{Z}_2$.

$$\mathbf{Z}_1 : \mathbf{Z}_2 = \begin{bmatrix} 1 & 0 & 1 \\ 0 & 2 & 2 \\ 1 & 2 & 3 \end{bmatrix} \left(\begin{bmatrix} 1 & 0 & 1 \\ 0 & 2 & 2 \\ 1 & 2 & 3 \end{bmatrix} + \begin{bmatrix} 1 & 0 & 1 \\ 0 & 1 & 1 \\ 1 & 1 & 2 \end{bmatrix} \right)^{\dagger} \begin{bmatrix} 1 & 0 & 1 \\ 0 & 1 & 1 \\ 1 & 1 & 2 \end{bmatrix}$$

$$= \frac{1}{324} \begin{bmatrix} 1 & 0 & 1 \\ 0 & 2 & 2 \\ 1 & 2 & 3 \end{bmatrix} \begin{bmatrix} 84 & -60 & 24 \\ -60 & 66 & 6 \\ 24 & 6 & 30 \end{bmatrix} \begin{bmatrix} 1 & 0 & 1 \\ 0 & 1 & 1 \\ 1 & 1 & 2 \end{bmatrix}$$

$$= \begin{bmatrix} 1/2 & 0 & 1/2 \\ 0 & 2/3 & 2/3 \\ 1/2 & 2/3 & 7/6 \end{bmatrix}.$$

The generalized inverse can be computed easily by several methods. The reader is encouraged to verify that the z_{kj} values obtained from $\mathbf{Z}_1 : \mathbf{Z}_2$ agree with those obtained by direct computation from (9).

4. Shorted matrices

The generalized inverse appears in situations other than just parallel connections

Suppose that one is interested in a 3-port network N with impedance matrix $\mathbf{Z}$. Now short out port 3 to produce a new 3-port network, N′ and denote its impedance matrix by $\mathbf{Z}'$. Since v_3 is always zero in N′ we must have $z'_{31} = z'_{32} = z'_{33} = 0$, that is, the bottom row of $\mathbf{Z}'$ is zero. If N is a reciprocal network, then the third column of $\mathbf{Z}$ must also consist of zeros. $\mathbf{Z}'$ would then have the form

$$\mathbf{Z}' = \begin{bmatrix} z'_{11} & z'_{12} & 0 \\ z'_{21} & z'_{22} & 0 \\ 0 & 0 & 0 \end{bmatrix}. \tag{1}$$

The obvious question is: What is the relationship between the z_{kj} and the z'_{kj}? The answer, which at first glance is probably not obvious, is:

Proposition 5.4.1 *Suppose that* N *is a resistive n-port network with impedance matrix* $\mathbf{Z}$. *Partition* $\mathbf{Z}$ *as* $\mathbf{Z} = \left[\begin{array}{c|c} \mathbf{Z}_{11} & \mathbf{Z}_{12} \\ \hline \mathbf{Z}_{21} & \mathbf{Z}_{22} \end{array}\right]$; $\mathbf{Z}_{22} \in \mathbb{C}^{s \times s}$, $1 \le s \le n$. *Then* $\mathbf{Z}' = \left[\begin{array}{c|c} \mathbf{Z}_{11} - \mathbf{Z}_{12}\mathbf{Z}_{22}^{\dagger}\mathbf{Z}_{21} & \mathbf{0} \\ \hline \mathbf{0} & \mathbf{0} \end{array}\right]$ *is the impedance matrix of the network N′ formed by shorting the last s ports of* N *if* N *is reciprocal.*

Proof Write $\mathbf{i} = \begin{bmatrix} \mathbf{i}_o \\ \mathbf{i}_s \end{bmatrix}$, $\mathbf{v} = \begin{bmatrix} \mathbf{v}_o \\ \mathbf{v}_s \end{bmatrix}$ where $\mathbf{i}_s, \mathbf{v}_s \in \mathbb{C}^s$. Then $\mathbf{v} = \mathbf{Z}\mathbf{i}$ may be written as

$$\begin{aligned} \mathbf{v}_o &= \mathbf{Z}_{11}\mathbf{i}_o + \mathbf{Z}_{12}\mathbf{i}_s \\ \mathbf{v}_s &= \mathbf{Z}_{21}\mathbf{i}_o + \mathbf{Z}_{22}\mathbf{i}_s \end{aligned} \tag{2}$$

Suppose now that the last s ports of N are shorted. We must determine the matrix $\mathbf{X}$ such that $\mathbf{v}_o = \mathbf{X}\mathbf{i}_o$. Since the last s ports are shorted, $\mathbf{v}_s = \mathbf{0}$. Thus the second equation of (2) becomes $\mathbf{Z}_{22}\mathbf{i}_s = -\mathbf{Z}_{21}\mathbf{i}_o$.
Hence

$$\mathbf{i}_s = -\mathbf{Z}_{22}^{\dagger}\mathbf{Z}_{21}\mathbf{i}_o + [\mathbf{I} - \mathbf{Z}_{22}^{\dagger}\mathbf{Z}_{22}]\mathbf{i}_s = -\mathbf{Z}_{22}^{\dagger}\mathbf{Z}_{21}\mathbf{i}_o + \mathbf{h} \text{ where } \mathbf{h} \in N(\mathbf{Z}_{22}). \tag{3}$$

If $(\mathbf{Z}\mathbf{i}, \mathbf{i}) = 0$, then $\mathbf{Z}\mathbf{i} = \mathbf{0}$ since $\mathbf{Z} \ge \mathbf{0}$. Thus $N(\mathbf{Z}_{22}) \subseteq N(\mathbf{Z}_{12})$. (Consider $\mathbf{i}$ with $\mathbf{i}_o = \mathbf{0}$). Substituting equation (3) into the first equation of (2) now gives $\mathbf{v}_o = \mathbf{Z}_{11}\mathbf{i}_o + \mathbf{Z}_{12}\mathbf{i}_s = \mathbf{Z}_{11}\mathbf{i}_o + \mathbf{Z}_{12}(-\mathbf{Z}_{22}^{\dagger}\mathbf{Z}_{21}\mathbf{i}_o + \mathbf{h})) = \mathbf{Z}_{11}\mathbf{i}_o - \mathbf{Z}_{12}\mathbf{Z}_{22}^{\dagger}\mathbf{Z}_{21}\mathbf{i}_o = (\mathbf{Z}_{11} - \mathbf{Z}_{12}\mathbf{Z}_{22}^{\dagger}\mathbf{Z}_{21})\mathbf{i}_o$ as desired. The zero blocks appear in the $\mathbf{Z}'$ matrix for the same reason that zeros appeared in the special case (1). ■

$\mathbf{Z}'$ is sometimes referred to as a *shorted matrix*. Properties of shorted matrices often correspond to physical properties of the circuit. We will mention one. Others are developed in the exercises along with a generalization of the definition of shorted matrix.

Suppose that $\mathbf{Z}, \mathbf{Z}'$ are as in Proposition 1. Then $\mathbf{Z} \ge \mathbf{Z}' \ge \mathbf{0}$. This corresponds to the physical fact that a short circuit can only lower resistance of a network and not increase it.

It is worth noting that in the formula for $\mathbf{Z}'$ in Proposition 1 a weaker type of inverse than the generalized inverse would have sufficed. However, as we will see later, it would then have been more difficult to show that $\mathbf{Z} \geq \mathbf{Z}'$. In the parallel sum the Penrose conditions are needed and a weaker inverse would not have worked.

5. Other uses of the generalized inverse

The applications of the generalized inverse in Sections 3 and 4 were chosen partly because of their uniqueness. There are other uses of the generalized inverse which are more routine.

For example, suppose that we have an n-port network N with impedance matrix $\mathbf{Z}$. Then $\mathbf{v} = \mathbf{Zi}$. It might be desirable to be able to produce a particular output $\mathbf{v}_0$. In that case we would want to solve $\mathbf{v}_0 = \mathbf{Zi}$. If $\mathbf{v}_0 \notin R(\mathbf{Z})$, then we must seek approximate solutions. This would be a least squares problem as discussed in Chapter 2. $\mathbf{Z}^\dagger \mathbf{v}$ would correspond to that least squares solution which requires the least current input (in the sense that $\|\mathbf{i}\|$ is minimized).

Of course, this approach would work for inputs and outputs other than just current inputs and voltage outputs. The only requirements are that with respect to the new variables the network has the properties of homogeneity and superposition otherwise we cannot get a linear system of equations unless a first order approximation is to be taken. In practice, there should also be a way to compute the matrix of coefficients of the system of equations.

Another use of the generalized inverse is in minimizing *quadratic forms* subject to linear constraints. Recall that a quadratic form is a function ϕ from $\mathbb{C}^n$ to $\mathbb{C}$ of the form $\phi(\mathbf{x}) = (\mathbf{Ax}, \mathbf{x})$, $\mathbf{A} \in \mathbb{C}^{n \times n}$, for a fixed $\mathbf{A}$. The instantaneous power dissipated by a circuit, the instantaneous value of the energy stored in the inductors (if any are in the circuit), and the instantaneous value of the energy stored in the capacitors, may all be written in the form $(\mathbf{Ai}, \mathbf{i})$ where $\mathbf{i}$ is a vector made up of the *loop currents.* A description of loop currents and how they can be used to get a system of equations describing a network may be found in Chapter III of Huelsman. His book provides a very good introduction to and amplification of the ideas presented in this chapter.

6. Exercises

For Exercises (1)–(8) assume that $\mathbf{A}, \mathbf{B}, \mathbf{C}, \mathbf{D} \geq \mathbf{0}$ are $n \times n$ hermitian matrices.

1. Show that $\mathbf{A}:\mathbf{B} = \mathbf{B}:\mathbf{A}$.
2. Show that $\mathbf{A}:\mathbf{B} \geq \mathbf{0}$.
3. Prove that $R(\mathbf{A}:\mathbf{B}) = R(\mathbf{A}) \cap R(\mathbf{B})$.
4. Prove that $(\mathbf{A}:\mathbf{B}):\mathbf{C} = \mathbf{A}:(\mathbf{B}:\mathbf{C})$.

*5. Prove that $\operatorname{Tr}(\mathbf{A}:\mathbf{B}) \leq (\operatorname{Tr}\mathbf{A}):(\operatorname{Tr}\mathbf{B})$.

*6. Prove that $\det(\mathbf{A}:\mathbf{B}) \leq (\det\mathbf{A}):(\det\mathbf{B})$.

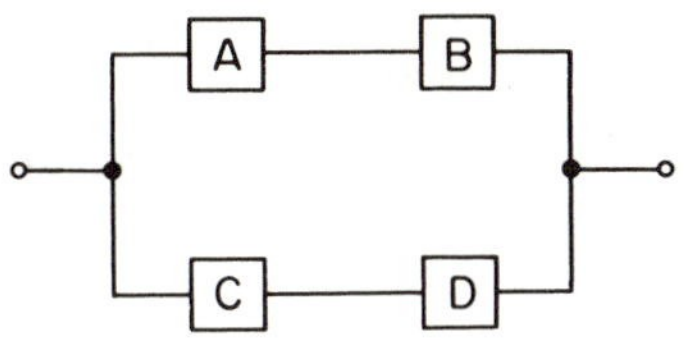

Fig. 5.13

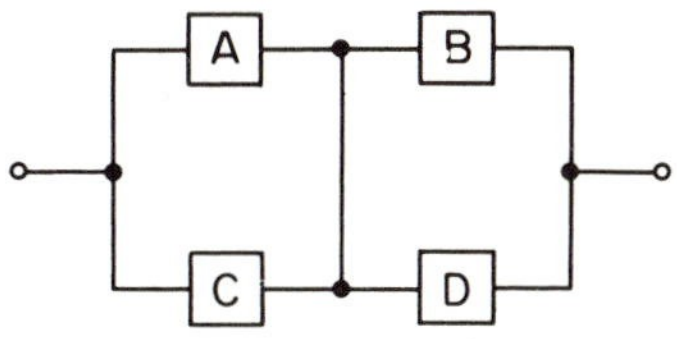

Fig. 5.14

7. a. Prove that if $\mathbf{A} \geq \mathbf{B}$, then $\mathbf{A}:\mathbf{C} \geq \mathbf{B}:\mathbf{C}$.
   b. Formulate the physical analogue of 7a in terms of impedances.

*8. Show that $(\mathbf{A}+\mathbf{B}):(\mathbf{C}+\mathbf{D}) \geq \mathbf{A}:\mathbf{C}+\mathbf{B}:\mathbf{D}$. (This corresponds to the assertion that Fig. 5.13 has more impedance than Fig. 5.14 since the latter has more paths.)

For Exercises (9)–(14) assume that $\underset{\sim}{\mathbf{E}}$ is an hermitian positive semi-definite operator on $\mathbb{C}^n$.

Pick a subspace $M \subseteq \mathbb{C}^n$. Let B be an orthonormal basis consisting of first an orthonormal basis for M and then one for $M^{\perp}$. With respect to B, $\underset{\sim}{\mathbf{E}}$ has the matrix $\mathbf{E} = \begin{bmatrix} \mathbf{A} & \mathbf{B} \\ \mathbf{B}^* & \mathbf{C} \end{bmatrix}$, where $\mathbf{E} \geq \mathbf{0}$. Define $\mathbf{E}_M = \begin{bmatrix} \mathbf{A}-\mathbf{B}\mathbf{C}^{\dagger}\mathbf{B}^* & \mathbf{0} \\ \mathbf{0} & \mathbf{0} \end{bmatrix}$.

9. Prove that $\mathbf{E} \geq \mathbf{E}_M$.
10. Show that if $\mathbf{D}$ is hermitian positive semi-definite and $\mathbf{E} \geq \mathbf{D} \geq \mathbf{0}$ and $R(\mathbf{D}) \subseteq M$, $\mathbf{D} \leq \mathbf{E}_M$.

*11. Prove that $\mathbf{E}_M = \lim_{n \to \infty} \mathbf{E}:n\mathbf{P}_M$.

For the next three exercises $\mathbf{F}$ is another hermitian positive semi-definite matrix partitioned with respect to B just like $\mathbf{E}$ was.

12. Suppose that $\mathbf{E} \geq \mathbf{F} \geq \mathbf{0}$. Show that $\mathbf{E}_M \geq \mathbf{F}_M \geq \mathbf{0}$.
13. Prove that $(\mathbf{E}+\mathbf{F})_M \geq \mathbf{E}_M + \mathbf{F}_M$.
14. Determine when equality holds in Exercise 13.

Let L, M be subspaces. Prove $\mathbf{P}_{L \cap M} = 2(\mathbf{P}_L : \mathbf{P}_M)$.

7. References and further reading

A good introduction to the use of matrix theory in circuit theory is [44]. Our Section 2 is a very condensed version of the development there. In particular, Huelsman discusses how to handle more general networks than ours with the port notation by the use of 'grounds' or reference nodes. Matrices other than the impedance matrix are also discussed in detail.

Many papers have been published on n-ports. The papers of Cederbaum, together with their bibliographies, will get the interested reader started. Two of his papers are listed in the bibliography at the end of this book [27], [28].

The parallel sum of matrices has been studied by Anderson and colleagues in [2], [3], [4], [5] and his thesis. Exercises (1)–(8) come from [3] while (9)–(14) are propositions and lemmas from [2]. The theory of shorted operators is extended to operators on Hilbert space in [5]. In [4] the operations of ordinary and parallel addition are treated as special cases of a more general type of matrix addition. The minimization of quadratic forms is discussed in [10]. The authors of [42] use the generalized inverse to minimize quadratic forms and eliminate 'unwanted variables'.

The reader interested in additional references is referred to the bibliographies of the above and in particular [4].

6
(i,j,k)-generalized inverses and linear estimation

1. Introduction

We have seen in the earlier chapters that the generalized inverse $\mathbf{A}^\dagger$ of $\mathbf{A}\in\mathbb{C}^{m\times n}$, although useful, has some shortcomings. For example: computation of $\mathbf{A}^\dagger$ can be difficult, $\mathbf{A}^\dagger$ is lacking in desirable spectral properties, and the generalized inverse of a product is not necessarily the product of the generalized inverses in reverse order.

It seems reasonable that in order to define a generalized inverse which overcomes one or more of these deficiencies, one must expect to give up something. The importance one attaches to the various types of generalized inverses will depend on the particular applications which one has in mind. For some applications the properties which are lost will not be nearly as important as those properties which are gained.

From a theoretical point of view, the definition and properties of the generalized inverse defined in Chapter 1 are probably more elegant than those of this chapter. However, the concepts of this chapter are considered by many to be more practical than those of the previous chapters.

2. Definitions

Recall that $\mathbf{L}(\mathbb{C}^n, \mathbb{C}^m)$ denotes the set of linear transformations from $\mathbb{C}^n$ into $\mathbb{C}^m$. For $\underset{\sim}{\mathbf{A}}\in\mathbf{L}(\mathbb{C}^n, \mathbb{C}^m)$, $\underset{\sim}{\mathbf{A}}^\dagger$ was defined in Chapter 1 as follows. $\mathbb{C}^n$ was decomposed into the direct sum of $N(\underset{\sim}{\mathbf{A}})$ and $N(\underset{\sim}{\mathbf{A}})^\perp$. $\underset{\sim}{\mathbf{A}}_1$ denoted the restriction of $\underset{\sim}{\mathbf{A}}$ to $N(\underset{\sim}{\mathbf{A}})^\perp$, so that $\underset{\sim}{\mathbf{A}}_1$ was a one to one mapping of $N(\underset{\sim}{\mathbf{A}})^\perp$ onto $R(\underset{\sim}{\mathbf{A}})$. $\underset{\sim}{\mathbf{A}}^\dagger$ was then defined to be

$$\underset{\sim}{\mathbf{A}}^\dagger\mathbf{x} \equiv \begin{cases} \underset{\sim}{\mathbf{A}}_1^{-1}\mathbf{x} & \text{if } \mathbf{x}\in R(\underset{\sim}{\mathbf{A}}) \\ \mathbf{0} & \text{if } \mathbf{x}\in R(\underset{\sim}{\mathbf{A}})^\perp \end{cases}$$

Instead of considering orthogonal complements of $N(\underset{\sim}{\mathbf{A}})$ and $R(\underset{\sim}{\mathbf{A}})$, one could consider any pair of complementary subspaces and obtain a linear transformation which could be considered as a generalized inverse for $\underset{\sim}{\mathbf{A}}$.

Definition 6.2.1 (*Functional Definition*) *Let* $\underset{\sim}{\mathbf{A}} \in \mathbf{L}(\mathbb{C}^n, \mathbb{C}^m)$ *and let* N *and* R *be complementary subspaces of* $N(\underset{\sim}{\mathbf{A}})$ *and* $R(\underset{\sim}{\mathbf{A}})$, *that is,* $\mathbb{C}^n = N(\underset{\sim}{\mathbf{A}}) \dot{+} N$ *and* $\mathbb{C}^m = R(\underset{\sim}{\mathbf{A}}) + R$. *Let* $\underset{\sim}{\mathbf{A}}_1 = \underset{\sim}{\mathbf{A}}|_N$ (*i.e.* $\underset{\sim}{\mathbf{A}}$ *restricted to* N. *Note that* $\underset{\sim}{\mathbf{A}}_1$ *is a one to one mapping of* N *onto* $R(\underset{\sim}{\mathbf{A}})$ *so that* $\underset{\sim}{\mathbf{A}}_1^{-1} : R(\underset{\sim}{\mathbf{A}}) \to N$ *exists*). *For* $\mathbf{x} \in \mathbb{C}^m$, *let* $\mathbf{x} = \mathbf{r}_1 + \mathbf{r}_2$ *where* $\mathbf{r}_1 \in R(\underset{\sim}{\mathbf{A}})$ *and* $\mathbf{r}_2 \in R$. *The function* $\underset{\sim}{\mathbf{G}}_{N,R}$ *defined by*

$$\underset{\sim}{\mathbf{G}}_{N,R}\mathbf{x} \equiv \underset{\sim}{\mathbf{A}}_1^{-1}\mathbf{r}_1$$

is either called the (N, R)-generalized inverse for $\underset{\sim}{\mathbf{A}}$ *or a* prescribed range/null space generalized inverse for $\underset{\sim}{\mathbf{A}}$. *For a given* N *and* R, $\underset{\sim}{\mathbf{G}}_{N,R}$ *is a uniquely defined linear transformation from* $\mathbb{C}^m$ *into* $\mathbb{C}^n$. *Therefore, for* $\mathbf{A} \in \mathbb{C}^{m \times n}$, $\mathbf{A}$ *induces a function* $\underset{\sim}{\mathbf{A}}$ *and we can define* $\mathbf{G}_{N,R} \in \mathbb{C}^{n \times m}$ *to be the matrix of* $\underset{\sim}{\mathbf{G}}_{N,R}$ (*with respect to the standard basis*).

In the terminology of Definition 1, $\mathbf{A}^\dagger$ is the $(R(\mathbf{A}^*), N(\mathbf{A}^*))$-generalized inverse for $\mathbf{A}$. In order to avoid confusion, we shall henceforth refer to $\mathbf{A}^\dagger$ as the *Moore–Penrose inverse of* $\mathbf{A}$.

In Chapter 1 three equivalent definitions for $\mathbf{A}^\dagger$ were given; a functional definition, a projective definition (Moore's definition), and an algebraic definition (Penrose's definition). We can construct analogous definitions for the (N, R)-generalized inverse. It will be assumed throughout this section that N, R are as in Definition 1.

The projection operators which we will be dealing with in this chapter will not necessarily be orthogonal projectors. So as to avoid confusion, the following notation will be adopted.

Notation. To denote the *oblique projector* whose range is M and whose null space is N we shall use the symbol $\mathbf{P}_{M,N}$. The symbol $\mathbf{P}_M$, will denote, as before, the *orthogonal projector* whose range is M and whose null space is $M^\perp$.

Starting with Definition 1 it is straightforward to arrive at the following two alternative characterizations of $\mathbf{G}_{N,R}$.

Definition 6.2.2 (*Projective Definition*) *For* $\mathbf{A} \in \mathbb{C}^{m \times n}$, $\mathbf{G}_{N,R}$ *is called the* (N, R)-*generalized inverse for* $\mathbf{A}$ *if* $R(\mathbf{G}_{N,R}) = N$, $N(\mathbf{G}_{N,R}) = R$, $\mathbf{A}\mathbf{G}_{N,R} = \mathbf{P}_{R(\mathbf{A}),R}$, and $\mathbf{G}_{N,R}\mathbf{A} = \mathbf{P}_{N,N(\mathbf{A})}$.

Definition 6.2.3 (*Algebraic Definition*) *For* $\mathbf{A} \in \mathbb{C}^{m \times n}$, $\mathbf{G}_{N,R}$ *is the* (N, R)-*generalized inverse for* $\mathbf{A}$ *if* $R(\mathbf{G}_{N,R}) = N$, $N(\mathbf{G}_{N,R}) = R$, $\mathbf{A}\mathbf{G}_{N,R}\mathbf{A} = \mathbf{A}$, *and* $\mathbf{G}_{N,R}\mathbf{A}\mathbf{G}_{N,R} = \mathbf{G}_{N,R}$.

Theorem 6.2.1 *The functional, the projective, and the algebraic definitions of the* (N, R)-*generalized inverse are equivalent.*

Proof We shall first prove that Definition 1 is equivalent to Definition 2 and then that Definition 2 is equivalent to Definition 3.

If $\underset{\sim}{\mathbf{G}}_{N,R}$ satisfies the conditions of Definition 1, then it is clear that $R(\underset{\sim}{\mathbf{G}}_{N,R}) = N$ and $N(\underset{\sim}{\mathbf{G}}_{N,R}) = R$. But $\underset{\sim}{\mathbf{A}}\underset{\sim}{\mathbf{G}}_{N,R}$ is the identity function on $R(\underset{\sim}{\mathbf{A}})$ and the zero function on R. Thus $\mathbf{A}\mathbf{G}_{N,R} = \mathbf{P}_{R(\mathbf{A}),R}$. Similarly $\mathbf{G}_{N,R}\mathbf{A} = \mathbf{P}_{N,N(\mathbf{A})}$, and Definition 1 implies Definition 2. Conversely, if $\mathbf{G}_{N,R}$ satisfies the conditions of Definition 2, then N and R must be complementary subspaces for $N(\mathbf{A})$ and $R(\mathbf{A})$ respectively. It also follows that $\underset{\sim}{\mathbf{A}}\underset{\sim}{\mathbf{G}}_{N,R}$ must be the identity on $R(\underset{\sim}{\mathbf{A}})$ and zero on R while $\underset{\sim}{\mathbf{G}}_{N,R}\underset{\sim}{\mathbf{A}}$ is the identity on N and zero on $N(\mathbf{A})$. Thus $\underset{\sim}{\mathbf{G}}_{N,R}$ satisfies the conditions of Definition 1, and hence Definition 2 implies Definition 1. That Definition 2 implies Definition 3 is clear. To complete the proof, we need only to show that Definition 3 implies Definition 2. Assuming $\mathbf{G}$ satisfies the conditions of Definition 3, we obtain $(\mathbf{AG})^2 = \mathbf{AG}$ and $(\mathbf{GA})^2 = \mathbf{GA}$ so that $\mathbf{AG}$ and $\mathbf{GA}$ are projectors. Furthermore, $R(\mathbf{AG}) \subseteq R(\mathbf{A}) = R(\mathbf{AGA}) \subseteq R(\mathbf{AG})$ and $R(\mathbf{GA}) \subseteq R(\mathbf{G}) = R(\mathbf{GAG}) \subseteq R(\mathbf{GA})$. Thus $R(\mathbf{AG}) = R(\mathbf{A})$ and $R(\mathbf{GA}) = R(\mathbf{G}) = N$. Likewise, it is a simple matter to show that $N(\mathbf{AG}) = N(\mathbf{G}) = R$ and $N(\mathbf{GA}) = N(\mathbf{A})$, so that $\mathbf{AG} = \mathbf{P}_{R(\mathbf{A}),R}$ and $\mathbf{GA} = \mathbf{P}_{N,N(\mathbf{A})}$, which are the conditions of Definition 2 and the proof is complete. ■

Corollary 6.2.1 *For* $\mathbf{A} \in \mathbb{C}^{m \times n}$, *the class of all prescribed range/null space generalized inverses for* $\mathbf{A}$ *is precisely the set*

$$\{\mathbf{X} \in \mathbb{C}^{n \times m} \mid \mathbf{AXA} = \mathbf{A} \text{ and } \mathbf{XAX} = \mathbf{X}\}, \tag{1}$$

i.e. those matrices which satisfy the first and second *Penrose conditions.*

The definition of a prescribed range/null space inverse was formulated as an extension of the Moore–Penrose inverse with no particular applications in mind. Let us now be a bit more practical and look at a problem of fundamental importance. Consider a system of linear equations written as $\mathbf{Ax} = \mathbf{b}$. If $\mathbf{A}$ is square and non-singular, then one of the characteristics of $\mathbf{A}^{-1}$ is that for every $\mathbf{b}$, $\mathbf{A}^{-1}\mathbf{b}$ is the solution. In order to generalize this property, one might ask for $\mathbf{A} \in \mathbb{C}^{m \times n}$, what are the characteristics of a matrix $\mathbf{G} \in \mathbb{C}^{n \times m}$ such that $\mathbf{Gb}$ is a solution of $\mathbf{Ax} = \mathbf{b}$ for every $\mathbf{b} \in \mathbb{C}^m$ for which $\mathbf{Ax} = \mathbf{b}$ is consistent? That is, what are the characteristics of $\mathbf{G}$ if

$$\mathbf{AGb} = \mathbf{b} \text{ for every } \mathbf{b} \in R(\mathbf{A})? \tag{2}$$

In loose terms, we are asking what do the 'equation solving generalized inverses of $\mathbf{A}$' look like? This is easy to answer. Since $\mathbf{AGb} = \mathbf{b}$ for every $\mathbf{b} \in R(\mathbf{A})$, it is clear that $\mathbf{AGA} = \mathbf{A}$. Conversely, suppose that $\mathbf{G}$ satisfies $\mathbf{AGA} = \mathbf{A}$. For every $\mathbf{b} \in R(\mathbf{A})$ there exists an $\mathbf{x}_b \in \mathbb{C}^n$ such that $\mathbf{Ax}_b = \mathbf{b}$. Therefore, $\mathbf{AGb} = \mathbf{AGAx}_b = \mathbf{Ax}_b = \mathbf{b}$ for every $\mathbf{b} \in R(\mathbf{A})$. Below is a formal statement of our observations.

Theorem 6.2.2 *For* $\mathbf{A} \in \mathbb{C}^{m \times n}$, $\mathbf{G} \in \mathbb{C}^{n \times m}$ *has the property that* $\mathbf{Gb}$ *is a solution of* $\mathbf{Ax} = \mathbf{b}$ *for every* $\mathbf{b} \in \mathbb{C}^m$ *for which* $\mathbf{Ax} = \mathbf{b}$ *is consistent if and only if*

$$\mathbf{G} \in \{\mathbf{X} \in \mathbb{C}^{n \times m} \mid \mathbf{AXA} = \mathbf{A}\}. \tag{3}$$

Thus the 'equation solving generalized inverses for $\mathbf{A}$' are precisely those which satisfy the *first* Penrose condition of Definition 1.1.3.

Let us now be more particular. Suppose we seek $\mathbf{G}\in\mathbb{C}^{n\times m}$ such that, in addition to being an 'equation solving inverse' in the sense of (2), we also require that for each $\mathbf{b}\in R(\mathbf{A})$, $\|\mathbf{Gb}\| < \|\mathbf{z}\|$ for all $\mathbf{z}\neq\mathbf{Gb}$ and $\mathbf{z}\in\{\mathbf{x}\,|\,\mathbf{Ax}=\mathbf{b}\}$. That is, for each $\mathbf{b}\in R(\mathbf{A})$ we want $\mathbf{Gb}$ to be the solution of minimal norm. On the basis of Theorem 2.1.1 we may restate our objective as follows. For each $\mathbf{b}\in R(\mathbf{A})$ we require that $\mathbf{Gb}=\mathbf{A}^{\dagger}\mathbf{b}$. Therefore, $\mathbf{G}$ must satisfy the equation

$$\mathbf{GA}=\mathbf{A}^{\dagger}\mathbf{A}=\mathbf{P}_{R(\mathbf{A}^*)}, \tag{4}$$

which is equivalent to

$$\mathbf{AGA}=\mathbf{A} \text{ and } (\mathbf{GA})^*=\mathbf{GA}. \tag{5}$$

The equivalence of (4) and (5) is left to the exercises. Suppose now that $\mathbf{G}$ is any matrix which satisfies (5). All of the above implications are reversible so that $\mathbf{Gb}$ must be the solution of minimal norm. Below is a formal statement of what we have just proven.

Theorem 6.2.3 *For* $\mathbf{A}\in\mathbb{C}^{m\times n}$ *and* $\mathbf{b}\in R(\mathbf{A})$, $\mathbf{A}(\mathbf{Gb})=\mathbf{b}$ *and* $\|\mathbf{Gb}\|<\|\mathbf{z}\|$ *for all* $\mathbf{z}\neq\mathbf{Gb}$ *and* $\mathbf{z}\in\{\mathbf{x}\,|\,\mathbf{Ax}=\mathbf{b}\}$ *if and only if*

$$\mathbf{G}\in\{\mathbf{X}\in\mathbb{C}^{n\times m}\,|\,\mathbf{AXA}=\mathbf{A} \text{ and } (\mathbf{XA})^*=\mathbf{XA}\}.$$

We define the term *minimum norm generalized inverse* to be a matrix which satisfies the *first* and *fourth* Penrose conditions of Definition 1.1.3.

Let us now turn our attention to inconsistent systems of equations. As in Chapter 2, the statement $\mathbf{Ax}=\mathbf{b}$ is to be taken as an *open statement* and the set of vectors $\{\mathbf{z}\,|\,\mathbf{Az}$ is equal to $\mathbf{b}\}$ (i.e. the 'solution set' for the open statement) may or may not be empty, depending on whether $\mathbf{b}\in R(\mathbf{A})$. When the solution set for $\mathbf{Ax}=\mathbf{b}$ is empty, we say $\mathbf{Ax}=\mathbf{b}$ is *inconsistent*. In dealing with inconsistent equations, a common practice is to seek a least squares solution as defined in Definition 2.1.1. As we saw in Theorem 2.1.2, $\mathbf{A}^{\dagger}\mathbf{b}$ is always a least squares solution of $\mathbf{Ax}=\mathbf{b}$. However, $\mathbf{A}^{\dagger}\mathbf{b}$ is a special least squares solution. It is the least squares solution of minimal norm. In some applications, one might settle for obtaining any least squares solution and not care about the one of minimal norm.

For $\mathbf{A}\in\mathbb{C}^{m\times n}$, let us try to determine the characteristics of a matrix $\mathbf{G}$ such that $\mathbf{Gb}$ is a least squares solution of $\mathbf{Ax}=\mathbf{b}$ for all $\mathbf{b}\in\mathbb{C}^m$. To begin with, we can infer from Corollary 2.1.1 that $\|\mathbf{AGb}-\mathbf{b}\|$ is minimal if and only if $\mathbf{AGb}=\mathbf{P}_{R(\mathbf{A})}\mathbf{b}=\mathbf{AA}^{\dagger}\mathbf{b}$. This being true for all $\mathbf{b}\in\mathbb{C}^m$ yields $\mathbf{AG}=\mathbf{AA}^{\dagger}$. But $\mathbf{AG}=\mathbf{AA}^{\dagger}$ is equivalent to $\mathbf{AGA}=\mathbf{A}$ and $(\mathbf{AG})^*=\mathbf{AG}$. The proof of this equivalence is left to the exercises. We formally state the above observations in the following theorem.

Theorem 6.2.4 *For* $\mathbf{A}\in\mathbb{C}^{m\times n}$, *for* $\mathbf{Gb}$ *is a least squares solution of* $\mathbf{Ax}=\mathbf{b}$ *for all* $\mathbf{b}\in\mathbb{C}^m$ *if and only if*

$$\mathbf{G}\in\{\mathbf{X}\in\mathbb{C}^{n\times m}\,|\,\mathbf{AXA}=\mathbf{A} \text{ and } (\mathbf{AX})^*=\mathbf{AX}\}.$$

We define the term *least squares generalized inverse* to be a matrix which satisfies the *first* and *third* Penrose conditions.

Looking at Corollary 1 and Theorems 2–4, one sees that each of the different types of $\mathbf{G}$ matrices discussed can be characterized as a solution to some subset of the Penrose conditions of Definition 1.1.3. To simplify our nomenclature we make the following standard definition.

Definition 6.2.4 For $\mathbf{A} \in \mathbb{C}^{m \times n}$, *a matrix* $\mathbf{G} \in \mathbb{C}^{n \times m}$ *is called an* (i, j, k)-*inverse for* $\mathbf{A}$ if $\mathbf{G}$ *satisfies the* ith, jth, *and* kth *Penrose conditions;*

(1) $\mathbf{AXA} = \mathbf{A}$,
(2) $\mathbf{XAX} = \mathbf{X}$,
(3) $(\mathbf{AX})^* = \mathbf{AX}$,
(4) $(\mathbf{XA})^* = \mathbf{XA}$.

Furthermore, the set of all (i, j, k)-*inverses for* $\mathbf{A}$ *will be denoted by* $\mathbf{A}\{i, j, k\}$.

For example, $\mathbf{G}$ is a $(1, 3)$-inverse for $\mathbf{A}$ if $\mathbf{AGA} = \mathbf{A}$ and $(\mathbf{AG})^* = \mathbf{AG}$. We write $\mathbf{G} \in \mathbf{A}\{1, 3\}$. Note that $\mathbf{G}$ may or may not satisfy either of the other two Penrose conditions. This notation requires one to pay particular attention to how the equations are ordered, but experience has shown this convention to be efficient and useful.

Notation. For $\mathbf{A} \in \mathbb{C}^{m \times n}$, the symbol $\mathbf{A}^-$ will be used to designate an arbitrary element of $\mathbf{A}\{1\}$.

The $(\cdot)^-$ notation is a convenience that must be treated with some care. It is frequently used to make statements which hold for the entire class $\mathbf{A}\{1\}$. For example, the statement $\operatorname{rank}(\mathbf{AA}^-) = \operatorname{rank}(\mathbf{A})$ is understood to mean that '$\operatorname{rank}(\mathbf{AG}) = \operatorname{rank}(\mathbf{A})$ for every $\mathbf{G} \in \mathbf{A}\{1\}$'. The phrase 'for every $\mathbf{G} \in \mathbf{A}\{1\}$' will always be implicit, unless otherwise stated, but generally will not appear. Because expressions involving the $(\cdot)^-$ notation are not always uniquely defined matrices, ambiguities can arise. For example, in investigating the possibility of a reverse order law for (1)-inverses, what should it mean to write $(\mathbf{AB})^- = \mathbf{B}^-\mathbf{A}^-$? It is better to avoid the $(\cdot)^-$ notation in situations of this type. At times it will be necessary to formulate statements more explicitly than the $(\cdot)^-$ notation allows.

Some authors have assigned special notations to different kinds of (i, j, k)-inverses. We will not do this. Since almost every application of (i, j, k)-inverses involves subsets of $\mathbf{A}\{1\}$ (the equation solvers) we will simply use the $(\cdot)^-$ together with a qualifying phrase. For example, we might say 'let $\mathbf{A}^-$ be a least squares inverse' when we wish to designate an arbitrary element of $\mathbf{A}\{1, 3\}$, or we can simply write 'let $\mathbf{A}^- \in \mathbf{A}\{1, 3\}$'.

The term 'generalized inverse' would be inappropriate if the related concepts did not coincide with the usual notion of matrix inverse in the special case that the matrix under consideration is non-singular. Note that if $\mathbf{A} \in \mathbb{C}^{n \times m}$ is non-singular, then $\mathbf{A}\{1\} = \{\mathbf{A}^{-1}\}$. Note also that $\mathbf{0} \in \mathbf{A}\{2\}$ even if $\mathbf{A}$ is non-singular.

Table 6.1 summarizes the information concerning the important types of (i, j, k)-inverses.

Let $\mathbf{A} \in \mathbb{C}^{m \times n}$

Table 6.1

Type of inverse	Terminology	Properties
(1)-inverse	An Equation Solving Inverse (Sometimes called a g-inverse)	$\mathbf{G} \in \mathbf{A}\{1\}$ if and only if $\mathbf{Gb}$ is a solution of $\mathbf{Ax} = \mathbf{b}$ for every $\mathbf{b} \in R(\mathbf{A})$
(1, 2)-inverse	A prescribed range/null space inverse. (The (N, R)-inverse) (Some authors have also called this a reflexive inverse)	If $\mathbf{G} \in \mathbf{A}\{1, 2\}$, then $N(\mathbf{A}) \dot{+} R(\mathbf{G}) = \mathbb{C}^n$ and $R(\mathbf{A}) \dot{+} N(\mathbf{G}) = \mathbb{C}^m$. That is, each (1, 2)-inverse defines complementary subspaces for $N(\mathbf{A})$ as well as $R(\mathbf{A})$. Conversely, each pair (N, R), where N and R are complements of $N(\mathbf{R})$ and $R(\mathbf{A})$ respectively, uniquely determines a (1, 2)-inverse, $\mathbf{G}_{N,R}$ with $R(\mathbf{G}_{N,R}) = N$ and $N(\mathbf{G}_{N,R}) = R$
(1, 3)-inverse	A Least Squares Inverse	$\mathbf{G} \in \mathbf{A}\{1, 3\}$ if and only if $\mathbf{Gb}$ is a least squares solution of $\mathbf{Ax} = \mathbf{b}$ for every $\mathbf{b} \in \mathbb{C}^m$
(1, 4)-inverse	A Minimum Norm Inverse	$\mathbf{G} \in \mathbf{A}\{1, 4\}$ if and only if $\mathbf{Gb}$ is the minimum norm solution of $\mathbf{Ax} = \mathbf{b}$ for every $\mathbf{b} \in R(\mathbf{A})$
(1, 2, 3, 4)-inverse	The Moore–Penrose Inverse, $\mathbf{A}^\dagger$	$\mathbf{A}\{1, 2, 3, 4\}$ contains exactly one element, $\mathbf{A}^\dagger$, which is the $(R(\mathbf{A}^*), N(\mathbf{A}^*))$-inverse for $\mathbf{A}$. $\mathbf{A}^\dagger\mathbf{b}$ is the minimal norm least squares solution of $\mathbf{Ax} = \mathbf{b}$. If $\mathbf{b} \in R(\mathbf{A})$, then $\mathbf{A}^\dagger\mathbf{b}$ is the solution of minimal norm

There are, of course, several possible (i, j, k)-inverses which are not included in Table 6.1. However, they are of lesser importance and their properties can be inferred from those listed in the table. For example, a generalized inverse which will provide least squares solutions and whose range and null space are complements of $N(\mathbf{A})$ on $R(\mathbf{A})$, respectively, must clearly be a (1, 2, 3)-inverse.

3. (1)-Inverses

As already mentioned, the important types of (i, j, k)-inverses are usually members of $\mathbf{A}\{1\}$. Therefore, we will take some time to discuss (1)-inverses.

All of the facts listed below are self-evident and are presented here for completeness.

Theorem 6.3.1 *If* $\mathbf{A} \in \mathbb{C}^{m \times n}$, *then*

(i) $rank(\mathbf{A}^-) \geq rank(\mathbf{A})$,
(ii) $rank(\mathbf{AA}^-) = rank(\mathbf{A}^-\mathbf{A}) = rank(\mathbf{A})$,
(iii) $(\mathbf{A}^-)^* \in \mathbf{A}^*\{1\}$,
(iv) *For non-singular* $\mathbf{P}$ *and* $\mathbf{Q}$, $\mathbf{Q}^{-1}\mathbf{A}^-\mathbf{P}^{-1} \in (\mathbf{PAQ})\{1\}$.
(v) *If* $\mathbf{A}$ *has full column rank, then* $\mathbf{A}^-\mathbf{A} = \mathbf{I}_n$.
(vi) *If* $\mathbf{A}$ *has full row rank, then* $\mathbf{AA}^- = \mathbf{I}_m$.
(vii) *If* $\mathbf{P}$ *has full column rank and* $\mathbf{Q}$ *has full row rank, then* $\mathbf{Q}^-\mathbf{A}^-\mathbf{P}^- \in (\mathbf{PAQ})\{1\}$,

(viii) $\left[\begin{array}{c|c} \mathbf{A}^- & \mathbf{0} \\ \hline \mathbf{0} & \mathbf{B}^- \end{array}\right]$ *is a* (1)*-inverse for* $\left[\begin{array}{c|c} \mathbf{A} & \mathbf{0} \\ \hline \mathbf{0} & \mathbf{B} \end{array}\right]$.

(ix) *If* $\mathbf{A}$ *is hermitian, then there exists a hermitian* (1)*-inverse for* $\mathbf{A}$. (*For example*, $\mathbf{A}^\dagger$)

(x) *If* $\mathbf{A}$ *is positive* (*negative*) *semi-definite, then there exists a positive* (*negative*) *semi-definite* (1)*-inverse for* $\mathbf{A}$. (*For example*, $\mathbf{A}^\dagger$)

Theorem 6.3.2 *For* $\mathbf{A}, \mathbf{X}, \mathbf{B}, \mathbf{C}$ *conformable matrices, the matrix equation* $\mathbf{AXB} = \mathbf{C}$ *has a solution if and only if* $\mathbf{AA}^-\mathbf{CB}^-\mathbf{B} = \mathbf{C}$*; in which case the set of all solutions is given by*

$$\{\mathbf{X}\} = \{\mathbf{A}^-\mathbf{CB}^- + \mathbf{H} - \mathbf{A}^-\mathbf{AHBB}^- \mid \mathbf{H} \text{ arbitrary}\}. \tag{1}$$

Proof If $\mathbf{AXB} = \mathbf{C}$ is consistent then there exists an $\mathbf{X}_o$ such that $\mathbf{C} = \mathbf{AX}_o\mathbf{B}$. Thus $\mathbf{AA}^-\mathbf{CB}^-\mathbf{B} = \mathbf{AA}^-\mathbf{AX}_o\mathbf{BB}^-\mathbf{B} = \mathbf{AX}_o\mathbf{B} = \mathbf{C}$. If $\mathbf{C} = \mathbf{AA}^-\mathbf{CB}^-\mathbf{B}$, then $\mathbf{X} = \mathbf{A}^-\mathbf{CB}^-$ is a particular solution. To prove (1), note first that for every $\mathbf{H}, \mathbf{A}^-\mathbf{CB}^- + \mathbf{H} - \mathbf{A}^-\mathbf{AHBB}^-$ is a solution of $\mathbf{AXB} = \mathbf{C}$. Given a particular solution $\mathbf{X}_o$, there exists an $\mathbf{H}_o$ such that $\mathbf{X}_o = \mathbf{A}^-\mathbf{CB}^- + \mathbf{H}_o - \mathbf{A}^-\mathbf{AH}_o\mathbf{BB}^-$ (by the consistency condition), take $\mathbf{H}_o = \mathbf{X}_o$. ■

Corollary 6.3.1 *For* $\mathbf{A} \in \mathbb{C}^{m \times n}$, $\mathbf{b} \in R(\mathbf{A})$, *the set of all solutions of* $\mathbf{Ax} = \mathbf{b}$ *can be written as*

$$\{\mathbf{x}\} = \{\mathbf{A}^-\mathbf{b} + (\mathbf{I} - \mathbf{A}^-\mathbf{A})\mathbf{h} \mid \mathbf{h} \in \mathbb{C}^n\}.$$

Likewise, for $\mathbf{X} \in \mathbb{C}^{n \times k}$, $\mathbf{c} \in R(\mathbf{X}^*)$, *the solution set of* $\mathbf{c}^* = \mathbf{l}^*\mathbf{X}$ *is*

$$\mathbf{l}^* = \{\mathbf{c}^*\mathbf{X}^- + \mathbf{h}^*(\mathbf{I} - \mathbf{XX}^-) \mid \mathbf{h} \in \mathbb{C}^n\}.$$

Corollary 6.3.2 *For* $\mathbf{A} \in \mathbb{C}^{m \times n}$, $\mathbf{X} \in \mathbb{C}^{n \times p}$, $\mathbf{Y} \in \mathbb{C}^{q \times m}$, *the set of solutions to* $\mathbf{AX} = \mathbf{0}$ *is given by*

$$\{\mathbf{X}\} = \{(\mathbf{I} - \mathbf{A}^-\mathbf{A})\mathbf{H} \mid \mathbf{H} \in \mathbb{C}^{n \times p}\}$$

and the set of solutions to $\mathbf{YA} = \mathbf{0}$ *is given by*

$$\{\mathbf{Y}\} = \{\mathbf{K}(\mathbf{I} - \mathbf{AA}^-) \mid \mathbf{K} \in \mathbb{C}^{q \times m}\}.$$

Theorem 6.3.3 *If* $\mathbf{A} \in \mathbb{C}^{m \times n}$, *then* $\mathbf{A}\{1\}$ *can be characterized by either of the following:*

$$\mathbf{A}\{1\} = \{\mathbf{A}^- + \mathbf{H} - \mathbf{A}^-\mathbf{AHAA}^- \mid \mathbf{H} \in \mathbb{C}^{n \times m}\} \tag{2}$$

$$\mathbf{A}\{1\} = \{\mathbf{A}^- + \mathbf{H}(\mathbf{I} - \mathbf{AA}^-) + (\mathbf{I} - \mathbf{A}^-\mathbf{A})\,\mathbf{K} \mid \mathbf{H}, \mathbf{K} \in \mathbb{C}^{n \times m}\}. \tag{3}$$

Proof To prove (2), note that $\mathbf{AXA} = \mathbf{A}$ is always consistent since $\mathbf{AA}^\dagger\mathbf{A} = \mathbf{A}$. The result in (2) now follows from Theorem 2. To prove (3), note that $\mathbf{A}^- + \mathbf{H}(\mathbf{I} - \mathbf{AA}^-) + (\mathbf{I} - \mathbf{A}^-\mathbf{A})\mathbf{K} \in \mathbf{A}\{1\}$ for all $\mathbf{H}, \mathbf{K} \in \mathbb{C}^{n \times m}$ and that for any particular element $\mathbf{A}_o^- \in \mathbf{A}\{1\}$, there exist an $\mathbf{H}_o$ and $\mathbf{K}_o$ such that $\mathbf{A}_o^- = \mathbf{A}^- + \mathbf{H}_o(\mathbf{I} - \mathbf{A}\mathbf{A}^-) + (\mathbf{I} - \mathbf{A}^-\mathbf{A})\mathbf{K}_o$, namely $\mathbf{H}_o = \mathbf{A}_o^- - \mathbf{A}^-$ and $\mathbf{K}_o = \mathbf{A}_o^-\mathbf{A}\mathbf{A}^-$. ■

Since (1)-inverses provide solutions to consistent systems, it seems only natural to investigate the possibility of using (1)-inverses to provide a common solution, if one exists, to two systems. Let $\mathbf{A} \in \mathbb{C}^{m \times n}$, $\mathbf{B} \in \mathbb{C}^{p \times n}$, $\mathbf{a} \in \mathbb{C}^m$, $\mathbf{b} \in \mathbb{C}^p$ and consider the two systems

$$\mathbf{A}\mathbf{x} = \mathbf{a} \text{ and } \mathbf{B}\mathbf{x} = \mathbf{b}. \tag{4}$$

The problem is to find an $\mathbf{x} \in \mathbb{C}^n$ which satisfies both. This problem is clearly equivalent to solving the partitioned system

$$\begin{bmatrix} \mathbf{A} \\ \hdashline \mathbf{B} \end{bmatrix} \mathbf{x} = \begin{bmatrix} \mathbf{a} \\ \hdashline \mathbf{b} \end{bmatrix}. \tag{5}$$

If (1)-inverses could be obtained for the partitioned matrix $\mathbf{M} = \begin{bmatrix} \mathbf{A} \\ \hdashline \mathbf{B} \end{bmatrix}$, then the system (4), could be completely analysed. The following result provides the solution to this problem.

Theorem 6.3.4 *For* $\mathbf{A} \in \mathbb{C}^{m \times n}$, $\mathbf{B} \in \mathbb{C}^{p \times n}$, *a* (1)-*inverse for the partitioned matrix* $\mathbf{M} = \begin{bmatrix} \mathbf{A} \\ \hdashline \mathbf{B} \end{bmatrix}$ *is given by*

$$\mathbf{G}_\mathrm{M} = [(\mathbf{I} - (\mathbf{I} - \mathbf{A}^-\mathbf{A})(\mathbf{B}(\mathbf{I} - \mathbf{A}^-\mathbf{A}))^-\mathbf{B})\mathbf{A}^- \mid (\mathbf{I} - \mathbf{A}^-\mathbf{A})(\mathbf{B}(\mathbf{I} - \mathbf{A}^-\mathbf{A}))^-]. \tag{6}$$

Furthermore, for $\mathbf{C} \in \mathbb{C}^{m \times r}$, *a* (1)-*inverse for the partitioned matrix* $\mathbf{N} = [\mathbf{A} \mid \mathbf{C}]$ *is given by*

$$\mathbf{G}_\mathrm{N} = \begin{bmatrix} \mathbf{A}^-(\mathbf{I} - \mathbf{C}((\mathbf{I} - \mathbf{A}\mathbf{A}^-)\mathbf{C})^-(\mathbf{I} - \mathbf{A}\mathbf{A}^-)) \\ \hdashline ((\mathbf{I} - \mathbf{A}\mathbf{A}^-)\mathbf{C})^-(\mathbf{I} - \mathbf{A}\mathbf{A}^-) \end{bmatrix}. \tag{7}$$

If $(\cdot)^-$ is taken to mean (1, 2)-inverse, then $\mathbf{G}_\mathrm{M} \in \mathbf{M}\{1,2\}$ and $\mathbf{G}_\mathrm{N} \in \mathbf{N}\{1,2\}$. The representation for the entire class $\mathbf{M}\{1\}$ as well as $\mathbf{N}\{1\}$ can be given in terms of $\mathbf{G}_\mathrm{M}$ and $\mathbf{G}_\mathrm{N}$. We have chosen not to present these representations.

Theorem 4 is proven by simply verifying the defining equations are satisfied.

We can now say something about (4).

Theorem 6.3.5 *Let* $\mathbf{A} \in \mathbb{C}^{m \times n}$, $\mathbf{B} \in \mathbb{C}^{p \times n}$, $\mathbf{a} \in R(\mathbf{A})$, $\mathbf{b} \in R(\mathbf{B})$, *and let* $\mathbf{x}_a$ *and* $\mathbf{x}_b$ *be any two particular solutions for the systems*

$$\mathbf{A}\mathbf{x} = \mathbf{a} \text{ and } \mathbf{B}\mathbf{x} = \mathbf{b} \tag{8}$$

respectively. Let $\mathbf{F} = \mathbf{B}(\mathbf{I} - \mathbf{A}^-\mathbf{A})$. *The following statements are equivalent.*

The two systems in (8) *possess a common solution.* (9)

$$\mathbf{B}\mathbf{x}_a - \mathbf{b} \in R(\mathbf{F}) = R(\mathbf{B}/N(\mathbf{A})) \tag{10}$$

$$\mathbf{x}_a - \mathbf{x}_b \in N(\mathbf{A}) + N(\mathbf{B}). \tag{11}$$

Furthermore, when a common solution exists, a particular common solution is given by

$$\mathbf{x}_c = (\mathbf{I} - (\mathbf{I} - \mathbf{A}^-\mathbf{A})\mathbf{F}^-\mathbf{B})\mathbf{x}_a + (\mathbf{I} - \mathbf{A}^-\mathbf{A})\mathbf{F}^-\mathbf{b} \tag{12}$$

and the set of all common solutions can be written as

$$\{\mathbf{x}_c + (\mathbf{I} - \mathbf{A}^-\mathbf{A})(\mathbf{I} - \mathbf{F}^-\mathbf{F})\mathbf{h} \,|\, \mathbf{h} \in \mathbb{C}^n\}. \tag{13}$$

Proof The chain of implication to be proven is (11) ⇒(10) ⇒(9) ⇒(11).
(11) ⇒(10) : Suppose (11) holds. Then $\mathbf{x}_a - \mathbf{x}_b = \mathbf{n}_a + \mathbf{n}_b$, $\mathbf{n}_a \in N(\mathbf{A})$, $\mathbf{n}_b \in N(\mathbf{B})$ so that $\mathbf{B}\mathbf{x}_a - \mathbf{b} = \mathbf{B}(\mathbf{x}_a - \mathbf{x}_b) = \mathbf{B}\mathbf{n}_a \in R(\mathbf{F})$.

(10) ⇒(9): If (10) holds, then the vector $\mathbf{x}_c$ of (12) is a common solution since $\mathbf{A}\mathbf{x}_c = \mathbf{A}\mathbf{x}_a = \mathbf{a}$, and

$$\mathbf{B}\mathbf{x}_c = \mathbf{B}\mathbf{x}_a - \mathbf{F}\mathbf{F}^-\mathbf{B}\mathbf{x}_a + \mathbf{F}\mathbf{F}^-\mathbf{b}. \tag{14}$$

Now (10) yields that $\mathbf{F}\mathbf{F}^-(\mathbf{B}\mathbf{x}_a - \mathbf{b}) = \mathbf{B}\mathbf{x}_a - \mathbf{b}$, or $\mathbf{B}\mathbf{x}_a - \mathbf{F}\mathbf{F}^-\mathbf{B}\mathbf{x}_a = \mathbf{b} - \mathbf{F}\mathbf{F}^-\mathbf{b}$. Therefore, (14) becomes $\mathbf{B}\mathbf{x}_c = \mathbf{b}$.

(9) ⇒(11): If there exists a common solution then the two solution sets must intersect. That is, $\{\mathbf{x}_a + N(\mathbf{A})\} \cap \{\mathbf{x}_b + N(\mathbf{B})\} \neq \emptyset$. Thus there exist vectors $\mathbf{n}_a \in N(\mathbf{A})$ and $\mathbf{n}_b \in N(\mathbf{B})$ such that $\mathbf{x}_a + \mathbf{n}_a = \mathbf{x}_b + \mathbf{n}_b$, and (11) follows. To obtain the set of all solutions, use the fact that they can be written as

$$\left\{\mathbf{x}_c + N\left(\begin{bmatrix}\mathbf{A}\\ \mathbf{B}\end{bmatrix}\right)\right\} = \left\{\mathbf{x}_c + \left(\mathbf{I} - \begin{bmatrix}\mathbf{A}\\ \mathbf{B}\end{bmatrix}^-\begin{bmatrix}\mathbf{A}\\ \mathbf{B}\end{bmatrix}\right)\mathbf{h} \,\middle|\, \mathbf{h} \in \mathbb{C}^n\right\}$$
$$= \left\{\mathbf{x}_c + \left(\mathbf{I} - \mathbf{G}_{\mathrm{M}}\begin{bmatrix}\mathbf{A}\\ \mathbf{B}\end{bmatrix}\right)\mathbf{h}\right\}$$

where $\mathbf{G}_{\mathrm{M}}$ is given in (6). Now,

$\mathbf{I} - \mathbf{G}_{\mathrm{M}}\begin{bmatrix}\mathbf{A}\\ \mathbf{B}\end{bmatrix} = \mathbf{I} - (\mathbf{A}^-\mathbf{A} - (\mathbf{I} - \mathbf{A}^-\mathbf{A})\mathbf{F}^-\mathbf{B}\mathbf{A}^-\mathbf{A}) - (\mathbf{I} - \mathbf{A}^-\mathbf{A})\mathbf{F}^-\mathbf{B} =$
$(\mathbf{I} - \mathbf{A}^-\mathbf{A})(\mathbf{I} - \mathbf{F}^-\mathbf{B} + \mathbf{F}^-\mathbf{B}\mathbf{A}^-\mathbf{A}) = (\mathbf{I} - \mathbf{A}^-\mathbf{A})(\mathbf{I} - \mathbf{F}^-\mathbf{B}(\mathbf{I} - \mathbf{A}^-\mathbf{A})) =$
$(\mathbf{I} - \mathbf{A}^-\mathbf{A})(\mathbf{I} - \mathbf{F}^-\mathbf{F})$, which gives (13). ■

We shall now present some results on finer partitions than those discussed in Theorem 4. Representations for (1)- and (1, 2)-inverses of matrices partitioned as

$$\mathbf{M} = \left[\begin{array}{c|c}\mathbf{A} & \mathbf{C}\\ \hline \mathbf{R} & \mathbf{D}\end{array}\right]$$

where **A**, **C**, **R**, and **D** are any conformable matrices will be given. First, we need a technical lemma and some notation.

Notation. For $\mathbf{T} \in \mathbb{C}^{m \times n}$ and $\mathbf{T}^- \in \mathbf{T}\{1\}$, $\mathbf{E}_{\mathrm{T}}$ and $\mathbf{F}_{\mathrm{T}}$ will denote the projectors $\mathbf{E}_{\mathrm{T}} = \mathbf{I} - \mathbf{T}\mathbf{T}^-$ and $\mathbf{F}_{\mathrm{T}} = \mathbf{I} - \mathbf{T}^-\mathbf{T}$.

Lemma 6.3.1 Let $\mathbf{N} \in \mathbb{C}^{m \times n}$ *such that* $\mathbf{N} = \begin{bmatrix}\mathbf{X} & \mathbf{Y}\\ \mathbf{W} & \mathbf{Z}\end{bmatrix}$, *and let* $\mathbf{X}^-, \mathbf{Y}^-$, *and* $\mathbf{W}^-$ *be* (1)- *or* (1, 2)-*inverses for* **X**, **Y** *and* **W**, *respectively, which satisfy* $\mathbf{X}^-\mathbf{Y}=\mathbf{0}$, $\mathbf{Y}^-\mathbf{X}=\mathbf{0}$, *and* $\mathbf{W}\mathbf{X}^-=\mathbf{0}$, $\mathbf{X}\mathbf{W}^-=\mathbf{0}$. *Then, depending on how* $(\cdot)^-$ *is interpreted, a* (1)- *or* (1, 2)-*inverse for* **N** *is given by*

$$\mathbf{N}^- = \left[\begin{array}{c|c} \mathbf{X}^- - \mathbf{W}^-\mathbf{Z}\mathbf{Y}^- & \mathbf{W}^- \\ \hline \mathbf{Y}^- & \mathbf{0} \end{array}\right] + \begin{bmatrix} \mathbf{W}^-\mathbf{Z} \\ -\mathbf{I} \end{bmatrix} \mathbf{Q}[\mathbf{Z}\mathbf{Y}^- \mid -\mathbf{I}]. \tag{15}$$

where $\mathbf{Q}=\mathbf{F}_{\mathbf{Y}}(\mathbf{E}_{\mathbf{W}}\mathbf{Z}\mathbf{F}_{\mathbf{Y}})^-\mathbf{E}_{\mathbf{W}}$.

Proof The proof amounts to showing that the defining relations are satisfied. Let $\mathbf{L}_1$ and $\mathbf{L}_2$ denote the first and second term of the sum on the right-hand side of (15), so that

$$\mathbf{N}\mathbf{L}_1\mathbf{N} = \left[\begin{array}{c|c} \mathbf{X} & \mathbf{Y} \\ \hline \mathbf{W} & \mathbf{E}_{\mathbf{W}}\mathbf{Z}\mathbf{Y}^-\mathbf{Y} + \mathbf{W}\mathbf{W}^-\mathbf{Z} \end{array}\right], \quad \mathbf{N}\mathbf{L}_2\mathbf{N} = \left[\begin{array}{c|c} \mathbf{0} & \mathbf{0} \\ \hline \mathbf{0} & \mathbf{E}_{\mathbf{W}}\mathbf{Z}\mathbf{Q}\mathbf{Z}\mathbf{F}_{\mathbf{Y}} \end{array}\right].$$

Now, $\mathbf{E}_{\mathbf{W}}\mathbf{Z}\mathbf{Q}\mathbf{Z}\mathbf{F}_{\mathbf{Y}} = \mathbf{E}_{\mathbf{W}}\mathbf{Z}\mathbf{F}_{\mathbf{Y}}$ and $\mathbf{W}\mathbf{W}^-\mathbf{Z} = \mathbf{Z} - \mathbf{E}_{\mathbf{W}}\mathbf{Z}$, and hence, $\mathbf{E}_{\mathbf{W}}\mathbf{Z}\mathbf{Y}^-\mathbf{Y} + \mathbf{W}\mathbf{W}^-\mathbf{Z} + \mathbf{E}_{\mathbf{W}}\mathbf{Z}\mathbf{Q}\mathbf{Z}\mathbf{F}_{\mathbf{Y}} = \mathbf{Z}$ so that $\mathbf{N}\mathbf{N}^-\mathbf{N}=\mathbf{N}$. If $(\cdot)^-$ is interpreted as meaning a (1, 2)-inverse, then a direct calculation shows that $\mathbf{L}_1\mathbf{N}\mathbf{L}_1 = \mathbf{L}_1$, $\mathbf{L}_1\mathbf{N}\mathbf{L}_2 = \mathbf{0}$, and $\mathbf{L}_2\mathbf{N}\mathbf{L}_1 = \mathbf{0}$. By using the fact that **Z** is a (1)-inverse for **Q**, it is also easy to verify that $\mathbf{L}_2\mathbf{N}\mathbf{L}_2 = \mathbf{L}_2$ and hence $\mathbf{N}^-\mathbf{N}\mathbf{N}^- = \mathbf{N}^-$. ■

In passing, we remark that there are three other forms of Lemma 1 which are possible by considering the following three sets of hypotheses:

$$\mathbf{Y}^-\mathbf{X}=\mathbf{0}, \quad \mathbf{X}^-\mathbf{Y}=\mathbf{0}, \quad \mathbf{Z}\mathbf{Y}^-=\mathbf{0} \text{ and } \mathbf{Y}\mathbf{Z}^-=\mathbf{0}. \tag{16}$$

$$\mathbf{W}^-\mathbf{Z}=\mathbf{0}, \quad \mathbf{Z}^-\mathbf{W}=\mathbf{0}, \quad \mathbf{X}\mathbf{W}^-=\mathbf{0} \text{ and } \mathbf{W}\mathbf{X}^-=\mathbf{0}. \tag{17}$$

$$\mathbf{Z}^-\mathbf{W}=\mathbf{0}, \quad \mathbf{W}^-\mathbf{Z}=\mathbf{0}, \quad \mathbf{Y}\mathbf{Z}^-=\mathbf{0} \text{ and } \mathbf{Z}\mathbf{Y}^-=\mathbf{0}. \tag{18}$$

By performing a permutation of rows, or columns, or both, and then applying Theorem 2.1, a representation which resembles (15) can be obtained for each of the previous cases.

With the aid of Lemma 1, it is now possible to develop a representation for a (1)- or (1, 2)-inverse of a completely general partitioned matrix.

Theorem 6.3.6 *Let* $\mathbf{M}\in\mathbb{C}^{m\times n}$ *denote the matrix*

$$\mathbf{M} = \begin{bmatrix} \mathbf{A} & \mathbf{C} \\ \mathbf{R} & \mathbf{D} \end{bmatrix}.$$

and let $\mathbf{Z}=\mathbf{D}-\mathbf{R}\mathbf{A}^-\mathbf{C}$, $\mathbf{Y}=\mathbf{E}_{\mathbf{A}}\mathbf{C}$, $\mathbf{W}=\mathbf{R}\mathbf{F}_{\mathbf{A}}$ *and* $\mathbf{Q}=\mathbf{F}_{\mathbf{Y}}(\mathbf{E}_{\mathbf{W}}\mathbf{Z}\mathbf{F}_{\mathbf{Y}})^-\mathbf{E}_{\mathbf{W}}$. *A* (1)- *or* (1, 2)-*inverse for* **M**, *depending on how* $(\cdot)^-$ *is interpreted, is given by*

$$\mathbf{M}^- = \left[\begin{array}{c|c} \mathbf{A}^- - \mathbf{A}^-\mathbf{C}\mathbf{Y}^-\mathbf{E}_{\mathbf{A}} - \mathbf{F}_{\mathbf{A}}\mathbf{W}^-\mathbf{R}\mathbf{A}^- - \mathbf{F}_{\mathbf{A}}\mathbf{W}^-\mathbf{Z}\mathbf{Y}^-\mathbf{E}_{\mathbf{A}} & \mathbf{F}_{\mathbf{A}}\mathbf{W}^- \\ \hline \mathbf{Y}^-\mathbf{E}_{\mathbf{A}} & \mathbf{0} \end{array}\right]$$
$$+ \begin{bmatrix} \mathbf{F}_{\mathbf{A}}\mathbf{W}^-\mathbf{Z} + \mathbf{A}^-\mathbf{C} \\ -\mathbf{I} \end{bmatrix} \mathbf{Q}[\mathbf{Z}\mathbf{Y}^-\mathbf{E}_{\mathbf{A}} + \mathbf{R}\mathbf{A}^- \mid -\mathbf{I}].$$

Proof For the moment, let $(\cdot)^-$ denote a (1, 2)-inverse and let **P**, **S** and **N** denote the matrices

$$\mathbf{P}=\begin{bmatrix}\mathbf{I} & \mathbf{0}\\ \mathbf{RA}^- & \mathbf{I}\end{bmatrix},\ \mathbf{S}=\begin{bmatrix}\mathbf{I} & \mathbf{A}^-\mathbf{C}\\ \mathbf{0} & \mathbf{I}\end{bmatrix}\ \text{and}\ \mathbf{N}=\begin{bmatrix}\mathbf{A} & \mathbf{Y}\\ \mathbf{W} & \mathbf{Z}\end{bmatrix}.$$

Then, $\mathbf{M}=\mathbf{PNS}$ and a (1,2)-inverse for **M** is given by

$$\mathbf{M}^-=\mathbf{S}^{-1}\mathbf{N}^-\mathbf{P}^{-1}. \tag{20}$$

For every (1,2)-inverse $\mathbf{Y}^-$ and $\mathbf{W}^-$, the matrices $\mathbf{G}=\mathbf{Y}^-\mathbf{E}_\mathbf{A}$ and $\mathbf{H}=\mathbf{F}_\mathbf{A}\mathbf{W}^-$ are (1,2)-inverses for **Y** and **W**, respectively, such that $\mathbf{GA}=\mathbf{0}$ and $\mathbf{AH}=\mathbf{0}$. Since it is also true that each (1,2)-inverse $\mathbf{A}^-$ satisfies $\mathbf{A}^-\mathbf{Y}=\mathbf{0}$ and $\mathbf{WA}^-=\mathbf{0}$, Lemma 1 is used to obtain a (1,2)-inverse for **N** as

$$\mathbf{N}^-=\left[\begin{array}{c|c}\mathbf{A}^- - \mathbf{F}_\mathbf{A}\mathbf{W}^-\mathbf{ZY}^-\mathbf{E}_\mathbf{A} & \mathbf{F}_\mathbf{A}\mathbf{W}^-\\ \hline \mathbf{Y}^-\mathbf{E}_\mathbf{A} & \mathbf{0}\end{array}\right]+\begin{bmatrix}\mathbf{F}_\mathbf{A}\mathbf{W}^-\mathbf{Z}\\ -\mathbf{I}\end{bmatrix}\mathbf{Q}[\mathbf{ZY}^-\mathbf{E}_\mathbf{A} \mid -\mathbf{I}]. \tag{21}$$

Using (21) with (20) yields (19), the desired result for (1, 2)-inverses. If $(\cdot)^-$ is interpreted as meaning only a (1)-inverse, it is a matter of direct computation to verify that the matrix (19) is still a (1)-inverse. ■

Observe that **M** may also be factored in three other ways as

$$\begin{aligned}\mathbf{M}&=\begin{bmatrix}\mathbf{I} & \mathbf{0}\\ \mathbf{DC}^- & \mathbf{I}\end{bmatrix}\begin{bmatrix}\mathbf{C} & \mathbf{E}_\mathbf{C}\mathbf{A}\\ \mathbf{DF}_\mathbf{C} & \mathbf{R}-\mathbf{DC}^-\mathbf{A}\end{bmatrix}\begin{bmatrix}\mathbf{C}^-\mathbf{A} & \mathbf{I}\\ \mathbf{I} & \mathbf{0}\end{bmatrix}\\ &=\begin{bmatrix}\mathbf{AR}^- & \mathbf{I}\\ \mathbf{I} & \mathbf{0}\end{bmatrix}\begin{bmatrix}\mathbf{R} & \mathbf{E}_\mathbf{R}\mathbf{D}\\ \mathbf{AF}_\mathbf{R} & \mathbf{C}-\mathbf{AR}^-\mathbf{D}\end{bmatrix}\begin{bmatrix}\mathbf{I} & \mathbf{R}^-\mathbf{D}\\ \mathbf{0} & \mathbf{I}\end{bmatrix}\\ &=\begin{bmatrix}\mathbf{CD}^- & \mathbf{I}\\ \mathbf{I} & \mathbf{0}\end{bmatrix}\begin{bmatrix}\mathbf{D} & \mathbf{E}_\mathbf{D}\mathbf{R}\\ \mathbf{CF}_\mathbf{D} & \mathbf{A}-\mathbf{CD}^-\mathbf{R}\end{bmatrix}\begin{bmatrix}\mathbf{D}^-\mathbf{R} & \mathbf{I}\\ \mathbf{I} & \mathbf{0}\end{bmatrix}.\end{aligned}$$

Coupling each of these with the appropriate form of Lemma 1 which is obtained from (16), (17) or (18), one can use the same method as in the proof of Theorem 6 to derive three other representations for $\mathbf{M}^-$ which resemble that obtained in Theorem 6.

Theorem 6 has several useful consequences.

Corollary 6.3.3 *If* $\mathbf{M}=\left[\begin{array}{c|c}\mathbf{A} & \mathbf{C}\\ \hline \mathbf{R} & \mathbf{D}\end{array}\right]$ *where* $R(\mathbf{C})\subseteq R(\mathbf{A})$ *and* $RS(\mathbf{R})\subseteq RS(\mathbf{A})$, *(*$RS(\cdot)$ *denotes the row space), then a particular* (1)- *or* (1, 2)-*inverse for* **M** *is given by*

$$\mathbf{M}^-=\left[\begin{array}{c|c}\mathbf{A}^-+\mathbf{A}^-\mathbf{C}(\mathbf{D}-\mathbf{RA}^-\mathbf{C})^-\mathbf{RA}^- & -\mathbf{A}^-\mathbf{C}(\mathbf{D}-\mathbf{RA}^-\mathbf{C})^-\\ \hline -(\mathbf{D}-\mathbf{RA}^-\mathbf{C})^-\mathbf{RA}^- & (\mathbf{D}-\mathbf{RA}^-\mathbf{C})^-\end{array}\right]$$

This gives the familiar form which occurs when the matrices **M** and **A** are non-singular and $(\cdot)^-$ is taken as $(\cdot)^{-1}$.

By using Lemma 3, we have the following.

Corollary 6.3.4 *If* $\mathbf{M} = \left[\begin{array}{c|c} \mathbf{A} & \mathbf{C} \\ \hline \mathbf{R} & \mathbf{D} \end{array}\right]$ *where* $\mathbf{A} \in \mathbb{C}^{r \times r}$ *and rank*$(\mathbf{M}) =$ *rank*$(\mathbf{A}) = r$, *then a particular* (1)- *or* (1, 2)-*inverse for* $\mathbf{M}$ *is given by*

$$\mathbf{M}^- = \left[\begin{array}{c|c} \mathbf{A}^- & \mathbf{0} \\ \hline \mathbf{0} & \mathbf{0} \end{array}\right].$$

In many applications, the matrices involved are either positive or negative semi-definite. In these cases, (1)-inverses of partitioned matrices are easy to find. Below we give the result for positive semi-definite matrices. The results for negative semi-definite matrices are left as exercises.

Corollary 6.3.5 *If* $\mathbf{S} = \left[\begin{array}{c|c} \mathbf{A} & \mathbf{C} \\ \hline \mathbf{C}^* & \mathbf{D} \end{array}\right]$ *is positive semi-definite, then a particular* (1)-*inverse for* $\mathbf{S}$ *is given by*

$$\mathbf{S}^- = \left[\begin{array}{c|c} \mathbf{A}^- + \mathbf{A}^-\mathbf{C}(\mathbf{D} - \mathbf{C}^*\mathbf{A}^-\mathbf{C})^-\mathbf{C}^*\mathbf{A}^- & -\mathbf{A}^-\mathbf{C}(\mathbf{D} - \mathbf{C}^*\mathbf{A}^-\mathbf{C})^- \\ \hline -(\mathbf{D} - \mathbf{C}^*\mathbf{A}^-\mathbf{C})^-\mathbf{C}^*\mathbf{A}^- & (\mathbf{D} - \mathbf{C}^*\mathbf{A}^-\mathbf{C})^- \end{array}\right]$$

Proof Since $\mathbf{S}$ is positive semi-definite, $\mathbf{S}$ can be written as

$$\mathbf{S} = [\mathbf{S}_1^* \,|\, \mathbf{S}_1^*]\left[\begin{array}{c} \mathbf{S}_1 \\ \hline \mathbf{S}_2 \end{array}\right] = \left[\begin{array}{c|c} \mathbf{S}_1^*\mathbf{S}_1 & \mathbf{S}_1^*\mathbf{S}_2 \\ \hline \mathbf{S}_2^*\mathbf{S}_1 & \mathbf{S}_2^*\mathbf{S}_2 \end{array}\right] = \left[\begin{array}{c|c} \mathbf{A} & \mathbf{C} \\ \hline \mathbf{C}^* & \mathbf{D} \end{array}\right].$$

Clearly, $R(\mathbf{C}) = R(\mathbf{S}_1^*\mathbf{S}_2) \subseteq R(\mathbf{S}_1^*\mathbf{S}_1) = R(\mathbf{A})$ and $RS(\mathbf{C}^*) = RS(\mathbf{S}_2^*\mathbf{S}_1) \subseteq RS(\mathbf{S}_1^*\mathbf{S}_1) = RS(\mathbf{A})$. The result now follows from Corollary 3. ■

Partitioned matrices of the form

$$\mathbf{B} = \left[\begin{array}{c|c} \mathbf{A}^*\mathbf{A} & \mathbf{X} \\ \hline \mathbf{X}^* & \mathbf{0} \end{array}\right]$$

occurred in our treatment of the constrained least squares problem of Section 6 in Chapter 3. They also occur in statistical applications. As we shall see in the next section, (1)-inverses of $\mathbf{B}$ can also be used to present a unified treatment of the subject of linear models. Theorem 6 provides a (1)-inverse of $\mathbf{B}$ which will be used in the next section.

Corollary 6.3.6 *Let* $\mathbf{V} \in \mathbb{C}^{n \times n}$ *be positive semi-definite,* $\mathbf{X} \in \mathbb{C}^{n \times r}$, $\mathbf{B} = \left[\begin{array}{c|c} \mathbf{V} & \mathbf{X} \\ \hline \mathbf{X}^* & \mathbf{0} \end{array}\right]$. *A particular* (1)-*inverse for* $\mathbf{B}$ *is given by*

$$\mathbf{B}^- = \left[\begin{array}{c|c} \mathbf{0} & \mathbf{X}^{-*} \\ \hline \mathbf{X}^- & -\mathbf{X}^-\mathbf{V}\mathbf{X}^{-*} \end{array}\right] + \left[\begin{array}{c} \mathbf{I} \\ \hline -\mathbf{X}^-\mathbf{V} \end{array}\right]\mathbf{Q}[\mathbf{I} \,|\, -\mathbf{V}\mathbf{X}^{-*}]$$

where $\mathbf{Q} = \mathbf{F}_{\mathbf{X}^*}(\mathbf{E}_{\mathbf{X}}\mathbf{V}\mathbf{F}_{\mathbf{X}^*})^-\mathbf{E}_{\mathbf{X}}$.

Theorem 6 can also be used to represent the rank of a large partitioned matrix in terms of ranks of matrices of lower order.

Theorem 6.3.7 *Let* $\mathbf{M} \in \mathbb{C}^{m \times n}$ *denote the matrix* $\mathbf{M} = \begin{bmatrix} \mathbf{A} & \mathbf{C} \\ \mathbf{R} & \mathbf{D} \end{bmatrix}$. *The rank of* $\mathbf{M}$ *is given by* $rank(\mathbf{M}) = rank(\mathbf{A}) + rank(\mathbf{Y}) + rank(\mathbf{W}) + rank(\mathbf{U})$, *where* $\mathbf{Y} = \mathbf{E_A C}$, $\mathbf{W} = \mathbf{RF_A}$, *and* $\mathbf{U} = \mathbf{E_W}(\mathbf{D} - \mathbf{RA^- C})\mathbf{F_Y} = \mathbf{E_W Z F_Y}$. *Moreover, the expressions* $rank(\mathbf{Y})$, $rank(\mathbf{W})$, *and* $rank(\mathbf{U})$ *do not depend upon which* (1)-*inverses are used.*

Proof For every (1)-inverse $\mathbf{M}^-$ of $\mathbf{M}$, the product $\mathbf{MM}^-$ is idempotent. Using Theorem 6 we compute $\mathbf{MM}^-$ as

$$\mathbf{MM}^- = \left[\begin{array}{c|c} \mathbf{AA}^- + \mathbf{YY}^-\mathbf{E_A} & \mathbf{0} \\ \hline ********* & \mathbf{WW}^- + \mathbf{UU}^-\mathbf{E_W} \end{array}\right],$$

so that (Tr = trace)

$$\begin{aligned} \operatorname{rank}(\mathbf{M}) &= \operatorname{rank}(\mathbf{MM}^-) = \operatorname{Tr}(\mathbf{MM}^-) \\ &= \operatorname{Tr}(\mathbf{AA}^-) + \operatorname{Tr}(\mathbf{YY}^-\mathbf{E_A}) + \operatorname{Tr}(\mathbf{WW}^-) + \operatorname{Tr}(\mathbf{UU}^-\mathbf{E_W}). \end{aligned} \tag{22}$$

Because $\mathbf{E_A Y} = \mathbf{Y}$ and $\mathbf{E_W U} = \mathbf{U}$, and since $\mathbf{AA}^-$, $\mathbf{YY}^-$, $\mathbf{WW}^-$, and $\mathbf{UU}^-$ are idempotent, (22) becomes $\operatorname{rank}(\mathbf{M}) = \operatorname{rank}(\mathbf{A}) + \operatorname{rank}(\mathbf{Y}) = \operatorname{rank}(\mathbf{W}) + \operatorname{rank}(\mathbf{U})$. To show that rank($\mathbf{Y}$), rank($\mathbf{W}$), and rank($\mathbf{U}$) are independent of the (1)-inverses used, let $\mathbf{G}_1$ and $\mathbf{G}_2$ be two (1)-inverses for $\mathbf{A}$ and let $\mathbf{E}_1 = \mathbf{I} - \mathbf{AG}_1$, $\mathbf{E}_2 = \mathbf{I} - \mathbf{AG}_2$, $\mathbf{Y}_1 = \mathbf{E}_1\mathbf{C}$ and $\mathbf{Y}_2 = \mathbf{E}_2\mathbf{C}$. Because $\mathbf{E}_1\mathbf{E}_2 = \mathbf{E}_1$ and $\mathbf{E}_2\mathbf{E}_1 = \mathbf{E}_2$, it follows that $\mathbf{Y}_1 = \mathbf{E}_1\mathbf{Y}_2$ and $\mathbf{Y}_2 = \mathbf{E}_2\mathbf{Y}_1$, so that $\operatorname{rank}(\mathbf{Y}_1) = \operatorname{rank}(\mathbf{Y}_2)$. Similar remarks may be made about $\mathbf{W}$. From the first part of the theorem, $\operatorname{rank}(\mathbf{U}) = \operatorname{rank}(\mathbf{M}) - \operatorname{rank}(\mathbf{A}) - \operatorname{rank}(\mathbf{Y}) - \operatorname{rank}(\mathbf{W})$, so that rank($\mathbf{U}$) is also constant with respect to the (1)-inverses used. ■

By performing row permutations, column permutations, or both, one can obtain three more forms of Theorem 7.

Corollary 6.3.7 *For the matrix* $\mathbf{M}$ *of Theorem* 6.3.7,

$$rank(\mathbf{M}) \leq rank(\mathbf{A}) + rank(\mathbf{C}) + rank(\mathbf{R}) + rank(\mathbf{Z})$$

where $\mathbf{Z} = \mathbf{D} - \mathbf{RA^- C}$, $\mathbf{A}^- \in \mathbf{A}\{1\}$.

Proof The inequality follows directly from Theorem 7 since $\mathbf{Y} = \mathbf{E_A C}$, $\mathbf{W} = \mathbf{RF_A}$, and $\mathbf{U} = \mathbf{E_W Z F_Y}$. ■

Corollary 6.3.8 *If* $rank(\mathbf{M}) = rank(\mathbf{A})$, *then*

$$R(\mathbf{C}) \subseteq R(\mathbf{A}), \quad RS(\mathbf{R}) \subseteq RS(\mathbf{A}), \text{ and } \mathbf{D} = \mathbf{RA^- C} \tag{23}$$

for every $\mathbf{A}^-$. *Conversely, if the three conditions of* (23) *hold for some* $\mathbf{A}^-$, *then* $rank(\mathbf{M}) = rank(\mathbf{A})$.

Proof If $\operatorname{rank}(\mathbf{M}) = \operatorname{rank}(\mathbf{A})$, then $\mathbf{Y} = \mathbf{0}$, $\mathbf{W} = \mathbf{0}$, and $\mathbf{U} = \mathbf{0}$. $\mathbf{Y} = \mathbf{0}$ implies $R(\mathbf{C}) \subseteq R(\mathbf{A})$ and $\mathbf{W} = \mathbf{0}$ implies $RS(\mathbf{R}) \subseteq RS(\mathbf{A})$. Since $\mathbf{Y} = \mathbf{0}$ and $\mathbf{W} = \mathbf{0}$, it follows that $\mathbf{U} = \mathbf{Z}$ and $\mathbf{U} = \mathbf{0}$ implies $\mathbf{D} = \mathbf{RA^- C}$. Conversely, the three conditions of (23) imply $\mathbf{Y} = \mathbf{0}$, $\mathbf{W} = \mathbf{0}$, and $\mathbf{U} = \mathbf{0}$. ■

Again, there are three additional forms of Corollaries 7 and 8 which can be proven. For example:

Corollary 6.3.9 *If* $R(\mathbf{C}) \subseteq R(\mathbf{A})$ *and* $RS(\mathbf{R}) \subseteq RS(\mathbf{A})$, *then* $rank(\mathbf{M}) = rank(\mathbf{A}) + rank(\mathbf{D} - \mathbf{RA}^{-}\mathbf{C})$ *for every* $\mathbf{A}^{-}$. *Conversely, if* $rank(\mathbf{M}) = rank(\mathbf{A}) + rank(\mathbf{D} - \mathbf{RA}^{-}\mathbf{C})$ *for some* $\mathbf{A}^{-}$, *then* $R(\mathbf{C}) \subseteq (\mathbf{A})$ *and* $RS(\mathbf{R}) \subseteq RS(\mathbf{A})$.

We now turn to the slightly more specialized problem of block triangular matrices. That is, either $\mathbf{R} = \mathbf{0}$ or $\mathbf{C} = \mathbf{0}$. We shall limit the discussion to *upper* block triangular matrices. For each statement concerning upper block triangular matrices, there is a corresponding statement about lower block triangular matrices easily proved using transposes.

Let $\mathbf{T} = \left[\begin{array}{c|c} \mathbf{A} & \mathbf{C} \\ \hline \mathbf{0} & \mathbf{D} \end{array}\right]$, $\mathbf{A} \in \mathbb{C}^{m \times n}$, $\mathbf{C} \in \mathbb{C}^{m \times r}$, $\mathbf{D} \in \mathbb{C}^{p \times r}$

From Theorem 6 one gets that:

Theorem 6.3.8 *A* (1)*-inverse for* $\mathbf{T}$ *is always given by*

$$\mathbf{T}^{-} = \left[\begin{array}{c|c} \mathbf{A}^{-} & -\mathbf{A}^{-}\mathbf{C}\mathbf{D}^{-} \\ \hline \mathbf{0} & \mathbf{D}^{-} \end{array}\right] + \left[\begin{array}{c} -\mathbf{A}^{-}\mathbf{C} \\ \hline \mathbf{I} \end{array}\right] \mathbf{Q} [\mathbf{I} \mid -\mathbf{C}\mathbf{D}^{-}] \tag{24}$$

where $\mathbf{Q} = \mathbf{F}_{\mathrm{D}}(\mathbf{E}_{\mathrm{A}}\mathbf{C}\mathbf{F}_{\mathrm{D}})^{-}\mathbf{E}_{\mathrm{A}}$. *Furthermore, if* $(\cdot)^{-}$ *represents any* (1, 2)*-inverse, then* (24) *yields a* (1, 2)*-inverse for* $\mathbf{T}$.
From Theorem 7, we have that:

Theorem 6.3.9 *For every choice of* (1)*-inverses in* $\mathbf{E}_{\mathrm{A}}$ *and* $\mathbf{F}_{\mathrm{D}}$, $rank(\mathbf{T}) = rank(\mathbf{A}) + rank(\mathbf{D}) + rank(\mathbf{E}_{\mathrm{A}}\mathbf{C}\mathbf{F}_{\mathrm{D}})$.

4. Applications to the theory of linear estimation

Once one realizes that generalized inverses can be used to provide expressions for solutions (or least squares solutions) to a linear system of algebraic equations, it is only natural to use this tool in connection with the statistical theory of linear estimation. Indeed, the popularity of generalized inverses during the last two decades was, in large part, due to the interest statisticians exhibited for the subject. Much of the early theory of generalized inverses was developed by statisticians with specific applications relating to linear estimation in mind.

One advantage of introducing generalized inverses into the theory of linear estimation is that a unified theory can be presented which draws no distinction between full rank and rank deficient models or between models with singular variance matrices as opposed to those with non-singular variance matrices.

We will confine our discussion to applications involving problems of linear estimation. A complete treatment of how generalized inverses are utilized in statistical applications would require another book almost the size of this one.

In Chapter 2 the application of the Moore–Penrose inverse to least squares problems was presented in a way that avoided the introduction of statistical terminology. However, in this section we will assume the reader is familiar with standard statistical terminology and some of the basic concepts pertaining to statistical models.

We will analyse the linear model $\mathbf{y} = \mathbf{Xb} + \mathbf{e}$ where $\mathbf{y}$ is an $(n \times 1)$ vector of observable random variables, $\mathbf{X}$ is an $(n \times k)$ matrix of known constants, $\mathbf{b}$ is a $(k \times 1)$ unknown vector of parameters, and $\mathbf{e}$ is a $(n \times 1)$ vector of non-observable random variables with zero expectation and variance matrix $\mathrm{E}(\mathbf{ee}^*) = \mathrm{Var}(\mathbf{y}) = \sigma^2\mathbf{V}$. Here $\mathbf{V}$ is positive semi-definite and known, but σ is unknown. We will denote this model by $(\mathbf{y}, \mathbf{Xb}, \sigma^2\mathbf{V})$. If $\mathrm{rank}(\mathbf{X}) = k$ and $\mathbf{V} = \mathbf{I}$, then the celebrated Gauss–Markov theorem guarantees that the least squares solution $\hat{\mathbf{b}} = \mathbf{X}^\dagger\mathbf{y} = (\mathbf{X}^*\mathbf{X})^{-1}\mathbf{X}^*\mathbf{y}$ provides the minimum variance linear unbiased estimate of $\mathbf{b}$.

However, in the general case where $\mathbf{X}$ is possibly rank deficient or $\mathbf{V}$ is possibly singular, there may be no unbiased estimate of $\mathbf{b}$. Then only certain linear functions of $\mathbf{b}$ are unbiasedly estimable. The problem is to obtain minimum variance linear unbiased estimates for estimable linear functions of $\mathbf{b}$ as well as an unbiased estimate for σ^2. Throughout this section, all matrices will have real entries and $(\cdot)^*$ denotes the transpose.

When $\mathbf{V}$ is singular, there are some natural restrictions on $\mathbf{y}$ as well as $\mathbf{b}$. In order to derive these restrictions, as well as other results, we will need to make frequent use of the following fact.

Lemma 6.4.1 *For the model* $(\mathbf{y}, \mathbf{Xb}, \sigma^2\mathbf{V})$ *and for* $\mathbf{l}_1, \mathbf{l}_2 \in \mathbb{R}^n$ *and* $\alpha_1, \alpha_2 \in \mathbb{R}$, $\mathrm{Cov}(\mathbf{l}_1^*\mathbf{y} + \alpha_1, \mathbf{l}_2^*\mathbf{y} + \alpha_2) = \sigma^2\mathbf{l}_1^*\mathbf{V}\mathbf{l}_2$ *and* $\mathrm{Var}(\mathbf{l}^*\mathbf{y} + \alpha) = \sigma^2\mathbf{l}^*\mathbf{V}\mathbf{l}$.

Proof $\mathrm{Cov}(\mathbf{l}_1^*\mathbf{y} + \alpha_1, \mathbf{l}_2^*\mathbf{y} + \alpha_2) = \mathrm{E}[\mathbf{l}_1^*\mathbf{y} - \mathrm{E}(\mathbf{l}_1^*\mathbf{y})][\mathbf{l}_2^* - \mathrm{E}(\mathbf{l}_2^*\mathbf{y})] = \mathrm{E}([\mathbf{l}_1^*(\mathbf{y} - \mathbf{Xb})][\mathbf{l}_2^*(\mathbf{y} - \mathbf{Xb})]) = \mathrm{E}[\mathbf{l}_1^*\mathbf{ee}^*\mathbf{l}_2] = \sigma^2\mathbf{l}_1^*\mathbf{V}\mathbf{l}_2$. That $\mathrm{Var}(\mathbf{l}^*\mathbf{y} + \alpha) = \sigma^2\mathbf{l}^*\mathbf{V}\mathbf{l}$ follows by taking $\mathbf{l}_1 = \mathbf{l}_2$. ■

We now investigate the restrictions which are naturally present when $\mathbf{V}$ is singular.

Since $\mathbf{V}$ is positive semi-definite, there exists an orthogonal matrix $\mathbf{S}$ such that $\mathbf{S}^*\mathbf{VS}$ is the diagonal matrix,

$$\mathbf{S}^*\mathbf{VS} = \begin{bmatrix} \lambda_1^2 & & & & & 0 \\ & \lambda_2^2 & & & & \\ & & \ddots & & & \\ & & & \lambda_r^2 & & \\ & & & & 0 & \\ & & & & & \ddots \\ 0 & & & & & 0 \end{bmatrix} = \begin{bmatrix} \mathbf{D}^2 & \mathbf{0} \\ \mathbf{0} & \mathbf{0} \end{bmatrix} \text{ where } \lambda_i \neq 0 \text{ and}$$

$r = \mathrm{rank}(\mathbf{V})$. If $\mathbf{T} = \mathbf{S}\begin{bmatrix} \mathbf{D}^{-1} & \mathbf{0} \\ \mathbf{0} & \mathbf{I}_{n-r} \end{bmatrix}$, then it is easy to see that the model $(\mathbf{y}, \mathbf{Xb}, \sigma^2\mathbf{V})$ is equivalent to the model

$$\left(\mathbf{T}^*\mathbf{y}, \mathbf{T}^*\mathbf{Xb}, \begin{bmatrix} \sigma^2\mathbf{I}_r & \mathbf{0} \\ \mathbf{0} & \mathbf{0} \end{bmatrix}\right). \tag{1}$$

Let $\mathbf{S}$ be partitioned as $\mathbf{S} = [\mathbf{P} \,|\, \mathbf{Q}]$ where $\mathbf{P}$ is $(n \times r)$. Then (1) can be written in the equivalent form

$$\left\{\begin{array}{l}\mathbf{D}^{-1}\mathbf{P}^*\mathbf{y} = \mathbf{D}^{-1}\mathbf{P}^*\mathbf{X}\mathbf{b} + \mathbf{D}^{-1}\mathbf{P}^*\mathbf{e} \\ \quad\;\; \mathbf{Q}^*\mathbf{y} = \mathbf{Q}^*\mathbf{X}\mathbf{b} + \mathbf{Q}^*\mathbf{e}\end{array}\right\}. \tag{2}$$

It follows that $\mathrm{Var}(\mathbf{Q}^*\mathbf{e}) = \mathbf{0}$ and $\mathrm{E}(\mathbf{Q}^*\mathbf{e}) = \mathbf{0}$ so that

$$\mathbf{Q}^*\mathbf{y} = \mathbf{Q}^*\mathbf{X}\mathbf{b} \text{ with probability 1.} \tag{3}$$

Equation (3) is just a set of linear restrictions on $\mathbf{b}$. If the linear system in (3) is assumed to be consistent, then the model $(\mathbf{y}, \mathbf{X}\mathbf{b}, \sigma^2\mathbf{V})$ is called a *consistent model.* Therefore, the model (2), and hence $(\mathbf{y}, \mathbf{X}\mathbf{b}, \sigma^2\mathbf{V})$, is equivalent to a restricted model of the form

$$(\tilde{\mathbf{y}},\ \tilde{\mathbf{X}}\mathbf{b} \mid \mathbf{R}\mathbf{b} = \mathbf{f}, \sigma^2\mathbf{I}) \tag{4}$$

where $\tilde{\mathbf{y}} = \mathbf{D}^{-1}\mathbf{P}^*\mathbf{y}$, $\tilde{\mathbf{X}} = \mathbf{D}^{-1}\mathbf{P}^*\mathbf{X}$, $\tilde{\mathbf{e}} = \mathbf{D}^{-1}\mathbf{P}^*\mathbf{e}$, $\mathbf{R} = \mathbf{Q}^*\mathbf{X}$, and $\mathbf{f} = \mathbf{Q}^*\mathbf{y}$. The notation $(\tilde{\mathbf{y}}, \tilde{\mathbf{X}}\mathbf{b} \mid \mathbf{R}\mathbf{b} = \mathbf{f}, \sigma^2\mathbf{I})$ has the obvious interpretation

$$\tilde{\mathbf{y}} = \tilde{\mathbf{X}}\mathbf{b} + \tilde{\mathbf{e}}, \text{ for } \mathbf{b} \text{ such that } \mathbf{R}\mathbf{b} = \mathbf{f},\ \mathrm{Var}(\tilde{\mathbf{e}}) = \sigma^2\mathbf{I}.$$

In the sequel, we will always assume all models we write are consistent.

Definition 6.4.1 *A linear function* $\mathbf{c}^*\mathbf{b}$ *is said to be* linearly unbiasedly estimable *under* $(\mathbf{y}, \mathbf{X}\mathbf{b}, \sigma^2\mathbf{V})$ *if there exists a vector* $\mathbf{l} \in \mathbb{R}^n$ *and scalar* $\alpha \in \mathbb{R}$ *such that*

$$\mathrm{E}(\mathbf{l}^*\mathbf{y} + \alpha) = \mathbf{c}^*\mathbf{b} \text{ for all } \mathbf{b} \text{ such that } \mathbf{Q}^*\mathbf{X}\mathbf{b} = \mathbf{Q}^*\mathbf{y}$$

where $\mathbf{Q}^*$ *is as in* (2). *Whenever we use the term 'estimable' in the sequel, we will mean* linearly unbiasedly *estimable.*

We are now in a position to characterize those vectors $\mathbf{c}$ such that $\mathbf{c}^*\mathbf{b}$ is unbiasedly estimable.

Theorem 6.4.1 *The function* $\mathbf{c}^*\mathbf{b}$ *is estimable under* $(\mathbf{y}, \mathbf{X}\mathbf{b}, \sigma^2\mathbf{V})$ *if and only if* $\mathbf{c} \in R(\mathbf{X}^*)$.

Proof From Corollary 3.1, $\mathbf{Q}^*\mathbf{X}\mathbf{b} = \mathbf{Q}^*\mathbf{y}$ if and only if

$$\mathbf{b} \in \mathscr{S} = \{(\mathbf{Q}^*\mathbf{X})^-\mathbf{Q}^*\mathbf{y} + [\mathbf{I} - (\mathbf{Q}^*\mathbf{X})^-(\mathbf{Q}^*\mathbf{X})]\mathbf{h} \mid \mathbf{h} \in \mathbb{R}^k\}.$$

There exists $\mathbf{l} \in \mathbb{R}^n$ and $\alpha \in \mathbb{R}$ such that

$$\mathrm{E}(\mathbf{l}^*\mathbf{y} + \alpha) = \mathbf{c}^*\mathbf{b},\ \mathbf{b} \in \mathscr{S}$$

$$\Leftrightarrow \mathbf{l}^*\mathbf{X}\mathbf{b} + \alpha = \mathbf{c}^*\mathbf{b},\ \mathbf{b} \in \mathscr{S}$$

$$\Leftrightarrow (\mathbf{c}^* - \mathbf{l}^*\mathbf{X})\mathbf{b} = \alpha,\ \mathbf{b} \in \mathscr{S}$$

$$\Leftrightarrow (\mathbf{c}^* - \mathbf{l}^*\mathbf{X})[(\mathbf{Q}^*\mathbf{X})^-\mathbf{Q}^*\mathbf{y} + (\mathbf{I} - (\mathbf{Q}^*\mathbf{X})^-(\mathbf{Q}^*\mathbf{X}))\mathbf{h}] = \alpha,\ \mathbf{h} \in \mathbb{R}^k$$

$$\Leftrightarrow \left\{\begin{array}{l}\alpha = (\mathbf{c}^* - \mathbf{l}^*\mathbf{X})(\mathbf{Q}^*\mathbf{X})^-\mathbf{Q}^*\mathbf{y} \\ (\mathbf{c}^* - \mathbf{l}^*\mathbf{X})[\mathbf{I} - (\mathbf{Q}^*\mathbf{X})^-(\mathbf{Q}^*\mathbf{X})] = \mathbf{0}\end{array}\right.$$

$$\Leftrightarrow \begin{cases} \alpha = (\mathbf{c}^* - \mathbf{l}^*\mathbf{X})(\mathbf{Q}^*\mathbf{X})^- \mathbf{Q}^*\mathbf{y} \\ \mathbf{c}^* = [\mathbf{l}^* + (\mathbf{c}^* - \mathbf{l}^*\mathbf{X})(\mathbf{Q}^*\mathbf{X})^- \mathbf{Q}^*]\mathbf{X} \end{cases}$$

$$\Leftrightarrow \begin{cases} \alpha = \mathbf{d}^*\mathbf{y} \\ \mathbf{c}^* = [\mathbf{l}^* + \mathbf{d}^*]\mathbf{X}, \end{cases} \quad \mathbf{d}^* = (\mathbf{c}^* - \mathbf{l}^*\mathbf{X})(\mathbf{Q}^*\mathbf{X})^- \mathbf{Q}^*. \quad \blacksquare \tag{5}$$

There are several important consequences of Theorem 1.

Corollary 6.4.1 *A linear function* $\mathbf{l}^*\mathbf{y} + \alpha$ *is an unbiased estimate of* $\mathbf{c}^*\mathbf{b}$ *under* $(\mathbf{y}, \mathbf{Xb}, \sigma^2\mathbf{V})$ *if and only if there exists a vector* $\mathbf{d}$ *such that* $\alpha = \mathbf{y}^*\mathbf{d}$, *with probability* 1, *and* $\mathbf{X}^*(\mathbf{l} + \mathbf{d}) = \mathbf{c}$.

In some of the literature on linear estimation, it is often stated that if $\mathbf{l}$ is a vector such that $\mathbf{l}^*\mathbf{y}$ is an unbiased estimate for $\mathbf{c}^*\mathbf{b}$ under $(\mathbf{y}, \mathbf{Xb}, \sigma^2\mathbf{V})$, then $\mathbf{X}^*\mathbf{l} = \mathbf{c}$. This is wrong. Corollary 4.1 shows that $\mathbf{X}^*\mathbf{l} = \mathbf{c}$, is sufficient but *not* necessary for $\mathbf{l}^*\mathbf{y}$ to be an unbiased estimate for $\mathbf{c}^*\mathbf{b}$ under $(\mathbf{y}, \mathbf{Xb}, \sigma^2\mathbf{V})$.

We next need a result which will tell us what the entire set of linear unbiased estimates for $\mathbf{c}^*\mathbf{b}$ looks like.

Theorem 6.4.2 *Suppose* $\mathbf{c}^*\mathbf{b}$ *is estimable under* $(\mathbf{y}, \mathbf{Xb}, \sigma^2\mathbf{V})$. *The set of linear unbiased estimates for* $\mathbf{c}^*\mathbf{b}$ *is given by*

$$\mathrm{U} = \{\psi \mid \psi = \xi^*\mathbf{y} \textit{ where } \xi^*\mathbf{X} = \mathbf{c}^*\}$$

Proof Suppose first that $\psi \in \mathrm{U}$. Then $\psi = \xi^*\mathbf{y}$ for some ξ^* such that $\mathbf{c}^* = \xi^*\mathbf{X}$. Clearly, $\mathrm{E}(\psi) = \xi^*\mathbf{Xb} = \mathbf{c}^*\mathbf{b}$. Conversely, suppose ψ is a linear unbiased estimate for $\mathbf{c}^*\mathbf{b}$. Then, by definition, there exists a vector $\mathbf{l}$ and a scalar α such that $\psi = \mathbf{l}^*\mathbf{y} + \alpha$. Since ψ is unbiased, Corollary 4.1 guarantees that there is a vector $\mathbf{d}$ such that $\mathbf{c}^* = (\mathbf{l} + \mathbf{d})^*\mathbf{X}$ and $\mathbf{d}^*\mathbf{y} = \alpha$ (with probability 1) so that

$$\psi = \mathbf{l}^*\mathbf{y} + \alpha = \mathbf{l}^*\mathbf{y} + \mathbf{d}^*\mathbf{y} = (\mathbf{l} + \mathbf{d})^*\mathbf{y}.$$

Therefore, ψ has the form $\xi^*\mathbf{y}$ with $\xi^*\mathbf{X} = \mathbf{c}^*$ and hence $\psi \in \mathrm{U}$. $\blacksquare$

As an immediate consequence of Theorem 2, we have the following.

Corollary 6.4.2 *Suppose that* $\mathbf{c}^*\mathbf{b}$ *is estimable under* $(\mathbf{y}, \mathbf{Xb}, \sigma^2\mathbf{V})$. $\mathbf{l}$ *is a vector such* $\mathbf{l}^*\mathbf{y}$ *is an unbiased estimate of* $\mathbf{c}^*\mathbf{b}$ *if and only if there exists a vector* ξ *such that* $\mathbf{l}^*\mathbf{y} = \xi^*\mathbf{y}$ *where* $\xi^*\mathbf{X} = \mathbf{c}^*$.

Using Corollary 6.3.1, the set of unbiased estimates for $\mathbf{c}^*\mathbf{b}$ given in Theorem 2 can be written in terms of any element of $\mathbf{X}\{1\}$.

Corollary 6.4.3 *Suppose* $\mathbf{c}^*\mathbf{b}$ *is estimable under* $(\mathbf{y}, \mathbf{Xb}, \sigma^2\mathbf{V})$ *and* $\mathbf{X}^- \in \mathbf{X}\{1\}$. *The set of linear unbiased estimates for* $\mathbf{c}^*\mathbf{b}$ *is given by*

$$\mathrm{U} = \{\mathbf{c}^*\mathbf{X}^-\mathbf{y} + \mathbf{h}^*\mathbf{y} \mid \mathbf{h} \in N(\mathbf{X}^*)\}.$$

We now address the problem of finding a form for the minimum variance linear unbiased estimate of an estimable function. From

Theorem 2 and Lemma 1, we know that when $\mathbf{c}^*\mathbf{b}$ is estimable under $(\mathbf{y}, \mathbf{Xb}, \sigma^2\mathbf{V})$, ψ is a linear unbiased estimate for $\mathbf{c}^*\mathbf{b}$ which has minimum variance if and only if there is a vector ξ such that $\psi = \xi^*\mathbf{y}$ and the minimum

$$\min_{\mathbf{z}\in\mathscr{S}} \mathbf{z}^*\mathbf{V}\mathbf{z} \text{ where } \mathscr{S} = \{\mathbf{p} \mid \mathbf{X}^*\mathbf{p} = \mathbf{c}\} \tag{6}$$

is attained at $\mathbf{z} = \xi$. Since $\mathbf{V}$ is positive semi-definite, there exists a matrix $\mathbf{A}$ such that $\mathbf{V} = \mathbf{A}^*\mathbf{A}$. Thus (6) is equivalent to

$$\min_{\mathbf{z}\in\mathscr{S}} \|\mathbf{Az}\| \text{ where } \mathscr{S} = \{\mathbf{p} \mid \mathbf{X}^*\mathbf{p} = \mathbf{c}\}. \tag{7}$$

This is a constrained least squares problem of the type studied in Section 3.6. Theorem 3.6.1 guarantees that the minimum (7) is attained at a vector $\mathbf{z}_0$ if and only if there is a vector λ_0 such that

$$\begin{bmatrix} \mathbf{z}_0 \\ \lambda_0 \end{bmatrix} \text{ is a least squares solution of } \left[\begin{array}{c|c} \mathbf{V} & \mathbf{X} \\ \hline \mathbf{X}^* & \mathbf{0} \end{array}\right] \begin{bmatrix} \mathbf{z} \\ \lambda \end{bmatrix} = \begin{bmatrix} \mathbf{0} \\ \mathbf{c} \end{bmatrix}. \tag{8}$$

Let $\mathbf{B} = \left[\begin{array}{c|c} \mathbf{V} & \mathbf{X} \\ \hline \mathbf{X}^* & \mathbf{0} \end{array}\right]$. Using Corollary 3.6, it is easy to see that $\begin{bmatrix} \mathbf{0} \\ \mathbf{c} \end{bmatrix} \in R(\mathbf{B})$ because for the $\mathbf{B}^-$ of Corollary 3.6,

$$\mathbf{B}\mathbf{B}^-\begin{bmatrix} \mathbf{0} \\ \mathbf{c} \end{bmatrix} = \begin{bmatrix} \mathbf{0} \\ (\mathbf{X}^-\mathbf{X})^*\mathbf{c} \end{bmatrix} = \begin{bmatrix} \mathbf{0} \\ \mathbf{c} \end{bmatrix}, \tag{9}$$

by Theorem 6.4.1. Thus we can make an even stronger statement than (8) by replacing the phrase 'least squares solution' in (8) with the word 'solution' to give the following result.

Theorem 6.4.3 *When* $\mathbf{c}^*\mathbf{b}$ *is estimable under* $(\mathbf{y}, \mathbf{Xb}, \sigma^2\mathbf{V})$, ψ *is a linear unbiased estimate of* $\mathbf{c}^*\mathbf{b}$ *which has minimum variance if and only if there are vectors* ξ *and* λ *such that* $\psi = \xi^*\mathbf{y}$ *and*

$$\left[\begin{array}{c|c} \mathbf{V} & \mathbf{X} \\ \hline \mathbf{X}^* & \mathbf{0} \end{array}\right] \begin{bmatrix} \xi \\ \lambda \end{bmatrix} = \begin{bmatrix} \mathbf{0} \\ \mathbf{c} \end{bmatrix}. \tag{10}$$

A useful equivalent formulation of Theorem 3 is as follows.

Theorem 6.4.4 *When* $\mathbf{c}^*\mathbf{b}$ *is estimable under* $(\mathbf{y}, \mathbf{Xb}, \sigma^2\mathbf{V})$, ψ *is a linear unbiased estimate of* $\mathbf{c}^*\mathbf{b}$ *which has minimum variance if and only if there is a vector* ξ *such that each of the following conditions hold.*

(i) $\psi = \xi^*\mathbf{y}$
(ii) $\mathbf{X}^*\xi = \mathbf{c}$
(iii) $\mathbf{V}\xi \in R(\mathbf{X})$

The linear unbiased estimate of $\mathbf{c}^*\mathbf{b}$ which has minimum variance is unique in the following sense.

Theorem 6.4.5 *Suppose that* $\mathbf{c}^*\mathbf{b}$ *is estimable under* $(\mathbf{y}, \mathbf{Xb}, \sigma^2\mathbf{V})$. *If*

ψ_1 and ψ_2 are both linear unbiased estimates of $\mathbf{c}^*\mathbf{b}$ *which have minimum variance, then* $\psi_1 = \psi_2$ *with probability* 1.

Proof From Theorem 4 we know that there exist vectors ξ_1 and ξ_2 such that

$$\psi_1 - \psi_2 = (\xi_1 - \xi_2)^*\mathbf{y}, \tag{11}$$

$$(\xi_1 - \xi_2)^*\mathbf{X} = \mathbf{0}^*, \text{ and} \tag{12}$$

$$\mathbf{V}(\xi_1 - \xi_2) \in R(\mathbf{X}). \tag{13}$$

From (13) we know there is a vector $\mathbf{h}$ such that $\mathbf{V}(\xi_1 - \xi_2) = \mathbf{Xh}$ so that we can use Lemma 1 together with (12) to obtain

$$\mathrm{Var}(\psi_1 - \psi_2) = \sigma^2(\xi_1 - \xi_2)^*\mathbf{V}(\xi_1 - \xi_2) = \sigma^2(\xi_1 - \xi_2)^*\mathbf{Xh} = 0.$$

Thus, there is a constant κ such that $\psi_1 - \psi_2 = \kappa$ with probability 1. However, (11) and (12) imply that $\kappa = \mathrm{E}(\psi_1 - \psi_2) = \mathrm{E}[(\xi_1 - \xi_2)^*\mathbf{y}] = (\xi_1 - \xi_2)^*\mathbf{Xb} = 0.$ ■

In light of this result, we make the following definition.

Definition 6.4.2 *When* $\mathbf{c}^*\mathbf{b}$ *is estimable under* $(\mathbf{y}, \mathbf{Xb}, \sigma^2\mathbf{V})$, ψ *is called the* best linear unbiased estimate (*or* BLUE) *for* $\mathbf{c}^*\mathbf{b}$ *when* ψ *is the unique linear unbiased estimate for* $\mathbf{c}^*\mathbf{b}$ *with minimum variance.*

If ψ is the BLUE for $\mathbf{c}^*\mathbf{b}$, then ψ is unique. There are, however, generally infinitely many vectors ξ satisfying the conditions of Theorem 4 which can give rise to ψ. Although there may be a slight theoretical interest in representing all of the ξ's associated with the BLUE ψ, this is usually not the problem of prime concern. The important problem is to obtain some formula for ψ. If any one particular ξ satisfying the conditions of Theorem 4 can be determined, then the problem of finding the BLUE of $\mathbf{c}^*\mathbf{b}$ is considered to be solved. It is clear from Theorem 3 that knowledge of any (1)-inverse of the matrix $\left[\begin{array}{c|c} \mathbf{V} & \mathbf{X} \\ \hline \mathbf{X}^* & \mathbf{0} \end{array}\right]$ can provide a representation for the BLUE of $\mathbf{c}^*\mathbf{b}$. Such a (1)-inverse can also provide other valuable information. Before pursuing this further, we need the following definition.

Definition 6.4.3 *For the linear model* $(\mathbf{y}, \mathbf{Xb}, \sigma^2\mathbf{V})$, *let* $\mathbf{B}$ *denote the matrix.*

$$\mathbf{B} = \left[\begin{array}{c|c} \mathbf{V} & \mathbf{X} \\ \hline \mathbf{X}^* & \mathbf{0} \end{array}\right] \quad (\mathbf{V} \text{ is } n \times n \text{ and } \mathbf{X} \text{ is } n \times k).$$

An $n \times n$ matrix is said to be a $\mathbf{B}_{11}$-matrix if it appears as the upper left hand block in some $\mathbf{B}^- \in \mathbf{B}\{1\}$. Likewise, those $n \times k$ matrices which appear as an upper right hand block in some $\mathbf{B}^-$ are called $\mathbf{B}_{12}$-matrices; those $k \times n$ matrices which appear in the lower left hand corner of some $\mathbf{B}^-$ are called $\mathbf{B}_{21}$-matrices; and those $k \times k$ matrices which appear in the lower right hand corner of some $\mathbf{B}^-$ are called $\mathbf{B}_{22}$-matrices.

A somewhat amazing fact about $\mathbf{B}_{ij}$-matrices is that each class is completely independent of every other class in the sense that if $\mathbf{Q}$ is any $\mathbf{B}_{11}$-matrix, $\mathbf{U}$ is any $\mathbf{B}_{12}$-matrix, $\mathbf{L}$ is any $\mathbf{B}_{21}$-matrix, and $\mathbf{T}$ is any $\mathbf{B}_{22}$-matrix, then the composite matrix $\left[\begin{array}{c|c} \mathbf{Q} & \mathbf{U} \\ \hline \mathbf{L} & \mathbf{T} \end{array}\right]$ is always a (1)-inverse for $\mathbf{B}$. Furthermore, the $\mathbf{B}_{ij}$-matrices can be computed independently of one another. This means that each block which appears in a $\mathbf{B}^-$ can be calculated as a separate entity without regard to any other block which might appear in the same $\mathbf{B}^-$ (or any other $\mathbf{B}^-$). In order to establish these facts, we need some preliminary lemmas.

Lemma 6.4.2 *Let* $\mathbf{E} = \mathbf{I} - \mathbf{X}\mathbf{X}^\dagger$. *A matrix* $\mathbf{Q}$ *is a* $\mathbf{B}_{11}$*-matrix if and only if* $\mathbf{Q}$ *satisfies the four equations*

$$\mathbf{E}(\mathbf{V} - \mathbf{V}\mathbf{Q}\mathbf{V}) = \mathbf{0}, \tag{14}$$

$$\mathbf{V}\mathbf{Q}\mathbf{X} = \mathbf{0}, \tag{15}$$

$$\mathbf{X}^*\mathbf{Q}\mathbf{V} = \mathbf{0}, \tag{16}$$

$$\mathbf{X}^*\mathbf{Q}\mathbf{X} = \mathbf{0}. \tag{17}$$

Proof Suppose first that $\mathbf{Q}$ is a $\mathbf{B}_{11}$-matrix. Then there must exist matrices $\mathbf{W}_{12}$, $\mathbf{W}_{21}$, and $\mathbf{W}_{22}$ such that $\begin{bmatrix} \mathbf{Q} & \mathbf{W}_{12} \\ \mathbf{W}_{21} & \mathbf{W}_{22} \end{bmatrix} \in \mathbf{B}\{1\}$. This implies, by direct multiplication, that (i) $\mathbf{V}\mathbf{Q}\mathbf{V} + \mathbf{X}\mathbf{W}_{21}\mathbf{V} + \mathbf{V}\mathbf{W}_{12}\mathbf{X}^* + \mathbf{X}\mathbf{W}_{22}\mathbf{X}^* = \mathbf{V}$; (ii) $\mathbf{V}\mathbf{Q}\mathbf{X} + \mathbf{X}\mathbf{W}_{21}\mathbf{X} = \mathbf{X}$; (iii) $\mathbf{X}^*\mathbf{Q}\mathbf{V} + \mathbf{X}^*\mathbf{W}_{12}\mathbf{X}^* = \mathbf{X}^*$; and (iv) $\mathbf{X}^*\mathbf{Q}\mathbf{X} = \mathbf{0}$. Note that (iv) is equation (17). To establish (15), multiply (ii) on the left by $\mathbf{X}^*\mathbf{Q}^*$ and use (iv) to obtain $\mathbf{X}^*\mathbf{Q}^*\mathbf{V}\mathbf{Q}\mathbf{X} = \mathbf{0}$. Since $\mathbf{V}$ is positive semi-definite, $\mathbf{V} = \mathbf{A}^*\mathbf{A}$ for some $\mathbf{A}$ and it is easy to see that (15) follows. Equation (16) follows in a similar manner. Equations (ii) and (iii) now degenerate to

$$\mathbf{X}\mathbf{W}_{21}\mathbf{X} = \mathbf{X} \text{ and } \mathbf{X}^*\mathbf{W}_{12}\mathbf{X}^* = \mathbf{X}^*. \tag{18}$$

To establish (14), notice that for every vector $\mathbf{h}$, and $\mathbf{B}^- \in \mathbf{B}\{1\}$, (9) guarantees that $\mathbf{B}\mathbf{B}^-\begin{bmatrix} \mathbf{0} \\ \mathbf{X}^*\mathbf{h} \end{bmatrix} = \begin{bmatrix} \mathbf{0} \\ \mathbf{X}^*\mathbf{h} \end{bmatrix}$. From this it follows that $\mathbf{V}\mathbf{W}_{12}\mathbf{X}^* + \mathbf{X}\mathbf{W}_{22}\mathbf{X}^* = \mathbf{0}$ so that (i) becomes

$$\mathbf{V} - \mathbf{V}\mathbf{Q}\mathbf{V} = \mathbf{X}\mathbf{W}_{21}\mathbf{V}. \tag{19}$$

Equation (14) is obtained by multiplying (19) on the left by $\mathbf{E}$. Conversely, if $\mathbf{Q}$ satisfies (14)–(17), then

$$\left[\begin{array}{c|c} \mathbf{Q} & (\mathbf{I} - \mathbf{Q}\mathbf{V})\mathbf{X}^{\dagger *} \\ \hline \mathbf{X}^\dagger(\mathbf{I} - \mathbf{V}\mathbf{Q}) & \mathbf{X}^\dagger(\mathbf{V}\mathbf{Q}\mathbf{V} - \mathbf{V})\mathbf{X}^{\dagger *} \end{array}\right] \in \mathbf{B}\{1\}$$

and $\mathbf{Q}$ is a $\mathbf{B}_{11}$-matrix. ■

Lemma 6.4.3 *The term* $\mathbf{VQV}$ *is invariant for all* $\mathbf{B}_{11}$*-matrices* $\mathbf{Q}$. *Moreover, for every* $\mathbf{B}_{11}$*-matrix* $\mathbf{Q}$, $\mathbf{VQV} = \mathbf{A}^*(\mathbf{AE})(\mathbf{AE})^\dagger\mathbf{A}$ *where* $\mathbf{E} = \mathbf{I} - \mathbf{XX}^\dagger$ *and* $\mathbf{V} = \mathbf{A}^*\mathbf{A}$.

Proof Suppose $\mathbf{Q}$ is a $\mathbf{B}_{11}$-matrix so that (14)–(17) hold. By direct multiplication, along with (15) and (16), and the fact that $\mathbf{X}^\dagger = (\mathbf{X}^*\mathbf{X})^\dagger\mathbf{X}^* = \mathbf{X}^*(\mathbf{XX}^*)^\dagger$, it can be verified that $\mathbf{VEQEV} = \mathbf{VQV}$. Let $\mathbf{K} = \mathbf{EVE}$ and $\mathbf{G} = \mathbf{QKK}^\dagger + \mathbf{K}^\dagger\mathbf{KQ} - \mathbf{Q}$. It now follows from (14) that $\mathbf{Q} \in \mathbf{K}\{1\}$ and hence $\mathbf{G} \in \mathbf{K}\{1\}$. Now observe that $\mathbf{Q}$ can be written as

$$\mathbf{Q} = \mathbf{Q}(\mathbf{I} - \mathbf{KK}^\dagger)\mathbf{E} + \mathbf{E}(\mathbf{I} - \mathbf{K}^\dagger\mathbf{K})\mathbf{Q} + \mathbf{EGE} \tag{20}$$

since (15) and (16) imply $\mathbf{KQE} = \mathbf{KQ}$ and $\mathbf{EQK} = \mathbf{QK}$. But from $\mathbf{K} = \mathbf{K}^*$, it follows that $\mathbf{KK}^\dagger = \mathbf{K}^\dagger\mathbf{K}$ and $(\mathbf{I} - \mathbf{KK}^\dagger)\mathbf{EV} = \mathbf{0} = \mathbf{VE}(\mathbf{I} - \mathbf{K}^\dagger\mathbf{K})$. Therefore, (20) together with $\mathbf{KK}^\dagger\mathbf{EV} = \mathbf{EV}$ and $\mathbf{VEK}^\dagger\mathbf{K} = \mathbf{VE}$ yields $\mathbf{VQV} = \mathbf{VEGEV} = \mathbf{VEK}^\dagger\mathbf{KGKK}^\dagger\mathbf{EV}$. Now use the fact that $\mathbf{G} \in \mathbf{K}\{1\}$ to obtain $\mathbf{VQV} = \mathbf{VEK}^\dagger\mathbf{EV} = \mathbf{VE}(\mathbf{EA}^*\mathbf{AE})^\dagger\mathbf{EV} = \mathbf{A}^*\mathbf{AE}(\mathbf{AE})^\dagger(\mathbf{AE})^{*\dagger}\mathbf{EA}^*\mathbf{A} = \mathbf{A}^*(\mathbf{AE})(\mathbf{AE})^\dagger\mathbf{A}$. ■

Lemma 6.4.4 *Let* $\mathbf{D} = \mathbf{A}^*[\mathbf{I} - (\mathbf{AE})(\mathbf{AE})^\dagger]\mathbf{A}$ *where* $\mathbf{A}$ *and* $\mathbf{E}$ *are as in* Lemma 3. (*By virtue of Lemma* 3, $\mathbf{D} = \mathbf{V} - \mathbf{VQV}$ *where* $\mathbf{Q}$ *can be any* $\mathbf{B}_{11}$*-matrix.*) *Each of the following statements hold.*

$$\mathbf{U} \text{ is a } \mathbf{B}_{12}\text{-matrix if and only if } \mathbf{U} \in \mathbf{X}^*\{1\} \text{ and } \mathbf{VUX}^* = \mathbf{D}. \tag{21}$$

$$\mathbf{L} \text{ is } \mathbf{B}_{21}\text{-matrix if and only if } \mathbf{L} \in \mathbf{X}\{1\} \text{ and } \mathbf{XLV} = \mathbf{D}. \tag{22}$$

$$\mathbf{T} \text{ is } \mathbf{B}_{22}\text{-matrix if and only if } \mathbf{XTX}^* = -\mathbf{D}. \tag{23}$$

Proof of (21). Suppose first that $\mathbf{U} \in \mathbf{X}^*\{1\}$ where $\mathbf{VUX}^* = \mathbf{D}$. To see that $\mathbf{U}$ is a $\mathbf{B}_{12}$-matrix, let $\mathbf{Q} = \mathbf{E}(\mathbf{EVE})^\dagger\mathbf{E}$ and verify that

$$\mathbf{M}_U = \left[\begin{array}{c|c} \mathbf{Q} & \mathbf{U} \\ \hline \mathbf{X}^\dagger(\mathbf{I} - \mathbf{VQ}) & -\mathbf{X}^\dagger\mathbf{DX}^{\dagger *} \end{array}\right] \in \mathbf{B}\{1\}$$

by observing that Corollary 3.6 implies that $\mathbf{Q}$ is a $\mathbf{B}_{11}$-matrix so that (14)–(17) can be used. Conversely, suppose that $\mathbf{U}$ is a $\mathbf{B}_{12}$-matrix. This means there exist matrices $\mathbf{Q}$, $\mathbf{L}$ and $\mathbf{T}$ such that $\left[\begin{array}{c|c} \mathbf{Q} & \mathbf{U} \\ \hline \mathbf{L} & \mathbf{T} \end{array}\right] \in \mathbf{B}\{1\}$.
From (18) we have that $\mathbf{U} \in \mathbf{X}^*\{1\}$. The fact that $\mathbf{VUX}^* = \mathbf{V} - \mathbf{VQV} = \mathbf{D}$ follows from (19). Thus (21) holds.

The same type of argument is used to prove (22), (23) except in place of $\mathbf{M}_U$, one uses

$$\mathbf{M}_L = \left[\begin{array}{c|c} \mathbf{Q} & (\mathbf{I} - \mathbf{QV})\mathbf{X}^{\dagger *} \\ \hline \mathbf{L} & -\mathbf{X}^\dagger\mathbf{DX}^{\dagger *} \end{array}\right] \text{ for (22) and}$$

$$\mathbf{M}_T = \left[\begin{array}{c|c} \mathbf{Q} & (\mathbf{I} - \mathbf{QV})\mathbf{X}^{\dagger *} \\ \hline \mathbf{X}^\dagger(\mathbf{I} - \mathbf{VQ}) & \mathbf{T} \end{array}\right] \text{ for (23).} \quad ■$$

By combining the results of Lemmas 2–4 one arrives at the following important result concerning the independence of the various classes of $\mathbf{B}_{ij}$-matrices.

Theorem 6.4.6 *If* $\mathbf{G}_{11}$ *is any* $\mathbf{B}_{11}$*-matrix,* $\mathbf{G}_{12}$ *is any* $\mathbf{B}_{12}$*-matrix,* $\mathbf{G}_{21}$ *is any* $\mathbf{B}_{21}$*-matrix and* $\mathbf{G}_{22}$ *is any* $\mathbf{B}_{22}$*-matrix, then the composite matrix* $\begin{bmatrix} \mathbf{G}_{11} & \mathbf{G}_{12} \\ \mathbf{G}_{21} & \mathbf{G}_{22} \end{bmatrix}$ *is a* (1)*-inverse for* $\mathbf{B}$*. Furthermore, the matrices* $\mathbf{G}_{ij}$ *can be computed independently of each other. The equations on which such calculations must be based are given in* (14)–(17) *and* (21)–(23).

Although it is not necessary to know a $\mathbf{B}_{11}$-matrix in order to compute $\mathbf{B}_{12}$-, $\mathbf{B}_{21}$-, or $\mathbf{B}_{22}$-matrices, knowledge of any $\mathbf{B}_{11}$-matrix can be useful since the matrix $\mathbf{D}$ of Lemma 4 is then readily available. Once $\mathbf{D}$ is known, a set of $\mathbf{B}_{12}$-, $\mathbf{B}_{21}$-, and $\mathbf{B}_{22}$-matrices can be easily computed.

The importance of the different $\mathbf{B}_{ij}$-matrices in linear estimation is given in the following fundamental theorem.

Theorem 6.4.7 *If* $\mathbf{c}^*\mathbf{b}$ *is estimable under* $(\mathbf{y}, \mathbf{Xb}, \sigma^2\mathbf{V})$*, then each of the following is true.*

If $\mathbf{G}_{12}$ *is any* $\mathbf{B}_{12}$*-matrix, then the* BLUE *of* $\mathbf{c}^*\mathbf{b}$ *is given by*

$$\psi = \mathbf{c}^*\mathbf{G}_{12}^*\mathbf{y}. \tag{24}$$

If $\mathbf{G}_{21}$ *is any* $\mathbf{B}_{21}$*-matrix, then the* BLUE *of* $\mathbf{c}^*\mathbf{b}$ *is also given by*

$$\psi = \mathbf{c}^*\mathbf{G}_{21}\mathbf{y}. \tag{24$'$}$$

Suppose $\mathbf{c}_1^*\mathbf{b}$ *and* $\mathbf{c}_2^*\mathbf{b}$ *are both estimable with* BLUE's ψ_1 *and* ψ_2*, respectively. If* $\mathbf{G}_{22}$ *is any* $\mathbf{B}_{22}$*-matrix, then* $\mathrm{Cov}(\psi_1, \psi_2)$

$$= -\sigma^2\mathbf{c}_1^*\mathbf{G}_{22}\mathbf{c}_2 = -\sigma^2\mathbf{c}_2^*\mathbf{G}_{22}\mathbf{c}_1. \tag{25}$$

If $\mathbf{G}_{22}$ *is any* $\mathbf{B}_{22}$*-matrix and* ψ *is the* BLUE *of* $\mathbf{c}^*\mathbf{b}$*, then*

$$\mathrm{Var}(\psi) = -\sigma^2\mathbf{c}^*\mathbf{G}_{22}\mathbf{c}. \tag{26}$$

If $\mathbf{G}_{11}$ *is any* $\mathbf{B}_{11}$*-matrix, then an unbiased estimator for* σ^2 *is given by*

$$\gamma^{-1}\mathbf{y}^*\mathbf{G}_{11}\mathbf{y} \text{ where } \gamma = \mathrm{Tr}(\mathbf{G}_{11}\mathbf{V}). \tag{27}$$

Proof of (24): If $\mathbf{G}_{12}$ is a $\mathbf{B}_{12}$-matrix, then there exist matrices $\mathbf{Q}, \mathbf{L}$, and $\mathbf{T}$ such that $\begin{bmatrix} \mathbf{Q} & \mathbf{G}_{12} \\ \mathbf{L} & \mathbf{T} \end{bmatrix} \in \mathbf{B}\{1\}$. Theorem 3 guarantees that the BLUE for $\mathbf{c}^*\mathbf{b}$ is given by $\psi = \xi^*\mathbf{y}$ where ξ satisfies (10). Therefore, a solution for $\begin{bmatrix} \xi \\ \lambda \end{bmatrix}$ is $\mathbf{B}^-\begin{bmatrix} \mathbf{0} \\ \mathbf{c} \end{bmatrix}$ for any $\mathbf{B}^-$. Thus one solution for ξ is $\mathbf{G}_{12}\mathbf{c}$, and hence $\psi = \mathbf{c}^*\mathbf{G}_{12}^*\mathbf{y}$.

Proof of (24′): If $\mathbf{G}_{21}$ is a $\mathbf{B}_{21}$-matrix, then by Lemma 4, $\mathbf{G}_{21}^*$ is a $\mathbf{B}_{12}$-matrix, (24′) now follows from (24).

Proof of (25): If $\mathbf{G}_{22}$ is a $\mathbf{B}_{22}$-matrix, then there exist matrices $\mathbf{Q}, \mathbf{U}$ and $\mathbf{L}$ such that $\begin{bmatrix} \mathbf{Q} & \mathbf{U} \\ \mathbf{L} & \mathbf{G}_{22} \end{bmatrix} \in \mathbf{B}\{1\}$. From Theorem 1, we know that there exists a vectors $\mathbf{h}_1$ and $\mathbf{h}_2$ such that $\mathbf{X}^*\mathbf{h}_1 = \mathbf{c}_1$ and $\mathbf{X}^*\mathbf{h}_2 = \mathbf{c}_2$. Use (24) together with Lemma 1 to obtain

$$\begin{aligned}
\mathrm{Cov}(\psi_1, \psi_2) &= \mathrm{Cov}(\mathbf{c}_1^*\mathbf{U}^*\mathbf{y}, \mathbf{c}_2^*\mathbf{U}^*\mathbf{y}) = \sigma^2\mathbf{c}_1^*\mathbf{U}^*\mathbf{V}\mathbf{U}\mathbf{c}_2 \\
&= \sigma^2\mathbf{h}_1^*\mathbf{X}\mathbf{U}^*\mathbf{V}\mathbf{U}\mathbf{X}^*\mathbf{h}_1 \\
&= \sigma^2\mathbf{h}_1^*\mathbf{X}\mathbf{U}^*\mathbf{D}\mathbf{h}_2 && \text{(from (21))} \\
&= -\sigma^2\mathbf{h}_1^*\mathbf{X}\mathbf{U}^*\mathbf{X}\mathbf{G}_{22}\mathbf{X}^*\mathbf{h}_2 && \text{(from (23))} \\
&= -\sigma^2\mathbf{h}_1^*\mathbf{X}\mathbf{G}_{22}\mathbf{X}^*\mathbf{h}_2 && \text{(from (21) since } \mathbf{U}\in\mathbf{X}^*\{1\}) \\
&= -\sigma^2\mathbf{c}_1^*\mathbf{G}_{22}\mathbf{c}_2.
\end{aligned}$$

The fact that $\mathrm{Cov}(\psi_1, \psi_2)$ is also given by $-\sigma^2\mathbf{c}_2^*\mathbf{G}_{22}\mathbf{c}_1$ is immediate.

Proof of (26): This is obtained from (25) by taking $\mathbf{c}_1 = \mathbf{c}_2$.

Proof of (27): If $\mathbf{G}_{11}$ is a $\mathbf{B}_{11}$-matrix, then
$\mathbf{y}^*\mathbf{G}_{11}\mathbf{y} = (\mathbf{X}\mathbf{b}-\mathbf{e})^*\mathbf{G}_{11}(\mathbf{X}\mathbf{b}-\mathbf{e}) = \mathbf{b}^*\mathbf{X}^*\mathbf{G}_{11}\mathbf{X}\mathbf{b} - 2\mathbf{b}^*\mathbf{X}^*\mathbf{G}_{11}\mathbf{e} + \mathbf{e}^*\mathbf{G}_{11}\mathbf{e}$.
Using (17), together with the fact that $E(\mathbf{e}) = 0$, yields

$$\begin{aligned}
E(\mathbf{y}^*\mathbf{G}_{11}\mathbf{y}) &= E(\mathbf{e}^*\mathbf{G}_{11}\mathbf{e}) = E[\mathrm{Tr}(\mathbf{G}_{11}\mathbf{e}\mathbf{e}^*)] \\
&= \mathrm{Tr}[E(\mathbf{G}_{11}\mathbf{e}\mathbf{e}^*)] = \sigma^2\mathrm{Tr}(\mathbf{G}_{11}\mathbf{V}). \quad ■
\end{aligned}$$

In (27) we made the assumption that $\mathrm{Tr}(\mathbf{G}_{11}\mathbf{V}) \neq 0$. It can be shown that $\mathrm{Tr}(\mathbf{G}_{11}\mathbf{V}) = 0$ if and only if $R(\mathbf{V}) \subseteq R(\mathbf{X})$, which is clearly a pathological situation. The details are left as exercises.

Theorem 7 shows that once any element of $\mathbf{B}\{1\}$ is known, the problem of inference from a general linear model is completely solved and the problem of inference is thus reduced to the calculation of specific $\mathbf{B}_{ij}$-matrices.

Actually, knowledge of any $\mathbf{B}_{11}$-matrix together with any element of $\mathbf{X}\{1\}$ will suffice in order to produce the quantities of Theorem 7 (i.e. *a priori* knowledge of a $\mathbf{B}_{12}$-, $\mathbf{B}_{21}$- or a $\mathbf{B}_{22}$-matrix is not necessary).

Theorem 6.4.8 *If* $\mathbf{c}^*\mathbf{b}$ *is estimable under* $(\mathbf{y}, \mathbf{X}\mathbf{b}, \sigma^2\mathbf{V})$ *and if* $\mathbf{Q}$ *is any* $\mathbf{B}_{11}$*-matrix and* $\mathbf{X}^-$ *is any element of* $\mathbf{X}\{1\}$*, then each of the following is true.*

(i) *The* BLUE *of* $\mathbf{c}^*\mathbf{b}$ *is* $\psi = \mathbf{c}^*\mathbf{X}^-(\mathbf{I}-\mathbf{V}\mathbf{Q})\mathbf{y}$.

(ii) *If* ψ *is the* BLUE *of* $\mathbf{c}^*\mathbf{b}$, *then* $\mathrm{Var}(\psi) = \sigma^2\mathbf{c}^*\mathbf{X}^-(\mathbf{V}-\mathbf{V}\mathbf{Q}\mathbf{V})\mathbf{X}^{-*}\mathbf{c}$.
$= \sigma^2\mathbf{c}^*\mathbf{X}^-\mathbf{D}\mathbf{X}^{-*}\mathbf{c}$.

(iii) *If* ψ_1 *and* ψ_2 *are the* BLUE's *of* $\mathbf{c}_1^*\mathbf{b}$ *and* $\mathbf{c}_2^*\mathbf{b}$, *respectively, (assuming each are estimable) then*
$\mathrm{Cov}(\psi_1, \psi_2) = \sigma^2\mathbf{c}_1^*\mathbf{X}^-(\mathbf{V}-\mathbf{V}\mathbf{Q}\mathbf{V})\mathbf{X}^{-*}\mathbf{c}_2$.
$= \sigma^2\mathbf{c}_1^*\mathbf{X}^-\mathbf{D}\mathbf{X}^{-*}\mathbf{c}_2$

(iv) *An unbiased estimation for* σ^2 *is* $\gamma^{-1}\mathbf{y}^*\mathbf{Q}\mathbf{y}$ *where* $\gamma = \mathrm{Tr}(\mathbf{Q}\mathbf{V})$.

Proof If $\mathbf{Q}$ is a $\mathbf{B}_{11}$-matrix then (14)–(17) hold and it is not difficult to show that

$$\left[\begin{array}{c|c} \mathbf{Q} & (\mathbf{I}-\mathbf{Q}\mathbf{V})\mathbf{X}^{-*} \\ \hline \mathbf{X}^-(\mathbf{I}-\mathbf{V}\mathbf{Q}) & \mathbf{X}^-(\mathbf{V}\mathbf{Q}\mathbf{V}-\mathbf{V})\mathbf{X}^{-*} \end{array}\right] \in \mathbf{B}\{1\}.$$

The desired result now follows from Theorem 7. ■

If any $\mathbf{X}^-$ is known, then a $\mathbf{B}_{11}$-matrix is always available via the formula $\mathbf{Q} = (\mathbf{I} - \mathbf{XX}^-)^*[(\mathbf{I} - \mathbf{XX}^-)\mathbf{V}(\mathbf{I} - \mathbf{XX}^-)^*]^-(\mathbf{I} - \mathbf{XX}^-)$. However, if a $\mathbf{B}_{11}$-matrix is known, then computing $\mathbf{X}^-$ is unnecessary. The next result shows that once a $\mathbf{B}_{11}$-matrix is known, then all one needs is any solution $\mathbf{l}^*$ of the system $\mathbf{l}^*\mathbf{X} = \mathbf{c}$.

Theorem 6.4.9 *If $\mathbf{c}^*\mathbf{b}$ is estimable under $(\mathbf{y}, \mathbf{Xb}, \sigma^2\mathbf{V})$ and $\mathbf{l}^*$ is any solution of $\mathbf{c}^* = \mathbf{l}^*\mathbf{X}$ and $\mathbf{Q}$ is any $\mathbf{B}_{11}$-matrix, then each of the following is true.*

(i) *The* BLUE *of* $\mathbf{c}^*\mathbf{b}$ *is* $\psi = \mathbf{l}^*(\mathbf{I} - \mathbf{VQ})\mathbf{y}$

(ii) *If* ψ *is the* BLUE *of* $\mathbf{c}^*\mathbf{b}$, *then* $\mathrm{Var}(\psi) = \sigma^2\mathbf{l}^*(\mathbf{V} - \mathbf{VQV})\mathbf{l} = \sigma^2\mathbf{l}^*\mathbf{D}\mathbf{l}$.

(iii) *If* ψ_1 *and* ψ_2 *are* BLUE's *of* $\mathbf{c}_1^*\mathbf{b}$ *and* $\mathbf{c}_2^*\mathbf{b}$, *respectively, then* $\mathrm{Cov}(\psi_1, \psi_2) = \sigma^2\mathbf{l}_1^*(\mathbf{V} - \mathbf{VQV})\mathbf{l} = \sigma^2\mathbf{l}_1^*\mathbf{D}\mathbf{l}_2$ *where* $\mathbf{l}_1^*\mathbf{X} = \mathbf{c}_1^*$ *and* $\mathbf{l}_2^*\mathbf{X} = \mathbf{c}_2^*$.

Proof We know from Theorem 1 that $\mathbf{l}^*\mathbf{X} = \mathbf{c}^*$ is always consistent so that $\mathbf{c}^* = \mathbf{c}^*\mathbf{XX}^\dagger$. For a particular solution, $\mathbf{l}_o^*$, there is always a particular member $\mathbf{X}_o^- \in \mathbf{X}\{1\}$ such that $\mathbf{l}_o^* = \mathbf{c}^*\mathbf{X}_o^-$, namely $\mathbf{X}_o^- = \mathbf{X}^\dagger + \mathbf{c}^{*\dagger}\mathbf{l}_o^* - \mathbf{c}^{*\dagger}\mathbf{c}^*\mathbf{X}^\dagger$. The desired conclusions now follow from Theorem 8. ■

We conclude by considering the special, but important, case when $\mathbf{V}$ is non-singular. It is a simple exercise to show that $(\mathbf{X}^*\mathbf{V}^{-1}\mathbf{X})^-\mathbf{X}^*\mathbf{V}^{-1} \in \mathbf{X}\{1\}$. It is then easy to use Lemma 2 to show that $[\mathbf{V}^{-1} - \mathbf{V}^{-1}\mathbf{X}(\mathbf{X}^*\mathbf{V}^{-1}\mathbf{X})^- \times \mathbf{X}^*\mathbf{V}^{-1}]$ is a $\mathbf{B}_{11}$-matrix. Therefore, $\mathbf{D} = \mathbf{V} - \mathbf{VQV} = \mathbf{X}(\mathbf{X}^*\mathbf{V}^{-1}\mathbf{X})^-\mathbf{X}^*$ and it is clear from Lemma 4 that $(\mathbf{X}^*\mathbf{V}^{-1}\mathbf{X})^-\mathbf{X}^*\mathbf{V}^{-1}$ is a $\mathbf{B}_{21}$-matrix and $-(\mathbf{X}^*\mathbf{V}^{-1}\mathbf{X})^-$ is a $\mathbf{B}_{22}$-matrix.

These observations together with (7) give the following useful result.

Corollary 6.4.4 *If $\mathbf{c}^*\mathbf{b}$ is estimable under $(\mathbf{y}, \mathbf{Xb}, \sigma^2\mathbf{V})$ and $\mathbf{V}$ is non-singular, then each of the following hold.*

(i) *The* BLUE *of* $\mathbf{c}^*\mathbf{b}$ *is* $\psi = \mathbf{c}^*(\mathbf{X}^*\mathbf{V}^{-1}\mathbf{X})^-\mathbf{X}^*\mathbf{V}^{-1}\mathbf{y}$.

(ii) $\mathrm{Var}(\psi) = \sigma^2\mathbf{c}^*(\mathbf{X}^*\mathbf{V}^{-1}\mathbf{X})^-\mathbf{c}$ *and* $\mathrm{Cov}(\psi_1, \psi_2) = \sigma^2\mathbf{c}_1^*(\mathbf{X}^*\mathbf{V}^{-1}\mathbf{X})^-\mathbf{c}_2$ *where* ψ_1 *and* ψ_2 *are the* BLUE's *for* $\mathbf{c}_1^*\mathbf{b}$ *and* $\mathbf{c}_2^*\mathbf{b}$.

(iii) *An unbiased estimator for* σ^2 *is given by* $\gamma^{-1}\mathbf{y}^*[\mathbf{V}^{-1} - \mathbf{V}^{-1}\mathbf{X}(\mathbf{X}^*\mathbf{V}^{-1}\mathbf{X})^-\mathbf{X}^*\mathbf{V}^{-1}]\,\mathbf{y}$ *where* $\gamma = n\text{-rank}(\mathbf{X})$.

Perhaps the most common situation encountered is when $\mathbf{V} = \mathbf{I}$, in which case we have the following.

Corollary 6.4.5 *If $\mathbf{c}^*\mathbf{b}$ is estimable under $(\mathbf{y}, \mathbf{Xb}, \sigma^2\mathbf{I})$, then each of the following hold.*

(i) *The* BLUE *of* $\mathbf{c}^*\mathbf{b}$ *is* $\psi = \mathbf{c}^*(\mathbf{X}^*\mathbf{X})^-\mathbf{X}^*\mathbf{y}$.

(ii) $\mathrm{Var}(\psi) = \sigma^2\mathbf{c}^*(\mathbf{X}^*\mathbf{X})^-\mathbf{c}$ *and* $\mathrm{Cov}(\psi_1, \psi_2) = \sigma^2\mathbf{c}_1^*(\mathbf{X}^*\mathbf{X})^-\mathbf{c}_2$ *where* ψ_1 *and* ψ_2 *are the* BLUE's *for* $\mathbf{c}_1^*\mathbf{b}$ *and* $\mathbf{c}_2^*\mathbf{b}$.

(iii) *An unbiased estimator for* σ^2 *is given by* $\gamma^{-1}\mathbf{y}^*[\mathbf{I} - \mathbf{X}(\mathbf{X}^*\mathbf{X})^-\mathbf{X}^*]\mathbf{y}$ *where* $\gamma = n\text{-rank}(\mathbf{X})$.

Notice that $(\mathbf{X}^*\mathbf{X})^-\mathbf{X}^* \in \mathbf{X}\{1,3\}$ so that $(\mathbf{X}^*\mathbf{X})^-\mathbf{X}^*\mathbf{y}$ is just another way of representing any least squares solution of $\mathbf{Xb} = \mathbf{y}$. Also $\mathbf{c}^*(\mathbf{X}^*\mathbf{X})^-$ is just any solution, $\mathbf{l}^*$, of $\mathbf{c}^* = \mathbf{l}^*(\mathbf{X}^*\mathbf{X})$. Thus Corollary 4 can also be stated in terms of solutions of $\mathbf{c}^* = \mathbf{l}^*(\mathbf{X}^*\mathbf{X})$ or in terms of least squares solutions of $\mathbf{Xb} = \mathbf{y}$. Similar remarks can be made about the results in Corollary 3 because $\mathbf{c}^*(\mathbf{X}^*\mathbf{V}^{-1}\mathbf{X})^-$ represents any solution of $\mathbf{c}^* = \mathbf{l}^*(\mathbf{X}^*\mathbf{V}^{-1}\mathbf{X})$ and $(\mathbf{X}^*\mathbf{V}^{-1}\mathbf{X})^-\mathbf{X}^*\mathbf{V}^{-1}\mathbf{y}$ represents any weighted least squares solution of $\mathbf{Xb} = \mathbf{y}$. (By a weighted least squares solution of $\mathbf{Xb} = \mathbf{y}$, we mean any vector $\mathbf{z}$ such that $\|\mathbf{Xz} - \mathbf{y}\|^2_{\mathbf{V}^{-1}} = (\mathbf{Xz} - \mathbf{y})^*\mathbf{V}^{-1}(\mathbf{Xz} - \mathbf{y})$ is minimized, or equivalently, any solution of the weighted normal equations $\mathbf{X}^*\mathbf{V}^{-1}\mathbf{Xz} = \mathbf{X}^*\mathbf{V}^{-1}\mathbf{y}$.)

In conclusion, we note that not only are linear models with singular variance matrices representable as restricted linear models but that restricted linear models $(\mathbf{y}, \mathbf{Xb} \mid \mathbf{Rb} = \mathbf{f}, \sigma^2\mathbf{V})$ are just special cases of linear models where the variance matrix is singular. Indeed, one can always write

$$\tilde{\mathbf{x}} = \begin{bmatrix} \mathbf{X} \\ \mathbf{R} \end{bmatrix}, \tilde{\mathbf{y}} = \begin{bmatrix} \mathbf{y} \\ \mathbf{f} \end{bmatrix} \text{ and } \tilde{\mathbf{V}} = \left[\begin{array}{c|c} \mathbf{V} & \mathbf{0} \\ \hline \mathbf{0} & \mathbf{0} \end{array}\right]$$

and it is clear that the restricted model $(\mathbf{y}, \mathbf{Xb} \mid \mathbf{Rb} = \mathbf{f}, \sigma^2\mathbf{V})$ is equivalent to $(\tilde{\mathbf{y}}, \tilde{\mathbf{X}}\mathbf{b}, \sigma^2\tilde{\mathbf{V}})$ where $\tilde{\mathbf{V}}$ is singular.

5. Exercises

Verify each of the following assertions.

1. $(\mathbf{A}^- \otimes \mathbf{B}^-) \in (\mathbf{A} \otimes \mathbf{B})\{1\}$ where $\mathbf{A} \otimes \mathbf{B}$ denotes the Kronecker product of $\mathbf{A}$ and $\mathbf{B}$.
2. Let $\mathbf{G} = \mathbf{U}(\mathbf{VAU})^-\mathbf{V}$ and let $\text{rank}(\mathbf{A}) = r$. Each of the following is true. (a) $\mathbf{G} \in \mathbf{A}\{1\}$ iff $\text{rank}(\mathbf{VAU}) = r$. (b) $\mathbf{G} \in \mathbf{A}\{2\}$ and $R(\mathbf{G}) = R(\mathbf{U})$ iff $\text{rank}(\mathbf{VAU}) = \text{rank}(\mathbf{U})$. (c) $\mathbf{G} \in \mathbf{A}\{2\}$ and $N(\mathbf{G}) = N(\mathbf{V})$ iff $\text{rank}(\mathbf{VAU}) = \text{rank}(\mathbf{V})$. (d) $\mathbf{G}$ is a $(R(\mathbf{U}), N(\mathbf{V}))$-inverse for $\mathbf{A}$ iff $\text{rank}(\mathbf{U}) = \text{rank}(\mathbf{V}) = \text{rank}(\mathbf{VAU}) = r$.
3. If $\text{rank}(\mathbf{A}^*\mathbf{VA}) = \text{rank}(\mathbf{A})$, then $\mathbf{A}(\mathbf{A}^*\mathbf{VA})^-(\mathbf{A}^*\mathbf{VA}) = \mathbf{A}$ and $(\mathbf{A}^*\mathbf{VA})(\mathbf{A}^*\mathbf{VA})^-\mathbf{A}^* = \mathbf{A}^*$.
4. Verify $\mathbf{A}(\mathbf{A}^*\mathbf{A})^-\mathbf{A}^* = \mathbf{A}\mathbf{A}^\dagger$.
5. If $R(\mathbf{C}) \subseteq R(\mathbf{A})$ and $RS(\mathbf{R}) \subseteq RS(\mathbf{A})$, then $\mathbf{RA}^-\mathbf{C}$ is invariant over $\mathbf{A}\{1\}$.
6. $\mathbf{A}^-\mathbf{A}\mathbf{A}^- \in \mathbf{A}\{1,2\}$.
7. For $\mathbf{r} \in \mathbb{C}^{1 \times n}$, $\mathbf{rA}^-\mathbf{A} = \mathbf{r} \Leftrightarrow \mathbf{r} \in RS(\mathbf{A})$
8. For $\mathbf{G} \in \mathbf{A}\{1\}$, the following statements are equivalent: (a) $\mathbf{G} \in \mathbf{A}\{1,2\}$, (b) $\text{rank}(\mathbf{A}) = \text{rank}(\mathbf{G})$, (c) $\mathbf{G} = \mathbf{G}_1\mathbf{A}\mathbf{G}_2$ for some $\mathbf{G}_1, \mathbf{G}_2 \in \mathbf{A}\{1\}$.
9. If $\mathbf{A} \in \mathbb{C}^{m \times n}$ has rank r, then there exists $\mathbf{G}_i \in \mathbf{A}\{1\}$ such that $\text{rank}(\mathbf{G}_i) = r + i$, $i = 0, 1, 2, \ldots, \min(m, n)$.
10. If $\mathbf{A} \in \mathbb{C}^{n \times n}$ and $\mathbf{P}$ is a non-singular matrix such that $\mathbf{PA} = \mathbf{H}$ where $\mathbf{H}$ is the Hermite canonical form for $\mathbf{A}$, then $\mathbf{P} \in \mathbf{A}\{1\}$.
11. For every $\mathbf{A} \in \mathbb{C}^{m \times n}$, $\mathbf{B} \in \mathbb{C}^{n \times p}$, there exists a $\mathbf{G} \in \mathbf{A}\{1\}$ and $\mathbf{F} \in \mathbf{B}\{1\}$ such $\mathbf{FG} \in (\mathbf{AB})\{1\}$.

12. Let $K^{m \times n}$ denotes the set of $m \times n$ matrices with integer entries. If $\mathbf{A} \in K^{m \times n}$, then there exists $\mathbf{G} \in \mathbf{A}\{1,2\}$ such that $\mathbf{AG} \in K^{m \times m}$ and $\mathbf{GA} \in K^{n \times n}$. (Hint: Consider $\mathbf{G} = \mathbf{QS}^{\dagger}\mathbf{P}$ where $\mathbf{PAQ} = \mathbf{S}$ and $\mathbf{S}$ is the Smith canonical form with $\mathbf{P}$ and $\mathbf{Q}$ being non-singular.)
13. Let $\mathbf{A} \in K^{m \times n}$ and $\mathbf{b} \in K^{m \times 1}$ where $\mathbf{b} \in R(\mathbf{A})$. Let $\mathbf{G} = \mathbf{QS}^{\dagger}\mathbf{P}$ (as described above). $\mathbf{Ax} = \mathbf{b}$ has an integer solution if and only if $\mathbf{Gb} \in K^{n \times 1}$; in which case the general integer solution is given by $\mathbf{x} = \mathbf{Gb} + (\mathbf{I} - \mathbf{GA})\mathbf{h}$, $\mathbf{h} \in K^{n \times 1}$.
14. Let $\mathbf{P}^*$ and $\mathbf{Q}^*$ be permutation matrices such that $\mathbf{M} = \mathbf{P}^*\begin{bmatrix} \mathbf{A} & \mathbf{C} \\ \mathbf{R} & \mathbf{D} \end{bmatrix}\mathbf{Q}^*$ where $\operatorname{rank}(\mathbf{M}) = \operatorname{rank}(\mathbf{A}_{r \times r}) = r$ and let $\mathbf{T} = \mathbf{A}^{-1}\mathbf{C}$, $\mathbf{S} = \mathbf{RA}^{-1}$. Then

    (i) $\mathbf{Q}\left[\begin{array}{c|c} \mathbf{A}^{-1} & \mathbf{0} \\ \hline \mathbf{0} & \mathbf{0} \end{array}\right]\mathbf{P} \in \mathbf{M}\{1,2\}$,

    (ii) $\mathbf{Q}\begin{bmatrix} \mathbf{A}^{-1} \\ \mathbf{0} \end{bmatrix}(\mathbf{I} + \mathbf{S}^*\mathbf{S})^{-1}[\mathbf{I} \mid \mathbf{S}^*]\mathbf{P} \in \mathbf{M}\{1,2,3\}$,

    (iii) $\mathbf{Q}\begin{bmatrix} \mathbf{I} \\ \mathbf{T}^* \end{bmatrix}(\mathbf{I} + \mathbf{TT}^*)^{-1}[\mathbf{A}^{-1} \mid \mathbf{0}]\mathbf{P} \in \mathbf{M}\{1,2,4\}$,

    (iv) $\mathbf{Q}\begin{bmatrix} \mathbf{I} \\ \mathbf{T}^* \end{bmatrix}(\mathbf{I} + \mathbf{TT}^*)^{-1}\mathbf{A}^{-1}(\mathbf{I} + \mathbf{S}^*\mathbf{S})[\mathbf{I} \mid \mathbf{S}^*]\mathbf{P} = \mathbf{M}^{\dagger}$.

15. Let $\mathbf{P}$ and $\mathbf{Q}$ be non-singular matrices and let $\mathbf{A}$ be an $r \times r$ non-singular matrix such that $\mathbf{M} = \mathbf{P}^{-1}\begin{bmatrix} \mathbf{A} & \mathbf{0} \\ \mathbf{0} & \mathbf{0} \end{bmatrix}\mathbf{Q}^{-1}$.

    Let $\mathbf{G} = \mathbf{Q}\begin{bmatrix} \mathbf{Z} & \mathbf{U} \\ \mathbf{V} & \mathbf{W} \end{bmatrix}\mathbf{P}$. Then

    (i) $\mathbf{G} \in \mathbf{M}\{1\}$ iff $\mathbf{Z} = \mathbf{A}^{-1}$.
    (ii) $\mathbf{G} \in \mathbf{M}\{1,2\}$ iff $\mathbf{Z} = \mathbf{A}^{-1}$ and $\mathbf{W} = \mathbf{VAU}$.
    (iii) $\mathbf{G} \in \mathbf{M}\{1,2,3\}$ iff $\mathbf{Z} = \mathbf{A}^{-1}$, $\mathbf{U} = -\mathbf{A}^{-1}\mathbf{P}_1\mathbf{P}_2^{\dagger}$, $\mathbf{W} = -\mathbf{VP}_1\mathbf{P}_2^{\dagger}$ where $\mathbf{P} = \begin{bmatrix} \mathbf{P}_1 \\ \mathbf{P}_2 \end{bmatrix}$
    (iv) $\mathbf{G} \in \mathbf{M}\{1,2,4\}$ iff $\mathbf{Z} = \mathbf{A}^{-1}$, $\mathbf{V} = -\mathbf{Q}_2^{\dagger}\mathbf{Q}_1\mathbf{A}^{-1}$, $\mathbf{W} = -\mathbf{Q}_2^{\dagger}\mathbf{Q}_1\mathbf{U}$ where $\mathbf{Q} = [\mathbf{Q}_1 \mid \mathbf{Q}_2]$
    (v) $\mathbf{G} = \mathbf{M}^{\dagger}$ iff $\mathbf{Z} = \mathbf{A}^{-1}$, $\mathbf{U} = -\mathbf{A}^{-1}\mathbf{P}_1\mathbf{P}_2^{\dagger}$, $\mathbf{V} = -\mathbf{Q}_2^{\dagger}\mathbf{Q}_1\mathbf{A}^{-1}$, and $\mathbf{W} = \mathbf{Q}_2^{\dagger}\mathbf{Q}_1\mathbf{A}^{-1}\mathbf{P}_1\mathbf{P}_2^{\dagger}$

16. For $\mathbf{A} \in \mathbb{C}^{m \times n}$, $\mathbf{c} \in \mathbb{C}^m$, $\mathbf{d} \in \mathbb{C}^n$, let $\mathbf{E} = \mathbf{I} - \mathbf{AA}^-$, $\mathbf{F} = \mathbf{I} - \mathbf{A}^-\mathbf{A}$, and $\beta = 1 + \mathbf{d}^*\mathbf{A}^-\mathbf{c}$.

    A (1)-inverse for $\mathbf{A} + \mathbf{cd}^*$ is given by one of the following:

    (i) $(\mathbf{A} + \mathbf{cd}^*)^- = \mathbf{A}^- - \dfrac{\mathbf{A}^-\mathbf{cc}^*\mathbf{E}^*\mathbf{E}}{\mathbf{c}^*\mathbf{E}^*\mathbf{Ec}} - \dfrac{\mathbf{FF}^*\mathbf{dd}^*\mathbf{A}^-}{\mathbf{d}^*\mathbf{FF}^*\mathbf{d}} + \beta\dfrac{\mathbf{FF}^*\mathbf{dc}^*\mathbf{E}^*\mathbf{E}}{(\mathbf{c}^*\mathbf{E}^*\mathbf{Ec})(\mathbf{d}^*\mathbf{FF}^*\mathbf{d})}$ when $\mathbf{c} \notin R(\mathbf{A})$, $\mathbf{d} \notin R(\mathbf{A}^*)$.

(ii) $(\mathbf{A}+\mathbf{cd}^*)^- = \mathbf{A}^- - \dfrac{\mathbf{FF}^*\mathbf{dd}^*\mathbf{A}^-}{\mathbf{d}^*\mathbf{FF}^*\mathbf{d}}$ when $\beta = 0$, $\mathbf{c}\in R(\mathbf{A})$, $\mathbf{d}\notin R(\mathbf{A}^*)$

(iii) $(\mathbf{A}+\mathbf{cd}^*)^- = \mathbf{A}^- - \beta^{-1}\mathbf{A}^-\mathbf{cd}^*\mathbf{A}^-$ when $\beta \neq 0$ and either $\mathbf{c}\in R(\mathbf{A})$ or $\mathbf{d}\in R(\mathbf{A}^*)$.

(iv) $(\mathbf{A}+\mathbf{cd}^*)^- = \mathbf{A}^- - \dfrac{\mathbf{A}^-\mathbf{cc}^*\mathbf{E}^*\mathbf{E}}{\mathbf{c}^*\mathbf{E}^*\mathbf{Ec}}$ when $\beta = 0$, $\mathbf{c}\notin R(\mathbf{A})$, $\mathbf{d}\in R(\mathbf{A}^*)$

(v) $(\mathbf{A}+\mathbf{cd}^*)^- = \mathbf{A}^-$ when $\beta = 0$, $\mathbf{c}\in R(\mathbf{A})$, $\mathbf{d}\in R(\mathbf{A}^*)$.

17. $\mathbf{A}^\dagger = \mathbf{A}^*(\mathbf{A}^*\mathbf{AA}^*)^-\mathbf{A}^*$.
18. $\mathbf{G}\in\mathbf{A}\{2\}$ iff there exist a pair of orthogonal projections $\mathbf{P}$ and $\mathbf{Q}$ such that $\mathbf{G} = (\mathbf{PAQ})^\dagger$.
19. $\mathbf{G}\in\mathbf{A}\{1,2\}$ iff there exist $\mathbf{A}_1^-, \mathbf{A}_2^-\in\mathbf{A}\{1\}$ such that $\mathbf{G} = \mathbf{A}_1^-\mathbf{AA}_2^-$.
20. $\mathbf{A}\{1,3\} = \{\mathbf{A}^\dagger + (\mathbf{I}-\mathbf{A}^-\mathbf{A})\mathbf{H}\,|\,\mathbf{H}$ is arbitrary$\}$, $\mathbf{A}\{1,4\} = \{\mathbf{A}^\dagger + \mathbf{K}(\mathbf{I}-\mathbf{AA}^-)\,|\,\mathbf{K}$ is arbitrary$\}$.
21. Let $\mathbf{A}$ be $n\times n$. If $\mathbf{P}$ is a non-singular matrix such that $\mathbf{PA}^*\mathbf{A}$ is in Hermite form, then $\mathbf{PA}^*\in\mathbf{A}\{1,3\}$.
22. Let M be a subspace and let $\mathbf{P} = \mathbf{P}_M$ and $\mathbf{P}_\perp = \mathbf{P}_{M^\perp}$. The constrained system $\mathbf{Ax}+\mathbf{y} = \mathbf{b}$, $\mathbf{x}\in M$, $\mathbf{y}\in M^\perp$ is consistent iff $\mathbf{b}\in R(\mathbf{AP}+\mathbf{P}_\perp)$; in which case the solutions are $\mathbf{x} = \mathbf{P}(\mathbf{AP}+\mathbf{P}_\perp)^-\mathbf{b}$ and $\mathbf{y} = \mathbf{b}-\mathbf{Ax}$. When $(\mathbf{AP}+\mathbf{P}_\perp)^{-1}$ exists, the matrix $\mathbf{G} = \mathbf{P}(\mathbf{AP}+\mathbf{P}_\perp)^{-1}\mathbf{b}$ is called the *Bott–Duffin inverse*.
23. $(\mathbf{AP}+\mathbf{P}_\perp)^{-1}$ exists iff $(\mathbf{K}^*\mathbf{AK})^{-1}$ exists where the columns of $\mathbf{K}$ form a basis for M. The Bott–Duffin inverse is $\mathbf{G} = \mathbf{K}(\mathbf{K}^*\mathbf{AK})^{-1}\mathbf{K}^*$.
24. When it exists, the Bott–Duffin inverse is the $(M, M^\perp)$-inverse of $\mathbf{PAP}$.
25. Let $\mathbf{A}_{m\times n}$ denote an incidence matrix of a directed graph consisting of m nodes $\{\mathrm{N}_1, \mathrm{N}_2, \ldots, \mathrm{N}_m\}$ and n directed paths $\{\mathrm{P}_1, \mathrm{P}_2, \ldots, \mathrm{P}_n\}$ between nodes. That is, $a_{ij} = 1$ if P_j is a path directed *away from* N_i, $a_{ij} = -1$ if P_j is a path directed *into* N_i, and $a_{ij} = 0$ if P_j is a path which neither leads away from or into N_i. Suppose the graph is connected (i.e. every pair of nodes is connected by some sequence of paths.) If $\mathbf{G}\in\mathbf{A}\{1,3\}$, then $\mathbf{I}-\mathbf{AG} = \dfrac{1}{m}\mathbf{J}$ where $\mathbf{J}$ is a matrix of 1's.
26. If $\mathbf{A}$ is the incidence matrix of a connected di-graph, then $\operatorname{rank}(\mathbf{A}) = m-1$, where $m =$ number of nodes.
27. Let $\mathbf{W}$ be a positive definite matrix and let $\|\cdot\|_{\mathbf{W}}$ be the norm associated with $\mathbf{W}$ (i.e. $\|\mathbf{x}\|_{\mathbf{W}}^2 = \mathbf{x}^*\mathbf{Wx}$). $\mathbf{G}$ is a matrix such that $\mathbf{x} = \mathbf{Gb}$ is a weighted least squares solution ($\|\mathbf{Ax}-\mathbf{b}\|_{\mathbf{W}}$ is minimal) for every $\mathbf{b}$ iff $\mathbf{G}$ satisfies $\mathbf{AGA} = \mathbf{A}$ and $(\mathbf{WAG})^* = \mathbf{WAG}$ (A weighted (1, 3)-inverse).
28. Let $\mathbf{V}$ be positive definite. $\mathbf{Gy}$ is a $\mathbf{V}^{-1}$-least squares solution of $\mathbf{Xb} = \mathbf{y}$ for all $\mathbf{y}$ iff $\mathbf{G}$ is a $\mathbf{B}_{21}$-matrix.
29. Let $\mathbf{V}$ be positive definite and suppose $\mathbf{Ax} = \mathbf{b}$ is consistent. $\mathbf{G}$ is a matrix such that $\mathbf{x} = \mathbf{Gb}$ is the minimal $\mathbf{V}$-norm solution of $\mathbf{Ax} = \mathbf{b}$ for all $\mathbf{b}\in R(\mathbf{A})$ iff $\mathbf{G}$ satisfies $\mathbf{AGA} = \mathbf{A}$ and $(\mathbf{VGA})^* = \mathbf{VGA}$ (A weighted (1, 4)-inverse).

30. (Weighted Moore–Penrose inverse) $\mathbf{AGA} = \mathbf{A}$, $\mathbf{GAG} = \mathbf{G}$, $(\mathbf{WAG})^* = \mathbf{WAG}$, and $(\mathbf{VGA})^* = \mathbf{VGA}$, iff for all $\mathbf{b}$, $\mathbf{Gb}$ is the $\mathbf{W}$-least squares solution of $\mathbf{Ax} = \mathbf{b}$ which has minimal $\mathbf{V}$-norm. Moreover, there exists a unique solution for $\mathbf{G}$ which can be expressed as $\mathbf{G} = \mathbf{V}^{-1/2} \times (\mathbf{W}^{1/2}\mathbf{A}\mathbf{V}^{-1/2})^\dagger \mathbf{W}^{1/2} = \mathbf{V}^{-1}\mathbf{A}^*\mathbf{WA}(\mathbf{A}^*\mathbf{WAV}^{-1}\mathbf{A}^*\mathbf{WA})^-\mathbf{A}^*\mathbf{W}$.
31. Let $\mathbf{V}$ be positive semi-definite and let $\|\mathbf{x}\|_{\mathbf{V}}$ denote the semi-norm $(\mathbf{x}^*\mathbf{Vx})^{1/2}$. A vector ξ is called a minimum $\mathbf{V}$-semi-norm solution of the system $\mathbf{X}^*\mathbf{z} = \mathbf{c}$, $\mathbf{c} \in R(\mathbf{X}^*)$ if $\mathbf{X}^*\xi = \mathbf{c}$ and $\|\xi\|_{\mathbf{V}}$ is minimal among all solutions. The following statements are equivalent.

    (i) $\mathbf{Gc}$ is a minimum $\mathbf{V}$-semi-norm for every $\mathbf{c} \in R(\mathbf{X}^*)$.
    (ii) $\mathbf{G}$ is a $\mathbf{B}_{12}$-matrix.
    (iii) $\mathbf{G} \in \mathbf{X}^*\{1\}$ and $\mathbf{XG}^*\mathbf{V} = \mathbf{VGX}^*$. (The same as in Exercise 29).

32. Let $\mathbf{W}$ be positive semi-definite. $\mathbf{G}$ is a matrix such that $\mathbf{Gb}$ is a $\mathbf{W}$-least squares solution of $\mathbf{Ax} = \mathbf{b}$, for all $\mathbf{b}$, iff $\mathbf{A}^*\mathbf{WAG} = \mathbf{A}^*\mathbf{W}$. This last equation is equivalent to the two conditions $\mathbf{WAGA} = \mathbf{WA}$ and $(\mathbf{WAG})^* = \mathbf{WAG}$. (Notice that $\mathbf{G}$ is not necessarily in $\mathbf{A}\{1\}$, as was the case in Exercise 27.)
33. Let $\mathbf{V}$ and $\mathbf{W}$ be positive semi-definite. $\mathbf{G}$ is a matrix such that $\mathbf{Gb}$ is a minimal $\mathbf{V}$-semi-norm $\mathbf{W}$-least squares solution of $\mathbf{Ax} = \mathbf{b}$ iff $\mathbf{G}$ satisfies the four conditions $\mathbf{WAGA} = \mathbf{WA}$, $\mathbf{VGAG} = \mathbf{VG}$, $(\mathbf{WAG})^* = \mathbf{WAG}$, and $(\mathbf{VGA})^* = \mathbf{VGA}$. (If $\mathbf{V}$ is positive definite, there exists a unique solution for $\mathbf{G}$. If $\mathbf{V}$ is just semi-definite, $\mathbf{G}$ may not be unique.)
34. If $\mathbf{V}$ and $\mathbf{W}$ are positive semi-definite and $\mathbf{X} = \mathbf{A}^*\mathbf{WA}$, then every $\mathbf{B}_{12}$-matrix satisfies the four conditions of Exercise 33.
35. If $\mathbf{Q}$ is any $\mathbf{B}_{11}$-matrix, then $\mathrm{Tr}(\mathbf{QV}) = \mathrm{rank}[\mathbf{V} \mid \mathbf{X}] - \mathrm{rank}[\mathbf{X}]$. Furthermore, $\mathrm{Tr}(\mathbf{QV}) = 0$ iff $R(\mathbf{V}) \subseteq R(\mathbf{X})$ iff $\mathbf{0}$ is a $\mathbf{B}_{11}$-matrix.
36. $(\mathbf{V} + \mathbf{XX}^*)^-\mathbf{X}[\mathbf{X}^*(\mathbf{V} + \mathbf{XX}^*)^-\mathbf{X}]^-$ is always a $\mathbf{B}_{12}$-matrix. If $R(\mathbf{X}) \subseteq R(\mathbf{V})$, then $\mathbf{V}^-\mathbf{X}(\mathbf{X}^*\mathbf{V}^-\mathbf{X})^-$ is a $\mathbf{B}_{12}$-matrix.
37. The matrix $\begin{bmatrix} \mathbf{V} & \mathbf{X} \\ \mathbf{X}^* & \mathbf{0} \end{bmatrix}$ is non-singular iff $\mathrm{rank}(\mathbf{V}_{n \times n}) = n$ and $\mathrm{rank}(\mathbf{X}_{n \times k}) = k$.
38. If $\mathbf{M}$ is any matrix such that $R(\mathbf{V} + \mathbf{XMX}^*) = R([\mathbf{V} \mid \mathbf{X}])$ and if $\mathbf{W} = (\mathbf{V} + \mathbf{XMX}^*)^-$, then $\mathbf{L} = (\mathbf{X}^*\mathbf{WX})^-\mathbf{X}^*\mathbf{W}$ is a $\mathbf{B}_{21}$-matrix, $\mathbf{L}^*$ is a $\mathbf{B}_{12}$-matrix, $\mathbf{W}(\mathbf{I} - \mathbf{XL})$ is a $\mathbf{B}_{11}$-matrix, and $(\mathbf{X}^*\mathbf{WX})^- - \mathbf{M}$ is a $\mathbf{B}_{22}$-matrix.
39. The following statements are equivalent. (i) The invariant term $\mathbf{D}$ of Lemma 4.4 is the zero matrix. (ii) $\mathbf{0}$ is a $\mathbf{B}_{22}$-matrix. (iii) $R(\mathbf{V}) = R(\mathbf{V}|_{N(\mathbf{X}^*)})$. (iv) $\mathrm{rank}(\mathbf{V}) = \mathrm{Tr}(\mathbf{VQ})$ where $\mathbf{Q}$ is any $\mathbf{B}_{11}$-matrix. (v) $R(\mathbf{V}) \cap R(\mathbf{X}) = \mathbf{0}$.
40. (Use of 2-inverses in a generalized Newton's Method.) Let $\mathbf{x}_0 \in \mathbb{C}^n$ and let $\mathbf{B}(\mathbf{x}_0, r)$ be the open ball of radius r centred at $\mathbf{x}_0$. Let $\mathbf{f}$ be a function $\mathbf{f}: \mathbf{B}(\mathbf{x}_0, r) \to \mathbb{C}^m$ and let $\mathbf{J}(\mathbf{x}) \in \mathbb{C}^{m \times n}$, $\mathbf{X}(\mathbf{x}) \in \mathbb{C}^{n \times m}$ be defined for $\mathbf{x} \in \overline{\mathbf{B}(\mathbf{x}_0, \mathbf{r})}$ where $\mathbf{X}(\mathbf{x}) \in \mathbf{J}(\mathbf{x})\{2\}$. Suppose δ, ε, and γ are constants such

that the following hold:

(i) $\|\mathbf{f}(\mathbf{u}) - \mathbf{f}(\mathbf{v}) - \mathbf{J}(\mathbf{v})(\mathbf{u} - \mathbf{v})\| \leq \varepsilon \|\mathbf{u} - \mathbf{v}\|$, for $\mathbf{u}, \mathbf{v} \in B(\mathbf{x}_o, r)$ with $\mathbf{u} - \mathbf{v} \in R(\mathbf{X}(\mathbf{v}))$.

(ii) $\|(\mathbf{X}(\mathbf{u}) - \mathbf{X}(\mathbf{v}))\mathbf{f}(\mathbf{v})\| \leq \gamma \|\mathbf{u} - \mathbf{v}\|$, for $\mathbf{u}, \mathbf{v} \in B(\mathbf{x}_o, r)$.

(iii) $\varepsilon \|\mathbf{X}(\mathbf{u})\| + \gamma \leq \delta < 1$ for $\mathbf{u} \in B(\mathbf{x}_o, r)$

(iv) $\|\mathbf{X}(\mathbf{x}_o)\| \, \|\mathbf{f}(\mathbf{x}_o)\| < (1 + \delta)r$.

Then the sequence $\mathbf{x}_{k+1} = \mathbf{x}_k - \mathbf{X}(\mathbf{x}_k)\mathbf{f}(\mathbf{x}_k)$ converges to a point $\mathbf{p} \in \overline{B(\mathbf{x}_o, r)}$ which is a solution of $\mathbf{X}(\mathbf{p})\mathbf{f}(\mathbf{x}) = \mathbf{0}$. (If $\mathbf{X}(\mathbf{p})$ has full column rank, then $\mathbf{p}$ is a solution of $\mathbf{f}(\mathbf{x}) = \mathbf{0}$.)

41. For any choices of (1)-inverses for $\mathbf{X}, \mathbf{X}^*$, and $\mathbf{K} = \mathbf{E}_X \mathbf{V} \mathbf{F}_{X^*}$ where $\mathbf{E}_X = \mathbf{I} - \mathbf{X}\mathbf{X}^-$, $\mathbf{F}_{X^*} = \mathbf{I} - \mathbf{X}^{*-}\mathbf{X}^*$, the following statements are true. If $\mathbf{Q}$ is a $\mathbf{B}_{11}$-matrix, there exist matrices $\mathbf{Z}_1, \mathbf{Z}_2, \mathbf{G}$ such that $\mathbf{G} \in \mathbf{K}\{1\}$ and (\$) $\mathbf{Q} = \mathbf{Z}_1(\mathbf{I} - \mathbf{K}\mathbf{K}^-)\mathbf{E}_X + \mathbf{F}_{X^*}(\mathbf{I} - \mathbf{K}^-\mathbf{K})\mathbf{Z}_2 + \mathbf{F}_{X^*}\mathbf{G}\mathbf{E}_X$. Conversely, for every pair $\mathbf{Z}_1, \mathbf{Z}_2$, and every $\mathbf{G} \in \mathbf{K}\{1\}$, the matrix $\mathbf{Q}$ in (\$) is a $\mathbf{B}_{11}$-matrix.

7
The Drazin inverse

1. Introduction

In the previous chapters, the Moore–Penrose inverse and the other (i,j,k)-inverses were discussed in some detail. A major characteristic of the (i,j,k)-inverses is the fact that they provide some type of solution, or least squares solution, for a system of linear algebraic equations. That is, they are 'equation solving' inverses.

However, we also saw that there are some desirable properties that the (i,j,k)-inverses do not usually possess. For example, if $\mathbf{A}, \mathbf{B} \in \mathbb{C}^{n \times n}$, then there is no class, $C(i,j,k)$, of (i,j,k)-inverses for $\mathbf{A}$ and $\mathbf{B}$ such that $\mathbf{A}^-, \mathbf{B}^- \in C(i,j,k)$ implies any of the following:

(i) $\mathbf{A}\mathbf{A}^- = \mathbf{A}^-\mathbf{A}$,
(ii) $(\mathbf{A}^-)^p = (\mathbf{A}^p)^-$ for positive integers p,
(iii) $\lambda \in \sigma(\mathbf{A}) \Leftrightarrow \lambda^\dagger \in \sigma(\mathbf{A}^-)$,
(iv) $\mathbf{A}^{p+1}\mathbf{A}^- = \mathbf{A}^p$ for positive integers p, or if $\mathbf{A}$ is similar to $\mathbf{B}$ via the similarity transformation $\mathbf{P}$, then $\mathbf{A}^-$ is similar to $\mathbf{B}^-$ via $\mathbf{P}$.

Depending on the intended applications, it might be desirable to give up the algebraic equation solving properties the (1)-inverses possess in exchange for a generalized inverse which possesses some other 'inverse-like' properties. The Group and Drazin generalized inverses of this chapter will be of such a compromising nature. In many ways, they more closely resemble the true non-singular inverse than do the (i,j,k)-inverses. They will possess all of the above mentioned properties.

Although the Drazin inverse will not provide solutions of linear algebraic equations, it will provide solutions for systems of linear differential equations and linear difference equations as will be shown in Chapter 9.

Up to this point the underlying field has always been taken to be the field of complex numbers. Although this was not always necessary, the complex numbers provided the most natural setting for the development of the Moore–Penrose inverse as well as most of the other (i,j,k)-inverse. To extend the concepts of the previous chapters to matrices over different fields is somewhat artificial. One soon finds that the kind of field needed in order to obtain analogous results must possess properties which mimic those of the complex numbers.

There is nothing special about the complex numbers when it comes to defining the Drazin inverse. However, many of the results in the latter part of this chapter depend on the taking of limits. Rather than get into a technical discussion of the type of topology needed on the field, we shall merely note that *almost* all our results extend to arbitrary fields.

The Group inverse, as we shall see later, is just a special case of the Drazin inverse. However, because the Group inverse appears in some interesting applications, (see Chapter 8) we consider it as a separate entity.

2. Definitions

The Drazin inverse will only be defined for *square* matrices. Just as was the case when defining the (i, j, k)-inverses, there are at least two different approaches possible when formulating the definition. These are the functional or geometric definition and the algebraic definition. The algebraic definition was first given by M. P. Drazin in 1958 in the setting of an abstract ring. We will give both definitions and then show that they are equivalent. Before doing this, some preliminary geometrical facts are needed. Throughout this chapter, we adopt the convention that $\mathbf{0}^{\circ} = \mathbf{I}$.

Lemma 7.2.1 *Let $\underset{\sim}{\mathbf{A}}$ be a linear transformation on $\mathbb{C}^n$. There exists a non-negative integer k such that $\mathbb{C}^n = R(\underset{\sim}{\mathbf{A}}^k) \dot{+} N(\underset{\sim}{\mathbf{A}}^k)$.*

Proof Let k be the smallest non-negative integer such that $R(\underset{\sim}{\mathbf{A}}^{\circ}) \supset R(\underset{\sim}{\mathbf{A}}) \supset \ldots \supset R(\underset{\sim}{\mathbf{A}}^{k-1}) \supset R(\underset{\sim}{\mathbf{A}}^k) = R(\underset{\sim}{\mathbf{A}}^{k+1}) = R(\underset{\sim}{\mathbf{A}}^{k+2}) = \ldots$ or equivalently, $N(\underset{\sim}{\mathbf{A}}^{\circ}) \subset N(\underset{\sim}{\mathbf{A}}) \subset \ldots \subset N(\underset{\sim}{\mathbf{A}}^{k-1}) \subset N(\underset{\sim}{\mathbf{A}}^k) = N(\underset{\sim}{\mathbf{A}}^{k+1}) = N(\underset{\sim}{\mathbf{A}}^{k+2}) = \ldots$. Suppose that $\mathbf{x} \in R(\underset{\sim}{\mathbf{A}}^k) \cap N(\underset{\sim}{\mathbf{A}}^k)$. Then there exists a $\mathbf{z} \in \mathbb{C}^n$ such that $\underset{\sim}{\mathbf{A}}^k\mathbf{z} = \mathbf{x}$. Thus, $\underset{\sim}{\mathbf{A}}^{2k}\mathbf{z} = \underset{\sim}{\mathbf{A}}^k\mathbf{x} = \mathbf{0}$, so that $\mathbf{z} \in N(\underset{\sim}{\mathbf{A}}^{2k}) = N(\underset{\sim}{\mathbf{A}}^k)$. Thus $\mathbf{x} = \mathbf{0}$. Suppose, that $\operatorname{rank}(\underset{\sim}{\mathbf{A}}^k) = r$ so that $\dim[N(\underset{\sim}{\mathbf{A}}^k)] = n - r$. If $\{\mathbf{v}_1 \ldots, \mathbf{v}_r\}$ is a basis for $R(\underset{\sim}{\mathbf{A}}^k)$ and if $\{\mathbf{v}_{r+1}, \ldots, \mathbf{v}_n\}$ is a basis for $N(\underset{\sim}{\mathbf{A}}^k)$, it is easy to show that $\{\mathbf{v}_1, \ldots, \mathbf{v}_r, \mathbf{v}_{r+1}, \ldots, \mathbf{v}_n\}$ is a basis for $\mathbb{C}^n$. ■

The number k which was introduced in Lemma 1 will be very important.

Definition 7.2.1 *Let $\underset{\sim}{\mathbf{A}}$ be a linear transformation on $\mathbb{C}^n$. The smallest non-negative integer k such that $\mathbb{C}^n = R(\underset{\sim}{\mathbf{A}}^k) \dot{+} N(\underset{\sim}{\mathbf{A}}^k)$, or equivalently, the smallest non-negative integer k such that $\operatorname{rank}(\underset{\sim}{\mathbf{A}}^k) = \operatorname{rank}(\underset{\sim}{\mathbf{A}}^{k+1})$, is called* the index of $\underset{\sim}{\mathbf{A}}$ *and is denoted by* $\operatorname{Ind}(\underset{\sim}{\mathbf{A}})$.

Note that if $\underset{\sim}{\mathbf{A}}$ is invertible, $\operatorname{Ind}(\underset{\sim}{\mathbf{A}}) = 0$. Also $\operatorname{Ind}(\mathbf{0}) = 1$.

Several different characterizations of the index will be developed in the sequel.

Lemma 7.2.2 *If $\underset{\sim}{\mathbf{A}}$ is a linear transformation on $\mathbb{C}^n$ and* $\operatorname{Ind}(\underset{\sim}{\mathbf{A}}) = k$, *then $\underset{\sim}{\mathbf{A}}_1 = \underset{\sim}{\mathbf{A}}|_{R(\underset{\sim}{\mathbf{A}}^k)}$ (i.e. $\underset{\sim}{\mathbf{A}}$ restricted to $R(\underset{\sim}{\mathbf{A}}^k)$) is an invertible linear transformation on $R(\underset{\sim}{\mathbf{A}}^k)$.*

Proof $\underset{\sim}{\mathbf{A}}\,R(\underset{\sim}{\mathbf{A}}^k) = R(\underset{\sim}{\mathbf{A}}^{k+1}) = R(\underset{\sim}{\mathbf{A}}^k)$. ■

We now formulate a definition of the Drazin inverse of a linear transformation on $\mathbb{C}^n$.

Definition 7.2.2 (*Functional Definition.*) *Let $\underset{\sim}{\mathbf{A}}$ be a linear transformation on $\mathbb{C}^n$ such that* $\mathrm{Ind}(\underset{\sim}{\mathbf{A}}) = k$. *Let $\mathbf{x} \in \mathbb{C}^n$ and write $\mathbf{x} = \mathbf{u} + \mathbf{v}$ where $\mathbf{u} \in R(\underset{\sim}{\mathbf{A}}^k)$ and $\mathbf{v} \in N(\underset{\sim}{\mathbf{A}}^k)$. Let $\underset{\sim}{\mathbf{A}}_1 = \underset{\sim}{\mathbf{A}}\big|_{R(\underset{\sim}{\mathbf{A}}^k)}$. The linear transformation defined by $\underset{\sim}{\mathbf{A}}^{\mathrm{D}}\mathbf{x} = \underset{\sim}{\mathbf{A}}_1^{-1}\mathbf{u}$ is called the* Drazin inverse of $\underset{\sim}{\mathbf{A}}$. *For $\mathbf{A} \in \mathbb{C}^{n \times n}$, let $\underset{\sim}{\mathbf{A}}$ be the linear transformation induced on $\mathbb{C}^n$ by $\mathbf{A}$. The Drazin inverse, $\mathbf{A}^{\mathrm{D}}$, of $\mathbf{A}$ is defined to be the matrix of $\underset{\sim}{\mathbf{A}}^{\mathrm{D}}$ with respect to the standard basis.*

Theorem 7.2.1 (*The Canonical Form Representation For $\mathbf{A}$ and $\mathbf{A}^{\mathrm{D}}$.*) *If $\mathbf{A} \in \mathbb{C}^{n \times n}$ is such that $\mathrm{Ind}(\mathbf{A}) = k > 0$, then there exists a non-singular matrix $\mathbf{P}$ such that*

$$\mathbf{A} = \mathbf{P}\begin{bmatrix} \mathbf{C} & \mathbf{0} \\ \mathbf{0} & \mathbf{N} \end{bmatrix}\mathbf{P}^{-1} \tag{1}$$

where $\mathbf{C}$ is non-singular and $\mathbf{N}$ is nilpotent of index k. Furthermore, if $\mathbf{P}, \mathbf{C}$ and $\mathbf{N}$ are any matrices satisfying the above conditions, then

$$\mathbf{A}^{\mathrm{D}} = \mathbf{P}\begin{bmatrix} \mathbf{C}^{-1} & \mathbf{0} \\ \mathbf{0} & \mathbf{0} \end{bmatrix}\mathbf{P}^{-1}. \tag{2}$$

Proof Let $\underset{\sim}{\mathbf{A}}$ be the linear transformation induced on $\mathbb{C}^n$ by $\mathbf{A}$. Let $\mathrm{B} = \{\mathbf{v}_1, \mathbf{v}_2, \ldots, \mathbf{v}_r, \mathbf{v}_{r+1}, \ldots, \mathbf{v}_n\}$ be the basis for $\mathbb{C}^n$ constructed in the proof of Lemma 1 so that $\{\mathbf{v}_1, \ldots, \mathbf{v}_r\}$ is a basis for $R(\underset{\sim}{\mathbf{A}}^k)$ and $\{\mathbf{v}_{r+1}, \ldots, \mathbf{v}_n\}$ is a basis for $N(\underset{\sim}{\mathbf{A}}^k)$. Since $R(\underset{\sim}{\mathbf{A}}^k)$, $N(\underset{\sim}{\mathbf{A}}^k)$ are invariant subspaces for $\underset{\sim}{\mathbf{A}}$ and $\underset{\sim}{\mathbf{A}}^k(N(\underset{\sim}{\mathbf{A}}^k)) = \{\mathbf{0}\}$, we have the block form for $\mathbf{A}$ if $\mathbf{P} = [\mathbf{v}_1, \ldots, \mathbf{v}_n]$. The form for $\mathbf{A}^{\mathrm{D}}$ follows from the definition of $\mathbf{A}^{\mathrm{D}}$ if $\mathbf{P}$ is as specified. However if $\hat{\mathbf{P}}, \hat{\mathbf{C}}, \hat{\mathbf{N}}$ are such that (1) holds, and $\hat{\mathbf{C}}$ is non-singular, and $\hat{\mathbf{N}}^k = \mathbf{0}$, then the first r columns of $\hat{\mathbf{P}}$ are a basis for $R(\underset{\sim}{\mathbf{A}}^k)$ while the remaining columns are a basis for $N(\underset{\sim}{\mathbf{A}}^k)$. Thus (2) for any $\mathbf{P}, \mathbf{C}$ or $\mathbf{N}$ by Definition 2. ■

The algebraic definition of $\mathbf{A}^{\mathrm{D}}$, or $\underset{\sim}{\mathbf{A}}^{\mathrm{D}}$, is as follows.

Definition 7.2.3 (*Algebraic Definition.*) *If $\mathbf{A} \in \mathbb{C}^{n \times n}$ with $\mathrm{Ind}(\mathbf{A}) = k$ and if $\mathbf{A}^{\mathrm{D}} \in \mathbb{C}^{n \times n}$ is such that*

$$\mathbf{A}^{\mathrm{D}}\mathbf{A}\mathbf{A}^{\mathrm{D}} = \mathbf{A}^{\mathrm{D}}, \tag{3}$$

$$\mathbf{A}\mathbf{A}^{\mathrm{D}} = \mathbf{A}^{\mathrm{D}}\mathbf{A}, \text{ and} \tag{4}$$

$$\mathbf{A}^{k+1}\mathbf{A}^{\mathrm{D}} = \mathbf{A}^k, \tag{5}$$

then $\mathbf{A}^{\mathrm{D}}$ is called the Drazin inverse of $\mathbf{A}$.

Theorem 7.2.2 *For $\mathbf{A} \in \mathbb{C}^{n \times n}$, the functional definition of $\mathbf{A}^{\mathrm{D}}$ is equivalent to the algebraic definition of $\mathbf{A}^{\mathrm{D}}$.*

Proof Write $\mathbf{A}$ as in (1). That $\mathbf{A}^{\mathrm{D}}$ satisfies (3), (4) and (5) is trivial. Suppose then that $\mathbf{X}$ satisfies (3), (4) and (5).

Now $\mathbf{P}^{-1}\mathbf{X}\mathbf{P} = \begin{bmatrix} \mathbf{X}_{11} & \mathbf{X}_{12} \\ \mathbf{X}_{21} & \mathbf{X}_{22} \end{bmatrix}$ where $\mathbf{X}_{11}$ and $\mathbf{C}$ are the same size.

From (5) we have $\mathbf{C}^{k+1}\mathbf{X}_{11} = \mathbf{C}^k$, $\mathbf{C}^{k+1}\mathbf{X}_{12} = \mathbf{0}$. Thus $\mathbf{X}_{11} = \mathbf{C}^{-1}$ and $\mathbf{X}_{12} = \mathbf{0}$. But also $\mathbf{X}\mathbf{A}^{k+1} = \mathbf{A}^k$ by (4) and (5). Thus $\mathbf{X}_{21} = \mathbf{0}$.

There remains to show that $\mathbf{X}_{22} = \mathbf{0}$. From (3) and (4) we have $\mathbf{N}(\mathbf{X}_{22})^2 = \mathbf{X}_{22}$. Thus $\mathbf{N}^{k-1}\mathbf{X}_{22} = \mathbf{N}^k(\mathbf{X}_{22})^2 = \mathbf{0}$. But then $\mathbf{N}^{k-2}\mathbf{X}_{22} = \mathbf{N}^{k-1}(\mathbf{X}_{22})^2 = \mathbf{0}$. Continuing in this manner gives $\mathbf{X}_{22} = \mathbf{0}$ as desired. ■

Notice that $\mathbf{A}^{\mathrm{D}}$ exists and is unique for all $\mathbf{A} \in \mathbb{C}^{n \times n}$, since the functional definition is constructive in nature. Some important facts that are evident either from the definitions or from the above proof are listed in the following corollary.

Corollary 7.2.1 *If* $\mathbf{A} \in \mathbb{C}^{n \times n}$ *and* $\operatorname{Ind}(\mathbf{A}) = k$, *then*

(i) $R(\mathbf{A}^{\mathrm{D}}) = R(\mathbf{A}^k)$,
(ii) $N(\mathbf{A}^{\mathrm{D}}) = N(\mathbf{A}^k)$,
(iii) $\mathbf{A}\mathbf{A}^{\mathrm{D}} = \mathbf{A}^{\mathrm{D}}\mathbf{A} = \mathbf{P}_{R(\mathbf{A}^k), N(\mathbf{A}^k)}$,
(iv) $(\mathbf{I} - \mathbf{A}\mathbf{A}^{\mathrm{D}}) = (\mathbf{I} - \mathbf{A}^{\mathrm{D}}\mathbf{A}) = \mathbf{P}_{N(\mathbf{A}^k), R(\mathbf{A}^k)}$,
(v) *for a non-negative integer* p, $\mathbf{A}^{p+1}\mathbf{A}^{\mathrm{D}} = \mathbf{A}^p$ *if and only if* $p \geq k$, *and*
(vi) *if* $\mathbf{A}$ *is non-singular, then* $\mathbf{A}^{\mathrm{D}} = \mathbf{A}^{-1}$.

The number $k = \operatorname{Ind}(\mathbf{A})$ was used in the algebraic definition. Actually, any non-negative integer $p, p \geq k$, could have been used.

Theorem 7.2.3 *Let* $\mathbf{A} \in \mathbb{C}^{n \times n}$ *be such that* $\operatorname{Ind}(\mathbf{A}) = k$. *If* p *is a non-negative integer and* $\mathbf{X} \in \mathbb{C}^{n \times n}$ *is such that* $\mathbf{XAX} = \mathbf{X}$, $\mathbf{AX} = \mathbf{XA}$, *and* $\mathbf{A}^{p+1}\mathbf{X} = \mathbf{A}^p$, *then* $p \geq k$ *and* $\mathbf{X} = \mathbf{A}^{\mathrm{D}}$.

Proof $\mathbf{A}^{p+1}\mathbf{X} = \mathbf{A}^p$ implies that $R(\mathbf{A}^{p+1}) = R(\mathbf{A}^p)$ so that $p \geq k$. Write $p = k + i$. Then $(\mathbf{A}^{\mathrm{D}})^i\mathbf{A}^{k+1+i}\mathbf{X} = (\mathbf{A}^{\mathrm{D}})^i\mathbf{A}^{k+i}$. This reduces to $\mathbf{A}^{k+1}\mathbf{X} = \mathbf{A}^k$. Thus $\mathbf{X}$ satisfies the conditions of the algebraic definition of $\mathbf{A}^{\mathrm{D}}$. ■

Something that should immediately strike one's attention when looking at Definition 3 is that $\mathbf{A}^{\mathrm{D}}$ is not always a (1)-inverse for $\mathbf{A}$. This, of course, means that $\mathbf{A}^{\mathrm{D}}$ is not an 'equation solver'. That is, if $\mathbf{A} \in \mathbb{C}^{n \times n}$, $\mathbf{b} \in \mathbb{C}^n$, and $\mathbf{Ax} = \mathbf{b}$ is a consistent system of algebraic equations, then $\mathbf{A}^{\mathrm{D}}\mathbf{b}$ may not be a solution. In fact, $\mathbf{A}^{\mathrm{D}}\mathbf{b}$ is a solution of $\mathbf{Ax} = \mathbf{b}$ if and only if $\mathbf{b} \in R(\mathbf{A}^k)$ where $k = \operatorname{Ind}(\mathbf{A})$.

There are special cases when $\mathbf{A}^{\mathrm{D}}$ is a (1)-inverse for $\mathbf{A} \in \mathbb{C}^{n \times n}$.

Theorem 7.2.4 *For* $\mathbf{A} \in \mathbb{C}^{n \times n}$, $\mathbf{A}\mathbf{A}^{\mathrm{D}}\mathbf{A} = \mathbf{A}$ *if and only if* $\operatorname{Ind}(\mathbf{A}) \leq 1$.

Proof If $\operatorname{Ind}(\mathbf{A}) = 0$, then $\mathbf{A}^{\mathrm{D}} = \mathbf{A}^{-1}$ and $\mathbf{A}\mathbf{A}^{\mathrm{D}}\mathbf{A} = \mathbf{A}$. Suppose that $\operatorname{Ind}(\mathbf{A}) \geq 1$. Then relative to (1), (2) we have $\mathbf{A}\mathbf{A}^{\mathrm{D}}\mathbf{A} = \mathbf{A}$ if and only if $\mathbf{0} = \mathbf{N}$. But $\mathbf{0} = \mathbf{N}$ if and only if $\operatorname{Ind}(\mathbf{A}) = 1$. ■

The special case when $\mathrm{Ind}(\mathbf{A}) \le 1$ gives rise to what is known as the Group inverse. Notice that in this case, (5) can be rewritten as $\mathbf{AA^DA} = \mathbf{A}$.

Definition 7.2.4 *If* $\mathbf{A} \in \mathbb{C}^{n \times n}$ *is such that* $\mathrm{Ind}(\mathbf{A}) \le 1$, *then the Drazin inverse of* $\mathbf{A}$ *is called the* Group inverse *of* $\mathbf{A}$ *and is denoted by* $\mathbf{A}^{\#}$. *When it exists,* $\mathbf{A}^{\#}$ *is characterized as the unique matrix satisfying the three equations* $\mathbf{AA}^{\#}\mathbf{A} = \mathbf{A}$, $\mathbf{A}^{\#}\mathbf{AA}^{\#} = \mathbf{A}^{\#}$, *and* $\mathbf{AA}^{\#} = \mathbf{A}^{\#}\mathbf{A}$.

The following theorem makes it clear why the term 'group inverse' is used.

Theorem 7.2.5 *A linear transformation* $\underset{\sim}{\mathbf{A}}$ *on* $\mathbb{C}^n$ *which has rank r belongs to a multiplicative group, G, of linear transformations on* $\mathbb{C}^n$ *if and only if* $\mathrm{Ind}(\underset{\sim}{\mathbf{A}}) \le 1$. *Furthermore, if* $\underset{\sim}{\mathbf{A}} \in G$ *and if* $\underset{\sim}{\mathbf{A}}^g \in G$ *is the multiplicative inverse of* $\underset{\sim}{\mathbf{A}}$ *within G, then* $\underset{\sim}{\mathbf{A}}^g = \underset{\sim}{\mathbf{A}}^{\#}$.

Proof If $\underset{\sim}{\mathbf{A}}$ is in a multiplicative group G, then there exists an $\underset{\sim}{\mathbf{A}}^g \in G$ such that $\underset{\sim}{\mathbf{A}}^2\underset{\sim}{\mathbf{A}}^g = \underset{\sim}{\mathbf{A}}$, $\underset{\sim}{\mathbf{A}}\underset{\sim}{\mathbf{A}}^g = \underset{\sim}{\mathbf{A}}^g\underset{\sim}{\mathbf{A}}$, and $\underset{\sim}{\mathbf{A}}^g\underset{\sim}{\mathbf{A}}\underset{\sim}{\mathbf{A}}^g = \underset{\sim}{\mathbf{A}}^g$. Then $\underset{\sim}{\mathbf{A}}^g = \underset{\sim}{\mathbf{A}}^{\mathrm{D}}$ by Theorem 3 Also $\mathrm{Ind}(\underset{\sim}{\mathbf{A}}) \le 1$. Conversely, suppose $\mathrm{Ind}(\underset{\sim}{\mathbf{A}}) \le 1$. Then with $\mathbf{P}$ defined as in (1),

$$G = \left\{ \mathbf{P}\begin{bmatrix} \mathbf{X} & \mathbf{0} \\ \mathbf{0} & \mathbf{0} \end{bmatrix}\mathbf{P}^{-1} \,\middle|\, \mathbf{X} \in \mathbb{C}^{r \times r}, r = \mathrm{rank}(\mathbf{C}) \right\},$$

is a multiplicative group containing $\underset{\sim}{\mathbf{A}}$. ■

As a special case of Theorem 1 (or Theorem 5) we have the following.

Corollary 7.2.2 *For* $\mathbf{A} \in \mathbb{C}^{n \times n}$, $\mathbf{A}^{\#}$ *exists if and only if there exist non-singular matrices* $\mathbf{P}$ *and* $\mathbf{C}$ *such that* $\mathbf{A} = \mathbf{P}\begin{bmatrix} \mathbf{C} & \mathbf{0} \\ \mathbf{0} & \mathbf{0} \end{bmatrix}\mathbf{P}^{-1}$.

The following is a simple example of a group of singular matrices.

Example 7.2.1 Consider the following subset of $\mathbb{R}^{n \times n}$.

$$G = \left\{ \mathbf{A} \in \mathbb{R}^{n \times n} \,\middle|\, \mathbf{A} = \alpha\mathbf{J}, \text{ where } 0 \ne \alpha \in \mathbb{R} \text{ and } \mathbf{J} = \begin{bmatrix} 1 & 1 & \dots & 1 \\ 1 & 1 & \dots & 1 \\ \vdots & \vdots & & \vdots \\ 1 & 1 & \dots & 1 \end{bmatrix} \right\}$$

It is clear that G is a multiplicative group. The multiplicative identity in G is $\mathbf{E} = \dfrac{1}{n}\mathbf{J}$. If $\mathbf{A} = \alpha\mathbf{J}$, $\alpha \ne 0$, then the group inverse of $\mathbf{A}$ is $\mathbf{A}^{\#} = \dfrac{1}{\alpha n^2}\mathbf{J}$.

Another algebraic characterization of $\mathbf{A}^{\mathrm{D}}$ is illustrated in the heuristic diagram of Fig. 7.1. $\mathbb{C}^{n \times n}$ is a semi-group and the G's (one for each idempotent), are the maximal subgroups of $\mathbb{C}^{n \times n}$. Clearly, $\{G_i\}$ is a disjoint family but not a partition of $\mathbb{C}^{n \times n}$. If $\mathrm{Ind}(\mathbf{A}) \le 1$, then, as pointed out earlier, $\mathbf{A} \in G_i$, for some i, and $\mathbf{A}^{\mathrm{D}} = \mathbf{A}^{\#}$ is just the inverse of $\mathbf{A}$ with respect to the group G_i. If $\mathrm{Ind}(\mathbf{A}) = k > 1$, then it is not difficult to show that k can

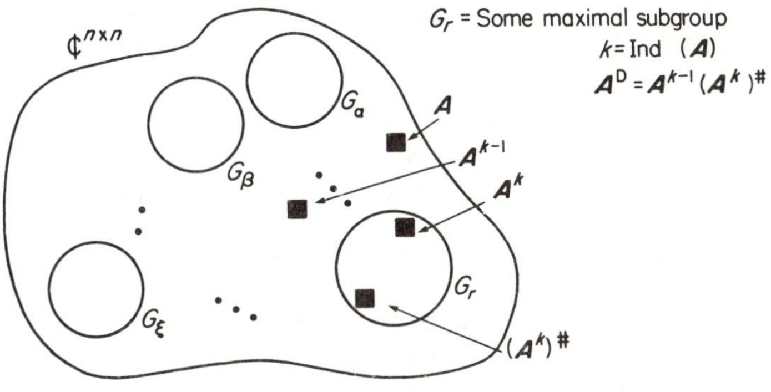

Fig. 7.1

be characterized as that number such that $\mathbf{A}^k \in G_r$, for some r, but $\mathbf{A}^{k-1} \notin G_r$, for all r. Thus, $\mathbf{A}^k$ has an inverse, $\mathbf{X}$, within G_r. The Drazin inverse of $\mathbf{A}$ is simply $\mathbf{A}^{\mathrm{D}} = \mathbf{A}^{k-1}\mathbf{X} = \mathbf{A}^{k-1}(\mathbf{A}^k)^\#$.

Suppose that the Jordan form of $\mathbf{A}$ is

$$J_{\mathrm{A}} = \left[\begin{array}{ccccc|ccccc} \mathbf{J}_1 & \mathbf{0} & \mathbf{0} & \dots & \mathbf{0} & \mathbf{0} & & \dots & & \mathbf{0} \\ \mathbf{0} & \mathbf{J}_2 & \mathbf{0} & \dots & \mathbf{0} & \mathbf{0} & & \dots & & \mathbf{0} \\ \vdots & & & & \vdots & \vdots & & & & \vdots \\ \mathbf{0} & \mathbf{0} & \mathbf{0} & \dots & \mathbf{J}_t & \mathbf{0} & & \dots & & \mathbf{0} \\ \hline \mathbf{0} & & \dots & & \mathbf{0} & \mathbf{J}_{t+1} & \mathbf{0} & \mathbf{0} & \dots & \mathbf{0} \\ \mathbf{0} & & \dots & & \mathbf{0} & \mathbf{0} & \mathbf{J}_{t+2} & \mathbf{0} & \dots & \mathbf{0} \\ \vdots & & & & \vdots & \vdots & & & & \vdots \\ \mathbf{0} & & \dots & & \mathbf{0} & \mathbf{0} & & \dots & & \mathbf{J}_k \end{array}\right].$$

(If a field other than $\mathbb{C}$ is used one still gets (1) but not possibly J_{A}).
If the Jordan blocks, $\mathbf{J}_i$, are arranged so that the diagonal elements of $\mathbf{J}_1, \mathbf{J}_2, \dots, \mathbf{J}_t$ are non-zero and the diagonal elements of $\mathbf{J}_{t+1}, \mathbf{J}_{t+2}, \dots, \mathbf{J}_k$ are zeros, then the matrices $\mathbf{C}$ and $\mathbf{N}$ in Theorem 1 may be taken to be

$$\mathbf{C} = \begin{bmatrix} \mathbf{J}_1 & \mathbf{0} & \mathbf{0} & \dots & \mathbf{0} \\ \mathbf{0} & \mathbf{J}_2 & \mathbf{0} & \dots & \mathbf{0} \\ \vdots & & \ddots & & \\ \mathbf{0} & \mathbf{0} & \mathbf{0} & \dots & \mathbf{J}_t \end{bmatrix} \text{ and } \mathbf{N} = \begin{bmatrix} \mathbf{J}_{t+1} & \mathbf{0} & \mathbf{0} & \dots & \mathbf{0} \\ \mathbf{0} & \mathbf{J}_{t+2} & \mathbf{0} & \dots & \mathbf{0} \\ \vdots & & \ddots & & \\ \mathbf{0} & \mathbf{0} & \mathbf{0} & \dots & \mathbf{J}_k \end{bmatrix}.$$

Theorem 1 will be fundamental in the development of the theory of the Drazin inverse. However, Theorem 1 also has a practical side. One may use this theorem to compute the Drazin inverse.

Algorithm 7.2.1 Computation of $\mathbf{A}^{\mathrm{D}}$ where $\mathbf{A} \in \mathbb{C}^{n \times n}$ and $\mathrm{Ind}(\mathbf{A}) = k$.

(I) Let p be an integer such that $p \geq k$. (p can always be taken to equal to n if no smaller value can be determined.) If $\mathbf{A}^p = \mathbf{0}$, then $\mathbf{A}^{\mathrm{D}} = \mathbf{0}$. Thus assume $\mathbf{A}^p \neq \mathbf{0}$.

(II) Row reduce $\mathbf{A}^p$ to its Hermite echelon form, $\mathbf{H}_{\mathbf{A}^p}$. (See Definition 1.3.2.) The sequence of reducing matrices need not be saved.

(III) By noting the position of the non-zero diagonal elements in $\mathbf{H}_{\mathbf{A}^p}$, select the distinguished columns from $\mathbf{A}^p$ and call them $\mathbf{v}_1, \mathbf{v}_2, \ldots, \mathbf{v}_r$. (This is a basis for $R(\mathbf{A}^k)$.)

(IV) Form the matrix $\mathbf{I} - \mathbf{H}_{\mathbf{A}^p}$ and save its non-zero columns. Call them $\mathbf{v}_{r+1}, \mathbf{v}_{r+2}, \ldots, \mathbf{v}_n$. (This is a basis for $N(\mathbf{A}^k)$.)

(V) Construct the non-singular matrix $\mathbf{P} = [\mathbf{v}_1 | \ldots | \mathbf{v}_r | \mathbf{v}_{r+1} | \ldots | \mathbf{v}_n]$.

(VI) Compute $\mathbf{P}^{-1}$

(VII) Form the product $\mathbf{P}^{-1}\mathbf{AP}$. This matrix will be in the form $\mathbf{P}^{-1}\mathbf{AP} = \begin{bmatrix} \mathbf{C} & \mathbf{0} \\ \mathbf{0} & \mathbf{N} \end{bmatrix}$ where $\mathbf{C}$ is non-singular and $\mathbf{N}$ is nilpotent.

(VIII) Compute $\mathbf{C}^{-1}$.

(IX) Compute $\mathbf{A}^{\mathrm{D}}$ by forming the product $\mathbf{A}^{\mathrm{D}} = \mathbf{P}\begin{bmatrix} \mathbf{C}^{-1} & \mathbf{0} \\ \mathbf{0} & \mathbf{0} \end{bmatrix}\mathbf{P}^{-1}$

Example 7.2.2 Let

$$\mathbf{A} = \begin{bmatrix} 2 & 0 & 0 \\ -1 & 1 & 1 \\ -1 & -1 & -1 \end{bmatrix}.$$

We shall find $\mathbf{A}^{\mathrm{D}}$ by using this algorithm.

(I) Since we don't know what $\mathrm{Ind}(\mathbf{A})$ is let $p = 3$. Then

$$\mathbf{A}^3 = \begin{bmatrix} 8 & 0 & 0 \\ -8 & 0 & 0 \\ 0 & 0 & 0 \end{bmatrix},$$

and

(II) $$\mathbf{H}_{\mathbf{A}^3} = \begin{bmatrix} 1 & 0 & 0 \\ 0 & 0 & 0 \\ 0 & 0 & 0 \end{bmatrix}.$$

(III) Thus $\mathbf{v}_1 = \begin{bmatrix} 8 \\ -8 \\ 0 \end{bmatrix}$ is a basis for $R(\mathbf{A}^k)$.

(IV) Now $\mathbf{I} - \mathbf{H}_{\mathbf{A}} = \begin{bmatrix} 0 & 0 & 0 \\ 0 & 1 & 0 \\ 0 & 0 & 1 \end{bmatrix}$, so that

$$\mathbf{v}_2 = \begin{bmatrix} 0 \\ 1 \\ 0 \end{bmatrix}, \mathbf{v}_3 = \begin{bmatrix} 0 \\ 0 \\ 1 \end{bmatrix} \text{ form a basis for } N(\mathbf{A}^k).$$

(V) This gives $\mathbf{P} = \begin{bmatrix} 8 & 0 & 0 \\ -8 & 1 & 0 \\ 0 & 0 & 1 \end{bmatrix}$,

(VI) $$\mathbf{P}^{-1} = \frac{1}{8}\begin{bmatrix} 1 & 0 & 0 \\ 8 & 8 & 0 \\ 0 & 0 & 8 \end{bmatrix},$$

and

(VII)

$$\mathbf{P}^{-1}\mathbf{A}\mathbf{P} = \left[\begin{array}{c|cc} 2 & 0 & 0 \\ \hline 0 & 1 & 1 \\ 0 & -1 & -1 \end{array}\right].$$

(VIII) Since $\mathbf{C} = 2$, $\mathbf{C}^{-1} = \frac{1}{2}$. We thus get

(IX)

$$\mathbf{A}^{\mathrm{D}} = \mathbf{P}\begin{bmatrix} \frac{1}{2} & 0 & 0 \\ 0 & 0 & 0 \\ 0 & 0 & 0 \end{bmatrix}\mathbf{P}^{-1} = \frac{1}{2}\begin{bmatrix} 1 & 0 & 0 \\ -1 & 0 & 0 \\ 0 & 0 & 0 \end{bmatrix}.$$

The next characterization of $\mathbf{A}^{\mathrm{D}}$ may be useful if one tries to formulate a definition for the Drazin inverse of a linear transformation on an infinite dimensional vector space [17].

For $\mathbf{A} \in \mathbb{C}^{n \times n}$, let C denote the class,

$$C = \{\mathbf{X} \in \mathbb{C}^{n \times n} \,|\, \mathbf{AX} = \mathbf{XA} \text{ and } \mathbf{XAX} = \mathbf{X}\}.$$

(Clearly, C is non-empty since $\mathbf{0} \in C$.) Define a partial ordering on C by $\mathbf{X}_1 \leq \mathbf{X}_2$ if and only if $\mathbf{X}_1\mathbf{A}\mathbf{X}_2 = \mathbf{X}_1 = \mathbf{X}_2\mathbf{A}\mathbf{X}_1$.

Theorem 7.2.6 $\mathbf{A}^{\mathrm{D}}$ *is the maximal element of* C.

Proof Suppose $\mathbf{X} \in C$. Then $\mathbf{X} = \mathbf{A}^n\mathbf{X}^{n+1}$ for $n = 1, 2, \ldots$ Thus, $R(\mathbf{X}) \subseteq R(\mathbf{A}^n)$ for each n. In particular, $R(\mathbf{X}) \subseteq R(\mathbf{A}^k)$ where $k = \mathrm{Ind}(\mathbf{A})$ so that $\mathbf{A}^{\mathrm{D}}\mathbf{A}\mathbf{X} = \mathbf{X}$. Furthermore, it is easy to see that $N(\mathbf{A}^n) \subseteq N(\mathbf{X})$ for every n. In particular, $N(\mathbf{A}^k) \subseteq N(\mathbf{X})$. It follows from this that $\mathbf{X}\mathbf{A}\mathbf{A}^{\mathrm{D}} = \mathbf{X}$. Therefore, $\mathbf{X} \leq \mathbf{A}^{\mathrm{D}}$ for every $\mathbf{X} \in C$. ■

3. Basic properties of the Drazin inverse

This section will present basic results about the Drazin inverse.

In Section 2, we saw that $\mathbf{A}^{\mathrm{D}}$ was not always a (1)-inverse for $\mathbf{A} \in \mathbb{C}^{n \times n}$. Though $\mathbf{A}\mathbf{A}^{\mathrm{D}}\mathbf{A}$ is usually unequal to $\mathbf{A}$, the product $\mathbf{A}\mathbf{A}^{\mathrm{D}}\mathbf{A} = \mathbf{A}^2\mathbf{A}^{\mathrm{D}}$ still plays an important role.

Definition 7.3.1 *For* $\mathbf{A} \in \mathbb{C}^{n \times n}$, *the product* $\mathbf{C}_{\mathrm{A}} = \mathbf{A}\mathbf{A}^{\mathrm{D}}\mathbf{A} = \mathbf{A}^2\mathbf{A}^{\mathrm{D}} = \mathbf{A}^{\mathrm{D}}\mathbf{A}^2$ *is called the* core of $\mathbf{A}$.

Intuitively, the 'core' of $\mathbf{A}$ should contain that which is basic to the structure of $\mathbf{A}$. If $\mathbf{C}_{\mathrm{A}}$ is removed from $\mathbf{A}$, then not much should remain.

The next theorem shows in what sense this is true.

Theorem 7.3.1 *If* $\mathbf{A} \in \mathbb{C}^{n \times n}$, *then* $\mathbf{A} - \mathbf{C}_{\mathrm{A}} = \mathbf{N}_{\mathrm{A}}$ *is a nilpotent matrix of index* $k = \mathrm{Ind}(\mathbf{A})$.

Proof The theorem is trivial if $\mathrm{Ind}(\mathbf{A}) = 0$. Thus assume $\mathrm{Ind}(\mathbf{A}) \geq 1$ and notice that $(\mathbf{N}_{\mathrm{A}})^k = (\mathbf{A} - \mathbf{A}\mathbf{A}^{\mathrm{D}}\mathbf{A})^k = (\mathbf{A}(\mathbf{I} - \mathbf{A}\mathbf{A}^{\mathrm{D}}))^k = \mathbf{A}^k(\mathbf{I} - \mathbf{A}\mathbf{A}^{\mathrm{D}}) = \mathbf{A}^k - \mathbf{A}^k = \mathbf{0}$. Since $\mathbf{A}^l - \mathbf{A}^{l+1}\mathbf{A}^{\mathrm{D}} \neq 0$ for $l < k$ we have $\mathrm{Ind}(\mathbf{N}_{\mathrm{A}}) = k$. ■

Definition 7.3.2 *For* $\mathbf{A}\in\mathbb{C}^{n\times n}$, *the matrix* $\mathbf{N}_{\mathbf{A}} = \mathbf{A} - \mathbf{C}_{\mathbf{A}} = (\mathbf{I} - \mathbf{A}\mathbf{A}^{\mathrm{D}})\mathbf{A}$ *is called the* nilpotent part of $\mathbf{A}$. *The decomposition* $\mathbf{A} = \mathbf{C}_{\mathbf{A}} + \mathbf{N}_{\mathbf{A}}$ *is called the* core-nilpotent decomposition of $\mathbf{A}$.

In terms of the canonical form representation of Theorem 2.1, we have the following.

Theorem 7.3.2 *If* $\mathbf{A}\in\mathbb{C}^{n\times n}$ *is written as* $\mathbf{A} = \mathbf{P}\begin{bmatrix}\mathbf{C} & \mathbf{0}\\ \mathbf{0} & \mathbf{N}\end{bmatrix}\mathbf{P}^{-1}$, *where* $\mathbf{P}$ *and* $\mathbf{C}$ *are non-singular and* $\mathbf{N}$ *is nilpotent of index* $k = \mathrm{Ind}(\mathbf{A})$, *then* $\mathbf{C}_{\mathbf{A}} = \mathbf{P}\begin{bmatrix}\mathbf{C} & \mathbf{0}\\ \mathbf{0} & \mathbf{0}\end{bmatrix}\mathbf{P}^{-1}$ *and* $\mathbf{N}_{\mathbf{A}} = \mathbf{P}\begin{bmatrix}\mathbf{0} & \mathbf{0}\\ \mathbf{0} & \mathbf{N}\end{bmatrix}\mathbf{P}^{-1}$.

The core-nilpotent decomposition of $\mathbf{A}$ is unique in the following sense.

Theorem 7.3.3 *For* $\mathbf{A}\in\mathbb{C}^{n\times n}$, $\mathbf{A}$ *has a unique decomposition as* $\mathbf{A} = \mathbf{X} + \mathbf{Y}$ *where* $\mathbf{XY} = \mathbf{YX} = \mathbf{0}$. $\mathrm{Ind}(\mathbf{X}) \le 1$, *and* $\mathbf{Y}$ *is nilpotent of index* $k = \mathrm{Ind}(\mathbf{A})$. *Moreover, this unique decomposition is given by* $\mathbf{A} = \mathbf{C}_{\mathbf{A}} + \mathbf{N}_{\mathbf{A}}$.

Proof Let $\mathbf{X}, \mathbf{Y}$ be as described in Theorem 3. If $\mathrm{Ind}(\mathbf{X}) = 0$, then $\mathbf{Y} = \mathbf{0}$ and $\mathbf{A}$ is invertible. Suppose then that $\mathrm{Ind}(\mathbf{X}) = 1$. Let $\mathbf{P}, \mathbf{C}$ be invertible matrices so that $\mathbf{X} = \mathbf{P}\begin{bmatrix}\mathbf{C} & \mathbf{0}\\ \mathbf{0} & \mathbf{0}\end{bmatrix}\mathbf{P}^{-1}$. Then $\mathbf{Y} = \mathbf{P}\begin{bmatrix}\mathbf{0} & \mathbf{0}\\ \mathbf{0} & \mathbf{Y}_2\end{bmatrix}\mathbf{P}^{-1}$ since $\mathbf{XY} = \mathbf{YX} = \mathbf{0}$ and $\mathbf{C}$ is invertible. Thus $\mathbf{Y}_2$ is nilpotent with $\mathrm{Ind}(\mathbf{Y}_2) = \mathrm{Ind}(\mathbf{A})$ since $\mathbf{Y}$ is. But $\mathbf{A} = \mathbf{X} + \mathbf{Y} = \mathbf{P}\begin{bmatrix}\mathbf{C} & \mathbf{0}\\ \mathbf{0} & \mathbf{Y}_2\end{bmatrix}\mathbf{P}^{-1}$ so that $\mathbf{X} = \mathbf{C}_{\mathbf{A}}$, $\mathbf{Y} = \mathbf{N}_{\mathbf{A}}$ by Theorem 2. ■

Corollary 7.3.1 *If* $\mathbf{A}\in\mathbb{C}^{n\times n}$ *and if p is a positive integer, then* $\mathbf{C}_{\mathbf{A}}^p = \mathbf{C}_{\mathbf{A}^p}$, $\mathbf{N}_{\mathbf{A}}^p = \mathbf{N}_{\mathbf{A}^p}$, *and* $\mathbf{A}^p = \mathbf{C}_{\mathbf{A}^p} + \mathbf{N}_{\mathbf{A}^p} = \mathbf{C}_{\mathbf{A}}^p + \mathbf{N}_{\mathbf{A}}^p$. *If* $p \ge \mathrm{Ind}(\mathbf{A})$, *then* $\mathbf{A}^p = \mathbf{C}_{\mathbf{A}}^p$.

The next lemma summarizes some of the basic relationships between $\mathbf{A}, \mathbf{C}_{\mathbf{A}}, \mathbf{N}_{\mathbf{A}}$, and $\mathbf{A}^{\mathrm{D}}$.

Lemma 7.3.1 *For* $\mathbf{A}\in\mathbb{C}^{n\times n}$, *the following statements are true.*

(i) $\mathrm{Ind}(\mathbf{A}^{\mathrm{D}}) = \mathrm{Ind}(\mathbf{C}_{\mathbf{A}}) = \begin{cases}1 & \text{if } \mathrm{Ind}(\mathbf{A}) \ge 1\\ 0 & \text{if } \mathrm{Ind}(\mathbf{A}) = 0\end{cases}$

(ii) $\mathbf{N}_{\mathbf{A}}\mathbf{C}_{\mathbf{A}} = \mathbf{C}_{\mathbf{A}}\mathbf{N}_{\mathbf{A}} = \mathbf{0}$.

(iii) $\mathbf{N}_{\mathbf{A}}\mathbf{A}^{\mathrm{D}} = \mathbf{A}^{\mathrm{D}}\mathbf{N}_{\mathbf{A}} = \mathbf{0}$.

(iv) $\mathbf{C}_{\mathbf{A}}\mathbf{A}\mathbf{A}^{\mathrm{D}} = \mathbf{A}\mathbf{A}^{\mathrm{D}}\mathbf{C}_{\mathbf{A}} = \mathbf{C}_{\mathbf{A}}$.

(v) $(\mathbf{A}^{\mathrm{D}})^{\mathrm{D}} = \mathbf{C}_{\mathbf{A}}$.

(vi) $\mathbf{A} = \mathbf{C}_{\mathbf{A}}$ *if and only if* $\mathrm{Ind}(\mathbf{A}) \le 1$.

(vii) $((\mathbf{A}^{\mathrm{D}})^{\mathrm{D}})^{\mathrm{D}} = \mathbf{A}^{\mathrm{D}}$.

(viii) $\mathbf{A}^{\mathrm{D}} = \mathbf{C}_{\mathbf{A}}^{\mathrm{D}}$.

(ix) $(\mathbf{A}^{\mathrm{D}})^* = (\mathbf{A}^*)^{\mathrm{D}}$. *(In the case of a general field, (*) is taken to mean transpose)*

There are cases when the Drazin inverse coincides with the Moore–Penrose inverse.

Theorem 7.3.4 *For* $\mathbf{A}\in\mathbb{C}^{n\times n}$, $\mathbf{A}^{\mathrm{D}}=\mathbf{A}^{\dagger}$ *if and only if* $\mathbf{A}$ *is an* EP *matrix,* (*See Chapter 7 for a discussion of* EP *matrices.*)

Proof If $\mathbf{A}$ is EP, then $\mathbf{AA}^{\dagger}=\mathbf{A}^{\dagger}\mathbf{A}$. Since $\mathbf{A}^{\dagger}$ is always a (1, 2)-inverse for $\mathbf{A}$, it follows that $\mathbf{A}^{\dagger}=\mathbf{A}^{\#}=\mathbf{A}^{\mathrm{D}}$. Conversely, if $\mathbf{A}^{\dagger}=\mathbf{A}^{\mathrm{D}}$, then $\mathbf{AA}^{\dagger}=\mathbf{AA}^{\mathrm{D}}=\mathbf{A}^{\mathrm{D}}\mathbf{A}=\mathbf{A}^{\dagger}\mathbf{A}$ so that $\mathbf{A}$ must be EP. ■

4. Spectral properties of the Drazin inverse

In what follows, $\sigma(\cdot)$ will always denote the spectrum, that is the set of eigenvalues. For a non-singular matrix $\mathbf{A}$, it is easily proven that $\lambda\in\sigma(\mathbf{A})$ if and only if $\lambda^{-1}\in\sigma(\mathbf{A}^{-1})$. Furthermore, $\mathbf{x}$ is an eigenvector for $\mathbf{A}$ corresponding to λ if and only if $\mathbf{x}$ is an eigenvector for $\mathbf{A}^{-1}$ corresponding to λ^{-1}.

Recall the definition of a generalized eigenvector.

Definition 7.4.1 *If* $\mathbf{A}\in\mathbb{C}^{n\times n}$ *and* $\mathbf{x}$ *is a non-zero vector such that there is a positive integer* p *and a scalar* $\lambda\in\sigma(\mathbf{A})$ *for which* $(\mathbf{A}-\lambda\mathbf{I})^{p}\mathbf{x}=\mathbf{0}$ *and* $(\mathbf{A}-\lambda\mathbf{I})^{p-1}\mathbf{x}\neq\mathbf{0}$, *then* $\mathbf{x}$ *is called a* generalized eigenvector for $\mathbf{A}$ of grade p.

An eigenvector of grade one is, of course, just an eigenvector. For a non-singular matrix $\mathbf{A}$, it is well known that $\mathbf{x}$ is a generalized eigenvector for $\mathbf{A}$ of grade p corresponding to $\lambda\in\sigma(\mathbf{A})$ if and only if $\mathbf{x}$ is a generalized eigenvector for $\mathbf{A}^{-1}$ of grade p corresponding to $\lambda^{-1}\in\sigma(\mathbf{A}^{-1})$. The next theorem shows that the same situation holds for Drazin inverses of singular matrices.

Theorem 7.4.1 *For* $\mathbf{A}\in\mathbb{C}^{n\times n}$ *such that* $\mathrm{Ind}(\mathbf{A})=k$, $\lambda\in\sigma(\mathbf{A})$ *if and only if* $\lambda^{\dagger}\in\sigma(\mathbf{A}^{\mathrm{D}})$. $\mathbf{x}$ *is a generalized eigenvector for* $\mathbf{A}$ *of grade* p *corresponding to* $\lambda\in\sigma(\mathbf{A})$, $\lambda\neq 0$, *if and only if* $\mathbf{x}$ *is a generalized eigenvector for* $\mathbf{A}^{\mathrm{D}}$ *of grade* p *corresponding to* $\lambda^{-1}\in\sigma(\mathbf{A}^{\mathrm{D}})$. *Furthermore,* $\mathbf{x}$ *is a generalized eigenvector for* $\mathbf{A}$ *corresponding to* $\lambda=0$ *if and only if* $\mathbf{x}\in N(\mathbf{A}^{k})=N(\mathbf{A}^{\mathrm{D}})$.

Proof If $\mathrm{Ind}(\mathbf{A})=0$ we are done. Suppose that $\mathbf{A}=\mathbf{P}\begin{bmatrix}\mathbf{C} & \mathbf{0}\\ \mathbf{0} & \mathbf{N}\end{bmatrix}\mathbf{P}^{-1}$, $\mathbf{x}=\mathbf{P}\begin{bmatrix}\mathbf{u}_1\\ \mathbf{u}_2\end{bmatrix}$. Then $\mathbf{x}$ is a generalized eigenvector for $\mathbf{A}$ of grade p for $\lambda\neq 0$ if and only if $\mathbf{u}_2=\mathbf{0}$ and $\mathbf{u}_1$ is a generalized eigenvector of grade p for $\mathbf{C}$. Since $\mathbf{C}$ is invertible and $\mathbf{A}^{\mathrm{D}}=\mathbf{P}\begin{bmatrix}\mathbf{C}^{-1} & \mathbf{0}\\ \mathbf{0} & \mathbf{0}\end{bmatrix}\mathbf{P}^{-1}$ we are done. The $\lambda=0$ case is obvious. ■

Corollary 7.4.2 *Let* $\mathbf{A}\in\mathbb{C}^{n\times n}$ *be such that* $\mathrm{Ind}(\mathbf{A})=k$. *If* $\mathbf{x}$ *is a generalized eigenvector for* $\mathbf{A}$ *corresponding to* $\lambda\neq 0$, *then* $\mathbf{x}\in R(\mathbf{A}^{k})$.

5. $\mathbf{A}^D$ as a polynomial in A

If $\mathbf{A}$ is a non-singular matrix, then it is easy to show that $\mathbf{A}^{-1}$ can be expressed as a polynomial in $\mathbf{A}$. This property does not carry over to the (i, j, k)-inverses. In particular, if $\mathbf{A}$ is square, then there may not exist a polynomial $p(x)$ such that $\mathbf{A}^{\dagger} = p(\mathbf{A})$. However, the Drazin inverse of $\mathbf{A}$ is always expressible as a polynomial in $\mathbf{A}$.

Theorem 7.5.1 *If* $\mathbf{A} \in \mathbb{C}^{n \times n}$, *then there exists a polynomial* $p(x)$ *such that* $\mathbf{A}^D = p(\mathbf{A})$.

Proof Use Theorem 2.1 and write $\mathbf{A}$ as $\mathbf{A} = \mathbf{P}\begin{bmatrix} \mathbf{C} & \mathbf{0} \\ \mathbf{0} & \mathbf{N} \end{bmatrix}\mathbf{P}^{-1}$ where $\mathbf{P}$ and $\mathbf{C}$ are non-singular and $\mathbf{N}$ is nilpotent of index $k = \operatorname{Ind}(\mathbf{A})$. Since $\mathbf{C}$ is non-singular, we know that there exists a polynomial $q(x)$ such that $\mathbf{C}^{-1} = q(\mathbf{C})$. Let $p(x)$ be the polynomial defined by $p(x) = x^k[q(x)]^{k+1}$. Then

$$p(\mathbf{A}) = \mathbf{A}^k[q(\mathbf{A})]^{k+1} = \mathbf{P}\begin{bmatrix} \mathbf{C}^k & \mathbf{0} \\ \mathbf{0} & \mathbf{0} \end{bmatrix}\begin{bmatrix} q(\mathbf{C}) & \mathbf{0} \\ \mathbf{0} & q(\mathbf{N}) \end{bmatrix}^{k+1}\mathbf{P}^{-1}$$

$$= \mathbf{P}\begin{bmatrix} \mathbf{C}^k[q(\mathbf{C})]^{k+1} & \mathbf{0} \\ \mathbf{0} & \mathbf{0} \end{bmatrix}\mathbf{P}^{-1} = \mathbf{P}\begin{bmatrix} \mathbf{C}^{-1} & \mathbf{0} \\ \mathbf{0} & \mathbf{0} \end{bmatrix}\mathbf{P}^{-1} = \mathbf{A}^D. \quad \blacksquare$$

The polynomial constructed in the proof of Theorem 1 is generally of much higher degree then is actually necessary. The next theorem shows how one might actually construct a polynomial $p(x)$ such that $p(\mathbf{A}) = \mathbf{A}^D$. Unlike Theorem 1 it uses the fact that $\mathbf{A}$ is a matrix over $\mathbb{C}$.

Theorem 7.5.2 [77] *Let* $\mathbf{A} \in \mathbb{C}^{n \times n}$. *Suppose that* $\{\lambda_0, \lambda_1, \lambda_2, \ldots, \lambda_t\}$ *are the distinct eigenvalues of* $\mathbf{A}$ *and* $\lambda_0 = 0$. *Let* m_i *denote the algebraic multiplicity of* λ_i *and* let $m = n - m_0 = m_1 + m_2 + \ldots + m_t$. *Let* $p(x)$ *be the polynomial of degree* $n - 1$ *such that* $p(x) = x^{m_0}(\alpha_0 + \alpha_1 x + \ldots + \alpha_{m-1}x^{m-1})$ *whose coefficients are the unique solutions of the following* $m \times m$ *system of linear equations.* ($(\cdot)^{(i)}$ *denotes the ith derivative with respect to x.*)

$$\begin{aligned} \frac{1}{\lambda_i} &= p(\lambda_i), \\ -\frac{1}{\lambda_i^2} &= p'(\lambda_i), \\ &\vdots \qquad\qquad \textit{for } i = 1, 2, \ldots, t, \\ \frac{(-1)^{m_i-1}(m_i-1)!}{\lambda_i^{m_i}} &= p^{(m_i-1)}(\lambda_i). \end{aligned}$$

Then $p(\mathbf{A}) = \mathbf{A}^D$.

Proof Since $\mathbf{A} \in \mathbb{C}^{n \times n}$, $\mathbf{A}$ is similar to a Jordan form. Write

$\mathbf{A} = \mathbf{T}\begin{bmatrix} \mathbf{J} & \mathbf{0} \\ \mathbf{0} & \mathbf{N} \end{bmatrix}\mathbf{T}^{-1}$, where $\mathbf{J}$ and $\mathbf{N}$ are the block diagonal matrices, $\mathbf{J} = \mathrm{Diag}[\mathbf{B}_1, \ldots, \mathbf{B}_h]$, $\mathbf{N} = \mathrm{Diag}[\mathbf{F}_1, \ldots, \mathbf{F}_g]$. Each $\mathbf{B}_j$ is an elementary Jordan block corresponding to a non-zero eigenvalue. That is, each $\mathbf{B}_j$ is of the form

$$\mathbf{B}_j = \begin{bmatrix} \lambda_l & 1 & 0 & 0 & \ldots & 0 & 0 \\ 0 & \lambda_l & 1 & 0 & \ldots & 0 & 0 \\ \vdots & & & & & & \\ 0 & 0 & 0 & 0 & \ldots & \lambda_l & 1 \\ 0 & 0 & 0 & 0 & \ldots & 0 & \lambda_l \end{bmatrix}_{s \times s}, \lambda_l \neq 0, \tag{1}$$

and $s \leq m_l$. Each $\mathbf{F}_j$ is an elementary Jordan block corresponding to a zero eigenvalue. That is, each $\mathbf{F}_j$ is of the form (1) with $\lambda_l = 0$. Clearly, $\mathbf{J}$ is non-singular and $\mathbf{N} \in \mathbb{C}^{m_0 \times m_0}$ is nilpotent of index $k = \mathrm{Ind}(\mathbf{A}) \leq m_0$.

Therefore, $\mathbf{A}^{\mathrm{D}} = \mathbf{T}\begin{bmatrix} \mathbf{J}^{-1} & \mathbf{0} \\ \mathbf{0} & \mathbf{0} \end{bmatrix}\mathbf{T}^{-1}$. Now, $p(\mathbf{A}) = \mathbf{T}\begin{bmatrix} p(\mathbf{J}) & \mathbf{0} \\ \mathbf{0} & p(\mathbf{N}) \end{bmatrix}\mathbf{T}^{-1}$ $= \mathbf{T}\begin{bmatrix} p(\mathbf{J}) & \mathbf{0} \\ \mathbf{0} & \mathbf{0} \end{bmatrix}\mathbf{T}^{-1}$, because $\mathbf{N}^{m_0} = \mathbf{0}$ implies $p(\mathbf{N}) = \mathbf{0}$. Since $p(\mathbf{J}) = \mathrm{Diag}[p(\mathbf{B}_1), \ldots, p(\mathbf{B}_h)]$, it suffices to show that $p(\mathbf{B}_j) = \mathbf{B}_j^{-1}$ for each j. But using (1), it is not difficult to verify that

$$p(\mathbf{B}_j) = \begin{bmatrix} p(\lambda_l) & \frac{p'(\lambda_l)}{1!} & \frac{p''(\lambda_l)}{2!} & \ldots & \frac{p^{(s-1)}(\lambda_l)}{(s-1)!} \\ 0 & p(\lambda_l) & \frac{p'(\lambda_l)}{1!} & \ldots & \vdots \\ \vdots & & & & \frac{p''(\lambda_l)}{2!} \\ \vdots & & & \ldots & \frac{p'(\lambda_l)}{1!} \\ 0 & 0 & 0 & 0 & p(\lambda_l) \end{bmatrix}_{s \times s}$$

$$= \begin{bmatrix} \frac{1}{\lambda_l} & \frac{-1}{\lambda_l^2} & \frac{1}{\lambda_l^3} & \ldots & \frac{(-1)^{s-1}}{\lambda_l^s} \\ 0 & \frac{1}{\lambda_l} & \frac{-1}{\lambda_l^2} & \ddots & \vdots \\ \vdots & \ddots & \ddots & \ddots & \frac{-1}{\lambda_l^2} \\ 0 & 0 & 0 \;\cdots\; 0 & \ddots & \frac{1}{\lambda_l} \end{bmatrix} = \mathbf{B}_j^{-1}.$$

Thus, $p(\mathbf{A}) = \mathbf{A}^{\mathrm{D}}$. ■

Theorem 2 can sometimes be useful in computing $\mathbf{A}^{\mathrm{D}}$. This is particularly true if m_o is large with respect to n. The following is an example where Theorem 2 can be used quite effectively.

Example 7.5.1 Let $\mathbf{A} = \begin{bmatrix} 2 & 4 & 6 & 5 \\ 1 & 4 & 5 & 4 \\ 0 & -1 & -1 & 0 \\ -1 & -2 & -3 & -3 \end{bmatrix}$. We shall use Theorem 2 to compute $\mathbf{A}^{\mathrm{D}}$. The first (and, in general, the most difficult) step is to compute the eigenvalues for $\mathbf{A}$. They are $\sigma(\mathbf{A}) = \{0, 0, 1, 1\}$. Thus, $m_o = 2$ and $m_1 = 2$ so that Theorem 2 implies that $\mathbf{A}^{\mathrm{D}}$ can be expressed as $\mathbf{A}^{\mathrm{D}} = \mathbf{A}^2(\alpha_o\mathbf{I} + \alpha_1\mathbf{A})$ since $p(x) = x^2(\alpha_o + \alpha_1 x)$. Now α_o and α_1 are the solutions of the system:

$$1 = p(1) = \alpha_o + \alpha_1$$
$$-1 = p'(1) = 2\alpha_o + 3\alpha_1$$

Therefore, $\alpha_o = 4$, $\alpha_1 = -3$, and

$$\mathbf{A}^{\mathrm{D}} = \mathbf{A}^2(4\mathbf{I} - 3\mathbf{A}) = \begin{bmatrix} 3 & -1 & 2 & 2 \\ 2 & 1 & 3 & 3 \\ -1 & 0 & -1 & -1 \\ -1 & 0 & -1 & -1 \end{bmatrix}.$$

For each $\mathbf{A} \in \mathbb{C}^{n \times n}$, there are two polynomials of special importance. These are the characteristic polynomial and the minimal polynomial. Let us examine each one. Consider first the minimal polynomial for $\mathbf{A}$,

$$m(x) = x^d + \alpha_{d-1}x^{d-1} + \ldots + \alpha_1 x + \alpha_o.$$

It is not difficult to show that $\mathbf{A}$ is non-singular if and only if $\alpha_o \neq 0$; in which case, $\mathbf{A}^{-1} = -\dfrac{1}{\alpha_o}(\mathbf{A}^{d-1} + \alpha_{d-1}\mathbf{A}^{d-2} + \ldots + \alpha_2\mathbf{A} + \alpha_1\mathbf{I})$. Now, assume $\mathbf{A}$ is singular so that $\alpha_o = 0$. Let i be the smallest number such that $0 = \alpha_o = \alpha_1 = \ldots = \alpha_{i-1}$, but $\alpha_i \neq 0$. This number, i, is sometimes called the *index of the zero eigenvalue*. The next theorem (valid in a general field) shows that the index of the zero eigenvalue of $\mathbf{A}$ is the same as the index of $\mathbf{A}$.

Theorem 7.5.3 *If* $\mathbf{A} \in \mathbb{C}^{n \times n}$, *and* $m(x) = x^d + \alpha_{d-1}x^{d-1} + \ldots + \alpha_i x^i$, *with* $\alpha_i \neq 0$, *is the minimal polynomial for* $\mathbf{A}$, *then* $i = \mathrm{Ind}(\mathbf{A})$.

Proof Use Theorem 2.1 and write $\mathbf{A}$ as $\mathbf{A} = \mathbf{P}\begin{bmatrix} \mathbf{C} & \mathbf{0} \\ \mathbf{0} & \mathbf{N} \end{bmatrix}\mathbf{P}^{-1}$ where $\mathbf{C}$ is non-singular and $\mathbf{N}$ is nilpotent of index $k = \mathrm{Ind}(\mathbf{A})$. Since $m(\mathbf{A}) = \mathbf{0}$, we can conclude that $\mathbf{0} = m(\mathbf{N}) = \mathbf{N}^d + \alpha_{d-1}\mathbf{N}^{d-1} + \ldots + \alpha_{i+1}\mathbf{N}^{i+1} + \alpha_i\mathbf{N}^i$ $= (\mathbf{N}^{d-i} + \alpha_{d-1}\mathbf{N}^{d-i-1} + \ldots + \alpha_i\mathbf{I})\mathbf{N}^i$. Since $(\mathbf{N}^{d-i} + \alpha_{d-1}\mathbf{N}^{d-i-1} + \alpha_i\mathbf{I})$

is invertible we have $\mathbf{N}^i = \mathbf{0}$. Hence $i \geq k$. Suppose that $k < i$. Then,

$$\mathbf{A}^{\mathrm{D}}\mathbf{A}^i = \mathbf{A}^{i-1}. \tag{2}$$

Write $m(x) = x^i q(x)$ so that $\mathbf{0} = m(\mathbf{A}) = \mathbf{A}^i q(\mathbf{A})$. Multiply both sides of this by $\mathbf{A}^{\mathrm{D}}$ and use (2) to obtain $\mathbf{0} = \mathbf{A}^{i-1}q(\mathbf{A})$. Thus, the polynomial $r(x) = x^{i-1}q(x)$ is such that $r(\mathbf{A}) = \mathbf{0}$ and $\deg[r(x)] < \deg[m(x)]$. This is a contradiction. Therefore, we can conclude that $k = i$. ■

Corollary 7.5.1 *Let* $\mathbf{A} \in \mathbb{C}^{n \times n}$, $k = \mathrm{Ind}(\mathbf{A})$, *and* m_o *denote the algebraic multiplicity of the zero eigenvalue. It is always the case that* $m_o \geq k$.

Proof The minimal polynomial, from Theorem 5.3, is $m(x) = x^k(x^{d-k} + \alpha_{d-1}x^{d-1-k} + \ldots + \alpha_{k+1}x + \alpha_k)$, $(\alpha_k \neq 0)$, and $m(x)$ must divide $p(x)$. ■

When one uses Theorem 2 to compute the Drazin inverse of $\mathbf{A} \in \mathbb{C}^{n \times n}$, it is necessary to compute each eigenvalue of $\mathbf{A}$ along with the multiplicities of each eigenvalue. Many times one can compute the coefficients of the characteristic polynomial for $\mathbf{A}$ easier than the eigenvalues. The following theorem shows how to obtain $\mathbf{A}^{\mathrm{D}}$ from the characteristic polynomial for $\mathbf{A}$.

Theorem 7.5.4 *Let* $\mathbf{A} \in \mathbb{C}^{n \times n}$ *and let* $k = \mathrm{Ind}(\mathbf{A})$. *Write the characteristic equation for* $\mathbf{A}$ *as* $0 = x^{m_o}(x^{n-m_o} + \beta_{n-1}x^{n-1-m_o} + \ldots + \beta_{m_o+1}x + \beta_{m_o}) = x^{m_o}q(x)$, $(\beta_{m_o} \neq 0)$. *Let*

$$r(x) = \begin{cases} -\dfrac{1}{\beta_{m_o}}(x^{n-m_o-1} + \beta_{n-1}x^{n-m_o-2} + \ldots + \beta_{m_o+1}), & \text{if } m_o < n \\ 0, & \text{if } m_o = n \end{cases} \tag{3}$$

Then, $\mathbf{A}^{\mathrm{D}} = \mathbf{A}^l[r(\mathbf{A})]^{l+1}$ *for each integer* $l \geq k$.

Proof If $m_o = n$, then $\mathbf{A}$ is nilpotent and $\mathbf{A}^{\mathrm{D}} = \mathbf{0}$, and the result is trivial. Thus assume $m_o < n$. Multiply both sides of $\mathbf{0} = \mathbf{A}^{m_o}q(\mathbf{A})$ by $(\mathbf{A}^{\mathrm{D}})^{m_o+1}$ to obtain $\mathbf{0} = \mathbf{A}^{\mathrm{D}}q(\mathbf{A})$. From this, it easily follows that $\mathbf{A}^{\mathrm{D}} = \mathbf{A}\mathbf{A}^{\mathrm{D}}r(\mathbf{A})$. By raising both sides to the $(l+1)$th power, we obtain $(\mathbf{A}^{\mathrm{D}})^{l+1} = \mathbf{A}\mathbf{A}^{\mathrm{D}}[r(\mathbf{A})]^{l+1}$. Multiplication on both sides of this by $\mathbf{A}^l$ yields $\mathbf{A}^{\mathrm{D}} = \mathbf{A}^l[r(\mathbf{A})]^{l+1}$. ■

Since the index of a matrix can never exceed its size, nor the number m_o of Corollary 1, we have the following.

Corollary 7.5.2 *For* $\mathbf{A} \in \mathbb{C}^{n \times n}$, $\mathbf{A}^{\mathrm{D}} = \mathbf{A}^n[r(\mathbf{A})]^{n+1} = \mathbf{A}^{m_o}[r(\mathbf{A})]^{m_o+1}$ *where* $r(x)$ *is the polynomial in Theorem* 4.

For $\mathbf{A} \in \mathbb{C}^{n \times n}$, the coefficients in the characteristic equation

$$x^n + \beta_{n-1}x^{n-1} + \beta_{n-2}x^{n-2} + \ldots + \beta_1 x + \beta_o = 0$$

for $\mathbf{A}$ can be computed recursively by the well known algorithm [43] (Tr denotes trace).

$$\beta_{n-j} = -\frac{1}{j}\mathrm{Tr}(\mathbf{A}\mathbf{S}_{j-1}) \tag{4}$$

where

$$\mathbf{S}_o = \mathbf{I} \text{ and } \mathbf{S}_j = \mathbf{A}\mathbf{S}_{j-1} + \beta_{n-j}\mathbf{I}. \tag{5}$$

Since $\mathbf{S}_j = \mathbf{A}^j + \beta_{n-1}\mathbf{A}^{j-1} + \ldots \beta_{n-j}\mathbf{I}$, the next result showing how this algorithm may be used to obtain the matrix $r(\mathbf{A})$ is immediate.

Theorem 7.5.5 *Let* $\mathbf{A}\in\mathbb{C}^{n\times n}$ *and let* $r(x)$ *be as in* (3). *If* $n = m_o$, *then* $\mathbf{A}^D = \mathbf{0}$. *If* $n > m_o$, *then* $r(\mathbf{A}) = -\dfrac{1}{\beta_{m_o}}\mathbf{S}_{n-m_o-1}$ *where* β_{m_o} *and* $\mathbf{S}_{m_o+1}$ *are computed from* (4) *and* (5). *Thus* $\mathbf{A}^D = -\dfrac{1}{\beta_{m_o}^{l+1}}\mathbf{A}^l\mathbf{S}_{n-m_o-1}^{l+1}$ *for each* $l \geq \text{Ind}(\mathbf{A})$.

Notice that if $\mathbf{S}_t = \mathbf{0}$, then $\mathbf{0} = \mathbf{S}_{t+1} = \mathbf{S}_{t+2} = \ldots = \mathbf{S}_{n-1}$ and $\beta_{n-t-1} = \beta_{n-t-2} = \ldots = \beta_o = 0$. Thus, it is easy to use Theorem 5 to obtain $r(\mathbf{A})$, and $\mathbf{A}^D$ as follows.

Algorithm 7.5.1 To compute $\mathbf{A}^D$ for $\mathbf{A}\in\mathbb{C}^{n\times n}$.

(I) Set $\mathbf{S}_o = \mathbf{I}$ and recursively compute $\mathbf{S}_j = \mathbf{A}\mathbf{S}_{j-1} + \beta_{n-j}\mathbf{I}$, $\beta_{n-j} = -\dfrac{1}{j}\text{Tr}(\mathbf{A}\mathbf{S}_{j-1})$, until some $\mathbf{S}_t = \mathbf{0}$, but $\mathbf{S}_{t-1} \neq \mathbf{0}$.

(II) Let u be that number such that $\beta_{n-u} \neq 0$ and $\beta_{n-u-1} = \beta_{n-u-2} = \ldots = \beta_{n-t-1} = 0$. (Notice that $n - u = m_o$, the algebraic multiplicity of the zero eigenvalue.)

(III) Let $l = n - u$ and compute $\mathbf{S}_{n-m_o-1}^{l+1} = \mathbf{S}_{u-1}^{l+1}$.

(IV) Compute $\mathbf{A}^D$ as $\mathbf{A}^D = -\dfrac{1}{\beta_l^{l+1}}\mathbf{A}^l\mathbf{S}_{u-1}^{l+1}$.

Note that not all of the computed $\mathbf{S}_j$'s must be saved. If $\beta_{n-j} \neq 0$, then $\mathbf{S}_{j-2}$ can be forgotten. However $\mathbf{S}_{j-1}$ needs to be saved until the next non-zero β appears. Also notice that this algorithm produces the value of the algebraic multiplicity of the zero eigenvalue for $\mathbf{A}$.

Example 7.5.2 Let

$$\mathbf{A} = \begin{bmatrix} 10 & -8 & 6 & -3 \\ 12 & -10 & 8 & -4 \\ 1 & -1 & 1 & 0 \\ -2 & 2 & -2 & 2 \end{bmatrix}.$$

We shall use Algorithm 1 to compute $\mathbf{A}^D$.

(I) Successive calculations give

$$\mathbf{S}_o = \mathbf{I}, \qquad \beta_3 = -\text{Tr}(\mathbf{A}\mathbf{S}_o) = -\text{Tr}(\mathbf{A}) = -3,$$

$$\mathbf{S}_1 = \mathbf{A}\mathbf{S}_o - 3\mathbf{I} = \begin{bmatrix} 7 & -8 & 6 & -3 \\ 12 & -13 & 8 & -4 \\ 1 & -1 & -2 & 0 \\ -2 & 2 & -2 & -1 \end{bmatrix},$$

$$\mathbf{AS}_1 = \begin{bmatrix} -14 & 12 & -10 & 5 \\ -20 & 18 & -16 & 8 \\ -4 & 4 & -4 & 1 \\ 4 & -4 & 4 & -4 \end{bmatrix}, \qquad \beta_2 = -\frac{1}{2}\mathrm{Tr}(\mathbf{AS}_1) = 2,$$

$$\mathbf{S}_2 = \mathbf{AS}_1 + 2\mathbf{I} = \begin{bmatrix} -12 & 12 & -10 & 5 \\ -20 & 20 & -16 & 8 \\ -4 & 4 & -2 & 1 \\ 4 & -4 & 4 & -2 \end{bmatrix},$$

$$\mathbf{AS}_2 = \begin{bmatrix} 4 & -4 & 4 & -2 \\ 8 & -8 & 8 & -4 \\ 4 & -4 & 4 & -2 \\ 0 & 0 & 0 & 0 \end{bmatrix}, \qquad \beta_1 = -\frac{1}{3}\mathrm{Tr}(\mathbf{AS}_2) = 0,$$

$$\mathbf{S}_3 = \mathbf{AS}_2,$$

$$\mathbf{AS}_3 = \begin{bmatrix} 0 & 0 & 0 & 0 \\ 0 & 0 & 0 & 0 \\ 0 & 0 & 0 & 0 \\ 0 & 0 & 0 & 0 \end{bmatrix}, \qquad \beta_o = -\frac{1}{4}\mathrm{Tr}(\mathbf{AS}_3) = 0,$$

and

$$\mathbf{S}_4 = \mathbf{AS}_3 = \mathbf{0}.$$

Therefore $t = 4$ in this example and the algebraic multiplicity of the zero eigenvalue is $m_o = 2$.

(II) Set $u = 2$.
(III) Set $l = 2$.
(IV) Compute $\mathbf{A}^{\mathrm{D}} = -\frac{1}{\beta_2^3}\mathbf{A}^2\mathbf{S}_1^3 = -\frac{1}{8}\mathbf{A}^2\mathbf{S}_1^3$ as follows. Since $\mathbf{S}_2 = \mathbf{AS}_1 + \beta_2\mathbf{I}$, we have that $\mathbf{AS}_2\mathbf{S}_1^2 = \mathbf{A}^2\mathbf{S}_1^3 + \beta_2\mathbf{AS}_1^2$. Write this as $\mathbf{A}^2\mathbf{S}_1^3 = [(\mathbf{AS}_2)\mathbf{S}_1 - \beta_2(\mathbf{AS}_1)]\mathbf{S}_1$. Since $\mathbf{AS}_2$ and $\mathbf{AS}_1$ have already been computed, only two matrix multiplications are necessary. This is more efficient than forming the product $\mathbf{A}^2\mathbf{S}_1^3$ directly. In general, one can always do something like this when this algorithm is used. Now

$$(\mathbf{AS}_2)\mathbf{S}_1 = \begin{bmatrix} -12 & 12 & -12 & 6 \\ -24 & 24 & -24 & 12 \\ -12 & 12 & -12 & 6 \\ 0 & 0 & 0 & 0 \end{bmatrix},$$

$$[(\mathbf{AS}_2)\mathbf{S}_1 - 2(\mathbf{AS}_1)] = \begin{bmatrix} 16 & -12 & 8 & -4 \\ 16 & -12 & 8 & -4 \\ -4 & 4 & -4 & 4 \\ -8 & 8 & -8 & 8 \end{bmatrix},$$

and

$$\mathbf{A}^2\mathbf{S}_1^3 = [(\mathbf{AS}_2)\mathbf{S}_1 - 2(\mathbf{AS}_1)]\mathbf{S}_1 = \begin{bmatrix} -16 & 12 & -8 & 4 \\ -16 & 12 & -8 & 4 \\ 8 & -8 & 8 & -8 \\ 16 & -16 & 16 & -16 \end{bmatrix}.$$

Therefore,

$$\mathbf{A}^{\mathrm{D}} = -\frac{1}{8}\mathbf{A}^2\mathbf{S}_1^3 = \begin{bmatrix} 2 & -\frac{3}{2} & 1 & -\frac{1}{2} \\ 2 & -\frac{3}{2} & 1 & -\frac{1}{2} \\ -1 & 1 & -1 & 1 \\ -2 & 2 & -2 & 2 \end{bmatrix}.$$

6. $\mathbf{A}^{\mathrm{D}}$ as a limit

It was previously shown how the Moore–Penrose inverse could be expressed by means of a limiting process. In this section, we will show how the Drazin inverse and the index of a square matrix can also be characterized in terms of a limiting process.

Whenever we consider the limit, as $z \to 0$, of an expression involving $(\mathbf{A} + z\mathbf{I})^{-1}$. $\mathbf{A} \in \mathbb{C}^{n \times n}$, we shall assume that $-z \notin \sigma(\mathbf{A})$.

Definition 7.6.1 *If* $\mathbf{A} \in \mathbb{C}^{n \times n}$, *and if* $\mathbf{C}_{\mathbf{A}}$ *and* $\mathbf{N}_{\mathbf{A}}$ *denote the core and nilpotent part of* $\mathbf{A}$, *respectively, then for integers* $m \geq -1$, *we define*

$$\mathbf{C}_{\mathbf{A}}^{(m)} = \mathbf{A}^{m+1}\mathbf{A}^{\mathrm{D}} = \begin{cases} \mathbf{A}^{\mathrm{D}}, & \text{if } m = -1 \\ \mathbf{AA}^{\mathrm{D}}, & \text{if } m = 0 \text{ and} \\ \mathbf{C}_{\mathbf{A}}^{m}, & \text{if } m \geq 1 \end{cases}$$

$$\mathbf{N}_{\mathbf{A}}^{(m)} = \begin{cases} \mathbf{0}, & \text{if } m = -1 \\ \mathbf{A}^m - \mathbf{C}_{\mathbf{A}}^{(m)}, & \text{if } m \geq 0, \end{cases} = \begin{cases} \mathbf{0}, & \text{if } m = -1 \\ \mathbf{I} - \mathbf{AA}^{\mathrm{D}}, & \text{if } m = 0. \\ \mathbf{N}_{\mathbf{A}}^{m}, & \text{if } m \geq 1 \end{cases}$$

Theorem 7.6.1 *Let* $\mathbf{A} \in \mathbb{C}^{n \times n}$ *and let* $\mathrm{Ind}(\mathbf{A}) = k$. *For every integer* $l \geq k$,

$$\mathbf{A}^{\mathrm{D}} = \lim_{z \to 0} (\mathbf{A}^{l+1} + z\mathbf{I})^{-1}\mathbf{A}^{l}. \tag{1}$$

For every non-negative integer l,

$$\mathbf{A}^{\mathrm{D}} = \lim_{z \to 0} (\mathbf{A}^{l+1} + z\mathbf{I})^{-1}\mathbf{C}_{\mathbf{A}}^{(l)}. \tag{2}$$

Proof If $k = 0$, then $\mathbf{A}$ is non-singular, and the result is evident. For $k > 0$, use Theorem 3.1 and write

$$(\mathbf{A}^{l+1} + z\mathbf{I})^{-1}\mathbf{C}_{\mathbf{A}}^{(l)} = \mathbf{P}\left[\begin{array}{c|c} (\mathbf{C}^{l+1} + z\mathbf{I})^{-1}\mathbf{C}^{l} & \mathbf{0} \\ \hline \mathbf{0} & \mathbf{0} \end{array}\right]\mathbf{P}^{-1}.$$

Since $\mathbf{C}$ is non-singular, and $\lim_{z \to 0} (\mathbf{C}^{l+1} + z\mathbf{I})^{-1}\mathbf{C}^{l} = \mathbf{C}^{-1}$, (2) is proven. If $l \geq k$, then $\mathbf{C}_{\mathbf{A}}^{(l)} = \mathbf{A}^{l+1}\mathbf{A}^{\mathrm{D}} = \mathbf{A}^{l}$ and (1) follows. ■

Since it is always true that $k \leq n$, we also have the following.

Corollary 7.7.1 *For* $\mathbf{A} \in \mathbb{C}^{n \times n}$, $\mathbf{A}^{\mathrm{D}} = \lim_{z \to 0} (\mathbf{A}^{n+1} + z\mathbf{I})^{-1}\mathbf{A}^n$.

The index of $\mathbf{A} \in \mathbb{C}^{n \times n}$ can also be characterized in terms of a limit. Before doing this, we need some preliminary results. The first is an obvious consequence of Theorem 3.1.

Lemma 7.6.1 *Let* $\mathbf{A} \in \mathbb{C}^{n \times n}$ *be a singular matrix. For a positive integer p,* $\mathrm{Ind}(\mathbf{A}^p) = 1$ *if and only if* $p \geq \mathrm{Ind}(\mathbf{A})$. *Equivalently, the smallest positive integer l for which* $\mathrm{Ind}(\mathbf{A}^l) = 1$ *is the index of* $\mathbf{A}$.

Lemma 7.6.2 *Let* $\mathbf{N} \in \mathbb{C}^{n \times n}$ *be a nilpotent matrix such that* $\mathrm{Ind}(\mathbf{N}) = k$. *For non-negative integers m and p, the limit*

$$\lim_{z \to 0} z^m(\mathbf{N} + z\mathbf{I})^{-1}\mathbf{N}^p \tag{3}$$

exists if and only if $m + p \geq k$. *When the limit exists, its value is given by*

$$\lim_{z \to 0} z^m(\mathbf{N} + z\mathbf{I})^{-1}\mathbf{N}^p = \begin{cases} (-1)^{m+1}\mathbf{N}^{m+p-1} & \text{if } m > 0 \\ \mathbf{0} & \text{if } m = 0 \end{cases} \tag{4}$$

Proof If $\mathbf{N} = \mathbf{0}$, then, from Lemma 2.1, we know that $k = 1$. The limit under consideration reduces to

$$\lim_{z \to 0} z^{m-1}\mathbf{0}^p = \begin{cases} \lim_{z \to 0} z^{m-1}\mathbf{I}, & \text{if } p = 0 \\ \mathbf{0}, & \text{if } p \geq 1. \end{cases}$$

It is evident this limit will exist if and only if either $p \geq 1$ or $m \geq 1$, which is equivalent to $m + p \geq 1$. Thus the result is established for $k = 1$. Assume now that $k \geq 1$, i.e. $\mathbf{N} \neq \mathbf{0}$. Since $(\mathbf{N} + z\mathbf{I})^{-1} = \sum_{i=0}^{k-1} (-1)^i \dfrac{\mathbf{N}^i}{z^{i+1}}$. We have

$$\begin{aligned} z^m(\mathbf{N} + z\mathbf{I})^{-1}\mathbf{N}^p = {} & z^{m-1}\mathbf{N}^p - z^{m-2}\mathbf{N}^{p+1} + \ldots + (-1)^{m-2}z\mathbf{N}^{m+p-2} \\ & + (-1)^{m-1}\mathbf{N}^{m+p-1} + \frac{(-1)^m\mathbf{N}^{m+p}}{z} + \ldots \\ & + \frac{(-1)^{k-1}\mathbf{N}^{p+k-1}}{z^{k-m}}. \end{aligned} \tag{5}$$

If $m + p \geq k$, then clearly the limit (3) exists. Conversely, if the limit (3) exists, then it can be seen from (5) that $\mathbf{N}^{m+p} = \mathbf{0}$ and hence $m + p \geq k$. ■

Theorem 7.6.2 *For* $\mathbf{A} \in \mathbb{C}^{n \times n}$ *where* $\mathrm{Ind}(\mathbf{A}) = k$ *and for non-negative integers m and p, the limit*

$$\lim_{z \to 0} z^m(\mathbf{A} + z\mathbf{I})^{-1}\mathbf{A}^p \tag{6}$$

exists if and only if $m + p \geq k$; *in which case the value of the limit is given by*

$$\lim_{z \to 0} z^m(\mathbf{A} + z\mathbf{I})^{-1}\mathbf{A}^p = \begin{cases} (-1)^{m-1}(\mathbf{I} - \mathbf{A}\mathbf{A}^{\mathrm{D}})\mathbf{A}^{m+p-1}, & \text{if } m > 0 \\ \mathbf{A}^{\mathrm{D}}\mathbf{A}^p, & \text{if } m = 0. \end{cases} \tag{7}$$

Proof If $k = 0$, the result is immediate. Assume $k \geq 1$ and use Theorem 2.1 to write $\mathbf{A} = \mathbf{P}\begin{bmatrix} \mathbf{C} & \mathbf{0} \\ \mathbf{0} & \mathbf{N} \end{bmatrix}\mathbf{P}^{-1}$ where $\mathbf{P}$ and $\mathbf{C}$ are non-singular and $\mathbf{N}$ is nilpotent of index k. Then

$$z^m(\mathbf{A} + z\mathbf{I})^{-1}\mathbf{A}^p = \mathbf{P}\left[\begin{array}{c|c} z^m(\mathbf{C} + z\mathbf{I})^{-1}\mathbf{C}^p & \mathbf{0} \\ \hline \mathbf{0} & z^m(\mathbf{N} + z\mathbf{I})^{-1}\mathbf{N}^p \end{array}\right]\mathbf{P}^{-1}. \tag{8}$$

Because $\mathbf{C}$ is non-singular, we always have

$$\lim_{z\to 0} z^m(\mathbf{C} + z\mathbf{I})^{-1}\mathbf{C}^p = \begin{cases} \mathbf{0}, & \text{if } m > 0 \\ \mathbf{C}^{p-1}, & \text{if } m = 0 \end{cases} \tag{9}$$

Thus the limit (6) exists if and only if the limit $\lim\limits_{z\to 0} z^m(\mathbf{N} + z\mathbf{I})^{-1}\mathbf{N}^p$ exists, which, by Lemma 2, is equivalent to saying $m + p \geq k$. The expression (7) is obtained from (8) by using (9) and (4). ■

There are some important corollaries to the above theorem. The first characterizes $\operatorname{Ind}(\mathbf{A})$ in terms of a limit.

Corollary 7.6.2 *For $\mathbf{A} \in \mathbb{C}^{n\times n}$, the following statements are equivalent*

(i) $\operatorname{Ind}(\mathbf{A}) = k$.
(ii) *k is the smallest non-negative integer such that the limit* $\lim\limits_{z\to 0} (\mathbf{A} + z\mathbf{I})^{-1}\mathbf{A}^k$ *exists.*
(iii) *k is the smallest non-negative integer such that the limit* $\lim\limits_{z\to 0} z^k(\mathbf{A} + z\mathbf{I})^{-1}$ *exists.*
(iv) *If* $\operatorname{Ind}(\mathbf{A}) = k$, *then* $\lim\limits_{z\to 0} (\mathbf{A} + z\mathbf{I})^{-1}\mathbf{A}^k = (\mathbf{A}\mathbf{A}^{\mathrm{D}})\mathbf{A}^{k-1} = \mathbf{C}_{\mathbf{A}}^{(k-1)}$
(v) *And when* $k > 0$, $\lim\limits_{z\to 0} z^k(\mathbf{A} + z\mathbf{I})^{-1} = (-1)^{k-1}(\mathbf{I} - \mathbf{A}\mathbf{A}^{\mathrm{D}})\mathbf{A}^{k-1} = \mathbf{N}_{\mathbf{A}}^{(k-1)}$.

Corollary 7.6.3 *For $\mathbf{A} \in \mathbb{C}^{n\times n}$ and for every integer $l \geq \operatorname{Ind}(\mathbf{A}) > 0$,* $\lim\limits_{z\to 0} (\mathbf{A} + z\mathbf{I})^{-1}(\mathbf{A}^l + z^l\mathbf{I}) = \mathbf{A}^{l-1}$.

Corollary 7.6.4 *For $\mathbf{A} \in \mathbb{C}^{n\times n}$, the following statements are equivalent.*

(i) $\operatorname{Ind}(\mathbf{A}) \leq 1$.
(ii) $\lim\limits_{z\to 0} (\mathbf{A} + z\mathbf{I})^{-1}\mathbf{A} = \mathbf{A}\mathbf{A}^{\#}$.
(iii) $\lim\limits_{z\to 0} z(\mathbf{A} + z\mathbf{I})^{-1} = \mathbf{I} - \mathbf{A}\mathbf{A}^{\#}$.

The index can also be characterized in terms of the limit (1).

Theorem 7.6.3 *For $\mathbf{A} \in \mathbb{C}^{n\times n}$, the smallest non-negative integer, l, such that*

$$\lim_{z\to 0} (\mathbf{A}^{l+1} + z\mathbf{I})^{-1}\mathbf{A}^l \tag{10}$$

exists is the index of $\mathbf{A}$.

Proof If $\operatorname{Ind}(\mathbf{A}) = 0$, then the existence of (10) is obvious. So suppose $\operatorname{Ind}(\mathbf{A}) = k \geq 1$. Using Theorem 2.1 we get

$$\mathbf{A} = \mathbf{P}\begin{bmatrix} \mathbf{C} & \mathbf{0} \\ \mathbf{0} & \mathbf{N} \end{bmatrix}\mathbf{P}^{-1}, \quad (\mathbf{A}^{l+1} + z\mathbf{I})^{-1}\mathbf{A}^{l}$$
$$= \mathbf{P}\left[\begin{array}{c|c} (\mathbf{C}^{l+1} + z\mathbf{I})^{-1}\mathbf{C}^{l} & \mathbf{0} \\ \hline \mathbf{0} & (\mathbf{N}^{l+1} + z\mathbf{I})^{-1}\mathbf{N}^{l} \end{array}\right]\mathbf{P}^{-1}.$$

The term $(\mathbf{C}^{l+1} + z\mathbf{I})^{-1}\mathbf{C}^{l}$ has a limit for all $l \geq 0$ since $\mathbf{C}$ is invertible. But $(\mathbf{N}^{l+1} + z\mathbf{I})^{-1}\mathbf{N}^{l} = \left(\mathbf{I} + \frac{\mathbf{N}^{l+1}}{z}\right)^{-1}\frac{\mathbf{N}^{l}}{z} = \left(\sum_{m=0}^{n} (-1)^m \frac{\mathbf{N}}{z^m} m(l+1)\right)\frac{\mathbf{N}^{l}}{z}$ which has a limit if and only if $\mathbf{N}^{l} = \mathbf{0}$. That is, $l \geq \operatorname{Ind}(\mathbf{A})$. ■

7. The Drazin inverse of a partitioned matrix

This section will investigate the Drazin inverse of matrices partitioned as $\mathbf{M} = \begin{bmatrix} \mathbf{A} & \mathbf{B} \\ \mathbf{D} & \mathbf{C} \end{bmatrix}$ where $\mathbf{A}$ and $\mathbf{C}$ are always assumed to be square.

Unfortunately, at the present time there is no known representation for $\mathbf{M}^{\mathrm{D}}$ with $\mathbf{A}, \mathbf{B}, \mathbf{C}, \mathbf{D}$ arbitrary. However, we can say something if either $\mathbf{D} = \mathbf{0}$ or $\mathbf{B} = \mathbf{0}$. In the following theorem, we assume $\mathbf{D} = \mathbf{0}$.

Theorem 7.7.1 *If* $\mathbf{M} = \begin{bmatrix} \mathbf{A} & \mathbf{B} \\ \mathbf{0} & \mathbf{C} \end{bmatrix} \in \mathbb{C}^{n\times n}$ *where* $\mathbf{A}$ *and* $\mathbf{C}$ *are square,* $k = \operatorname{Ind}(\mathbf{A})$, *and* $l = \operatorname{Ind}(\mathbf{C})$, *then* $\mathbf{M}^{\mathrm{D}} = \begin{bmatrix} \mathbf{A}^{\mathrm{D}} & \mathbf{X} \\ \mathbf{0} & \mathbf{C}^{\mathrm{D}} \end{bmatrix}$ *where*

$$\begin{aligned} \mathbf{X} &= (\mathbf{A}^{\mathrm{D}})^2\left[\sum_{i=0}^{n} (\mathbf{A}^{\mathrm{D}})^i\mathbf{B}\mathbf{C}^i\right](\mathbf{I} - \mathbf{C}\mathbf{C}^{\mathrm{D}}) \\ &\quad + (\mathbf{I} - \mathbf{A}\mathbf{A}^{\mathrm{D}})\left[\sum_{i=0}^{n} \mathbf{A}^i\mathbf{B}(\mathbf{C}^{\mathrm{D}})^i\right](\mathbf{C}^{\mathrm{D}})^2 - \mathbf{A}^{\mathrm{D}}\mathbf{B}\mathbf{C}^{\mathrm{D}} \\ &= (\mathbf{A}^{\mathrm{D}})^2\left[\sum_{i=0}^{l-1} (\mathbf{A}^{\mathrm{D}})^i\mathbf{B}\mathbf{C}^i\right](\mathbf{I} - \mathbf{C}\mathbf{C}^{\mathrm{D}}) \\ &\quad + (\mathbf{I} - \mathbf{A}\mathbf{A}^{\mathrm{D}})\left[\sum_{i=0}^{k-1} \mathbf{A}^i\mathbf{B}(\mathbf{C}^{\mathrm{D}})^i\right](\mathbf{C}^{\mathrm{D}})^2 - \mathbf{A}^{\mathrm{D}}\mathbf{B}\mathbf{C}^{\mathrm{D}}. \end{aligned} \tag{2}$$

(*We define* $\mathbf{0}^{\circ} = \mathbf{I}$)

Proof Expand the term $\mathbf{AX}$ as follows.

$$\begin{aligned} \mathbf{AX} = \sum_{0}^{l-1} (\mathbf{A}^{\mathrm{D}})^{i+1}\mathbf{B}\mathbf{C}^i - \sum_{0}^{l-1} (\mathbf{A}^{\mathrm{D}})^{i+1}\mathbf{B}\mathbf{C}^{i+1}\mathbf{C}^{\mathrm{D}} - \sum_{0}^{k-1} \mathbf{A}^{i+1}\mathbf{B}(\mathbf{C}^{\mathrm{D}})^{i+2} \\ - \sum_{0}^{k-1} \mathbf{A}^{\mathrm{D}}\mathbf{A}^{i+2}\mathbf{B}(\mathbf{C}^{\mathrm{D}})^{i+2} - \mathbf{A}\mathbf{A}^{\mathrm{D}}\mathbf{B}\mathbf{C} \end{aligned}$$

$$= \left(\mathbf{A}^{\mathrm{D}}\mathbf{B} + \sum_{0}^{l-2} (\mathbf{A}^{\mathrm{D}})^{i+2}\mathbf{B}\mathbf{C}^{i+1} \right) - \left(\mathbf{A}^{\mathrm{D}}\mathbf{B}\mathbf{C}\mathbf{C}^{\mathrm{D}} + \sum_{0}^{l-2} (\mathbf{A}^{\mathrm{D}})^{i+2}\mathbf{B}\mathbf{C}^{i+2}\mathbf{C}^{\mathrm{D}} \right)$$

$$+ \left(\sum_{1}^{k-1} \mathbf{A}^{i}\mathbf{B}(\mathbf{C}^{\mathrm{D}})^{i+1} + \mathbf{A}^{k}\mathbf{B}(\mathbf{C}^{\mathrm{D}})^{k+1} \right)$$

$$- \left(\sum_{1}^{k-1} \mathbf{A}^{\mathrm{D}}\mathbf{A}^{i+1}\mathbf{B}(\mathbf{C}^{\mathrm{D}})^{i+1} + \mathbf{A}^{k}\mathbf{B}(\mathbf{C}^{\mathrm{D}})^{k-1} \right) - \mathbf{A}\mathbf{A}^{\mathrm{D}}\mathbf{B}\mathbf{C}$$

$$= \mathbf{A}^{\mathrm{D}}\mathbf{B} + \sum_{0}^{l-2} (\mathbf{A}^{\mathrm{D}})^{i+2}\mathbf{B}\mathbf{C}^{i+1} - \mathbf{A}^{\mathrm{D}}\mathbf{B}\mathbf{C}\mathbf{C}^{\mathrm{D}} - \sum_{0}^{l-2} (\mathbf{A}^{\mathrm{D}})^{i+2}\mathbf{B}\mathbf{C}^{i+2}\mathbf{C}^{\mathrm{D}}$$

$$+ \sum_{1}^{k-1} \mathbf{A}^{i}\mathbf{B}(\mathbf{C}^{\mathrm{D}})^{i+1} - \sum_{1}^{k-1} \mathbf{A}^{\mathrm{D}}\mathbf{A}^{i+1}\mathbf{B}(\mathbf{C}^{\mathrm{D}})^{i+1} - \mathbf{A}\mathbf{A}^{\mathrm{D}}\mathbf{B}\mathbf{C}.$$

Now expand the term $\mathbf{XC}$ as follows.

$$\mathbf{XC} = \sum_{0}^{l-1} (\mathbf{A}^{\mathrm{D}})^{i+2}\mathbf{B}\mathbf{C}^{i+1} - \sum_{0}^{l-1} (\mathbf{A}^{\mathrm{D}})^{i+2}\mathbf{B}\mathbf{C}^{i+2}\mathbf{C}^{\mathrm{D}} + \sum_{0}^{k-1} \mathbf{A}^{i}\mathbf{B}(\mathbf{C}^{\mathrm{D}})^{i+1}$$

$$- \sum_{0}^{k-1} \mathbf{A}^{\mathrm{D}}\mathbf{A}^{i+1}\mathbf{B}(\mathbf{C}^{\mathrm{D}})^{i+1} - \mathbf{A}^{\mathrm{D}}\mathbf{B}\mathbf{C}^{\mathrm{D}}\mathbf{C}$$

$$= \left(\sum_{0}^{l-2} (\mathbf{A}^{\mathrm{D}})^{i+2}\mathbf{B}\mathbf{C}^{i+1} + (\mathbf{A}^{\mathrm{D}})^{l+1}\mathbf{B}\mathbf{C}^{l} \right) - \left(\sum_{0}^{l-2} (\mathbf{A}^{\mathrm{D}})^{i+2}\mathbf{B}\mathbf{C}^{i+2}\mathbf{C}^{\mathrm{D}} \right.$$

$$\left. + (\mathbf{A}^{\mathrm{D}})^{l+1}\mathbf{B}\mathbf{C}^{l} \right) + \left(\mathbf{B}\mathbf{C}^{\mathrm{D}} + \sum_{1}^{k-1} \mathbf{A}^{i}\mathbf{B}(\mathbf{C}^{\mathrm{D}})^{i+1} \right) - \left(\mathbf{A}^{\mathrm{D}}\mathbf{A}\mathbf{B}\mathbf{C}^{\mathrm{D}} \right.$$

$$\left. + \sum_{1}^{k-1} \mathbf{A}^{\mathrm{D}}\mathbf{A}^{i+1}\mathbf{B}(\mathbf{C}^{\mathrm{D}})^{i+1} \right) - \mathbf{A}^{\mathrm{D}}\mathbf{B}\mathbf{C}^{\mathrm{D}}\mathbf{C}.$$

It is easy to see that $\mathbf{AX} - \mathbf{XC} = \mathbf{A}^{\mathrm{D}}\mathbf{B} - \mathbf{B}\mathbf{C}^{\mathrm{D}}$, or $\mathbf{AX} + \mathbf{B}\mathbf{C}^{\mathrm{D}} = \mathbf{A}^{\mathrm{D}}\mathbf{B} + \mathbf{XC}$. From this it follows that $\begin{bmatrix} \mathbf{A} & \mathbf{B} \\ \mathbf{0} & \mathbf{C} \end{bmatrix} \begin{bmatrix} \mathbf{A}^{\mathrm{D}} & \mathbf{X} \\ \mathbf{0} & \mathbf{C}^{\mathrm{D}} \end{bmatrix} = \begin{bmatrix} \mathbf{A}^{\mathrm{D}} & \mathbf{X} \\ \mathbf{0} & \mathbf{C}^{\mathrm{D}} \end{bmatrix} \begin{bmatrix} \mathbf{A} & \mathbf{B} \\ \mathbf{0} & \mathbf{C} \end{bmatrix}$, so that condition (3) of Definition 2.3 is satisfied. To show that condition (2) holds, note that

$$\begin{bmatrix} \mathbf{A}^{\mathrm{D}} & \mathbf{X} \\ \mathbf{0} & \mathbf{C}^{\mathrm{D}} \end{bmatrix} \begin{bmatrix} \mathbf{A} & \mathbf{B} \\ \mathbf{0} & \mathbf{C} \end{bmatrix} \begin{bmatrix} \mathbf{A}^{\mathrm{D}} & \mathbf{X} \\ \mathbf{0} & \mathbf{C}^{\mathrm{D}} \end{bmatrix} = \left[\begin{array}{c|c} \mathbf{A}^{\mathrm{D}} & \mathbf{A}^{\mathrm{D}}\mathbf{A}\mathbf{X} + \mathbf{X}\mathbf{C}\mathbf{C}^{\mathrm{D}} + \mathbf{A}^{\mathrm{D}}\mathbf{B}\mathbf{C}^{\mathrm{D}} \\ \hline \mathbf{0} & \mathbf{C}^{\mathrm{D}} \end{array} \right].$$

Thus, it is only necessary to show that $\mathbf{A}^{\mathrm{D}}\mathbf{A}\mathbf{X} + \mathbf{X}\mathbf{C}\mathbf{C}^{\mathrm{D}} + \mathbf{A}^{\mathrm{D}}\mathbf{B}\mathbf{C}^{\mathrm{D}} = \mathbf{X}$. However, this is immediate from (2). Thus condition (2) of Definition 2.3 is satisfied. Finally, we will show that

$$\begin{bmatrix} \mathbf{A} & \mathbf{B} \\ \mathbf{0} & \mathbf{C} \end{bmatrix}^{n+2} \begin{bmatrix} \mathbf{A}^{\mathrm{D}} & \mathbf{X} \\ \mathbf{0} & \mathbf{C}^{\mathrm{D}} \end{bmatrix} = \begin{bmatrix} \mathbf{A} & \mathbf{B} \\ \mathbf{0} & \mathbf{C} \end{bmatrix}^{n+1}. \tag{3}$$

First notice that for any $p > 0$, $\begin{bmatrix} \mathbf{A} & \mathbf{B} \\ \mathbf{0} & \mathbf{C} \end{bmatrix}^{p} = \begin{bmatrix} \mathbf{A}^{p} & \mathbf{S}(p) \\ \mathbf{0} & \mathbf{C}^{p} \end{bmatrix}$, where $\mathbf{S}(p) = \sum_{i=0}^{p-1} \mathbf{A}^{p-1-i}\mathbf{B}\mathbf{C}^i$. Thus, since $n+2>k$ and $n+2>l$,

$$\begin{bmatrix} \mathbf{A} & \mathbf{B} \\ \mathbf{0} & \mathbf{C} \end{bmatrix}^{n+2} \begin{bmatrix} \mathbf{A}^{\mathrm{D}} & \mathbf{X} \\ \mathbf{0} & \mathbf{C}^{\mathrm{D}} \end{bmatrix} = \begin{bmatrix} \mathbf{A}^{n+1} & \mathbf{A}^{n+2}\mathbf{X} + \mathbf{S}(n+2)\mathbf{C}^{\mathrm{D}} \\ \mathbf{0} & \mathbf{C}^{n+1} \end{bmatrix}.$$

Therefore, it is only necessary to show that $\mathbf{A}^{n+2}\mathbf{X} + \mathbf{S}(n+2)\mathbf{C}^{\mathrm{D}} = \mathbf{S}(n+1)$. Observe first that since $l+k<n+1$, it must be the case that

$$\mathbf{A}^n(\mathbf{A}^{\mathrm{D}})^i = \mathbf{A}^{n-1} \text{ for } i = 1, 2, \ldots, l-1. \tag{4}$$

Thus,

$$\begin{aligned} \mathbf{A}^{n+2}\mathbf{X} &= \mathbf{A}^n\left[\sum_{i=0}^{l-1} (\mathbf{A}^{\mathrm{D}})^i\mathbf{B}\mathbf{C}^i\right](\mathbf{I} - \mathbf{C}\mathbf{C}^{\mathrm{D}}) - \mathbf{A}^{n+1}\mathbf{B}\mathbf{C}^{\mathrm{D}} \\ &= \left[\sum_{i=0}^{l-1} \mathbf{A}^{n-i}\mathbf{B}\mathbf{C}^i\right](\mathbf{I} - \mathbf{C}\mathbf{C}^{\mathrm{D}}) - \mathbf{A}^{n+1}\mathbf{B}\mathbf{C}^{\mathrm{D}} \\ &= \sum_{i=0}^{l-1} \mathbf{A}^{n-i}\mathbf{B}\mathbf{C}^i - \sum_{i=0}^{l-1} \mathbf{A}^{n-i}\mathbf{B}\mathbf{C}^{i+1}\mathbf{C}^{\mathrm{D}} - \mathbf{A}^{n+1}\mathbf{B}\mathbf{C}^{\mathrm{D}}. \end{aligned} \tag{5}$$

Now, $\mathbf{S}(n+2)\mathbf{C}^{\mathrm{D}} = \sum_{i=0}^{n+1} \mathbf{A}^{n+1-i}\mathbf{B}\mathbf{C}^i\mathbf{C}^{\mathrm{D}} = \sum_{i=0}^{l} \mathbf{A}^{n+1-i}\mathbf{B}\mathbf{C}^i\mathbf{C}^{\mathrm{D}}$

$$+ \sum_{i=l+1}^{n+1} \mathbf{A}^{n+1-i}\mathbf{B}\mathbf{C}^{i-1}.$$

By writing $\sum_{i=0}^{l} \mathbf{A}^{n+1-i}\mathbf{B}\mathbf{C}^i\mathbf{C}^{\mathrm{D}} = \mathbf{A}^{n+1}\mathbf{B}\mathbf{C}^{\mathrm{D}} + \sum_{i=1}^{l} \mathbf{A}^{n+1-i}\mathbf{B}\mathbf{C}^i\mathbf{C}^{\mathrm{D}}$

$$= \mathbf{A}^{n+1}\mathbf{B}\mathbf{C}^{\mathrm{D}} + \sum_{i=0}^{l-1} \mathbf{A}^{n-i}\mathbf{B}\mathbf{C}^{i+1}\mathbf{C}^{\mathrm{D}},$$

we obtain

$$\mathbf{S}(n+2)\mathbf{C}^{\mathrm{D}} = \mathbf{A}^{n+1}\mathbf{B}\mathbf{C}^{\mathrm{D}} + \sum_{i=0}^{l-1} \mathbf{A}^{n-i}\mathbf{B}\mathbf{C}^{i+1}\mathbf{C}^{\mathrm{D}} + \sum_{i=l+1}^{n+1} \mathbf{A}^{n+1-i}\mathbf{B}\mathbf{C}^{i-1}. \tag{6}$$

It is now easily seen from (5) and (6) that

$$\begin{aligned} \mathbf{A}^{n+2}\mathbf{X} + \mathbf{S}(n+2)\mathbf{C}^{\mathrm{D}} &= \sum_{i=0}^{l-1} \mathbf{A}^{n-i}\mathbf{B}\mathbf{C}^i + \sum_{i=l+1}^{n+1} \mathbf{A}^{n+1-i}\mathbf{B}\mathbf{C}^{i-1} \\ &= \sum_{i=0}^{l-1} \mathbf{A}^{n-i}\mathbf{B}\mathbf{C}^i + \sum_{i=l}^{n} \mathbf{A}^{n-i}\mathbf{B}\mathbf{C}^i \\ &= \sum_{i=0}^{n} \mathbf{A}^{n-i}\mathbf{B}\mathbf{C}^i = \mathbf{S}(n+1), \end{aligned}$$

which is the desired result. ■

By taking transposes we also have the following.

Corollary 7.7.1 *If* $\mathbf{L}\in\mathbb{C}^{n\times n}$ *is* $\mathbf{L}=\begin{bmatrix}\mathbf{C} & \mathbf{0}\\ \mathbf{B} & \mathbf{A}\end{bmatrix}$ *where* $\mathbf{A}$ *and* $\mathbf{C}$ *are square with* $\operatorname{Ind}(\mathbf{A})=k$ *and* $\operatorname{Ind}(\mathbf{C})=l$, *then* $\mathbf{L}^{\mathrm{D}}$ *is given by* $\mathbf{L}^{\mathrm{D}}=\begin{bmatrix}\mathbf{C}^{\mathrm{D}} & \mathbf{0}\\ \mathbf{X} & \mathbf{A}^{\mathrm{D}}\end{bmatrix}$ *where* $\mathbf{X}$ *is the matrix given in* (2).

There are many cases when one deals with a matrix in which two blocks are zero matrices.

Corollary 7.7.2 *Let*

$$\mathbf{M}_1=\begin{bmatrix}\mathbf{A} & \mathbf{B}\\ \mathbf{0} & \mathbf{0}\end{bmatrix}, \mathbf{M}_2=\begin{bmatrix}\mathbf{A} & \mathbf{0}\\ \mathbf{B} & \mathbf{0}\end{bmatrix}, \mathbf{M}_3=\begin{bmatrix}\mathbf{0} & \mathbf{0}\\ \mathbf{B} & \mathbf{A}\end{bmatrix}, \mathbf{M}_4=\begin{bmatrix}\mathbf{0} & \mathbf{B}\\ \mathbf{0} & \mathbf{A}\end{bmatrix}.$$

where $\mathbf{A}$ *is square and each* $\mathbf{M}_i$ *is square. Then,*

$$\mathbf{M}_1^{\mathrm{D}}=\begin{bmatrix}\mathbf{A}^{\mathrm{D}} & (\mathbf{A}^{\mathrm{D}})^2\mathbf{B}\\ \mathbf{0} & \mathbf{0}\end{bmatrix}, \mathbf{M}_2^{\mathrm{D}}=\begin{bmatrix}\mathbf{A}^{\mathrm{D}} & \mathbf{0}\\ \mathbf{B}(\mathbf{A}^{\mathrm{D}})^2 & \mathbf{0}\end{bmatrix}, \mathbf{M}_3^{\mathrm{D}}=\begin{bmatrix}\mathbf{0} & \mathbf{0}\\ (\mathbf{A}^{\mathrm{D}})^2\mathbf{B} & \mathbf{A}^{\mathrm{D}}\end{bmatrix},$$

and

$$\mathbf{M}_4^{\mathrm{D}}=\begin{bmatrix}\mathbf{0} & \mathbf{B}(\mathbf{A}^{\mathrm{D}})^2\\ \mathbf{0} & \mathbf{A}^{\mathrm{D}}\end{bmatrix}.$$

Each of these cases follows directly from Theorem 1 and Corollary 1.

The next result shows how $\operatorname{Ind}(\mathbf{M})$ is related to $\operatorname{Ind}(\mathbf{A})$ and $\operatorname{Ind}(\mathbf{C})$.

Theorem 7.7.2 *If* $\mathbf{M}=\begin{bmatrix}\mathbf{A} & \mathbf{B}\\ \mathbf{0} & \mathbf{C}\end{bmatrix}$, *with* $\mathbf{A}, \mathbf{C}$ *square, then* $\operatorname{Max}\{\operatorname{Ind}(\mathbf{A}), \operatorname{Ind}(\mathbf{C})\}\le\operatorname{Ind}(\mathbf{M})\le\operatorname{Ind}(\mathbf{A})+\operatorname{Ind}(\mathbf{C})$.

Proof By using (iii) of Corollary 6.2 we know that if $\operatorname{Ind}(\mathbf{M})=m$, then the limit

$$\lim_{z\to 0} z^m(\mathbf{M}+z\mathbf{I})^{-1} \tag{7}$$

exists. Since

$$z^m(\mathbf{M}+\mathbf{I})^{-1}=\left[\begin{array}{c|c} z^m(\mathbf{A}+z\mathbf{I})^{-1} & -z^m(\mathbf{A}+z\mathbf{I})^{-1}\mathbf{B}(\mathbf{C}+z\mathbf{I})^{-1}\\ \hline \mathbf{0} & z^m(\mathbf{C}+z\mathbf{I})^{-1}\end{array}\right], \tag{8}$$

one can see that the existence of the limit (7) implies that the limits $\lim_{z\to 0} z^m(\mathbf{A}+z\mathbf{I})^{-1}$ and $\lim_{z\to 0} z^m(\mathbf{C}+z\mathbf{I})^{-1}$ exist. From Corollary 6.2 we can conclude that $\operatorname{Ind}(\mathbf{A})\le m=\operatorname{Ind}(\mathbf{M})$ and $\operatorname{Ind}(\mathbf{C})\le m=\operatorname{Ind}(\mathbf{M})$, which establishes the first inequality of the theorem. On the other hand, if $\operatorname{Ind}(\mathbf{A})=k$ and $\operatorname{Ind}(\mathbf{C})=l$, then by Theorem 6.2 the limits $\lim_{z\to 0} z^{k+l}(\mathbf{A}+z\mathbf{I})^{-1}$, $\lim_{z\to 0} z^{k+l}(\mathbf{C}+z\mathbf{I})^{-1}$ and $\lim_{z\to 0} z^{k+l}(\mathbf{A}+z\mathbf{I})^{-1}\mathbf{B}(\mathbf{C}+z\mathbf{I})^{-1}=\lim_{z\to 0}[z^k(\mathbf{A}+z\mathbf{I})^{-1}]\mathbf{B}[z^l(\mathbf{C}+z\mathbf{I})^{-1}]$ each exist.

Thus $\lim_{z \to 0} z^{k+l}(\mathbf{M} + z\mathbf{I})^{-1}$ exists and $\text{Ind}(\mathbf{M}) \leq k + l$. ■

In the case when either **A** or **C** is non-singular, the previous theorem reduces to the following.

Corollary 7.7.3 *Let* $\mathbf{A} \in \mathbb{C}^{r \times r}$, $\mathbf{C} \in \mathbb{C}^{s \times s}$, *and* $\mathbf{M} = \begin{bmatrix} \mathbf{A} & \mathbf{B} \\ \mathbf{0} & \mathbf{C} \end{bmatrix}$. *If* **C** *is non-singular*, ($\text{Ind}(\mathbf{C}) = 0$), *then* $\text{Ind}(\mathbf{M}) = \text{Ind}(\mathbf{A})$. *Likewise, if* **A** *is non-singular, then* $\text{Ind}(\mathbf{M}) = \text{Ind}(\mathbf{C})$.

The case in which $\text{Ind}(\mathbf{M}) \leq 1$ is of particular interest and will find applications in the next chapter. The next theorem characterizes these matrices.

Theorem 7.7.3 *If* $\mathbf{A} \in \mathbb{C}^{r \times r}$, $\mathbf{C} \in \mathbb{C}^{s \times s}$ *and* $\mathbf{M} = \begin{bmatrix} \mathbf{A} & \mathbf{B} \\ \mathbf{0} & \mathbf{C} \end{bmatrix}$, *then* $\text{Ind}(\mathbf{M}) \leq 1$ *if and only if each of the following conditions is true:*

$$\text{Ind}(\mathbf{A}) \leq 1, \text{Ind}(\mathbf{C}) \leq 1, \tag{9}$$

and

$$(\mathbf{I} - \mathbf{A}\mathbf{A}^{\#})\mathbf{B}(\mathbf{I} - \mathbf{C}\mathbf{C}^{\#}) = \mathbf{0}. \tag{10}$$

Furthermore, when $\mathbf{M}^{\#}$ *exists, it is given by*

$$\mathbf{M}^{\#} = \left[\begin{array}{c|c} \mathbf{A}^{\#} & (\mathbf{A}^{\#})^2\mathbf{B}(\mathbf{I} - \mathbf{C}\mathbf{C}^{\#}) + (\mathbf{I} - \mathbf{A}\mathbf{A}^{\#})\mathbf{B}(\mathbf{C}^{\#})^2 - \mathbf{A}^{\#}\mathbf{B}\mathbf{C}^{\#} \\ \hline \mathbf{0} & \mathbf{C}^{\#} \end{array}\right]. \tag{11}$$

Proof Suppose first that $\text{Ind}(\mathbf{M}) \leq 1$. Then from Theorem 2, it follows that $\text{Ind}(\mathbf{A}) \leq 1$, $\text{Ind}(\mathbf{C}) \leq 1$ and (9) holds. Since $\text{Ind}(\mathbf{M}) \leq 1$, we know that $\mathbf{M}^{\mathrm{D}} = \mathbf{M}^{\#}$ is a (1)-inverse for **M**. Also, Theorem 5.1 guarantees that $\mathbf{M}^{\#}$ is a polynomial in **M** so that $\mathbf{M}^{\#}$ must be an upper block triangular (1)-inverse for **M**. Theorem 6.3.8 now implies that (10) must hold. Conversely, suppose that (9) and (10) hold. Then (9) implies that $\mathbf{A}^{\mathrm{D}} = \mathbf{A}^{\#}$ and $\mathbf{C}^{\mathrm{D}} = \mathbf{C}^{\#}$. Since $\mathbf{A}^{\#}$ and $\mathbf{C}^{\#}$ are (1)-inverses for **A** and **C**, (10) along with Theorem 6.3.8 implies that there exists an upper block triangular (1)-inverse for **M**. Theorem 6.3.9 implies that $\text{rank}(\mathbf{M}) = \text{rank}(\mathbf{A}) + \text{rank}(\mathbf{C})$. Similarly, $\text{rank}(\mathbf{M}^2) = \text{rank}(\mathbf{A}^2) + \text{rank}(\mathbf{C}^2) = \text{rank}(\mathbf{A}) + \text{rank}(\mathbf{C}) = \text{rank}(\mathbf{M})$ so that $\text{Ind}(\mathbf{M}) \leq 1$. The explicit form for $\mathbf{M}^{\#}$ given in (11) is a direct consequence of (2). ■

Corollary 7.7.4 *If* $\text{Ind}(\mathbf{A}) \leq 1$, $\text{Ind}(\mathbf{C}) \leq 1$, *and either* $R(\mathbf{B}) \subseteq R(\mathbf{A})$ *or* $N(\mathbf{C}) \subseteq N(\mathbf{B})$, *then* $\text{Ind}(\mathbf{M}) \leq 1$ *where* **M** *is as in Theorem* 3.

Proof $R(\mathbf{B}) \subseteq R(\mathbf{A})$ implies that $(\mathbf{I} - \mathbf{A}\mathbf{A}^{\#})\mathbf{B} = \mathbf{0}$ and $N(\mathbf{C}) \subseteq N(\mathbf{B})$ implies that $\mathbf{B}(\mathbf{I} - \mathbf{C}\mathbf{C}^{\#}) = \mathbf{0}$. ■

It is possible to generalize Theorem 3 to block triangular matrices of a general index.

Theorem 7.7.4 *Let* $\mathbf{A}\in\mathbb{C}^{r\times r}$, $\mathbf{C}\in\mathbb{C}^{s\times s}$ *both be singular (if either is non-singular, then Corollary 3 applies) and let* $\mathbf{M}=\begin{bmatrix}\mathbf{A} & \mathbf{B}\\ \mathbf{0} & \mathbf{C}\end{bmatrix}$. *For each positive integer p, let* $\mathbf{S}(p)=\sum_{i=0}^{p-1}\mathbf{A}^{p-1-i}\mathbf{B}\mathbf{C}^i$ *with the convention that* $\mathbf{0}^\circ=\mathbf{I}$. *Then,* $\mathrm{Ind}(\mathbf{M})\le m$ *if and only if each of the following conditions are true:*

$$\mathrm{Ind}(\mathbf{A})\le m, \tag{12}$$

$$\mathrm{Ind}(\mathbf{C})\le m, \text{ and} \tag{13}$$

$$(\mathbf{I}-\mathbf{A}\mathbf{A}^{\mathrm{D}})\mathbf{S}(m)(\mathbf{I}-\mathbf{C}\mathbf{C}^{\mathrm{D}})=\mathbf{0}. \tag{14}$$

Proof Notice that

$$\mathbf{M}^p=\left[\begin{array}{c|c}\mathbf{A}^p & \mathbf{S}(p)\\ \hline \mathbf{0} & \mathbf{C}^p\end{array}\right], \text{ for positive integers } p. \tag{15}$$

Assume first that $\mathrm{Ind}(\mathbf{M})\le m$. Then $\mathrm{Ind}(\mathbf{M}^m)=1$.
From Theorem 3 and the singularity of $\mathbf{A},\mathbf{C}$ we can conclude that

$$\mathrm{Ind}(\mathbf{A}^m)=1,\ \mathrm{Ind}(\mathbf{C}^m)=1, \text{ and} \tag{16}$$

$$(\mathbf{I}-\mathbf{A}^m(\mathbf{A}^m)^{\mathrm{D}})\mathbf{S}(m)(\mathbf{I}-\mathbf{C}^m(\mathbf{C}^m)^{\mathrm{D}})=\mathbf{0}. \tag{17}$$

Then (12), (13) hold by (16). Clearly, (17) reduces to (14). Conversely, suppose (12)–(14) hold. Then (16) and (17) hold. Theorem 3 now implies that $\mathrm{Ind}(\mathbf{M}^m)=1$. Therefore, $\mathrm{Ind}(\mathbf{M})\le 1$. ■

Lemma 7.7.1 *For* $\mathbf{T}\in\mathbb{C}^{n\times n}$, $\mathbf{R}\in\mathbb{C}^{r\times n}$, $\mathbf{S}\in\mathbb{C}^{n\times s}$, *such that rank*$(\mathbf{R})=n$ *and rank*$(\mathbf{S})=n$, *it is true that rank*$(\mathbf{RTS})=$ *rank*$(\mathbf{T})$.

Proof Note that $\mathbf{R}^\dagger\mathbf{R}=\mathbf{I}_n$ and $\mathbf{S}\mathbf{S}^\dagger=\mathbf{I}_n$. Thus $\mathrm{rank}(\mathbf{RTS})\le\mathrm{rank}(\mathbf{T})\le\mathrm{rank}(\mathbf{R}^\dagger\mathbf{RTS}\mathbf{S}^\dagger)\le\mathrm{rank}(\mathbf{RTS})$. ■

We now consider the Drazin inverse of a non-triangular partitioned matrix.

Theorem 7.7.5 *Let* $\mathbf{A}\in\mathbb{C}^{r\times r}$ *and* $\mathbf{M}=\begin{bmatrix}\mathbf{A} & \mathbf{B}\\ \mathbf{C} & \mathbf{D}\end{bmatrix}$. *If rank*$(\mathbf{M})=$ *rank*$(\mathbf{A})=r$, *then* $\mathrm{Ind}(\mathbf{M})=\mathrm{Ind}[\mathbf{A}(\mathbf{I}-\mathbf{QP})]+1=\mathrm{Ind}[(\mathbf{I}-\mathbf{QP})\mathbf{A}]+1$, *where* $\mathbf{P}=\mathbf{C}\mathbf{A}^{-1}$ *and* $\mathbf{Q}=\mathbf{A}^{-1}\mathbf{B}$.

Proof From Lemma 3.3.3, we have that $\mathbf{D}=\mathbf{C}\mathbf{A}^{-1}\mathbf{B}$ so that $\mathbf{M}=\begin{bmatrix}\mathbf{I}\\ \mathbf{P}\end{bmatrix}\mathbf{A}[\mathbf{I}\quad\mathbf{Q}]$. Thus, for every positive integer i,

$$\mathbf{M}^i=\begin{bmatrix}\mathbf{I}\\ \mathbf{P}\end{bmatrix}[\mathbf{A}(\mathbf{I}+\mathbf{QP})]^{i-1}\mathbf{A}[\mathbf{I}\quad\mathbf{Q}]=\begin{bmatrix}\mathbf{I}\\ \mathbf{P}\end{bmatrix}\mathbf{A}[(\mathbf{I}+\mathbf{QP})\mathbf{A}]^{i-1}[\mathbf{I}\quad\mathbf{Q}]. \tag{18}$$

Since $\begin{bmatrix} \mathbf{I} \\ \mathbf{P} \end{bmatrix}\mathbf{A}$ has full column rank and $[\mathbf{I} \quad \mathbf{Q}]$ has full row rank, we can conclude from Lemma 1 that $\mathrm{rank}(\mathbf{M}^{m+1}) = \mathrm{rank}([\mathbf{A}(\mathbf{I}+\mathbf{QP})]^m)$. Therefore, $\mathrm{rank}([\mathbf{A}(\mathbf{I}+\mathbf{QP})]^m) = \mathrm{rank}([\mathbf{A}(\mathbf{I}+\mathbf{QP})]^{m-1})$ if and only if $\mathrm{rank}(\mathbf{M}^{m+1}) = \mathrm{rank}(\mathbf{M})$. Hence $\mathrm{Ind}([\mathbf{A}(\mathbf{I}+\mathbf{QP})]) + 1 = \mathrm{Ind}(\mathbf{M})$. In a similar manner, one can show that $\mathrm{Ind}[(\mathbf{I}+\mathbf{QP})\mathbf{A}] + 1 = \mathrm{Ind}(\mathbf{M})$. ■

An immediate corollary is as follows.

Corollary 7.7.5 *For the situation of Theorem* 5, $\mathrm{Ind}(\mathbf{M}) = 1$ *if and only if* $(\mathbf{I}+\mathbf{QP})$ *is non-singular.*

The results we are developing now are not only useful in computing the index but also in computing $\mathbf{A}^{\mathrm{D}}$.

Theorem 7.7.6 *Let* $\mathbf{A} \in \mathbb{C}^{r\times r}$ *and* $\mathbf{M} = \begin{bmatrix} \mathbf{A} & \mathbf{B} \\ \mathbf{C} & \mathbf{D} \end{bmatrix}$. *If rank*$(\mathbf{M}) =$ *rank*$(\mathbf{A}) = r$, *then*

$$\mathbf{M}^{\mathrm{D}} = \begin{bmatrix} \mathbf{I} \\ \mathbf{CA}^{-1} \end{bmatrix}[(\mathbf{AS})^2]^{\mathrm{D}}\mathbf{A}[\mathbf{I} \,\vdots\, \mathbf{A}^{-1}\mathbf{B}] = \begin{bmatrix} \mathbf{I} \\ \mathbf{CA}^{-1} \end{bmatrix}\mathbf{A}[(\mathbf{SA})^2]^{\mathrm{D}}[\mathbf{I} \,\vdots\, \mathbf{A}^{-1}\mathbf{B}] \tag{19}$$

where $\mathbf{S} = \mathbf{I} + \mathbf{A}^{-1}\mathbf{BCA}^{-1}$.

Proof Let $\mathbf{R}$ denote the matrix $\mathbf{R} = \begin{bmatrix} \mathbf{I} \\ \mathbf{CA}^{-1} \end{bmatrix}[(\mathbf{AS})^2]^{\mathrm{D}}\mathbf{A}[\mathbf{I} \,\vdots\, \mathbf{A}^{-1}\mathbf{B}]$, and let $m = \mathrm{Ind}(\mathbf{M})$. By using (18), we obtain

$$\mathbf{M}^{m+1}\mathbf{R} = \begin{bmatrix} \mathbf{I} \\ \mathbf{CA}^{-1} \end{bmatrix}(\mathbf{AS})^{m+1}[(\mathbf{AS})^2]^{\mathrm{D}}\mathbf{A}[\mathbf{I} \,\vdots\, \mathbf{A}^{-1}\mathbf{B}].$$

By Theorem 5 we know $\mathrm{Ind}(\mathbf{AS}) = m - 1$ so that $(\mathbf{AS})^{m+1}[(\mathbf{AS})^2]^{\mathrm{D}} = (\mathbf{AS})^{m-1}$. Thus, it now follows that

$$\mathbf{M}^{m+1}\mathbf{R} = \begin{bmatrix} \mathbf{I} \\ \mathbf{CA}^{-1} \end{bmatrix}(\mathbf{AS})^{m-1}\mathbf{A}[\mathbf{I} \,\vdots\, \mathbf{A}^{-1}\mathbf{B}] = \mathbf{M}^m.$$

The facts that $\mathbf{MR} = \mathbf{RM}$ and $\mathbf{RMR}$ are easily verified and we have that $\mathbf{R}$ satisfies the algebraic definition for $\mathbf{M}^{\mathrm{D}}$. The second equality of (19) is similarly verified. ■

The case when $\mathrm{Ind}(\mathbf{M}) = 1$ is of particular interest.

Theorem 7.7.7 *Let* $\mathbf{A} \in \mathbb{C}^{r\times r}$ *and* $\mathbf{M} = \begin{bmatrix} \mathbf{A} & \mathbf{B} \\ \mathbf{C} & \mathbf{D} \end{bmatrix}$. *If rank*$(\mathbf{M}) =$ *rank*$(\mathbf{A}) = r$, *then* $\mathrm{Ind}(\mathbf{M}) = 1$ *if and only if* $\mathbf{S}^{-1}$ *exists where* $\mathbf{S} = \mathbf{I} + \mathbf{A}^{-1}\mathbf{BCA}^{-1}$. *When* $\mathrm{Ind}(\mathbf{M}) = 1$, $\mathbf{M}^{\#}$ *is given by*

$$\mathbf{M}^{\#} = \begin{bmatrix} \mathbf{I} \\ \mathbf{CA}^{-1} \end{bmatrix}(\mathbf{SAS})^{-1}[\mathbf{I} \,\vdots\, \mathbf{A}^{-1}\mathbf{B}] = \begin{bmatrix} \mathbf{I} \\ \mathbf{CA}^{-1} \end{bmatrix}\mathbf{S}^{-1}\mathbf{A}^{-1}\mathbf{S}^{-1}[\mathbf{I} \,\vdots\, \mathbf{A}^{-1}\mathbf{B}]$$

$$= \begin{bmatrix} \mathbf{A} \\ \mathbf{C} \end{bmatrix} (\mathbf{A}^2 + \mathbf{BC})^{-1} \mathbf{A} (\mathbf{A}^2 + \mathbf{BC})^{-1} [\mathbf{A} \mid \mathbf{B}].$$

This result follows directly from Corollary 5 and Theorem 6.

Before showing how some of these results concerning partitioned matrices can be used to obtain the Drazin inverse of a general matrix we need one more Lemma.

Lemma 7.7.2 *Consider a matrix of the form* $\mathbf{M} = \begin{bmatrix} \mathbf{A} & \mathbf{B} \\ \mathbf{0} & \mathbf{0} \end{bmatrix}$ *where* $\mathbf{A} \in \mathbb{C}^{r \times r}$. *Then,*

$$\operatorname{Ind}(\mathbf{A}) \le \operatorname{Ind}(\mathbf{M}) \le \operatorname{Ind}(\mathbf{A}) + 1. \tag{20}$$

Furthermore, suppose $\operatorname{rank}(\mathbf{M}) = r$. *Then* $\operatorname{Ind}(\mathbf{M}) = 1$ *if and only if* $\mathbf{A}$ *is non-singular.*

Proof (20) follows from Theorem 2. To prove the second statement of the lemma, first note that $\operatorname{Ind}(\mathbf{A}) = 0$ implies $\operatorname{Ind}(\mathbf{M}) = \operatorname{Ind}(\mathbf{0}) = 1$ by Corollary 3. Conversely, if $\operatorname{rank}(\mathbf{M}) = r$ and $\operatorname{Ind}(\mathbf{M}) = 1$, then $r = \operatorname{rank}(\mathbf{M}^2)$ or equivalently, $r = \operatorname{rank}([\mathbf{A}^2, \mathbf{AB}]) \le \operatorname{rank}(\mathbf{A}) \le r$. Thus $\operatorname{rank}(\mathbf{A}) = r$ so that $\mathbf{A}^{-1}$ exists. ■

The next theorem can be used to compute the group inverse in case $\operatorname{Ind}(\mathbf{M}) = 1$.

Theorem 7.7.8 *Let* $\mathbf{M} \in \mathbb{C}^{n \times n}$, $\mathbf{R} \in \mathbb{C}^{n \times n}$ *where* $\mathbf{R}$ *is a non-singular matrix such that* $\mathbf{RM} = \begin{bmatrix} \mathbf{A} & \mathbf{B} \\ \mathbf{0} & \mathbf{0} \end{bmatrix}$, $\mathbf{A} \in \mathbb{C}^{r \times r}$, *and* $r = \operatorname{rank}(\mathbf{M}) \le n$. *Write* $\mathbf{RMR}^{-1}$ *as*

$$\mathbf{RMR}^{-1} = \begin{bmatrix} \mathbf{A} & \mathbf{B} \\ \mathbf{0} & \mathbf{0} \end{bmatrix} \mathbf{R}^{-1} = \begin{bmatrix} \mathbf{U} & \mathbf{V} \\ \mathbf{0} & \mathbf{0} \end{bmatrix}$$ *where* $\mathbf{U} \in \mathbb{C}^{r \times r}$. *Then*

$$\mathbf{M}^{\mathrm{D}} = \mathbf{R}^{-1} \left[\begin{array}{c|c} \mathbf{U}^{\mathrm{D}} & (\mathbf{U}^{\mathrm{D}})^2 \mathbf{V} \\ \hline \mathbf{0} & \mathbf{0} \end{array} \right] \mathbf{R}.$$

If $\operatorname{Ind}(\mathbf{M}) = 1$, *then* $\mathbf{M}^{\#} = \mathbf{R}^{-1} \left[\begin{array}{c|c} \mathbf{U}^{-1} & \mathbf{U}^{-2} \mathbf{V} \\ \hline \mathbf{0} & \mathbf{0} \end{array} \right] \mathbf{R}.$

Proof Since $\operatorname{Ind}(\mathbf{M}) = \operatorname{Ind}(\mathbf{RMR}^{-1})$, we know from Lemma 2 that $\operatorname{Ind}(\mathbf{M}) = 1$ if and only if $\mathbf{U}^{-1}$ exists. The desired result now follows from Corollary 2. ■

Example 7.7.1 Let

$$\mathbf{I} = \begin{bmatrix} 1 & 2 & 1 \\ 2 & 4 & 2 \\ 1 & 0 & 0 \end{bmatrix}.$$

We shall calculate $\mathbf{M}^{\#}$ using Theorem 7. Row reduce $[\mathbf{M} \quad \mathbf{I}]$ to $[\mathbf{E}_{\mathbf{M}} \quad \mathbf{R}]$

where $\mathbf{E_M}$ is in row echelon form. ($\mathbf{R}$ is not unique.) Then,

$$\mathbf{E_M} = \begin{bmatrix} 1 & 0 & 0 \\ 0 & 1 & \frac{1}{2} \\ 0 & 0 & 0 \end{bmatrix}, \mathbf{R} = \begin{bmatrix} 0 & 0 & 1 \\ \frac{1}{2} & 0 & -\frac{1}{2} \\ -2 & 1 & 0 \end{bmatrix},$$

and $\mathbf{RM} = \mathbf{E_M}$. Now $\mathbf{R}^{-1} = \begin{bmatrix} 1 & 2 & 0 \\ 2 & 4 & 1 \\ 1 & 0 & 0 \end{bmatrix}$

so that $\mathbf{RMR}^{-1} = \left[\begin{array}{cc|c} 1 & 2 & 0 \\ \frac{5}{2} & 4 & 1 \\ \hline 0 & 0 & 0 \end{array}\right] = \left[\begin{array}{c|c} \mathbf{K} & \mathbf{V} \\ \hline \mathbf{0} & \mathbf{0} \end{array}\right]$.

Clearly, $\mathbf{U}^{-1}$ exists. This implies that $\mathrm{Ind}(\mathbf{M}) = 1$ and

$$\mathbf{M}^{\#} = \mathbf{R}^{-1}\left[\begin{array}{c|c} \mathbf{U}^{-1} & \mathbf{U}^{-2}\mathbf{V} \\ \hline \mathbf{0} & \mathbf{0} \end{array}\right]\mathbf{R} = \begin{bmatrix} -4 & 2 & 1 \\ -8 & 4 & 2 \\ 21 & -10 & -5 \end{bmatrix}.$$

From Lemma 6.1, we know that if $p \geq \mathrm{Ind}(\mathbf{M})$, then $\mathrm{Ind}(\mathbf{M}^p) = 1$. Thus, the above method could be applied to $\mathbf{M}^p$ to obtain $(\mathbf{M}^p)^{\#} = (\mathbf{M}^{\mathrm{D}})^p$. For a general matrix $\mathbf{M}$ with $p \geq \mathrm{Ind}(\mathbf{M})$, $\mathbf{M}^{\mathrm{D}}$ is then given by $\mathbf{M}^{\mathrm{D}} = (\mathbf{M}^{\mathrm{D}})^p\mathbf{M}^{p-1} = (\mathbf{M}^p)^{\#}\mathbf{M}^{p-1}$.

Another way Theorem 7 can be used to obtain $\mathbf{M}^{\mathrm{D}}$ for a matrix $\mathbf{M}$ of index greater than 1 is described below. Suppose $p \geq \mathrm{Ind}(\mathbf{M})$. Then $\mathrm{Ind}((\mathbf{RMR}^{-1})^p) = \mathrm{Ind}(\mathbf{RM}^p\mathbf{R}^{-1}) = \mathrm{Ind}(\mathbf{M}^p) = 1$. Thus if $\mathbf{RM} = \begin{bmatrix} \mathbf{A} & \mathbf{B} \\ \mathbf{0} & \mathbf{0} \end{bmatrix}$, then $\mathbf{RMR}^{-1} = \begin{bmatrix} \mathbf{S} & \mathbf{T} \\ \mathbf{0} & \mathbf{0} \end{bmatrix}$ and $(\mathbf{RMR}^{-1})^p = \mathbf{RM}^{-p}\mathbf{R}^{-1} = \begin{bmatrix} \mathbf{S}^p & \mathbf{S}^{p-1}\mathbf{T} \\ \mathbf{0} & \mathbf{0} \end{bmatrix}$.
It follows from (20) that $\mathrm{Ind}(\mathbf{S}^p) \leq 1$. Therefore one can use Theorem 7 to find $(\mathbf{S}^p)^{\#}$. (This is an advantage because $\mathbf{S}^p$ is a smaller size matrix.) Then,

$$\mathbf{M}^{\mathrm{D}} = (\mathbf{M}^p)^{\#}\mathbf{M}^{p-1} = \mathbf{R}^{-1}\left[\begin{array}{c|c} (\mathbf{S}^p)^{\#}\mathbf{S}^{p-1} & (\mathbf{S}^p)^{\#}\mathbf{S}^{p-2}\mathbf{T} \\ \hline \mathbf{0} & \mathbf{0} \end{array}\right]\mathbf{R}.$$

Finally, note that the singular value decomposition can be used in conjunction with Theorem 7. (See Chapter 12).

8. Other properties

It is possible to express the Drazin inverse in terms of any (1)-inverse.

Theorem 7.8.1 *If* $\mathbf{A} \in \mathbb{C}^{n \times n}$ *is such that* $\mathrm{Ind}(\mathbf{A}) = k$, *then for each integer*

$l \geq k$, and any (1)-inverse of $\mathbf{A}^{2l+1}$; $\mathbf{A}^{\mathrm{D}} = \mathbf{A}^l(\mathbf{A}^{2l+1})^{-}\mathbf{A}^l$. In particular, $\mathbf{A}^{\mathrm{D}} = \mathbf{A}^l(\mathbf{A}^{2l+1})^{\dagger}\mathbf{A}^l$.

Proof Let $\mathbf{A} = \mathbf{P}\begin{bmatrix} \mathbf{C} & \mathbf{0} \\ \mathbf{0} & \mathbf{N} \end{bmatrix}\mathbf{P}^{-1}$, $\mathbf{P}$ and $\mathbf{C}$ non-singular and $\mathbf{N}$ nilpotent. Then $\mathbf{A}^{2l+1} = \mathbf{P}\begin{bmatrix} \mathbf{C}^{2l+1} & \mathbf{0} \\ \mathbf{0} & \mathbf{0} \end{bmatrix}\mathbf{P}^{-1}$. If $\mathbf{X}$ is a (1)-inverse, then it is easy to see $\mathbf{X} = \mathbf{P}\begin{bmatrix} \mathbf{C}^{-2l-1} & \mathbf{X}_1 \\ \mathbf{X}_2 & \mathbf{X}_3 \end{bmatrix}\mathbf{P}^{-1}$ where $\mathbf{X}_1, \mathbf{X}_2, \mathbf{X}_3$ are arbitrary. That $\mathbf{A}^{\mathrm{D}} = \mathbf{A}^l\mathbf{X}\mathbf{A}^l$ is easily verified by multiplying the block matrices together. ∎

In Theorem 1.5, the Moore–Penrose inverse was expressed by using the full rank factorization of a matrix. The Drazin inverse can also be obtained by using the full rank factorization.

Theorem 7.8.2 *Suppose $\mathbf{A} \in \mathbb{C}^{n \times n}$ and perform a sequence of full rank factorizations:*
$\mathbf{A} = \mathbf{B}_1\mathbf{C}_1, \mathbf{C}_1\mathbf{B}_1 = \mathbf{B}_2\mathbf{C}_2, \mathbf{C}_2\mathbf{B}_2 = \mathbf{B}_3\mathbf{C}_3, \ldots$ so that $\mathbf{B}_i\mathbf{C}_i$ is a full rank factorization of $\mathbf{C}_{i-1}\mathbf{B}_{i-1}$, for $i = 2, 3, \ldots$. Eventually, there will be a pair of factors, $\mathbf{B}_k$ and $\mathbf{C}_k$, such that either $(\mathbf{C}_k\mathbf{B}_k)^{-1}$ exists or $\mathbf{C}_k\mathbf{B}_k = \mathbf{0}$. If k denotes the first integer for which this occurs, then

$$\mathrm{Ind}(\mathbf{A}) = \begin{cases} k & \text{when } (\mathbf{C}_k\mathbf{B}_k)^{-1} \text{ exist} \\ k+1 & \text{when } \mathbf{C}_k\mathbf{B}_k = \mathbf{0}. \end{cases}$$

When $\mathbf{C}_k\mathbf{B}_k$ is non-singular,

$$\text{rank}(\mathbf{A}^k) = \text{number of columns of } \mathbf{B}_k = \text{number of rows of } \mathbf{C}_k,$$

and

$$R(\mathbf{A}^k) = R(\mathbf{B}_1\mathbf{B}_2\cdots\mathbf{B}_k), \quad N(\mathbf{A}^k) = N(\mathbf{C}_k\mathbf{C}_{k-1}\cdots\mathbf{C}_1).$$

Moreover,

$$\mathbf{A}^{\mathrm{D}} = \begin{cases} \mathbf{B}_1\mathbf{B}_2\cdots\mathbf{B}_k(\mathbf{C}_k\mathbf{B}_k)^{-(k+1)}\mathbf{C}_k\mathbf{C}_{k-1}\cdots\mathbf{C}_1 & \text{when } (\mathbf{C}_k\mathbf{B}_k)^{-1} \text{ exists} \\ \mathbf{0} & \text{when } \mathbf{C}_k\mathbf{B}_k = \mathbf{0}. \end{cases}$$

Proof If $\mathbf{C}_i\mathbf{B}_i$ is $p \times p$ and has rank $q < p$, then $\mathbf{C}_{i+1}\mathbf{B}_{i+1}$ will be $q \times q$. That is, the size of $\mathbf{C}_{i+1}\mathbf{B}_{i+1}$ must be strictly less than that of $\mathbf{C}_i\mathbf{B}_i$ when $\mathbf{C}_i\mathbf{B}_i$ is singular. It follows that there eventually must be a pair of factors, $\mathbf{B}_k$ and $\mathbf{C}_k$, such that $\mathbf{C}_k\mathbf{B}_k$ is either non-singular or $\mathbf{0}$. Let k denote the first integer for which this occurs and write $\mathbf{A}^k = (\mathbf{B}_1\mathbf{C}_1)^k = \mathbf{B}_1(\mathbf{C}_1\mathbf{B}_1)^{k-1}\mathbf{C}_1 = \mathbf{B}_1(\mathbf{B}_2\mathbf{C}_2)^{k-1}\mathbf{C}_1 = \mathbf{B}_1\mathbf{B}_2(\mathbf{C}_2\mathbf{B}_2)^{k-2}\mathbf{C}_2\mathbf{C}_1 = \cdots = \mathbf{B}_1\mathbf{B}_2\cdots\mathbf{B}_{k-1} \times (\mathbf{B}_k\mathbf{C}_k)\mathbf{C}_{k-1}\mathbf{C}_{k-2}\cdots\mathbf{C}_1$, and $\mathbf{A}^{k+1} = \mathbf{B}_1\mathbf{B}_2\cdots\mathbf{B}_k(\mathbf{C}_k\mathbf{B}_k)\mathbf{C}_k\mathbf{C}_{k-1}\cdots\mathbf{C}_1$. Assume $\mathbf{C}_k\mathbf{B}_k$ is non-singular. If $\mathbf{B}_k$ is $p \times r$ and $\mathbf{C}_k$ is $r \times p$, then $\text{rank}(\mathbf{B}_k\mathbf{C}_k) = r$. Since $\mathbf{C}_k\mathbf{B}_k$ is $r \times r$ and non-singular, it follows that $\text{rank}(\mathbf{C}_k\mathbf{B}_k) = r = \text{rank}(\mathbf{B}_k\mathbf{C}_k)$ so that Lemma 7.1 guarantees that $\text{rank}(\mathbf{A}^{k+1}) = \text{rank}(\mathbf{C}_k\mathbf{B}_k) = \text{rank}(\mathbf{B}_k\mathbf{C}_k) = \text{rank}(\mathbf{A}^k)$. Since k is the smallest integer for which this holds, it must be the case that $\mathrm{Ind}(\mathbf{A}) = k$. The fact that $\mathrm{rank}(\mathbf{A}^k)$ = number of columns of $\mathbf{B}_k$ = number of rows of $\mathbf{C}_k$ is clear. By using the fact that the $\mathbf{B}_i$'s and $\mathbf{C}_i$'s are full rank factors, it is not difficult to see that $R(\mathbf{A}^k) = R(\mathbf{B}_1\mathbf{B}_2 \cdots \mathbf{B}_k)$ and $N(\mathbf{A}^k) = N(\mathbf{C}_k\mathbf{C}_{k-1} \cdots \mathbf{C}_1)$. If $\mathbf{C}_k\mathbf{B}_k = \mathbf{0}$, then it is clear that $\mathbf{A}$ must be nilpotent of index $k+1$. To prove the formula given for $\mathbf{A}^{\mathrm{D}}$ is valid, one simply verifies that the three conditions of the algebraic definition are satisfied. This is straightforward and is left as an exercise. ∎

There are several methods for performing full rank factorizations. One is the elimination scheme described in Algorithm 1.2. The others depend on orthogonalization techniques such as the modified Gram–Schmidt algorithm. Needless to say, the method chosen to perform the factorizations at each step can influence the final result.

Corollary 7.8.1 *If* $\mathbf{A} \in \mathbb{C}^{n\times n}$ *and if* $\mathbf{A} \neq \mathbf{0}$, *then* $\mathbf{A}^{\mathrm{D}} = \mathbf{A}^{n-1}\mathbf{B}(\mathbf{CB})^{-2}\mathbf{C} = \mathbf{B}(\mathbf{CB})^{-2}\mathbf{C}\mathbf{A}^{n-1}$ *where* $\mathbf{A}^n = \mathbf{BC}$ *is a full rank factorization.*

Corollary 7.8.2 *If* $\mathbf{A} \in \mathbb{C}^{n\times n}$ *is such that* $\mathrm{Ind}(\mathbf{A}) = 1$, *and* $\mathbf{A} = \mathbf{BC}$ *is a full rank factorization for* $\mathbf{A}$, *then* $\mathbf{A}^{\#} = \mathbf{B}(\mathbf{CB})^{-2}\mathbf{C}$.

Theorem 7.8.3 *If* $\mathbf{A} \in \mathbb{C}^{n\times n}$ *is such that rank*$(\mathbf{A}) = 1$, *then* $\mathbf{A}^{\mathrm{D}} = \mathbf{A}^{\#} = \dfrac{1}{[\mathrm{Tr}(\mathbf{A})]^2}\mathbf{A}$ *when* $\mathrm{Tr}(\mathbf{A}) \neq 0$ *and* $\mathbf{A}^{\mathrm{D}} = \mathbf{0}$ *when* $\mathrm{Tr}(\mathbf{A}) = 0$.

Proof If $\mathrm{rank}(\mathbf{A}) = 1$, then $\mathbf{A}$ can be written as $\mathbf{A} = \mathbf{cd}^*$ where $\mathbf{c} \in \mathbb{C}^n$ and $\mathbf{d} \in \mathbb{C}^n$. Now, $\mathrm{Tr}(\mathbf{A}) = \mathrm{Tr}(\mathbf{cd}^*) = \mathrm{Tr}(\mathbf{d}^*\mathbf{c}) = \mathbf{d}^*\mathbf{c}$. Thus $\mathbf{A}^2 = \mathbf{cd}^*\mathbf{cd}^* = \mathrm{Tr}(\mathbf{A})\mathbf{A}$. If $\mathrm{Tr}(\mathbf{A}) \neq 0$, then $R(\mathbf{A}^2) = R(\mathbf{A})$ so that $\mathrm{Ind}(\mathbf{A}) \leq 1$. The fact that

$\mathbf{A}^{\#} = \dfrac{1}{[\mathrm{Tr}(\mathbf{A})]^2}\mathbf{A}$ can now be deduced from Corollary 9.2, or else one can verify by direct computation the requirements of Definition 2.4. ∎

In general, the reverse order law does not hold for the Drazin inverse. That is, $(\mathbf{AB})^{\mathrm{D}} \neq \mathbf{B}^{\mathrm{D}}\mathbf{A}^{\mathrm{D}}$. In the case of the Moore–Penrose inverse, we saw that very strong conditions had to be placed on $\mathbf{A}$ and $\mathbf{B}$ in order to guarantee that $(\mathbf{AB})^{\dagger} = \mathbf{B}^{\dagger}\mathbf{A}^{\dagger}$. Even the commutativity of $\mathbf{A}$ and $\mathbf{B}$ is not strong enough to guarantee that $(\mathbf{AB})^{\dagger} = \mathbf{B}^{\dagger}\mathbf{A}^{\dagger}$. However, commutativity of $\mathbf{A}$ and $\mathbf{B}$ is enough to guarantee that $(\mathbf{AB})^{\mathrm{D}} = \mathbf{B}^{\mathrm{D}}\mathbf{A}^{\mathrm{D}}$.

Theorem 7.8.4 *If* $\mathbf{A}, \mathbf{B} \in \mathbb{C}^{n\times n}$ *are such that* $\mathbf{AB} = \mathbf{BA}$, *then*

(i) $(\mathbf{AB})^{\mathrm{D}} = \mathbf{B}^{\mathrm{D}}\mathbf{A}^{\mathrm{D}} = \mathbf{A}^{\mathrm{D}}\mathbf{B}^{\mathrm{D}}$,

(ii) $\mathbf{A}^{\mathrm{D}}\mathbf{B} = \mathbf{B}\mathbf{A}^{\mathrm{D}}$ and $\mathbf{A}\mathbf{B}^{\mathrm{D}} = \mathbf{B}^{\mathrm{D}}\mathbf{A}$.

In general,

(iii) $(\mathbf{AB})^{\mathrm{D}} = \mathbf{A}[(\mathbf{BA})^2]^{\mathrm{D}}\mathbf{B}$

even if $\mathbf{AB} \neq \mathbf{BA}$.

Proof Assume first that $\mathbf{AB} = \mathbf{BA}$. It follows from Theorem 6.1 that

$\mathbf{A}^{\mathrm{D}}$ is a polynomial in $\mathbf{A}$ and $\mathbf{B}^{\mathrm{D}}$ is a polynomial in $\mathbf{B}$. (i) and (ii) now are easily proven. Assume now that $\mathbf{A}$ and $\mathbf{B}$ do not necessarily commute. To prove (iii), let $\mathbf{Y} = \mathbf{A}[(\mathbf{BA})^2]^{\mathrm{D}}\mathbf{B}$. Clearly $\mathbf{YABY} = \mathbf{Y}$ and $\mathbf{ABY} = \mathbf{YAB} = \mathbf{A}(\mathbf{BA})^{\mathrm{D}}\mathbf{B}$. Let $k = \max\{\mathrm{Ind}(\mathbf{AB}), \mathrm{Ind}(\mathbf{BA})\}$. Then $(\mathbf{AB})^{k+2}\mathbf{Y} = (\mathbf{AB})^{k+2}\mathbf{A}(\mathbf{BA})^{2^{\mathrm{D}}}\mathbf{B} = (\mathbf{AB})^{k+1}\mathbf{ABA}(\mathbf{BA})^{2^{\mathrm{D}}}\mathbf{B} = (\mathbf{AB})^{k+1}\mathbf{A}(\mathbf{BA})^{\mathrm{D}}\mathbf{B} = \mathbf{A}(\mathbf{BA})^{k+1}(\mathbf{BA})^{\mathrm{D}}\mathbf{B} = \mathbf{A}(\mathbf{BA})^k\mathbf{B} = (\mathbf{AB})^{k+1}$. Therefore, by Theorem 2.2, $\mathbf{Y} = (\mathbf{AB})^{\mathrm{D}}$. ■

Corollary 7.8.3 *Let* $\mathbf{A}, \mathbf{B} \in \mathbb{C}^{n\times n}$ *be such that* $\mathbf{AB} = \mathbf{BA}$. *Then* $\mathrm{Ind}(\mathbf{AB}) \leq \max\{\mathrm{Ind}(\mathbf{A}), \mathrm{Ind}(\mathbf{B})\}$.

Given a solution to just one of the three defining conditions, $\mathbf{A}^{k+1}\mathbf{X} = \mathbf{A}^k$, one can construct $\mathbf{A}^{\mathrm{D}}$ from it.

Theorem 7.8.5 *Let* $\mathbf{A} \in \mathbb{C}^{n\times n}$. *Let* $\mathbf{B} \in \mathbb{C}^{n\times n}$ *be a matrix such that* $\mathbf{A}^{l+1}\mathbf{B} = \mathbf{A}^l$ *for some* $l \geq \mathrm{Ind}(\mathbf{A}) = k$. *Then* $\mathbf{A}^{\mathrm{D}} = \mathbf{A}^l\mathbf{B}^{l+1}$.

Proof If $\mathbf{A} = \mathbf{P}\begin{bmatrix} \mathbf{C} & \mathbf{0} \\ \mathbf{0} & \mathbf{N} \end{bmatrix}\mathbf{P}^{-1}$ and $\mathbf{A}^{l+1}\mathbf{B} = \mathbf{A}^l$, then

$$\mathbf{B} = \mathbf{P}\begin{bmatrix} \mathbf{C}^{-1} & \mathbf{0} \\ \mathbf{B}_1 & \mathbf{B}_2 \end{bmatrix}\mathbf{P}^{-1}.$$

Thus $\mathbf{A}^l\mathbf{B}^{l+1} = \mathbf{P}\begin{bmatrix} \mathbf{C}^l & \mathbf{0} \\ \mathbf{0} & \mathbf{0} \end{bmatrix}\begin{bmatrix} \mathbf{C}^{-l-1} & \mathbf{0} \\ \mathbf{X}_1 & \mathbf{X}_2 \end{bmatrix}\mathbf{P}^{-1} = \mathbf{P}\begin{bmatrix} \mathbf{C}^{-1} & \mathbf{0} \\ \mathbf{0} & \mathbf{0} \end{bmatrix}\mathbf{P}^{-1} = \mathbf{A}^{\mathrm{D}}$.

■

8

Applications of the Drazin inverse to the theory of finite Markov chains

1. Introduction and terminology

Let $\{X_t : t \in F \subseteq \mathbb{R}\}$ be an indexed set of random variables. If P is a probability measure such that

$$P(X_t \leq b \,|\, X_{t_i} = x_i, \text{ for } i = 1, \ldots, n) = P(X_t \leq b \,|\, X_{t_n} = x_n)$$

whenever $t_1 < t_2 < \ldots < t_n < t$, then the set $\{X_t : t \in F\}$ is called a *Markov process*. In other words, a Markov process is such that when the present state of the process is known, the probability of any future behaviour of the process is not changed by knowledge of its past behaviour.

If the index set F is countable and if the range of each X_k is the same finite set $\mathscr{S} = \{\mathscr{S}_1, \mathscr{S}_2, \ldots, \mathscr{S}_m\}$ (the elements of $\mathscr{S}$ are referred to as the *states* of the process), then the process is called a *finite Markov chain*.

It is convenient to think of X_k as being the outcome of the process on the kth step or trial. The probability of X_k being in state $\mathscr{S}_j$ given that X_{k-1} was in state $\mathscr{S}_i$ is $p_{ij}(k) = P(X_k = \mathscr{S}_j \,|\, X_{k-1} = \mathscr{S}_i)$. These numbers are called the *one-step transition probabilities*. If each of the one-step transition probabilities is independent of time, i.e. $p_{ij}(k) = p_{ij}$ for $k = 1, 2, 3 \ldots$, then the chain is said to be *homogeneous*.

In this chapter, we will confine our attention to finite homogeneous Markov chains and will use 'Markov chain' or 'chain' to denote a finite homogeneous Markov chain.

For an m-state chain, the matrix $\mathbf{T}$ whose (i,j)th entry is the one-step transition probability p_{ij} is called the *one-step transition matrix*, or simply the *transition matrix* of the chain.

An *ergodic set* (or class) is a set of states in which every state of the set is accessible from every other state of the set and no state outside the set is accessible from any state in the set. A *transient set* is a set of states in which every state of the set is accessible from every other state of the set, but some state outside the state is accessible from each state in the set. An *ergodic state* is a member of an ergodic set and a *transient state* is a member of a transient set.

An m-state Markov chain is said to be *ergodic* if the transition matrix of the chain is irreducible, or equivalently, its states form a single ergodic set. An ergodic chain is said to be *regular* if the transition matrix $\mathbf{T}$ has the property that there exists a positive integer p such that $\mathbf{T}^p > \mathbf{0}$. (For $\mathbf{X} \in \mathbb{R}^{m \times n}$, $\mathbf{X} > \mathbf{0}$ means each entry of $\mathbf{X}$ is positive.)

If an ergodic chain is not regular, then it is said to be *cyclic*. It can be shown that if an ergodic chain is cyclic, then each state can only be entered at periodic intervals.

A state is said to be *absorbing* if once it is entered, it can never be left. A chain is said to be an *absorbing chain* if it has at least one absorbing state and from every state it is possible to reach an absorbing state (but not necessarily in one step).

The theory of finite Markov chains provides one of the most beautiful and elegant applications of the theory of matrices. The classical theory of Markov chains did not include concepts relating to generalized inversion of matrices. In this chapter it will be demonstrated how the theory of generalized inverses can be used to unify the theory of finite Markov chains. It is the Drazin inverse rather than any of the (i, j, k)-inverses which must be used. Some types of (1)-inverses, including the Moore–Penrose inverse, can be 'forced' into the theory because of their equation solving abilities. However, they lead to cumbersome expressions which do little to enhance or unify the theory and provide no practical or computational advantage.

Throughout this chapter it is assumed that the reader is familiar with the classical theory as it is presented in the text by Kemeny and Snell [46].

All matrices used in this chapter are assumed to have only real entries so that $(\cdot)^*$ should be taken to mean transpose.

2. Introduction of the Drazin inverse into the theory of finite Markov chains

For an m-state chain whose transition matrix is $\mathbf{T}$, we will be primarily concerned with the matrix $\mathbf{A} = \mathbf{I} - \mathbf{T}$. Virtually everything that one wants to know about a chain can be extracted from $\mathbf{A}$ and its Drazin inverse. One of the most important reasons for the usefulness of the Drazin inverse is the fact that $\text{Ind}(\mathbf{A}) = 1$ for every transition matrix $\mathbf{T}$ so that the Drazin inverse is also the group inverse. This fact is obtainable from the classical theory of elementary divisors. However, we will present a different proof utilizing the theory of generalized inverses. After the theorem is proven, we will use the notation $\mathbf{A}^{\#}$ in place of $\mathbf{A}^{\mathrm{D}}$ in order to emphasize the fact that we are dealing with the group inverse.

Theorem 8.2.1 *If* $\mathbf{T} \in \mathbb{R}^{m \times m}$ *is any transition matrix (i.e.* $\mathbf{T}$ *is a stochastic matrix) and if* $\mathbf{A} = \mathbf{I} - \mathbf{T}$, *then* $\text{Ind}(\mathbf{A}) = 1$ *(i.e.* $\mathbf{A}^{\#}$ *exists).*

Proof The proof is in two parts. Part I is for the case when **T** is irreducible. Part II is for the case when **T** is reducible.

(I) If **T** is a stochastic matrix and **j** is a vector of 1's, then $\mathbf{Tj} = \mathbf{j}$ so that $1 \in \sigma(\mathbf{T})$. Since $\rho(\mathbf{T}) \leq \|\mathbf{T}\|_{0\infty} = 1$, (see page 211), it follows that $\rho(\mathbf{T}) = 1$. If **T** is irreducible, then the Perron–Frobenius Theorem implies that the eigenvalue 1 has algebraic multiplicity equal to one. Thus, $0 \in \sigma(\mathbf{A})$ with algebraic multiciplicity equal to one. Therefore, $\text{Ind}(\mathbf{A}) = 1$, which is equivalent to saying that $\mathbf{A}^{\#}$ exists by Theorem 7.2.4. ■

Before proving Part II of this theorem, we need the following fact.

Lemma 8.2.1 *If* $\mathbf{B} \geq \mathbf{0}$ *is irreducible and* $\mathbf{M} \geq \mathbf{0}$ *is a non-zero matrix such that* $\mathbf{B} + \mathbf{M} = \mathbf{S}$ *is a transition matrix, then* $\rho(\mathbf{B}) < 1$.

Proof Suppose the proposition is false. Then $\rho(\mathbf{B}) \geq 1$. However, since **S** is stochastic and $\mathbf{M} \geq \mathbf{0}$, it follows that $\|\mathbf{B}\|_{0\infty} \leq 1$. Thus, $\rho(\mathbf{B}) \leq \|\mathbf{B}\|_{0\infty} \leq 1$. Therefore, it must be the case that $1 = \rho(\mathbf{B}) = \rho(\mathbf{B}^*)$. The Perron–Frobenius Theorem implies that there exists a positive eigenvector, $\mathbf{v} > \mathbf{0}$, corresponding to the eigenvalue 1 for $\mathbf{B}^*$. Thus, $\mathbf{v} = \mathbf{B}^*\mathbf{v} = (\mathbf{S}^* - \mathbf{M}^*)\mathbf{v}$. By using the fact that $\mathbf{Sj} = \mathbf{j}$, we obtain $\mathbf{j}^*\mathbf{v} = \mathbf{j}^*\mathbf{S}^*\mathbf{v} - \mathbf{j}^*\mathbf{M}^*\mathbf{v} = \mathbf{j}^*\mathbf{v} - \mathbf{j}^*\mathbf{M}^*\mathbf{v}$. Therefore, $\mathbf{j}^*\mathbf{M}^*\mathbf{v} = 0$. However, this is impossible because $\mathbf{j} > \mathbf{0}$, $\mathbf{v} > \mathbf{0}$, $\mathbf{M} \geq \mathbf{0}$ and $\mathbf{M} \neq \mathbf{0}$. ■

We are now in a position to give the second part of the proof of Theorem 1.

(II) Assume now that the transition matrix is reducible. By a suitable permutation of the states, we can write

$$\mathbf{T} \sim \begin{bmatrix} \mathbf{X} & \mathbf{Y} \\ \mathbf{0} & \mathbf{Z} \end{bmatrix} \quad (\sim \text{ indicates equality after a suitable permutation has been performed})$$

where **X** and **Z** are square. If either **X** or **Z** is reducible, we can perform another permutation so that

$$\mathbf{T} \sim \begin{bmatrix} \mathbf{U} & \mathbf{V} & \mathbf{W} \\ \mathbf{0} & \mathbf{C} & \mathbf{D} \\ \mathbf{0} & \mathbf{0} & \mathbf{E} \end{bmatrix}.$$

If either **U**, **C** or **E**, is reducible, then another permutation is performed. Continuing in this manner, we eventually get

$$\mathbf{T} \sim \begin{bmatrix} \mathbf{B}_{11} & \mathbf{B}_{12} \cdots \mathbf{B}_{1n} \\ \mathbf{0} & \mathbf{B}_{22} \cdots \mathbf{B}_{2n} \\ \vdots & \ddots \\ \mathbf{0} & \mathbf{0} \quad \ldots \mathbf{B}_{nn} \end{bmatrix}$$

where $\mathbf{B}_{ii}$ is irreducible. If one or more rows of blocks are all zero except for the diagonal block (i.e. if there are subscripts i such that $\mathbf{B}_{ik} = \mathbf{0}$ for each $k \neq i$), then perform one last permutation and write

$$\mathbf{T} \sim \left[\begin{array}{cccc|cccc}
\mathbf{T}_{11} & \mathbf{T}_{12} & \cdots & \mathbf{T}_{1r} & \mathbf{T}_{1,r+1} & \mathbf{T}_{1,r+2} & \cdots & \mathbf{T}_{1,n} \\
\mathbf{0} & \mathbf{T}_{22} & \cdots & \mathbf{T}_{2r} & \mathbf{T}_{2,r+1} & \mathbf{T}_{2,r+2} & \cdots & \mathbf{T}_{2,n} \\
\vdots & & \ddots & & & & & \\
\mathbf{0} & \mathbf{0} & \cdots & \mathbf{T}_{rr} & \mathbf{T}_{r,r+1} & \mathbf{T}_{r,r+2} & \cdots & \mathbf{T}_{r,n} \\
\hline
\mathbf{0} & \mathbf{0}\ \mathbf{0} & \cdots & \mathbf{0} & \mathbf{T}_{r+1,r+1} & \mathbf{0} & \cdots & \mathbf{0} \\
\mathbf{0} & \mathbf{0}\ \mathbf{0} & \cdots & \mathbf{0} & \mathbf{0} & \mathbf{T}_{r+2,r+2} & \cdots & \mathbf{0} \\
\vdots & & & & & & \ddots & \\
\mathbf{0} & \mathbf{0}\ \mathbf{0} & \cdots & \mathbf{0} & \mathbf{0} & \mathbf{0} & \cdots & \mathbf{T}_{n,n}
\end{array}\right] \tag{1}$$

Each $\mathbf{T}_{ii}$ $(i = 1, 2, \ldots, n)$ is irreducible. From Part I of this proof, we know that $(\mathbf{I} - \mathbf{T}_{ii})^{\#}$ exists for every i. However, for $i = 1, 2, \ldots, r$, there is at least one index $k \neq i$ such that $\mathbf{T}_{ik} \neq \mathbf{0}$. It follows from Lemma 2.1 that $\rho(\mathbf{T}_{ii}) < 1$ for $i = 1, 2, \ldots r$. Therefore, $(\mathbf{I} - \mathbf{T}_{ii})^{-1}$ exists for $i = 1, 2, \ldots, r$. We can now conclude that there exists a permutation matrix $\mathbf{P}$ such that $\mathbf{A}$ can be written as

$$\mathbf{A} = \mathbf{P}^* \left[\begin{array}{c|c} \mathbf{G}_{11} & \mathbf{G}_{12} \\ \hline \mathbf{0} & \mathbf{G}_{22} \end{array}\right] \mathbf{P},$$

where $\mathbf{G}_{11}$ is non-singular and $\mathbf{G}_{22}^{\#}$ exists. It now follows from Theorem 7.7.3 that $\mathbf{A}^{\#}$ must exist and

$$\mathbf{A}^{\#} = \mathbf{P}^* \left[\begin{array}{c|c} \mathbf{G}_{11}^{-1} & \mathbf{G}_{11}^{-2}\mathbf{G}_{12}(\mathbf{I} - \mathbf{G}_{22}\mathbf{G}_{22}^{\#}) - \mathbf{G}_{11}^{-1}\mathbf{G}_{12}\mathbf{G}_{22}^{\#} \\ \hline \mathbf{0} & \mathbf{G}_{22}^{\#} \end{array}\right] \mathbf{P}.$$

Thus, the proof is complete. ■

Notice that for every transition matrix $\mathbf{T}$, it is always the case that $\mathbf{j} \in N(\mathbf{A}) = N(\mathbf{A}^{\#})$ so that $\mathbf{A}^{\#}\mathbf{j} = \mathbf{0}$. Furthermore, it is always the case that $(\mathbf{I} - \mathbf{A}\mathbf{A}^{\#})\mathbf{j} = \mathbf{j}$. We will frequently use these observations together with the following well-known Lemma, which we state without proof.

Lemma 8.2.2

(I) *Every transition matrix* $\mathbf{T}$ *is similar to a matrix of the form*

$$\left[\begin{array}{c|c} \mathbf{I}_k & \mathbf{0} \\ \hline \mathbf{0} & \mathbf{K} \end{array}\right], \text{ where } 1 \notin \sigma(\mathbf{K}).$$

(II) *If* $\mathbf{T}$ *is the transition matrix of an ergodic chain, then* $k = 1$, *i.e.* $\mathbf{T}$ *is similar to a matrix of the form*

$$\left[\begin{array}{c|c} 1 & \mathbf{0} \\ \hline \mathbf{0} & \mathbf{K} \end{array}\right], \ 1 \notin \sigma(\mathbf{K}).$$

(III) *If* $\mathbf{T}$ *is the transition matrix of a regular chain, then* $\lim_{n\to\infty} \mathbf{K}^n = \mathbf{0}$.

We are now in a position to relate the single expression $\mathbf{I} - \mathbf{A}\mathbf{A}^{\#}$ to various types of limiting processes which are frequently encountered in the theory of finite Markov chains.

Theorem 8.2.2 *Let* $\mathbf{T}$ *be the transition matrix of an m-state chain and let* $\mathbf{A} = \mathbf{I} - \mathbf{T}$. *Then*

$$\mathbf{I} - \mathbf{A}\mathbf{A}^{\#} = \begin{cases} \lim\limits_{n\to\infty} \dfrac{\mathbf{I} + \mathbf{T} + \mathbf{T}^2 + \ldots + \mathbf{T}^{n-1}}{n} & \textit{for every transition matrix.} \\ \lim\limits_{n\to\infty} (\alpha\mathbf{I} + (1-\alpha)\mathbf{T})^n & \textit{for every transition matrix } \mathbf{T} \textit{ and } 0 < \alpha < 1. \\ \lim\limits_{n\to\infty} \mathbf{T}^n & \textit{for every regular chain.} \\ \lim\limits_{n\to\infty} \mathbf{T}^n & \textit{for every absorbing chain.} \end{cases}$$

Proof For every transition matrix $\mathbf{T}$, we know from Lemma 2 that there exists a non-singular matrix $\mathbf{S}$ such that

$$\mathbf{T} = \mathbf{S}^{-1}\left[\begin{array}{c|c} \mathbf{I}_k & \mathbf{0} \\ \hline \mathbf{0} & \mathbf{K} \end{array}\right]\mathbf{S} \tag{2}$$

and $1 \notin \sigma(\mathbf{K})$. Therefore, $\mathbf{I} - \mathbf{K}$ is non-singular and

$$\mathbf{A} = \mathbf{S}^{-1}\left[\begin{array}{c|c} \mathbf{0} & \mathbf{0} \\ \hline \mathbf{0} & \mathbf{I} - \mathbf{K} \end{array}\right]\mathbf{S},\ \mathbf{A}^{\#} = \mathbf{S}^{-1}\left[\begin{array}{c|c} \mathbf{0} & \mathbf{0} \\ \hline \mathbf{0} & (\mathbf{I} - \mathbf{K})^{-1} \end{array}\right]\mathbf{S} \text{ and}$$

$$\mathbf{I} - \mathbf{A}\mathbf{A}^{\#} = \mathbf{S}^{-1}\left[\begin{array}{c|c} \mathbf{I}_k & \mathbf{0} \\ \hline \mathbf{0} & \mathbf{0} \end{array}\right]\mathbf{S}. \tag{3}$$

Assume first that $\mathbf{T}$ is the transition matrix of a regular chain. Then from Lemma 2 we know that $k = 1$ and $\lim\limits_{n\to\infty} \mathbf{K}^n = \mathbf{0}$. It is now clear that

$$\lim_{n\to\infty} \mathbf{T}^n = \mathbf{S}^{-1}\left[\begin{array}{c|c} 1 & \mathbf{0} \\ \hline \mathbf{0} & \mathbf{0} \end{array}\right]\mathbf{S} = \mathbf{I} - \mathbf{A}\mathbf{A}^{\#}.$$

Next, consider $\tilde{\mathbf{T}} = \alpha\mathbf{I} + (1-\alpha)\mathbf{T}$, $0 < \alpha < 1$. It is clear that $\tilde{\mathbf{T}}$ is a transition matrix whose eigenvalues are $\tilde{\lambda} = \alpha + (1-\alpha)\lambda$, $\lambda \in \sigma(\mathbf{T})$. Since each $\tilde{\lambda}$ is a convex combination of 1 and λ, it follows each $\tilde{\lambda}$ different from 1 is inside the unit circle (i.e. $|\tilde{\lambda}| < 1$ if $\tilde{\lambda} \neq 1$) so that by considering (2) it is clear that $\rho(\tilde{\mathbf{K}}) = \rho(\alpha\mathbf{I} + (1-\alpha)\mathbf{K}) < 1$ and

$$\mathbf{S}\left(\lim_{n\to\infty} \tilde{\mathbf{T}}^n\right)\mathbf{S}^{-1} = \lim_{n\to\infty} \left[\begin{array}{c|c} \mathbf{I} & \mathbf{0} \\ \hline \mathbf{0} & \alpha\mathbf{I} + (1-\alpha)\mathbf{K} \end{array}\right]^n$$

$$= \lim_{n\to\infty} \left[\begin{array}{c|c} \mathbf{I} & \mathbf{0} \\ \hline \mathbf{0} & \tilde{\mathbf{K}}^n \end{array}\right] = \left[\begin{array}{c|c} \mathbf{I} & \mathbf{0} \\ \hline \mathbf{0} & \mathbf{0} \end{array}\right] = \mathbf{S}(\mathbf{I} - \mathbf{A}\mathbf{A}^{\#})\mathbf{S}^{-1}.$$

Assume next that $\mathbf{T}$ is the transition matrix of an absorbing chain with exactly r absorbing states. Then there exists a permutation matrix $\mathbf{P}$ such that

$$\mathbf{T} = \mathbf{P}^*\left[\begin{array}{c|c} \mathbf{I}_r & \mathbf{0} \\ \hline \mathbf{R} & \mathbf{Q} \end{array}\right]\mathbf{P} \text{ and } \mathbf{T}^n = \mathbf{P}^*\left[\begin{array}{c|c} \mathbf{I}_r & \mathbf{0} \\ \hline \left(\sum_{k=0}^{n-1} \mathbf{Q}^k\right)\mathbf{R} & \mathbf{Q}^n \end{array}\right]\mathbf{P}.$$

For an absorbing chain, it is well-known that $\mathbf{I} + \mathbf{Q} + \mathbf{Q}^2 + \ldots = (\mathbf{I} - \mathbf{Q})^{-1}$ so that

$$\lim_{n\to\infty} \mathbf{T}^n = \mathbf{P}^*\left[\begin{array}{c|c} \mathbf{I}_r & \mathbf{0} \\ \hline (\mathbf{I} - \mathbf{Q})^{-1}\mathbf{R} & \mathbf{0} \end{array}\right]\mathbf{P}.$$

Since $\mathbf{A}$ has the form

$$\mathbf{A} = \mathbf{P}^*\left[\begin{array}{c|c} \mathbf{0} & \mathbf{0} \\ \hline -\mathbf{R} & \mathbf{I} - \mathbf{Q} \end{array}\right]\mathbf{P},$$

Theorem 7.7.3 yields

$$\mathbf{A}^\# = \mathbf{P}^*\left[\begin{array}{c|c} \mathbf{0} & \mathbf{0} \\ \hline -(\mathbf{I} - \mathbf{Q})^{-2}\mathbf{R} & (\mathbf{I} - \mathbf{Q})^{-1} \end{array}\right]\mathbf{P}.$$

Therefore,

$$\mathbf{I} - \mathbf{A}\mathbf{A}^\# = \mathbf{P}^*\left[\begin{array}{c|c} \mathbf{I}_r & \mathbf{0} \\ \hline (\mathbf{I} - \mathbf{Q})^{-1}\mathbf{R} & \mathbf{0} \end{array}\right]\mathbf{P} = \lim_{n\to\infty} \mathbf{T}^n. \tag{4}$$

Finally, assume that $\mathbf{T}$ is any transition matrix and is written in the form (2). Since $\mathbf{I} - \mathbf{K}$ is non-singular, we may write

$$\frac{\mathbf{I} + \mathbf{K} + \mathbf{K}^2 + \ldots + \mathbf{K}^{n-1}}{n} = \frac{(\mathbf{I} - \mathbf{K}^n)(\mathbf{I} - \mathbf{K})^{-1}}{n}. \tag{5}$$

By using (2) and (3) together with (5), it is a simple matter to verify that $\frac{\mathbf{I} + \mathbf{T} + \mathbf{T}^2 + \ldots + \mathbf{T}^{n-1}}{n} = \frac{(\mathbf{I} - \mathbf{T}^n)\mathbf{A}^\#}{n} + \mathbf{I} - \mathbf{A}\mathbf{A}^\#$. Since $\|\mathbf{T}^n\|_{0\infty} = 1$ for all n, it follows that $\lim_{n\to\infty} \frac{(\mathbf{I} - \mathbf{T}^n)\mathbf{A}^\#}{n} = \mathbf{0}$, and hence

$$\lim_{n\to\infty} \frac{\mathbf{I} + \mathbf{T} + \mathbf{T}^2 + \ldots + \mathbf{T}^{n-1}}{n} = \mathbf{I} - \mathbf{A}\mathbf{A}^\#. \quad \blacksquare$$

In the case of ergodic chains, the matrix $\mathbf{I} - \mathbf{A}\mathbf{A}^\#$ has a very special structure.

Theorem 8.2.3 *If $\mathbf{T}$ is the transition matrix of an m-state ergodic chain, then each row of $\mathbf{I} - \mathbf{A}\mathbf{A}^\#$ is the same vector $\mathbf{w}^* = [w_1, w_2, \ldots, w_m]$, which is the fixed probability (row) vector of $\mathbf{T}$.*

Proof This follows from the fact that $\lim_{n\to\infty} \mathbf{T}^n = \mathbf{I} - \mathbf{A}\mathbf{A}^\#$. $\blacksquare$

Throughout the rest of this chapter we will let **W** denote the limiting matrix, $\mathbf{I} - \mathbf{AA}^{\#}$.

3. Regular chains

Theorem 8.3.1 *For every transition matrix* **T**,

$$\mathbf{A}^{\#} = \lim_{n\to\infty} \sum_{k=0}^{n-1} \frac{n-k}{n}(\mathbf{T}^k - \mathbf{W}) = \mathbf{AA}^{\#} + \lim_{n\to\infty} \sum_{k=1}^{n-1} \frac{n-k}{n}(\mathbf{T}^k - \mathbf{W}).$$

If **T** *is the transition matrix of a regular chain, then the expression reduces to* $\mathbf{A}^{\#} = \sum_{k=0}^{\infty} (\mathbf{T}^k - \mathbf{W})$.

Proof Write **T** as in (I) of Lemma 2.2 so that

$$\sum_{k=0}^{n-1} \frac{n-k}{n}(\mathbf{T}^k - \mathbf{W}) = \mathbf{S}^{-1} \left[\begin{array}{c|c} \mathbf{0} & \mathbf{0} \\ \hline \mathbf{0} & \sum_{k=0}^{n-1} \frac{n-k}{n}\mathbf{K}^n \end{array}\right] \mathbf{S}.$$

Since $\mathbf{I} - \mathbf{K}$ is non-singular,

$$\sum_{k=0}^{n-1} \frac{n-k}{n}\mathbf{K}^k = (\mathbf{I} - \mathbf{K})^{-1} - (\mathbf{I} - \mathbf{K})^{-1}\mathbf{K}\left[\frac{1}{n}\sum_{k=0}^{n-1} \mathbf{K}^k\right].$$

From the first part of Theorem 2.2, it follows that

$$\lim_{n\to\infty} \frac{\mathbf{I} + \mathbf{K} + \ldots + \mathbf{K}^{n-1}}{n} = \mathbf{0}.$$

Therefore, $\lim_{n\to\infty} \sum_{k=0}^{n-1} \frac{n-k}{n}\mathbf{K}^k = (\mathbf{I} - \mathbf{K})^{-1}$, and hence

$$\lim_{n\to\infty} \sum_{k=0}^{n-1} \frac{n-k}{n}(\mathbf{T} - \mathbf{W})^k = \mathbf{S}^{-1} \left[\begin{array}{c|c} \mathbf{0} & \mathbf{0} \\ \hline \mathbf{0} & (\mathbf{I} - \mathbf{K})^{-1} \end{array}\right] \mathbf{S} = \mathbf{A}^{\#}.$$

The second equality in the first part of the theorem follows because $\mathbf{I} - \mathbf{W} = \mathbf{AA}^{\#}$. Assume now that **T** is the transition matrix of a regular chain. Write **T** as in Part (II) of Lemma 2.2. From Part (III) of that Lemma, we know that $\lim_{n\to\infty} \mathbf{K}^n = \mathbf{0}$ so that $\sum_{k=0}^{\infty} \mathbf{K}^k = (\mathbf{I} - \mathbf{K})^{-1}$. Therefore,

$$\sum_{k=0}^{\infty} (\mathbf{T}^k - \mathbf{W}) = \mathbf{S}^{-1} \left[\begin{array}{c|c} \mathbf{0} & \mathbf{0} \\ \hline \mathbf{0} & \sum_{k=0}^{\infty} \mathbf{K}^k \end{array}\right] \mathbf{S} = \mathbf{S}^{-1} \left[\begin{array}{c|c} \mathbf{0} & \mathbf{0} \\ \hline \mathbf{0} & (\mathbf{I} - \mathbf{K})^{-1} \end{array}\right] \mathbf{S} = \mathbf{A}^{\#}.$$

■

The second part of the proof of Theorem 2.3 provides an interpretation for each of the entries of $\mathbf{A}^{\#}$ for a regular chain.

Theorem 8.3.2 *Let* $\mathbf{T}$ *be the transition matrix of a regular chain and let* $\mathbf{N}^{(n)}$ *denote the matrix whose (i,j)th entry is the expected number of times the chain is in state* $\mathscr{S}_j$ *in the first n stages (the initial plus n − 1 stages) when the chain was initially in* $\mathscr{S}_i$. *Then* $\mathbf{A}^{\#} = \lim_{n\to\infty} (\mathbf{N}^{(n)} - n\mathbf{W})$.

Proof The result follows by combining the fact that $\mathbf{N}^{(n)} = \sum_{k=0}^{n} \mathbf{T}^k$ together with Lemma 3.1 since $\lim_{n\to\infty} (\mathbf{N}^{(n)} - n\mathbf{W}) = \lim_{n\to\infty} \left(\sum_{k=0}^{n-1} (\mathbf{T}^k - \mathbf{W}) \right) = \mathbf{A}^{\#}$. ∎

The (i,j)th entry, $\mathbf{A}^{\#}_{ij}$, of $\mathbf{A}^{\#}$ thus has the following meaning for a regular chain. For large n, the expected number of times in state $\mathscr{S}_j$, when initially in state $\mathscr{S}_i$, differs from nw_j by approximately $\mathbf{A}^{\#}_{ij}$ where w_j is the jth component of the fixed probability vector $\mathbf{w}^*$. In loose terms, one could write $\mathbf{N}^{(n)} \approx \mathbf{A}^{\#} + n(\mathbf{I} - \mathbf{A}\mathbf{A}^{\#})$ for large n. Furthermore, for large n, one can compare two starting states in terms of the elements of $\mathbf{A}^{\#}$ since $\lim_{n\to\infty} (\mathbf{N}^{(n)}_{ij} - \mathbf{N}^{(n)}_{kj}) = \mathbf{A}^{\#}_{ij} - \mathbf{A}^{\#}_{kj}$.

For an initial probability vector $\mathbf{p}^*$, the jth component, $(\mathbf{p}^*\mathbf{N}^{(n)})_j$, of $\mathbf{p}^*\mathbf{N}^{(n)}$ gives the expected number of times in state $\mathscr{S}_j$ in n stages. The following corollary provides a comparison of $(\mathbf{p}_1^*\mathbf{N}^{(n)})_j$ and $(\mathbf{p}_2^*\mathbf{N}^{(n)})_j$ for two different initial probability vectors.

Corollary 8.3.1 *Let* $\mathbf{T}$ *be the transition matrix of a regular chain, let* $\mathbf{p}^*$ *be an initial probability vector, and let* $\mathbf{w}^*$ *be the fixed probability vector for* $\mathbf{T}$. *Then* $\lim_{n\to\infty} (\mathbf{p}^*\mathbf{N}^{(n)} - n\mathbf{w}) = \mathbf{p}^*\mathbf{A}^{\#}$, *and for two initial vectors* $\mathbf{p}_1^*$ *and* $\mathbf{p}_2^*$, $\lim_{n\to\infty} (\mathbf{p}_1^*\mathbf{N}^{(n)} - \mathbf{p}_2^*\mathbf{N}^{(n)}) = (\mathbf{p}_1^* - \mathbf{p}_2^*)\mathbf{A}^{\#}$. *Furthermore,* $\mathrm{Tr}(\mathbf{A}^{\#}) = \lim_{n\to\infty} \sum_k [\mathbf{N}^{(n)}_{kk} - (\mathbf{p}^*\mathbf{N}^{(n)})_k] = \lim_{n\to\infty} (\mathrm{Tr}(\mathbf{N}^{(n)}) - \mathbf{p}^*\mathbf{N}^{(n)}\mathbf{j})$, *for every initial probability vector* $\mathbf{p}^*$.

Proof The first two limits are immediate. The third limit follows from the second and the fact that $\mathbf{A}^{\#}\mathbf{j} = \mathbf{0}$ since

$$\sum_k [\mathbf{N}^{(n)}_{kk} - (\mathbf{p}^*\mathbf{N}^{(n)})_k] = \sum_k [(\mathbf{e}_k^* - \mathbf{p}^*)\mathbf{N}^{(n)}\mathbf{e}_k] \to \sum_k [\mathbf{e}_k^*\mathbf{A}^{\#}\mathbf{e}_k - \mathbf{p}^*\mathbf{A}^{\#}\mathbf{e}_k$$
$$= \mathrm{Tr}(\mathbf{A}^{\#}) - \mathbf{p}^*\mathbf{A}^{\#}\mathbf{j} = \mathrm{Tr}(\mathbf{A}^{\#}). \quad ∎$$

4. Ergodic chains

In this section we will extend our attention to investigate ergodic chains in general. It will be shown that the matrix $\mathbf{A}^{\#}$ is the fundamental quantity in the theory of ergodic chains. Virtually everything that one would want to know about an ergodic chain can be determined by computing $\mathbf{A}^{\#}$.

We will begin by investigating the mean first passage times (i.e. the expected number of steps it takes to go from state $\mathscr{S}_i$ to $\mathscr{S}_j$ for the first time.)

Let $\mathbf{M}$ denote the matrix whose (i,j)th entry is the expected number of steps before entering state $\mathscr{S}_j$ for the first time after the initial state $\mathscr{S}_i$. $\mathbf{M}$ is called the *mean first passage matrix*.

For a square matrix $\mathbf{X}$, the diagonal matrix obtained by setting all off diagonal entries of $\mathbf{X}$ equal to zero is denoted by $\mathbf{X}_d$. $\mathbf{J}_m$ will denote the matrix of all 1's. If the size of $\mathbf{J}_m$ is understood from the context, then the subscript m will be omitted.

Theorem 8.4.1 *If* $\mathbf{T}$ *is the transition matrix of an m-state ergodic chain whose fixed probability vector is* $\mathbf{w}^* = [w_1, w_2, \dots, w_m]$, *then the mean first passage matrix is given by*

$$\mathbf{M} = (\mathbf{I} - \mathbf{A}^{\#} + \mathbf{J}\mathbf{A}_d^{\#})\mathbf{D} \tag{1}$$

where $\mathbf{D}$ *is the diagonal matrix*

$$\mathbf{D} = \begin{bmatrix} \frac{1}{w_1} & 0 & 0 \dots 0 \\ 0 & \frac{1}{w_2} & 0 \dots 0 \\ \vdots & & \\ 0 & 0 & 0 \dots \frac{1}{w_m} \end{bmatrix} = [(\mathbf{I} - \mathbf{A}\mathbf{A}^{\#})_d]^{-1}.$$

Proof It is known that the mean first passage matrix is the unique solution of the matrix equation

$$\mathbf{A}\mathbf{X} = \mathbf{J} - \mathbf{T}\mathbf{X}_d. \tag{2}$$

We simply verify that the right hand side of (1) satisfies the equation (2). Let $\mathbf{R}$ denote the right hand side of (1) and observe that $\mathbf{R}_d = \mathbf{D}$. Now, $\mathbf{A}\mathbf{R} = \mathbf{A}(\mathbf{I} - \mathbf{A}^{\#} + \mathbf{J}\mathbf{A}_d^{\#})\mathbf{D} = (\mathbf{A} - \mathbf{A}\mathbf{A}^{\#})\mathbf{D} = (\mathbf{A} - \mathbf{I} + \mathbf{W})\mathbf{D} = (\mathbf{W} - \mathbf{T})\mathbf{D} = \mathbf{J} - \mathbf{T}\mathbf{D} = \mathbf{J} - \mathbf{T}\mathbf{R}_d$. ■

Corollary 8.4.1 *Consider an ergodic chain whose fixed probability vector is* $\mathbf{w}^* = [w_1, w_2, \dots, w_m]$. *If the chain starts in state* $\mathscr{S}_i$, *then the expected number of steps taken before returning to state* $\mathscr{S}_i$ *for the first time is given by* $\mathbf{M}_{ii} = 1/(\mathbf{I} - \mathbf{A}\mathbf{A}^{\#})_{ii}$.

For an initial probability vector $\mathbf{p}^*$, the kth component, $(\mathbf{p}^*\mathbf{M})_k$, of $\mathbf{p}^*\mathbf{M}$ is the expected number of steps before entering state $\mathscr{S}_k$ for the first time. Consider the case of a regular chain that has gone through n steps before it is observed. If n is sufficiently large, the initial probability vector may be taken to be $\mathbf{p}^*(0) = \mathbf{w}^*$. In this case, $\mathbf{p}^*(t) = \mathbf{w}^*, t = 0, 1, 2, \dots$. When this situation occurs, we say that the chain is observed *in equilibrium*. The next theorem relates the diagonal elements of $\mathbf{A}^{\#}$ to the expected number of steps taken before entering state $\mathscr{S}_k$ when the initial distribution is $\mathbf{p}^*(0) = \mathbf{w}^*$.

Theorem 8.4.2 *If* $\mathbf{T}$ *is the transition matrix of an ergodic chain and* $\mathbf{w}^*$ *is the fixed probability vector of* $\mathbf{T}$, *then*

$$(\mathbf{w}^*\mathbf{M})_k = 1 + \frac{\mathbf{A}^{\#}_{kk}}{w_k}.$$

Proof From Theorem 4.1, we have $\mathbf{M} = \mathbf{D} - \mathbf{A}^{\#}\mathbf{D} + \mathbf{J}\mathbf{A}^{\#}_d\mathbf{D}$. Use this together with the fact that $\mathbf{w}^*\mathbf{A}^{\#} = \mathbf{0}$ to obtain $\mathbf{w}^*\mathbf{M} = \mathbf{w}^*\mathbf{D} + \mathbf{w}^*\mathbf{J}\mathbf{A}^{\#}_d\mathbf{D} = \mathbf{j}^*(\mathbf{I} + \mathbf{A}^{\#}_d\mathbf{D})$. Therefore, $(\mathbf{w}^*\mathbf{M})_k = 1 + (\mathbf{j}^*\mathbf{A}^{\#}_d\mathbf{D})_k = 1 + \dfrac{\mathbf{A}^{\#}_{kk}}{w_k}$. ■

The kth component of the vector $(\mathbf{p}^*_1 - \mathbf{p}^*_2)\mathbf{M}$ provides a comparison of the expected number of steps before entering state $\mathscr{S}_k$ for two different initial distributions $\mathbf{p}^*_1$ and $\mathbf{p}^*_2$. This expression is given below in terms of $\mathbf{A}^{\#}$.

Theorem 8.4.3 *For an ergodic chain and for two initial probability vectors* $\mathbf{p}^*_1$ *and* $\mathbf{p}^*_2$, $(\mathbf{p}^*_1 - \mathbf{p}^*_2)\mathbf{M} = (\mathbf{p}^*_1 - \mathbf{p}^*_2)(\mathbf{I} - \mathbf{A}^{\#})\mathbf{D}$.

Proof This follows from Theorem 4.1 since $\mathbf{p}^*_1\mathbf{J} = \mathbf{p}^*_2\mathbf{J} = \mathbf{j}^*$. ■

We now address ourselves to the problem of obtaining the variances of the first passage times. In Theorem 4.1, we saw how the matrix $\mathbf{A}^{\#}$ produces the expected first passage times. The following theorem shows that $\mathbf{A}^{\#}$ also produces the variances of the first passage times. For an ergodic chain, let $\mathbf{V}$ denote the matrix whose (i,j)-entry is the variance of the number of steps required to reach state $\mathscr{S}_j$ for the first time after the initial state $\mathscr{S}_i$.

Theorem 8.4.4. *For* $\mathbf{X} \in \mathbb{R}^{m \times m}$, *let* $\mathbf{X}_s$ *denote the matrix whose entries are the squares of the entries of* $\mathbf{X}$, *i.e.* $(\mathbf{X}_s)_{ij} = (\mathbf{X}_{ij})^2$. *The matrix of variances of the first passage times is given by* $\mathbf{V} = \mathbf{B} - \mathbf{M}_s$ *where* $\mathbf{M}$ *is the mean first passage matrix and*

$$\mathbf{B} = \mathbf{M}(2\mathbf{A}^{\#}_d\mathbf{D} + \mathbf{I}) + 2(\mathbf{A}^{\#}\mathbf{M} - \mathbf{J}[\mathbf{A}^{\#}\mathbf{M}]_d). \tag{3}$$

Proof If $\mathbf{V} = \mathbf{B} - \mathbf{M}_s$, then it is well known that $\mathbf{B}$ must be the unique solution of $\mathbf{A}\mathbf{B} = \mathbf{J} - \mathbf{T}\mathbf{B}_d + 2\mathbf{T}(\mathbf{M} - \mathbf{M}_d)$. From Theorem 4.1, we know that $\mathbf{M} - \mathbf{M}_d = \mathbf{M} - \mathbf{D} = -(\mathbf{A}^{\#} - \mathbf{J}\mathbf{A}^{\#}_d)\mathbf{D}$. Therefore $\mathbf{B}$ must be the unique solution of $\mathbf{A}\mathbf{B} = \mathbf{J} - \mathbf{T}\mathbf{B}_d - 2\mathbf{T}(\mathbf{A}^{\#} - \mathbf{J}\mathbf{A}^{\#}_d)\mathbf{D}$.

Let $\mathbf{R}$ denote the right hand side of (3) and observe that $\mathbf{R}_d = 2\mathbf{D}\mathbf{A}^{\#}_d\mathbf{D} + \mathbf{D}$. Now, use the facts that $\mathbf{A}\mathbf{J} = \mathbf{0}$, $\mathbf{A}\mathbf{M} = \mathbf{J} - \mathbf{T}\mathbf{D}$, $\mathbf{A}\mathbf{A}^{\#}\mathbf{M} = -\mathbf{A}^{\#}\mathbf{T}\mathbf{D}$, and $\mathbf{T}\mathbf{J} = \mathbf{J}$ to obtain $\mathbf{A}\mathbf{R} = (\mathbf{J} - \mathbf{T}\mathbf{D})(2\mathbf{A}^{\#}_d\mathbf{D} + \mathbf{I}) - 2\mathbf{A}^{\#}\mathbf{T}\mathbf{D} = \mathbf{J} - \mathbf{T}\mathbf{R}_d - 2\mathbf{T}(\mathbf{A}^{\#} - \mathbf{J}\mathbf{A}^{\#}_d)\mathbf{D}$. Therefore, (3) must be true. ■

5. Calculation of $\mathbf{A}^{\#}$ and $\mathbf{w}^*$ for an ergodic chain

Some practical methods for calculating the group inverse $\mathbf{G}^{\#}$ of a general square matrix $\mathbf{G}$ (provided of course that $\mathbf{G}^{\#}$ exists) were given in

Chapter 7. However, for the present situation we can take advantage of the special structure of $\mathbf{A} = \mathbf{I} - \mathbf{T}$ and devise an efficient algorithm by which to compute $\mathbf{A}^{\#}$.

Theorem 8.5.1 *Let* $\mathbf{T}$ *be the transition matrix of an m-state ergodic chain and let* $\mathbf{A} = \mathbf{I} - \mathbf{T}$. *Every principal submatrix of* $\mathbf{A}$ *of size* $k \times k$, $k = 1, 2, 3, \ldots, m-1$, *is non-singular. Furthermore, the inverse of each principal submatrix of* $\mathbf{A}$ *is a non-negative matrix.*

Proof Since $\mathbf{T}$ is the transition matrix of an ergodic chain, $\mathbf{T}$ is irreducible and $\rho(\mathbf{T}) = 1$. Let $\tilde{\mathbf{A}}$ be any $k \times k$ submatrix of $\mathbf{A}$, where $1 \leq k \leq m - 1$, so that $\tilde{\mathbf{A}} = \mathbf{I} - \tilde{\mathbf{T}}$ where $\tilde{\mathbf{T}}$ is a $k \times k$ principal submatrix of $\mathbf{T}$. It is well known that $\rho(\tilde{\mathbf{T}}) < \rho(\mathbf{T}) = 1$. Thus, $\tilde{\mathbf{A}}$ must be non-singular. For every matrix $\mathbf{B} \geq \mathbf{0}$, it is known that $\rho(\mathbf{B}) < 1$ if and only if $(\mathbf{I} - \mathbf{B})^{-1}$ exists and $(\mathbf{I} - \mathbf{B})^{-1} \geq \mathbf{0}$. Therefore, we can conclude that $(\tilde{\mathbf{A}})^{-1} \geq \mathbf{0}$ since $\tilde{\mathbf{T}} \geq \mathbf{0}$. ■

As a direct consequence of this theorem, we obtain the following result.

Corollary 8.5.1 *If* $\mathbf{T}$ *is the transition matrix of an m-state ergodic chain and we write* $\mathbf{A}$ *as*

$$\mathbf{A} = \mathbf{I} - \mathbf{T} = \left[\begin{array}{c|c} \mathbf{U} & \mathbf{c} \\ \hline \mathbf{d}^* & \alpha \end{array}\right]$$

where $\mathbf{U} \in \mathbb{R}^{(m-1)\times(m-1)}, \mathbf{c} \in \mathbb{R}^{m-1}, \mathbf{d} \in \mathbb{R}^{m-1}$, *and* $\alpha \in \mathbb{R}$, *then* $\mathbf{U}^{-1}$ *exists and* $\mathbf{U}^{-1} \geq \mathbf{0}$.

Using this result, it is now possible to give a useful formula for obtaining $\mathbf{A}^{\#}$.

Theorem 8.5.2 *For an m-state ergodic chain, write* $\mathbf{A}$ *as*

$$\mathbf{A} = \left[\begin{array}{c|c} \mathbf{U} & \mathbf{c} \\ \hline \mathbf{d}^* & \alpha \end{array}\right]$$

where $\mathbf{U} \in \mathbb{R}^{(m-1)\times(m-1)}, \mathbf{c} \in \mathbb{R}^{m-1}, \mathbf{d} \in \mathbb{R}^{m-1}$, *and* $\alpha \in \mathbb{R}$. *Adopt the following notation:* $\mathbf{h}^* = \mathbf{d}^*\mathbf{U}^{-1}, \delta = -\mathbf{h}^*\mathbf{U}^{-1}\mathbf{j}, \beta = 1 - \mathbf{h}^*\mathbf{j}$, *and* $\mathbf{F} = \mathbf{U}^{-1} - \dfrac{\delta}{\beta}\mathbf{I}$. *Then* $\delta > 0$, $\beta > 1$, *and* $\mathbf{A}^{\#}$ *is given by*

$$\mathbf{A}^{\#} = \left[\begin{array}{c|c} \mathbf{U}^{-1} + \dfrac{\mathbf{U}^{-1}\mathbf{j}\mathbf{h}^*\mathbf{U}^{-1}}{\delta} - \dfrac{\mathbf{F}\mathbf{j}\mathbf{h}^*\mathbf{F}}{\delta} & -\dfrac{\mathbf{F}\mathbf{j}}{\beta} \\ \hline \dfrac{\mathbf{h}^*\mathbf{F}}{\beta} & \dfrac{\delta}{\beta^2} \end{array}\right].$$

Proof To show $\delta > 0$ and $\beta > 1$, observe that $\mathbf{d}^* \leq \mathbf{0}$. From Corollary 5.1, it follows that $\mathbf{U}^{-1} \geq \mathbf{0}$ and hence $\mathbf{h}^* \leq \mathbf{0}$. However, $\mathbf{h}^*$ is not the zero vector. Otherwise, if $\mathbf{h}^* = \mathbf{0}$ then $\mathbf{d}^* = \mathbf{0}$, because $\mathbf{U}$ is non-singular, and this would imply that $\mathbf{T}$ is reducible, which is impossible. It now follows that $\delta > 0$ and $\beta > 1$. To show the validity of (1), first note that $(\mathbf{I} - \mathbf{j}\mathbf{h}^*)^{-1}$

$= \mathbf{I} + \mathbf{jh}^*/\beta$. Let $\mathbf{H} = (\mathbf{I} - \mathbf{jh}^*)^{-1}\mathbf{U}^{-1}(\mathbf{I} - \mathbf{jh}^*)^{-1}$ so that

$$\mathbf{H} = \mathbf{U}^{-1} + \frac{\mathbf{U}^{-1}\mathbf{jh}^*}{\beta} + \frac{\mathbf{jh}^*\mathbf{U}^{-1}}{\beta} - \frac{\delta\mathbf{jh}^*}{\beta^2}. \tag{2}$$

By a direct calculation, using (2), it is easy to see that $\mathbf{U}^{-1} + \dfrac{\mathbf{U}^{-1}\mathbf{jh}^*\mathbf{U}^{-1}}{\delta} - \dfrac{\mathbf{Fjh}^*\mathbf{F}}{\delta} = \mathbf{H}$. Likewise, use (2) to show that $\dfrac{\mathbf{h}^*\mathbf{F}}{\beta} = \mathbf{h}^*\mathbf{H}$, $\dfrac{\mathbf{Fj}}{\beta} = \mathbf{Hj}$, and $\dfrac{\delta}{\beta^2} = -\mathbf{h}^*\mathbf{Hj}$. Therefore, the matrix on the right hand side of (1) can be written as

$$\left[\begin{array}{c|c} \mathbf{H} & -\mathbf{Hj} \\ \hline \mathbf{h}^*\mathbf{H} & -\mathbf{h}^*\mathbf{Hj} \end{array}\right]. \tag{3}$$

Since $\operatorname{rank}(\mathbf{A}) = \operatorname{rank}(\mathbf{U}) = m - 1$, $\mathbf{A}$ can be written as

$$\mathbf{A} = \left[\begin{array}{c|c} \mathbf{U} & \mathbf{Uk} \\ \hline \mathbf{h}^*\mathbf{U} & \mathbf{h}^*\mathbf{Uk} \end{array}\right]$$

where $\mathbf{k} = \mathbf{U}^{-1}\mathbf{c}$. From Theorem 7.7.6, $\mathbf{A}^\#$ is given by

$$\mathbf{A}^\# = \left[\begin{array}{c|c} \mathbf{H} & \mathbf{Hk} \\ \hline \mathbf{h}^*\mathbf{H} & \mathbf{h}^*\mathbf{Hk} \end{array}\right]. \tag{4}$$

Because the row sums of $\mathbf{A}$ are all zero, it follows that $\mathbf{Uj} + \mathbf{c} = \mathbf{0}$ and hence $\mathbf{j} = -\mathbf{U}^{-1}\mathbf{c} = -\mathbf{k}$ so that the matrices in (3) and (4) are equal. ■

For an ergodic chain, if one desires to compute the fixed probability vector $\mathbf{w}^*$, one does not need to know the entire matrix $\mathbf{A}^\#$. As demonstrated in Theorem 2.3, knowledge of any single row of $\mathbf{A}^\#$ is sufficient. Theorem 5.2 provides an easily obtainable row, namely the last one, and thereby provides one with a relatively simple way of computing $\mathbf{w}^*$.

Theorem 8.5.3 *If* $\mathbf{T}$ *is the transition matrix of an m-state ergodic chain and* $\mathbf{A} = \mathbf{I} - \mathbf{T}$ *is partitioned as*

$$\mathbf{A} = \left[\begin{array}{c|c} \mathbf{U} & \mathbf{c} \\ \hline \mathbf{d}^* & \alpha \end{array}\right] = \left[\begin{array}{c|c} \mathbf{U} & -\mathbf{Uj} \\ \hline \mathbf{d}^* & -\mathbf{d}^*\mathbf{j} \end{array}\right]$$

where $\mathbf{U} \in \mathbb{R}^{(m-1)\times(m-1)}$, *then the fixed probability vector of* $\mathbf{T}$ *is given by*

$$\mathbf{w}^* = \frac{1}{\beta}\left[\begin{array}{c|c} -\mathbf{h}^* & 1 \end{array}\right] = \frac{1}{\beta}\left[\begin{array}{c|c} -\mathbf{d}^*\mathbf{U}^{-1} & 1 \end{array}\right]$$

where $\beta = 1 - \mathbf{h}^*\mathbf{j} = 1 - \mathbf{d}^*\mathbf{U}^{-1}\mathbf{j}$.

Proof From Theorem 2.3, $\mathbf{w}^* = \mathbf{e}_m^* - \mathbf{r}_m^*\mathbf{A}$, where $\mathbf{r}_m^*$ is the last row of $\mathbf{A}^\#$. From (1), $\mathbf{r}_m^* = \dfrac{1}{\beta}\left[\begin{array}{c|c} \mathbf{h}^*\mathbf{F} & \dfrac{\delta}{\beta} \end{array}\right]$ so that $\mathbf{r}_m^*\mathbf{A} = \dfrac{1}{\beta}\left[\begin{array}{c|c} \mathbf{h}^* & \mathbf{h}^*\mathbf{j} \end{array}\right]$. Therefore

$$\mathbf{w}^* = \mathbf{e}_m^* - \mathbf{r}_m^*\mathbf{A} = \frac{1}{\beta}\left[\begin{array}{c|c} -\mathbf{h}^* & 1 \end{array}\right].$$ ■

From a computational standpoint it is important to point out that when Theorem 5.3 is used to compute $\mathbf{w}^*$, it is not necessary to explicitly calculate $\mathbf{U}^{-1}$, just as it is not necessary to explicitly calculate $\mathbf{C}^{-1}$ in order to solve a non-singular system of linear equations $\mathbf{Cx} = \mathbf{b}$. Indeed, one may consider the vector $\mathbf{h}$ to be the solution of the non-singular system $\mathbf{U}^*\mathbf{x} = \mathbf{d}$ and proceed with the solution of the system by conventional methods.

Corollary 8.5.2 *For an ergodic chain,*

$$\mathbf{I} - \mathbf{AA}^{\#} = \frac{1}{\beta}\left[\begin{array}{c|c} -\mathbf{jh}^* & \mathbf{j} \\ \hline -\mathbf{h}^* & 1 \end{array}\right], \text{ and } \mathbf{AA}^{\#} = \left[\begin{array}{c|c} \mathbf{I} + \dfrac{\mathbf{jh}^*}{\beta} & \dfrac{-\mathbf{j}}{\beta} \\ \hline \dfrac{\mathbf{h}^*}{\beta} & 1 - 1/\beta \end{array}\right]$$

Proof The first result follows since $\mathbf{W} = \mathbf{I} - \mathbf{AA}^{\#}$; the second follows from the first. ■

As pointed out above, for an ergodic chain it is not necessary to explicitly compute $\mathbf{A}^{\#}$, or even $\mathbf{U}^{-1}$, in order to compute the fixed probability vector $\mathbf{w}^*$. But if one knows $\mathbf{A}^{\#}$ or $\mathbf{U}^{-1}$, then $\mathbf{w}^*$ is readily available. However, it is just the reverse that is often encountered in applications. That is, by theoretical considerations or perhaps by previous experience or experimentation, one knows what the fixed probability vector, $\mathbf{w}^*$, or the limiting matrix $\mathbf{W}$, has to be. In order to obtain some of the information about the chain which was discussed in the previous section, such as the mean first passage matrix, the matrix $\mathbf{A}^{\#}$ needs to be computed. It seems reasonable to try to use the already known information about $\mathbf{w}^*$, or $\mathbf{W}$ in order to obtain $\mathbf{A}^{\#}$ rather than starting from scratch and using only the knowledge of $\mathbf{A}$ and formula (1). The next theorem shows how this can be done.

Before stating that result, we give a very simple example of a situation where the fixed probability vector is known beforehand.

Example 8.5.1 Consider an m-state ergodic chain which is 'symmetric'. That is, the one-step probabilities satisfy $p_{ij} = p_{ji}$ for each $i, j = 1, 2, \ldots, m$. This implies that the transition matrix $\mathbf{T}$ is a doubly stochastic matrix. In particular, $\mathbf{j}^*\mathbf{T} = \mathbf{j}^*$. The vector $\mathbf{j}^* > \mathbf{0}$ is a fixed vector for $\mathbf{T}$ but it is not a probability vector. However this can easily be fixed by multiplying by $\dfrac{1}{m}$. Now,

$$\frac{1}{m}\mathbf{j}^* = \left(\frac{1}{m}\mathbf{j}^*\right)\mathbf{T}, \quad \frac{1}{m}\mathbf{j}^* > \mathbf{0}, \text{ and } \frac{1}{m}\mathbf{j}^*\mathbf{j} = 1.$$

Since the fixed probability vector is unique, it must be the case that $\mathbf{w}^* = \dfrac{1}{m}\mathbf{j}^*$. Thus a symmetric ergodic chain is an example (which occurs frequently) of a situation where $\mathbf{w}^*$ is known beforehand.

Let us now return to the problem of finding $\mathbf{A}^\#$ when $\mathbf{w}^*$, or $\mathbf{W}$ is already known. If $\mathbf{W}$ is known, then $\mathbf{AA}^\#$ is known since $\mathbf{AA}^\# = \mathbf{I} - \mathbf{W}$. The matrix $\mathbf{AA}^\#$ can be used to obtain $\mathbf{A}^\#$ as follows.

Theorem 8.5.4 *For an m-state ergodic chain with transition matrix* $\mathbf{T}$, *write* $\mathbf{A} = \mathbf{I} - \mathbf{T}$ *as*

$$\mathbf{A} = \left[\begin{array}{c|c} \mathbf{U} & \mathbf{c} \\ \hline \mathbf{d} & \alpha \end{array}\right]$$

where $\mathbf{U} \in \mathbb{R}^{(m-1)\times(m-1)}$. *The matrix* $\mathbf{A}^\#$ *is given by*

$$\mathbf{A}^\# = \mathbf{AA}^\# \left[\begin{array}{c|c} \mathbf{U}^{-1} & \mathbf{0} \\ \hline \mathbf{0} & 0 \end{array}\right] \mathbf{AA}^\#. \tag{5}$$

Proof If $\mathbf{A}^-$ is any (1)-inverse for $\mathbf{A}$ then $\mathbf{AA}^\#\mathbf{A}^-\mathbf{AA}^\# = \mathbf{A}^\#\mathbf{AA}^-\mathbf{AA}^\# = \mathbf{A}^\#\mathbf{AA}^\# = \mathbf{A}^\#$. But $\mathbf{X} = \left[\begin{array}{c|c} \mathbf{U}^{-1} & \mathbf{0} \\ \hline \mathbf{0} & 0 \end{array}\right]$ is a (1)-inverse for $\mathbf{A}$ since $\alpha = \mathbf{d}^*\mathbf{U}^{-1}\mathbf{c}$. ■

Frequently, one may wish to check a computed inverse. If the matrix under question is non-singular, one can compare the products of the original matrix by the computed inverse with the identity matrix. However, if the matrix $\mathbf{A}$ has index 1 and a computed $\mathbf{A}^\#$ is checked by comparing the product $\mathbf{AA}^\#\mathbf{A}$ with $\mathbf{A}$, $\mathbf{AA}^\#$ with $\mathbf{A}^\#\mathbf{A}$, and $\mathbf{A}^\#\mathbf{AA}^\#$ with $\mathbf{A}^\#$, then one has probably done more work doing the 'check' than in computing the original quantity. Since the number of arithmetic operations necessary to form the indicated matrix products is relatively large, it is possible that factors such as roundoff error can render the 'check' almost useless and leave the investigator totally unsure about the quantity he has computed. The next theorem provides an alternate means by which one can 'check' a computed $\mathbf{A}^\#$ for an ergodic chain.

For an ergodic chain, the fixed probability vector $\mathbf{w}^*$ (or equivalently, the limiting matrix $\mathbf{W}$) is either known from theoretical considerations, or else can usually be computed without much difficulty by simply proceeding as suggested in Theorem 5.3. Furthermore, iterative improvement techniques work well for producing very accurate numerical solutions for $\mathbf{w}^*$. For a general chain, one can use the second part of Theorem 2.2 to obtain $\mathbf{W}$. Once one has confidently obtained $\mathbf{w}^*$ (or $\mathbf{W}$), then a computed $\mathbf{A}^\#$ can be checked by using the following result.

Theorem 8.5.5 *For any chain with limiting matrix* $\mathbf{W}$, $\mathbf{A}^\#$ *can be characterized as the unique solution* $\mathbf{X}$ *of the two equations* $\mathbf{WX} = \mathbf{0}$ *and* $\mathbf{AX} = \mathbf{I} - \mathbf{W}$.

The proof is by direct substitution. If the chain is ergodic, then $\mathbf{WX} = \mathbf{0}$ can be replaced by $\mathbf{w}^*\mathbf{X} = \mathbf{0}^*$.

This theorem provides a method by which one can have some confidence in a computed value for $\mathbf{A}^{\#}$ since if $\mathbf{X}$ is a matrix such that $\mathbf{WX}$ is 'close' to $\mathbf{0}$ and $\mathbf{AX}$ is 'close' to $\mathbf{I}-\mathbf{W}$, then $\mathbf{X}$ must be 'close' to $\mathbf{A}^{\#}$. More precisely, we have the following.

Theorem 8.5.6 *For any chain, if* $\{\mathbf{X}_n\}$ *is a sequence of matrices such that* $\mathbf{WX}_n \to \mathbf{0}$ *and* $\mathbf{AX}_n \to \mathbf{I}-\mathbf{W}$, *then* $\mathbf{X}_n \to \mathbf{A}^{\#}$.

Proof $\mathbf{A}^{\#}-\mathbf{X}_n = \mathbf{AA}^{\#}(\mathbf{A}^{\#}-\mathbf{X}_n) - \mathbf{WX}_n = \mathbf{A}^{\#}[\mathbf{A}(\mathbf{A}^{\#}-\mathbf{X}_n)] - \mathbf{WX}_n \to \mathbf{0}$. ■

As with Theorem 5, if the chain is ergodic, then $\mathbf{WX}_n \to \mathbf{0}$ can be replaced by $\mathbf{w}^*\mathbf{X}_n \to \mathbf{0}^*$.

6. Non-ergodic chains and absorbing chains

In this section we will show that $\mathbf{A}^{\#}$ and $\mathbf{I}-\mathbf{AA}^{\#}$ can be useful in extracting information from non-ergodic chains. We first consider the problem of classifying the states as being either ergodic or transient. For chains with large numbers of states the problem of classifying the states, or equivalently, putting the transition matrix in a 'canonical' block triangular form is non-trivial. The following theorem shows how the projection $\mathbf{I}-\mathbf{AA}^{\#}$ can be used to classify the states. (As before $\mathbf{A}=\mathbf{I}-\mathbf{T}$ where $\mathbf{T}$ is the transition matrix.)

Theorem 8.6.1 *For a general chain, state* $\mathscr{S}_i$ *is a transient state if and only if the ith column of* $\mathbf{I}-\mathbf{AA}^{\#}$ *is entirely zero. Equivalently,* $\mathscr{S}_i$ *is an ergodic state if and only if the ith column contains at least one non-zero entry.*

Proof Perform the necessary permutations so that $\mathbf{T}$ has the form (1) of Section 2. Then all transient states are listed before any ergodic states, with the partition as indicated in (1) of Section 2. As argued in Theorem 2.1, $\mathbf{A}$ has the form

$$\mathbf{A} = \mathbf{P}^*\left[\begin{array}{c|c} \mathbf{G}_{11} & \mathbf{G}_{12} \\ \hline \mathbf{0} & \mathbf{G}_{22} \end{array}\right]\mathbf{P}$$

where $\mathbf{G}_{11}$ is non-singular, $\operatorname{Ind}(\mathbf{G}_{22})=1$, and $\mathbf{P}$ is a permutation matrix. By using Theorem 7.7.3, $\mathbf{I}-\mathbf{AA}^{\#}$ is seen to have the form

$$\mathbf{I}-\mathbf{AA}^{\#} = \mathbf{P}^*\left[\begin{array}{c|c} \mathbf{0} & \mathbf{X} \\ \hline \mathbf{0} & \mathbf{I}-\mathbf{G}_{22}\mathbf{G}_{22}^{\#} \end{array}\right]\mathbf{P}. \tag{1}$$

Furthermore, every column of $\mathbf{I}-\mathbf{G}_{22}\mathbf{G}_{22}^{\#}$ contains at least one non-zero entry since

$$\mathbf{G}_{22} = \begin{bmatrix} \mathbf{I}-\mathbf{T}_{r+1,r+1} & & \mathbf{0} \\ & \ddots & \\ \mathbf{0} & & \mathbf{I}-\mathbf{T}_{nn} \end{bmatrix} \tag{2}$$

and each $\mathbf{T}_{ii}$ $(i > r)$ is the transition matrix of an ergodic chain. Hence $\mathbf{I} - \mathbf{A}_{ii}\mathbf{A}_{ii}^{\#} = \mathbf{I} - (\mathbf{I} - \mathbf{T}_{ii})(\mathbf{I} - \mathbf{T}_{ii}^{\#}) > \mathbf{0}$ because it is the limiting matrix of the chain associated with $\mathbf{T}_{ii}$ and it is well known that the limiting matrix of an ergodic chain is strictly positive. ■

$\mathbf{I} - \mathbf{A}\mathbf{A}^{\#}$ can provide a distinction between the transient states and the ergodic states and it can completely solve the problem of determining the ergodic sets.

Theorem 8.6.2 *For a general chain, if states $\mathscr{S}_i$ and $\mathscr{S}_k$ belong to the same ergodic set, iff the ith and kth rows of $\mathbf{I} - \mathbf{A}\mathbf{A}^{\#}$ are equal.*

Proof Write the transition matrix in the form (1) of Section 2 so that (1) holds. The desired result follows from

$$\mathbf{I} - \mathbf{G}_{22}\mathbf{G}_{22}^{\#} = \begin{bmatrix} \mathbf{I} - \mathbf{A}_{r+1,r+1}\mathbf{A}_{r+1,r+1}^{\#} & & \mathbf{0} \\ & \ddots & \\ \mathbf{0} & & \mathbf{I} - \mathbf{A}_{nn}\mathbf{A}_{nn}^{\#} \end{bmatrix}$$

because each of the rows of $\mathbf{I} - \mathbf{A}_{ii}\mathbf{A}_{ii}^{\#}$ are identical $(i > r)$ since each $\mathbf{I} - \mathbf{A}_{ii}\mathbf{A}_{ii}^{\#}$ represents the limiting matrix of an ergodic chain. ■

For a general chain with more than one ergodic set, the elements of $\mathbf{I} - \mathbf{A}\mathbf{A}^{\#}$ can be used to obtain the probabilities of eventual absorption into any one particular ergodic set for each possible starting state.

Theorem 8.6.3 *For a general chain, let $[\mathscr{S}_k]$ denote the equivalence class (ergodic set) determined by the ergodic state $\mathscr{S}_k$. Let $\mathscr{T}_k$ denote the set of indices of those states which belong to $[\mathscr{S}_k]$. If $\mathscr{S}_i$ is a transient state, then*

$$\mathrm{P}\,(\textit{eventual absorption into } [\mathscr{S}_k] \mid \textit{initially in } \mathscr{S}_i) = \sum_{l \in \mathscr{T}_k} (\mathbf{I} - \mathbf{A}\mathbf{A}^{\#})_{il}.$$

Proof Permute the states so that $\mathbf{T}$ has the form (1) of Section 2. Replace each $\mathbf{T}_{ii}, i > r$, by an identity matrix and call the resulting matrix $\tilde{\mathbf{T}}$. From Theorem 2.1 we know that $\rho(\mathbf{T}_{ii}) < 1$ for $i = 1, 2, \ldots, r$ so that $\lim_{n \to \infty} \tilde{\mathbf{T}}^n$ exists and therefore must be given by $\lim_{n \to \infty} \tilde{\mathbf{T}}^n = \mathbf{I} - \tilde{\mathbf{A}}\tilde{\mathbf{A}}^{\#}$ where $\tilde{\mathbf{A}} = \mathbf{I} - \tilde{\mathbf{T}}$. This modified chain is clearly an absorbing chain and the probability of eventual absorption into state $\tilde{\mathscr{S}}_k$ when initially in $\tilde{\mathscr{S}}_i$ is given by $\left(\lim_{n \to \infty} \tilde{\mathbf{T}}^n\right)_{ik} = (\mathbf{I} - \tilde{\mathbf{A}}\tilde{\mathbf{A}}^{\#})_{ik}$. From this, it should be clear that in the original chain the probability of eventual absorption into the set $[\mathscr{S}_k]$ is simply the sum over $\mathscr{T}_k$ of the absorption probabilities. That is,

$$\mathrm{P}\,(\text{eventual absorption into } [\mathscr{S}_k] \mid \text{initially in } \mathscr{S}_i) = \sum_{l \in \mathscr{T}_k} (\mathbf{I} - \tilde{\mathbf{A}}\tilde{\mathbf{A}}^{\#})_{il}. \quad (3)$$

We must now show that the $(\tilde{\ })$ can be eliminated from (3). In order to do this, write $\mathbf{A}$ and $\tilde{\mathbf{A}}$ as

$$\tilde{\mathbf{A}} = \left[\begin{array}{c|c} \mathbf{G}_{11} & \mathbf{G}_{12} \\ \hline \mathbf{0} & \mathbf{0} \end{array}\right] \text{ and } \mathbf{A} = \left[\begin{array}{c|c} \mathbf{G}_{11} & \mathbf{G}_{12} \\ \hline \mathbf{0} & \mathbf{G}_{22} \end{array}\right].$$

Theorem 2.1 guarantees that $\mathbf{G}_{11}$ is non-singular and Theorem 7.7.3 yields

$$\mathbf{I} - \tilde{\mathbf{A}}\tilde{\mathbf{A}}^{\#} = \left[\begin{array}{c|c} \mathbf{0} & -\mathbf{G}_{11}^{-1}\mathbf{G}_{12} \\ \hline \mathbf{0} & \mathbf{I} \end{array}\right] \text{ and }$$

$$\mathbf{I} - \mathbf{A}\mathbf{A}^{\#} = \left[\begin{array}{c|c} \mathbf{0} & -\mathbf{G}_{11}^{-1}\mathbf{G}_{12}(\mathbf{I} - \mathbf{G}_{22}\mathbf{G}_{22}^{\#}) \\ \hline \mathbf{0} & \mathbf{I} - \mathbf{G}_{22}\mathbf{G}_{22}^{\#} \end{array}\right]$$

When $\mathbf{T}$ is in the form (1) of Section 2, the set of indicies $\mathscr{T}_k$ will be sequential. Let $\mathscr{T}_k = \{h, h+1, h+2, \ldots h+t\}$. Partition $\mathbf{I} - \tilde{\mathbf{A}}\tilde{\mathbf{A}}^{\#}$ and $\mathbf{I} - \mathbf{A}\mathbf{A}^{\#}$ as follows:

$$
\mathbf{I} - \tilde{\mathbf{A}}\tilde{\mathbf{A}}^{\#} =
\left[\begin{array}{ccc|ccccc}
\mathbf{0} & \cdots & \mathbf{0} & \mathbf{W}_{1,r+1} & \cdots & \widetilde{\mathbf{W}_{1q}} & \cdots & \mathbf{W}_{1n} \\
\vdots & & \vdots & & & & & \\
\mathbf{0} & \cdots & \mathbf{0} & \mathbf{W}_{p,r+1} & \cdots & \mathbf{W}_{pq} & \cdots & \mathbf{W}_{pn} \\
\vdots & & \vdots & & & & & \\
\mathbf{0} & \cdots & \mathbf{0} & \mathbf{W}_{r,r+1} & \cdots & \mathbf{W}_{rq} & \cdots & \mathbf{W}_{rn} \\
\hline
\mathbf{0} & \cdots & \mathbf{0} & \mathbf{I} & & \mathbf{0} & \cdots & \mathbf{0} \\
\vdots & & \vdots & & & & & \\
\mathbf{0} & \cdots & \mathbf{0} & \mathbf{0} & & \mathbf{I} & \cdots & \mathbf{0} \\
\vdots & & \vdots & & & & & \\
\mathbf{0} & \cdots & \mathbf{0} & \mathbf{0} & & \mathbf{0} & \cdots & \mathbf{I}
\end{array}\right]
\quad (4)
$$

columns $h, h+1, \ldots h+t$

$\}\leftarrow$ row i is in here

and

$$
\mathbf{I} - \mathbf{A}\mathbf{A}^{\#} =
\left[\begin{array}{ccc|ccccc}
\mathbf{0} \cdots \mathbf{0} & \mathbf{W}_{1,r+1}\mathbf{W}_{r+1,r+1} & \cdots & \widetilde{\mathbf{W}_{1q}\mathbf{W}_{qq}} & \cdots & \mathbf{W}_{1n}\mathbf{W}_{nn} \\
\mathbf{0} \cdots \mathbf{0} & \mathbf{W}_{p,r+1}\mathbf{W}_{r+1,r+1} & \cdots & \mathbf{W}_{pq}\mathbf{W}_{qq} & \cdots & \mathbf{W}_{pn}\mathbf{W}_{nn} \\
\mathbf{0} \cdots \mathbf{0} & \mathbf{W}_{r,r+1}\mathbf{W}_{r+1,r+1} & \cdots & \mathbf{W}_{rq}\mathbf{W}_{qq} & \cdots & \mathbf{W}_{rn}\mathbf{W}_{nn} \\
\hline
\mathbf{0} \cdots \mathbf{0} & \mathbf{W}_{r+1,r+1} & \cdots & \mathbf{0} & \cdots & \mathbf{0} \\
\mathbf{0} \cdots \mathbf{0} & \mathbf{0} & \cdots & \mathbf{W}_{qq} & \cdots & \mathbf{0} \\
\mathbf{0} \cdots \mathbf{0} & \mathbf{0} & \cdots & \mathbf{0} & \cdots & \mathbf{W}_{nn}
\end{array}\right]
\quad (5)
$$

columns $h, h+1, \ldots h+1$

row $\leftarrow i$ is in here

where $\mathbf{W}_{ii} = \mathbf{I} - \mathbf{A}_{ii}\mathbf{A}_{ii}^{\#}$. Suppose the ith row of $\mathbf{I} - \mathbf{A}\mathbf{A}^{\#}$ lies along the gth row of $\mathbf{W}_{pq}$. If P denotes the probability given in (3), then it is clear that P is given by

$$\mathrm{P} = g\text{th row sum of } \mathbf{W}_{pq}. \tag{6}$$

It is also evident that

$$\sum_{l \in \mathcal{T}_k} (\mathbf{I} - \mathbf{A}\mathbf{A}^{\#})_{il} = g\text{th row sum of } \mathbf{W}_{pq}\mathbf{W}_{qq}. \tag{7}$$

Since $\mathbf{W}_{qq}$ is the limiting matrix of $\mathbf{T}_{qq}$ and $\mathbf{T}_{qq}$ is the transition matrix of an ergodic chain, it follows that the rows of $\mathbf{W}_{qq}$ are identical and sum to 1. Therefore, the gth row sum of $\mathbf{W}_{pq}\mathbf{W}_{qq} = g$th row sum of $\mathbf{W}_{pq}$, and the desired result now follows by virtue of (6) and (7). ■

Theorem 8.6.4 *If $\mathcal{S}_i$ is a transient state and $\mathcal{S}_k$ is an ergodic state in an ergodic class $[\mathcal{S}_k]$ which is regular, then the limiting value for the nth step transition probability from $\mathcal{S}_i$ to $\mathcal{S}_k$ is given by $\lim_{n\to\infty} p_{ik}^{(n)} = (\mathbf{I} - \mathbf{A}\mathbf{A}^{\#})_{ik}$.*

Proof It is not hard to see that $\lim_{n\to\infty} p_{ik}^{(n)} = \alpha_{ik} w_{\mathcal{S}_k}$ where $\alpha_{ik} = \mathrm{P}$ (eventual absorption into $[\mathcal{S}_k] \mid$ initially in $\mathcal{S}_i$) and $w_{\mathcal{S}_k}$ is the component of the fixed probability vector associated with $[\mathcal{S}_k]$ corresponding to the state $\mathcal{S}_k$. Suppose $[\mathcal{S}_k]$ corresponds to $\mathbf{T}_{qq}$ when the transition matrix $\mathbf{T}$ is written in the form (1) of Section 2, and suppose the ith row of $\mathbf{I} - \mathbf{A}\mathbf{A}^{\#}$ lies along the gth row of the block $\mathbf{W}_{pq}\mathbf{W}_{qq}$ in (5). The kth column of $\mathbf{I} - \mathbf{A}\mathbf{A}^{\#}$ must therefore lie along one of the columns of $\mathbf{W}_{pq}\mathbf{W}_{qq}$, say the fth one, so that we can use (6) to obtain $(\mathbf{I} - \mathbf{A}\mathbf{A}^{\#})_{ik} = (\mathbf{W}_{pq}\mathbf{W}_{qq})_{gf} = (g\text{th row sum of } \mathbf{W}_{pq}) \times w_{\mathcal{S}_k} = \alpha_{ik} w_{\mathcal{S}_k}$. ■

The elements of $\mathbf{A}^{\#}$ itself contain important information about a general chain with more than one transient state.

Theorem 8.6.5 *If $\mathcal{S}_i$ and $\mathcal{S}_k$ are transient states, then $(\mathbf{A}^{\#})_{ik}$ is the expected number of times in $\mathcal{S}_k$ when initially in $\mathcal{S}_i$. Furthermore $\mathcal{S}_i$ and $\mathcal{S}_k$ belong to the same transient set iff $(\mathbf{A}^{\#})_{ik} > 0$ and $(\mathbf{A}^{\#})_{ki} > 0$*

Proof Permute the states so that $\mathbf{T}$ has the form (1) of Section 2 so that

$$\mathbf{T} = \left[\begin{array}{c|c} \mathbf{Q} & \mathbf{R} \\ \hline \mathbf{0} & \mathbf{E} \end{array}\right] \text{ and } \mathbf{A} = \left[\begin{array}{c|c} \mathbf{I} - \mathbf{Q} & -\mathbf{R} \\ \hline \mathbf{0} & \mathbf{I} - \mathbf{E} \end{array}\right]$$

where $\rho(\mathbf{Q}) < 1$. Notice that $\mathbf{T}_{ik} = \mathbf{Q}_{ik}$ because $\mathcal{S}_i$ and $\mathcal{S}_k$ are both transient states. By using the fact that $\left(\sum_{l=0}^{n-1} \mathbf{Q}^l\right)_{ik} = \left(\sum_{l=0}^{n-1} \mathbf{T}^l\right)_{ik}$ is the expected number of times in $\mathcal{S}_k$ in n steps when initially in $\mathcal{S}_i$, it is easy to see that the expected number of times in $\mathcal{S}_k$ when initially in $\mathcal{S}_i$ is

$$\lim_{n\to\infty} \left(\sum_{l=0}^{n-1} \mathbf{Q}^l\right)_{ik} = [(\mathbf{I} - \mathbf{Q})^{-1}]_{ik} = (\mathbf{A}^{\#})_{ik}. \quad ■$$

Theorem 8.6.6 *For a general chain, let $\mathcal{T}$ denote the set of indices* *corresponding to the transient states. If the chain is initially in the transient states $\mathscr{S}_i$, then $\sum_{k\in\mathscr{T}}(\mathbf{A}^{\#})_{ik}$ is the expected number of times the chain is in a transient state. (Expected number of times to reach an ergodic set.)*

Proof If a permutation is performed so that the transition matrix has the form (1) of Section 2, then

$$\mathbf{T}=\left[\begin{array}{c|c}\mathbf{Q} & \mathbf{R}\\ \hline \mathbf{0} & \mathbf{E}\end{array}\right], \rho(\mathbf{Q})<1.$$

If $\mathbf{Q}$ is $r\times r$, then $\mathscr{T}=\{1,2,\ldots,r\}$ and the previous theorem implies that $\sum_{k\in\mathscr{T}}(\mathbf{A}^{\#})_{ik}=\sum_{k=1}^{r}(\mathbf{I}-\mathbf{Q})^{-1}_{ik}$ is the expected number of times the chain is in a transient state when initially in $\mathscr{S}_i$. ■

Theorem 8.6.7 *If $\mathscr{S}_i$ and $\mathscr{S}_k$ are transient states, then*

$$(\mathbf{A}^{\#}[2\mathbf{A}_d^{\#}-\mathbf{I}]-\mathbf{A}_S^{\#})_{ik}=\textit{Variance of the number of times in } \mathscr{S}_k \textit{ when initially in } \mathscr{S}_i$$

and

$$\sum_{k\in\mathscr{T}}([2\mathbf{A}^{\#}-\mathbf{I}]\mathbf{A}^{\#}-\mathbf{A}_S^{\#})_{ik}=\textit{Variance of the number of times the chain is in a transient state when initially in } \mathscr{S}_i$$

where $\mathscr{T}$ is the set of indices corresponding to the transient states and $\mathbf{A}_d^{\#}$ and $\mathbf{A}_S^{\#}$ are as described in Theorems 4.1 *and* 4.4.

The proof is left as an exercise.

As direct corollaries of the above theorems, we obtain as special (but extremely useful) cases the following results about absorbing chains.

Corollary 8.6.1 *If $\mathbf{T}$ is the transition matrix for an absorbing chain, then the following statements are true.*

(i) *If $\mathscr{S}_k$ is an absorbing state, then $(\mathbf{I}-\mathbf{A}\mathbf{A}^{\#})_{ik}$ is the probability of being absorbed into $\mathscr{S}_k$ when initially in $\mathscr{S}_i$.*

(ii) *If $\mathscr{S}_i$ and $\mathscr{S}_k$ are non-absorbing states, then $(\mathbf{A}^{\#})_{ik}$ is the expected number of times in $\mathscr{S}_k$ when initially in $\mathscr{S}_i$.*

(iii) *If $\mathscr{T}$ is the set of indices corresponding to the non-absorbing states, then $\sum_{k\in\mathscr{T}}(\mathbf{A}^{\#})_{ik}$ is the expected number of steps until absorption when initially in the non-absorbing state $\mathscr{S}_i$.*

(iv) *If $\mathscr{S}_i$ and $\mathscr{S}_k$ are non-absorbing states, then $(\mathbf{A}^{\#}[2\mathbf{A}_d^{\#}-\mathbf{I}]-\mathbf{A}_S^{\#})_{ik}$ is the variance of the number of times in $\mathscr{S}_k$ when initially in $\mathscr{S}_i$.*

(v) *If $\mathscr{T}$ is the set of indices corresponding to the non-absorbing states, then $\sum_{k\in\mathscr{T}}([2\mathbf{A}^{\#}-\mathbf{I}]\mathbf{A}^{\#}-\mathbf{A}_S^{\#})_{ik}$ is the variance of the numbers of steps until absorption when initially in $\mathscr{S}_i$.*

In order to analyse a chain by utilizing the classical theory, it is always necessary to first permute or relabel the states so that the transition matrix assumes the canonical form (1) of Section 2. However, by analysing the chain using $\mathbf{A}^{\#}$ and $\mathbf{I} - \mathbf{A}\mathbf{A}^{\#}$, the problem of first permuting the states may be completely avoided since all results involving $\mathbf{A}^{\#}$ or $\mathbf{I} - \mathbf{A}\mathbf{A}^{\#}$ are independent of how the states are ordered or labelled. In fact, the results of this section *help* to perform a classification of states rather than requiring that a classification previously exist.

7. References and further reading

Almost any good text on probability theory treats the subject of finite Markov chains. However, not all authors use the tool of matrix theory for their development. In [34] and [66], the reader can find a good development of the subject in terms of matrix theory. In [46], the probabilistic approach is combined with the matrix theory approach. This text can provide the reader with all the needed background necessary to read this chapter.

Only the case where the state space is finite has been considered in this chapter. The industrious student might see what he can do with the subject when the state space is countably infinite, in which case one is dealing with infinite matrices. A good place to start is by reading [47]. See also [17].

9
Applications of the Drazin inverse

1. Introduction

The previous two chapters have developed the basic theory of the Drazin inverse and the applications of a special case, the group inverse, to Markov chains. This chapter will develop the application of the Drazin inverse to singular differential and difference equations.

We shall also discuss where these singular equations occur.

2. Applications of the Drazin inverse to linear systems of differential equations

In this section, we will be concerned with systems of first order linear differential equations of the form $\mathbf{A}\dot{\mathbf{x}}(t)+\mathbf{B}\mathbf{x}(t)=\mathbf{f}(t), \mathbf{x}(t_0)=\mathbf{c}\in\mathbb{C}^n$ where $\mathbf{A},\mathbf{B}\in\mathbb{C}^{n\times n}$ and $\mathbf{x}(t)$ and $\mathbf{f}(t)$ are vector valued functions of the real variable t, and $\mathbf{f}(t)$ is continuous in some interval containing t_0. If $\mathbf{A}$ is non-singular, then the classical theory applies and one has the following situation.

(I) The general solution of the homogeneous equation,

$$\mathbf{A}\dot{\mathbf{x}}(t)+\mathbf{B}\mathbf{x}(t)=\mathbf{0},$$

is given by $\mathbf{x}(t)=\mathrm{e}^{-\mathbf{A}^{-1}\mathbf{B}t}\mathbf{q}, \mathbf{q}\in\mathbb{C}^n$.

(II) The homogeneous initial value problem, $\mathbf{A}\dot{\mathbf{x}}(t)+\mathbf{B}\mathbf{x}(t)=\mathbf{0}, \mathbf{x}(t_0)=\mathbf{c}$, has the unique solution

$$\mathbf{x}(t)=\mathrm{e}^{-\mathbf{A}^{-1}\mathbf{B}(t-t_0)}\mathbf{c}.$$

(III) The general solution of the inhomogeneous equation $\mathbf{A}\dot{\mathbf{x}}(t)+\mathbf{B}\mathbf{x}(t)=\mathbf{f}(t)$, is given by $\mathbf{x}(t)=\mathrm{e}^{-\mathbf{A}^{-1}\mathbf{B}t}\mathbf{q}+\mathbf{A}^{-1}\mathrm{e}^{-\mathbf{A}^{-1}\mathbf{B}t}\int_a^t \mathrm{e}^{\mathbf{A}^{-1}\mathbf{B}s}\mathbf{f}(s)\mathrm{d}s$, $a\in\mathbb{R}, \mathbf{q}\in\mathbb{C}^n$.

(IV) The inhomogeneous initial value problem, $\mathbf{A}\dot{\mathbf{x}}(t)+\mathbf{B}\mathbf{x}(t)=\mathbf{f}(t)$, $\mathbf{x}(t_0)=\mathbf{c}$, has the unique solution $\mathbf{x}(t)=\mathrm{e}^{-\mathbf{A}^{-1}\mathbf{B}(t-t_0)}\mathbf{c}+\mathbf{A}^{-1}\int_{t_0}^t \mathrm{e}^{-\mathbf{A}^{-1}\mathbf{B}(t-s)} \times \mathbf{f}(s)\mathrm{d}s$.

In this section, we will examine what happens in each of these problems when $\mathbf{A}$ is a singular matrix.

When $\mathbf{A}$ is a singular matrix, things can happen that are impossible when $\mathbf{A}^{-1}$ exists. For example, the homogeneous initial value problem may be inconsistent, that is, there may not exist a solution. If there is a solution, it need not be unique. The following is a simple example that illustrates this fact.

Example 9.2.1 Let $\mathbf{A}=\begin{bmatrix}0 & 1\\ 0 & 0\end{bmatrix}$ and $\mathbf{B}=\begin{bmatrix}1 & 0\\ 0 & 1\end{bmatrix}$. Then the initial value problem $\mathbf{A}\dot{\mathbf{x}}(t)+\mathbf{B}\mathbf{x}(t)=\mathbf{0}, \mathbf{x}(0)=[1,1]^*$ clearly has no solution. If $\mathbf{A}=\begin{bmatrix}0 & 1 & 0\\ 0 & 0 & 0\\ 0 & 0 & 0\end{bmatrix}$ and $\mathbf{B}=\begin{bmatrix}0 & 1 & 1\\ 0 & 0 & 0\\ 0 & 0 & 0\end{bmatrix}$ and we impose the initial condition $\mathbf{x}(0)=[1,1,1]^*=\mathbf{c}$, then it is not difficult to see that the initial value problem $\mathbf{A}\dot{\mathbf{x}}(t)+\mathbf{B}\mathbf{x}(t)=\mathbf{0}, \mathbf{x}(0)=\mathbf{c}$, has infinitely many solutions. Notice that in each of the above examples, we even have that $\mathbf{AB}=\mathbf{BA}$.

The situations illustrated above motivate the following definitions.

Definition 9.2.1 *For* $\mathbf{A},\mathbf{B}\in\mathbb{C}^{n\times n}$ *and* $t_0\in\mathbb{R}$, *the vector* $\mathbf{c}\in\mathbb{C}^n$ *is said to be a* consistent initial vector associated with t_0 for the equation $\mathbf{A}\dot{\mathbf{x}}(t)+\mathbf{B}\mathbf{x}(t)=\mathbf{f}(t)$ *when the initial value problem* $\mathbf{A}\dot{\mathbf{x}}(t)+\mathbf{B}\mathbf{x}(t)=\mathbf{f}(t), \mathbf{x}(t_0)=\mathbf{c}$, *possesses at least one solution.*

Definition 9.2.2 *The equation* $\mathbf{A}\dot{\mathbf{x}}(t)+\mathbf{B}\mathbf{x}(t)=\mathbf{f}(t)$ *is said to be* tractable at the point t_0 *if the initial value problem* $\mathbf{A}\dot{\mathbf{x}}(t)+\mathbf{B}\mathbf{x}(t)=\mathbf{f}(t), \mathbf{x}(t_0)=\mathbf{c}$ *has a unique solution for each consistent initial vector*, $\mathbf{c}$, *associated with* t_0.

If the homogeneous equation $\mathbf{A}\dot{\mathbf{x}}(t)+\mathbf{B}\mathbf{x}(t)=\mathbf{0}$ is tractable at some point $t_0\in\mathbb{R}$, then it is tractable at every $t\in\mathbb{R}$. So we may simply say the equation is tractable.

Our goals are as follows.

(i) Characterize tractable homogeneous equations.
(ii) Provide, in closed form, the general solution of every tractable homogeneous equation.
(iii) Characterize the set of consistent initial vectors for tractable homogeneous equations.
(iv) Provide, in closed form, a particular solution for the inhomogeneous equation when the homogeneous equation is tractable.
(v) Characterize the set of consistent initial vectors associated with a point t_0 for an inhomogeneous equation when the homogeneous equation is tractable.
(vi) Provide, in closed form, the unique solution of $\mathbf{A}\dot{\mathbf{x}}(t)+\mathbf{B}\mathbf{x}(t)=\mathbf{f}(t)$, $\mathbf{x}(t_0)=\mathbf{c}$ where $\mathbf{c}$ is a consistent initial vector associated with t_0 and the differential equation is tractable.

The key to accomplishing (i)–(vi) is the following two results.

Theorem 9.2.1 *For* $\mathbf{A}, \mathbf{B} \in \mathbb{C}^{n \times n}$, *the homogeneous differential equation*

$$\mathbf{A}\dot{\mathbf{x}}(t) + \mathbf{B}\mathbf{x}(t) = \mathbf{0} \tag{1}$$

is tractable if and only if there exists a scalar $\lambda \in \mathbb{C}$ *such that* $(\lambda\mathbf{A} + \mathbf{B})^{-1}$ *exists.*

Lemma 9.2.1 *Let* $\mathbf{A}, \mathbf{B} \in \mathbb{C}^{n \times n}$. *Suppose there exists a* $\lambda \in \mathbb{C}$ *such that* $(\lambda\mathbf{A} + \mathbf{B})^{-1}$ *exists, and let*

$$\hat{\mathbf{A}}_\lambda = (\lambda\mathbf{A} + \mathbf{B})^{-1}\mathbf{A} \text{ and } \hat{\mathbf{B}}_\lambda = (\lambda\mathbf{A} + \mathbf{B})^{-1}\mathbf{B}. \tag{2}$$

Then $\hat{\mathbf{A}}_\lambda\hat{\mathbf{B}}_\lambda = \hat{\mathbf{B}}_\lambda\hat{\mathbf{A}}_\lambda$.

Theorem 1 shows that assuming $(\lambda\mathbf{A} + \mathbf{B})$ is invertible is a natural assumption. Lemma 1 means that we can assume for proof purposes that $\mathbf{A}$ and $\mathbf{B}$ commute. We shall prove the Lemma first.

Proof of Lemma 1. If there exists $\lambda \in \mathbb{C}$ such that $(\lambda\mathbf{A} + \mathbf{B})^{-1}$ exists, then $\lambda\hat{\mathbf{A}}_\lambda + \hat{\mathbf{B}}_\lambda = \mathbf{I}$. Thus $\hat{\mathbf{B}}_\lambda\hat{\mathbf{A}}_\lambda = \hat{\mathbf{A}}_\lambda\hat{\mathbf{B}}_\lambda$. ■

Proof of Theorem Suppose first that there exists $\lambda \in \mathbb{C}$ such that $(\lambda\mathbf{A} + \mathbf{B})^{-1}$ exists. Let $\hat{\mathbf{A}}_\lambda$ and $\hat{\mathbf{B}}_\lambda$ be as defined in (2). Clearly $\mathbf{A}\dot{\mathbf{x}}(t) + \mathbf{B}\mathbf{x}(t) = \mathbf{0}$ is tractable if and only if $\hat{\mathbf{A}}_\lambda\dot{\mathbf{x}}(t) + \hat{\mathbf{B}}_\lambda\mathbf{x}(t) = \mathbf{0}$ is. Taking a similarity we may write

$$\hat{\mathbf{A}}_\lambda = \begin{bmatrix} \mathbf{C} & \mathbf{0} \\ \mathbf{0} & \mathbf{N} \end{bmatrix}, \hat{\mathbf{B}}_\lambda = \begin{bmatrix} \mathbf{I} - \lambda\mathbf{C} & \mathbf{0} \\ \mathbf{0} & \mathbf{I} - \lambda\mathbf{N} \end{bmatrix} = \begin{bmatrix} \hat{\mathbf{B}}_1 & \mathbf{0} \\ \mathbf{0} & \hat{\mathbf{B}}_2 \end{bmatrix}, \mathbf{x}(t) = \begin{bmatrix} \mathbf{x}_1(t) \\ \mathbf{x}_2(t) \end{bmatrix} \tag{3}$$

since $\lambda\hat{\mathbf{A}}_\lambda + \hat{\mathbf{B}}_\lambda = \mathbf{I}$. Since $\mathbf{C}$ is invertible, $\mathbf{C}\dot{\mathbf{x}}_1(t) + (\mathbf{I} - \lambda\mathbf{C})\dot{\mathbf{x}}_2(t) = \mathbf{0}$ is tractable. Thus it suffices to show

$$\mathbf{N}\dot{\mathbf{x}}_2(t) + (\mathbf{I} - \lambda\mathbf{N})\mathbf{x}_2(t) = \mathbf{0} \text{ is tractable.} \tag{4}$$

Let $k = \operatorname{Ind}(\mathbf{N})$ and multiply (12) by $\mathbf{N}^{k-1}$. Then $(\mathbf{I} - \lambda\mathbf{N})\mathbf{N}^{k-1}\mathbf{x}_2(t) = \mathbf{0}$. Hence $\mathbf{N}^{k-1}\mathbf{x}_2(t) = \mathbf{0}$. Multiply (12) by $\mathbf{N}^{k-2}$. Then $\mathbf{N}^{k-1}\dot{\mathbf{x}}_2(t) + (\mathbf{I} - \lambda\mathbf{N})\mathbf{N}^{k-2}\mathbf{x}_2(t) = \mathbf{0}$ so that $\mathbf{N}^{k-2}\mathbf{x}_2(t) = \mathbf{0}$. Continuing in this manner we get that $\mathbf{x}_2(t) = \mathbf{0}$ and $\mathbf{N}\dot{\mathbf{x}}_2(t) + (\mathbf{I} - \lambda\mathbf{N})\mathbf{x}_2(t) = \mathbf{0}$ is trivially tractable.

Suppose now that $\mathbf{A}\dot{\mathbf{x}}(t) + \mathbf{B}\mathbf{x}(t) = \mathbf{0}$ is tractable. We need to show that there is a $\lambda \in \mathbb{C}$ such that $(\lambda\mathbf{A} + \mathbf{B})$ is invertible. Suppose that this is not true. Then $(\lambda\mathbf{A} + \mathbf{B})$ is singular for all $\lambda \in \mathbb{C}$. This means that for each $\lambda \in \mathbb{C}$, there is a vector $\mathbf{v}_\lambda \in \mathbb{C}^n$ such that

$$(\lambda\mathbf{A} + \mathbf{B})\mathbf{v}_\lambda = \mathbf{0} \text{ and } \mathbf{v}_\lambda \neq \mathbf{0}.$$

Let $\{\mathbf{v}_{\lambda_1}, \mathbf{v}_{\lambda_2}, \ldots, \mathbf{v}_{\lambda_s}\}$ be a finite linearly dependent set of such vectors. Let $\mathbf{x}_{\lambda_i}(\mathrm{t}) = \mathrm{e}^{\lambda_i t}\mathbf{v}_{\lambda_i}$ and let $\{\alpha_1, \alpha_2, \ldots, \alpha_s\} \subseteq \mathbb{C}$ be such that $\sum_{i=1}^{s} \alpha_i\mathbf{v}_{\lambda_i} = \mathbf{0}$, where not all the α_i's are 0. Then $\mathbf{z}(t) = \sum_{i=1}^{s} \alpha_i\mathbf{x}_{\lambda_i}(t)$ is not identically zero and is easily seen to be a solution of (1). However,

$$\mathbf{z}(0) = \sum_{i=1}^{s} \alpha_i \mathbf{v}_{\lambda_i} = \mathbf{0}.$$

Thus, there are two different solutions of (1), namely $\mathbf{z}(t)$ and $\mathbf{0}$, which satisfy the initial condition $\mathbf{x}(0) = \mathbf{0}$. Therefore, (1) is not tractable at $t = 0$, which contradicts our hypothesis. Hence, $(\lambda\mathbf{A} + \mathbf{B})^{-1}$ exists for some $\lambda \in \mathbb{C}$. ■

The next theorem will be used to show that most of our later development is independent of the scalar λ which is used in the expression $(\lambda\mathbf{A} + \mathbf{B})^{-1}$.

Theorem 9.2.2 *Suppose that* $\mathbf{A}, \mathbf{B} \in \mathbb{C}^{n \times n}$ *are such that there exists a* $\lambda \in \mathbb{C}$ *so that* $(\lambda\mathbf{A} + \mathbf{B})^{-1}$ *exists. Let* $\hat{\mathbf{A}}_\lambda = (\lambda\mathbf{A} + \mathbf{B})^{-1}\mathbf{A}$, $\hat{\mathbf{B}}_\lambda = (\lambda\mathbf{A} + \mathbf{B})^{-1}\mathbf{B}$, *and* $\hat{\mathbf{f}}_\lambda = (\lambda\mathbf{A} + \mathbf{B})^{-1}\mathbf{f}$ *for* $\mathbf{f} \in \mathbb{C}^n$. *For all* $\alpha, \mu \in \mathbb{C}$ *for which* $(\alpha\mathbf{A} + \mathbf{B})^{-1}$ *and* $(\mu\mathbf{A} + \mathbf{B})^{-1}$ *exist, the following statements are true.*

$$\hat{\mathbf{A}}_\alpha \hat{\mathbf{A}}_\alpha^{\mathrm{D}} = \hat{\mathbf{A}}_\mu \hat{\mathbf{A}}_\mu^{\mathrm{D}}. \tag{5}$$

$$\hat{\mathbf{A}}_\alpha^{\mathrm{D}} \hat{\mathbf{B}}_\alpha = \hat{\mathbf{A}}_\mu^{\mathrm{D}} \hat{\mathbf{B}}_\mu, \textit{ and } \hat{\mathbf{A}}_\alpha \hat{\mathbf{B}}_\alpha^{\mathrm{D}} = \hat{\mathbf{A}}_\mu \hat{\mathbf{B}}_\mu^{\mathrm{D}}. \tag{6}$$

$$\mathrm{Ind}(\hat{\mathbf{A}}_\alpha) = \mathrm{Ind}(\hat{\mathbf{A}}_\mu) \textit{ and } R(\hat{\mathbf{A}}_\alpha) = R(\hat{\mathbf{A}}_\mu). \tag{7}$$

$$\hat{\mathbf{A}}_\alpha^{\mathrm{D}} \hat{\mathbf{f}}_\alpha = \hat{\mathbf{A}}_\mu^{\mathrm{D}} \hat{\mathbf{f}}_\mu. \tag{8}$$

$$\hat{\mathbf{B}}_\alpha^{\mathrm{D}} \hat{\mathbf{f}}_\alpha = \hat{\mathbf{B}}_\mu^{\mathrm{D}} \hat{\mathbf{f}}_\mu. \tag{9}$$

Proof To prove (5), write

$$\begin{aligned}
\hat{\mathbf{A}}_\alpha^{\mathrm{D}} \hat{\mathbf{A}}_\alpha &= [(\alpha\mathbf{A} + \mathbf{B})^{-1}\mathbf{A}]^{\mathrm{D}} \hat{\mathbf{A}}_\alpha = [(\alpha\mathbf{A} + \mathbf{B})^{-1}(\mu\mathbf{A} + \mathbf{B})(\mu\mathbf{A} + \mathbf{B})^{-1}\mathbf{A}]^{\mathrm{D}} \hat{\mathbf{A}}_\alpha \\
&= [(\alpha\hat{\mathbf{A}}_\mu + \hat{\mathbf{B}}_\mu)^{-1}\hat{\mathbf{A}}_\mu]^{\mathrm{D}} \hat{\mathbf{A}}_\alpha \\
&= \hat{\mathbf{A}}_\mu^{\mathrm{D}}(\alpha\hat{\mathbf{A}}_\mu + \hat{\mathbf{B}}_\mu)\hat{\mathbf{A}}_\alpha \qquad \text{by Theorem 7.9.4.} \\
&= \hat{\mathbf{A}}_\mu^{\mathrm{D}}[(\mu\mathbf{A} + \mathbf{B})^{-1}(\alpha\mathbf{A} + \mathbf{B})]\hat{\mathbf{A}}_\alpha \\
&= \hat{\mathbf{A}}_\mu^{\mathrm{D}}(\mu\mathbf{A} + \mathbf{B})^{-1}(\alpha\mathbf{A} + \mathbf{B})(\alpha\mathbf{A} + \mathbf{B})^{-1}\mathbf{A} \\
&= \hat{\mathbf{A}}_\mu^{\mathrm{D}} \hat{\mathbf{A}}_\mu.
\end{aligned}$$

The proof of (6) is similar and is left as an exercise. To prove (7), write $\hat{\mathbf{A}}_\alpha = [(\alpha\mathbf{A} + \mathbf{B})^{-1}(\mu\mathbf{A} + \mathbf{B})](\mu\mathbf{A} + \mathbf{B})^{-1}\mathbf{A} = (\alpha\hat{\mathbf{A}}_\mu + \hat{\mathbf{B}}_\mu)^{-1}\hat{\mathbf{A}}_\mu = \hat{\mathbf{A}}_\mu(\alpha\hat{\mathbf{A}}_\mu + \hat{\mathbf{B}}_\mu)^{-1}$. Since $\hat{\mathbf{A}}_\mu$ and $\hat{\mathbf{B}}_\mu$ commute, it follows that for each positive integer m,

$$R(\hat{\mathbf{A}}_\alpha^m) = R(\hat{\mathbf{A}}_\mu^m).$$

Thus (7) follows. To prove (8), use the same technique used to prove (5) to obtain $\hat{\mathbf{A}}_\alpha^{\mathrm{D}}\hat{\mathbf{f}}_\alpha = [\hat{\mathbf{A}}_\mu^{\mathrm{D}}(\mu\mathbf{A} + \mathbf{B})^{-1}(\alpha\mathbf{A} + \mathbf{B})]\hat{\mathbf{f}}_\alpha = \hat{\mathbf{A}}_\mu^{\mathrm{D}}(\mu\mathbf{A} + \mathbf{B})^{-1}(\alpha\mathbf{A} + \mathbf{B}) \times (\alpha\mathbf{A} + \mathbf{B})^{-1}\mathbf{f} = \hat{\mathbf{A}}_\mu^{\mathrm{D}}\hat{\mathbf{f}}_\mu$. The proof of (9) is similar. ■

In view of the preceding theorem, we can now drop the subscript λ whenever the terms $\hat{\mathbf{A}}_\lambda\hat{\mathbf{A}}_\lambda^{\mathrm{D}}$, $\hat{\mathbf{A}}_\lambda^{\mathrm{D}}\hat{\mathbf{B}}_\lambda$, $R(\hat{\mathbf{A}}_\lambda)$, $\mathrm{Ind}(\hat{\mathbf{A}}_\lambda)$, $\hat{\mathbf{A}}_\lambda^{\mathrm{D}}\hat{\mathbf{f}}_\lambda$, and $\hat{\mathbf{B}}_\lambda^{\mathrm{D}}\hat{\mathbf{f}}_\lambda$ appear. We shall do so.

Let us return to the proof of Theorem 1. Recall that the original system was equivalent to the pair of equations

$$\mathbf{C}\dot{\mathbf{x}}_1(t) + \mathbf{B}_1\mathbf{x}_1(t) = \mathbf{0}, \quad \mathbf{C}\mathbf{B}_1 = \mathbf{B}_1\mathbf{C} \tag{10}$$

$$\mathbf{N}\dot{\mathbf{x}}_1(t) + \mathbf{B}_2\mathbf{x}_2(t) = \mathbf{0}, \quad \mathbf{N}\mathbf{B}_2 = \mathbf{B}_2\mathbf{N}, \tag{11}$$

$\mathbf{B}_2$ invertible, and the only solution of (11) was $\mathbf{x}_2(t) \equiv \mathbf{0}$. But (10) is consistent for any $\mathbf{x}_1(t_0)$ and the unique solution is $\mathbf{x}_1(t) = \exp(-\mathbf{C}^{-1}\mathbf{B}_1 \times (t - t_0))\mathbf{x}_1(t_0)$. Thus we have proved the first part of the next Theorem.

Theorem 9.2.3 *Suppose* $\mathbf{A}\dot{\mathbf{x}}(t) + \mathbf{B}\mathbf{x}(t) = \mathbf{0}$ *is tractable. Then the general solution is given by*

$$\mathbf{x}(t) = \mathrm{e}^{-\hat{\mathbf{A}}^{\mathrm{D}}\hat{\mathbf{B}}(t-t_0)}\hat{\mathbf{A}}\hat{\mathbf{A}}^{\mathrm{D}}\mathbf{q}, \quad \mathbf{q} \in \mathbb{C}^n. \tag{12}$$

A vector $\mathbf{c} \in \mathbb{C}^n$ *is a consistent initial vector for the homogeneous equation if and only if* $\mathbf{c} \in R(\hat{\mathbf{A}}^k) = R(\hat{\mathbf{A}}^{\mathrm{D}}\hat{\mathbf{A}})$.

Suppose that $\mathbf{f}(t)$ *is k-times continuously differentiable around* t_0. *Then the non-homogeneous equation* $\mathbf{A}\dot{\mathbf{x}}(t) + \mathbf{B}\mathbf{x}(t) = \mathbf{f}(t)$ *always possesses solutions and a particular solution is given by*

$$\mathbf{x}(t) = \mathrm{e}^{-\hat{\mathbf{A}}^{\mathrm{D}}\hat{\mathbf{B}}t}\int_{t_0}^{t} \mathrm{e}^{\hat{\mathbf{A}}^{\mathrm{D}}\hat{\mathbf{B}}s}\hat{\mathbf{A}}^{\mathrm{D}}\hat{\mathbf{f}}(s)\mathrm{d}s + (\mathbf{I} - \hat{\mathbf{A}}\hat{\mathbf{A}}^{\mathrm{D}})\sum_{i=0}^{k-1}(-1)^i[\hat{\mathbf{A}}\hat{\mathbf{B}}^{\mathrm{D}}]^i\hat{\mathbf{B}}^{\mathrm{D}}\hat{\mathbf{f}}^{(i)}(t). \tag{13}$$

Moreover, the expression (13) *is independent of* λ. *The general solution is given by*

$$\mathbf{x}(t) = \mathrm{e}^{-\hat{\mathbf{A}}^{\mathrm{D}}\hat{\mathbf{B}}(t-t_0)}\hat{\mathbf{A}}\hat{\mathbf{A}}^{\mathrm{D}}\mathbf{q} + \mathrm{e}^{-\hat{\mathbf{A}}^{\mathrm{D}}\hat{\mathbf{B}}t}\int_{t_0}^{t} \mathrm{e}^{\hat{\mathbf{A}}^{\mathrm{D}}\hat{\mathbf{B}}s}\hat{\mathbf{A}}^{\mathrm{D}}\hat{\mathbf{f}}(s)\mathrm{d}s$$

$$+ (\mathbf{I} - \hat{\mathbf{A}}\hat{\mathbf{A}}^{\mathrm{D}})\sum_{i=0}^{k-1}(-1)^i[\hat{\mathbf{A}}\hat{\mathbf{B}}^{\mathrm{D}}]^i\hat{\mathbf{B}}^{\mathrm{D}}\hat{\mathbf{f}}^{(i)}(t), \quad \mathbf{q} \in \mathbb{C}^n. \tag{14}$$

Let $\hat{\mathbf{w}} = (\mathbf{I} - \hat{\mathbf{A}}\hat{\mathbf{A}}^{\mathrm{D}})\sum_{i=0}^{k-1}(-1)^i(\hat{\mathbf{A}}\hat{\mathbf{B}}^{\mathrm{D}})^i\hat{\mathbf{B}}^{\mathrm{D}}\hat{\mathbf{f}}^{(i)}(t_0)$. *Then* $\hat{\mathbf{w}}$ *is independent of* λ. *A vector* $\mathbf{c} \in \mathbb{C}^n$ *is a consistent initial vector associated with* $t_0 \in \mathbb{R}$ *for the inhomogeneous equation if and only if* $\mathbf{c} \in \{\hat{\mathbf{w}} + R(\hat{\mathbf{A}}^k)\}$. *Furthermore, the inhomogeneous equation is tractable at* t_0 *and the unique solution of the initial value problem with* $\mathbf{x}(t_0) = \mathbf{c}$, $\mathbf{c}$ *a consistent initial vector associated with* t_0, *is given by* (14) *with* $\mathbf{q} = \mathbf{c}$.

Proof (14) will follow from (12) and (13). We have already shown (12). To see (13) let

$$\mathbf{x}_1(t) = \hat{\mathbf{A}}^{\mathrm{D}}\mathrm{e}^{-\hat{\mathbf{A}}^{\mathrm{D}}\hat{\mathbf{B}}t}\int_{0}^{t} \mathrm{e}^{\hat{\mathbf{A}}^{\mathrm{D}}\hat{\mathbf{B}}s}\mathbf{f}(s)\mathrm{d}s,$$

$$\mathbf{x}_2(t) = (\mathbf{I} - \hat{\mathbf{A}}\hat{\mathbf{A}}^{\mathrm{D}})\sum_{i=0}^{k-1}(-1)^i\hat{\mathbf{A}}^i(\hat{\mathbf{B}}^{\mathrm{D}})^{i+1}\hat{\mathbf{f}}^{(i)}(t),$$

where we have taken $t_0 = 0$ for notational convenience. We shall show that

$$\hat{\mathbf{A}}\dot{\mathbf{x}}_1(t) + \hat{\mathbf{B}}\mathbf{x}_1(t) = \hat{\mathbf{A}}\hat{\mathbf{A}}^{\mathrm{D}}\mathbf{f}(t) \text{ and} \tag{15}$$

$$\hat{\mathbf{A}}\dot{\mathbf{x}}_2(t) + \hat{\mathbf{B}}\mathbf{x}_2(t) = (\mathbf{I} - \hat{\mathbf{A}}\hat{\mathbf{A}}^{\mathrm{D}})\mathbf{f}(t). \tag{16}$$

To verify (15), note that $\hat{\mathbf{A}}\dot{\mathbf{x}}_1(t) = \hat{\mathbf{A}}[-\hat{\mathbf{A}}^{\mathrm{D}}\hat{\mathbf{B}}\mathbf{x}_1(t) + \hat{\mathbf{A}}^{\mathrm{D}}e^{-\hat{\mathbf{A}}^{\mathrm{D}}\hat{\mathbf{B}}t}e^{\hat{\mathbf{A}}^{\mathrm{D}}\hat{\mathbf{B}}t}\mathbf{f}(t)] = -\hat{\mathbf{A}}\hat{\mathbf{A}}^{\mathrm{D}}\hat{\mathbf{B}}\mathbf{x}_1(t) + \hat{\mathbf{A}}\hat{\mathbf{A}}^{\mathrm{D}}\mathbf{f}(t) = -\hat{\mathbf{B}}\mathbf{x}_1(t) + \hat{\mathbf{A}}\hat{\mathbf{A}}^{\mathrm{D}}\mathbf{f}(t)$, as desired. We now verify (16). $\hat{\mathbf{A}}\dot{\mathbf{x}}_2(t) = \hat{\mathbf{A}}(\mathbf{I} - \hat{\mathbf{A}}\hat{\mathbf{A}}^{\mathrm{D}})\sum_{i=0}^{k-1}(-1)^i\hat{\mathbf{A}}^i(\hat{\mathbf{B}}^{\mathrm{D}})^{i+1}\hat{\mathbf{f}}^{(i+1)}(t) = (\mathbf{I} - \hat{\mathbf{A}}\hat{\mathbf{A}}^{\mathrm{D}})\sum_{i=0}^{k-1}(-1)^i \times (\hat{\mathbf{A}}\hat{\mathbf{B}}^{\mathrm{D}})^{i+1}\hat{\mathbf{f}}^{(i+1)}(t) = (\mathbf{I} - \hat{\mathbf{A}}\hat{\mathbf{A}}^{\mathrm{D}})\sum_{i=1}^{k-1}(-1)^{i-1}(\hat{\mathbf{A}}\hat{\mathbf{B}}^{\mathrm{D}})^i\hat{\mathbf{f}}^{(i)}(t) = (\mathbf{I} - \hat{\mathbf{A}}\hat{\mathbf{A}}^{\mathrm{D}}) \times \hat{\mathbf{B}}\hat{\mathbf{B}}^{\mathrm{D}}\sum_{i=1}^{k-1}(-1)^{i-1}(\hat{\mathbf{A}}\hat{\mathbf{B}}^{\mathrm{D}})^i\hat{\mathbf{f}}^{(i)}(t) = (\mathbf{I} - \hat{\mathbf{A}}\hat{\mathbf{A}}^{\mathrm{D}})\hat{\mathbf{B}}\sum_{i=1}^{k-1}(-1)^{i-1}\hat{\mathbf{A}}^i(\hat{\mathbf{B}}^{\mathrm{D}})^{i+1}\hat{\mathbf{f}}^{(i)}(t) = (\mathbf{I} - \hat{\mathbf{A}}\hat{\mathbf{A}}^{\mathrm{D}})\hat{\mathbf{B}}(-\mathbf{x}_2(t) + \hat{\mathbf{B}}^{\mathrm{D}}\hat{\mathbf{f}}(t)) = -\hat{\mathbf{B}}\mathbf{x}_2(t) + (\mathbf{I} - \hat{\mathbf{A}}\hat{\mathbf{A}}^{\mathrm{D}})\hat{\mathbf{B}}\hat{\mathbf{B}}^{\mathrm{D}}\mathbf{f}(t) = -\hat{\mathbf{B}}\mathbf{x}_2(t) + (\mathbf{I} - \hat{\mathbf{A}}\hat{\mathbf{A}}^{\mathrm{D}})\hat{\mathbf{f}}(t)$ where the fact that $\hat{\mathbf{B}}, \hat{\mathbf{B}}^{\mathrm{D}}, \hat{\mathbf{A}}, \hat{\mathbf{A}}^{\mathrm{D}}$ commute has been used freely. Thus, $\mathbf{x}_1(t) + \mathbf{x}_2(t)$ is a particular solution as desired. The characterization of the consistent initial vectors for the inhomogeneous equation follow directly from (14). That the solutions are independent of λ follows from Theorem 2. ■

An important special case is when $\mathbf{B}$ is invertible. Then we may take $\lambda = 0$ and $\hat{\mathbf{A}} = \mathbf{B}^{-1}\mathbf{A}$, $\hat{\mathbf{B}} = \mathbf{I}$, $\hat{\mathbf{f}} = \mathbf{B}^{-1}\mathbf{f}$.

The Drazin inverse can sometimes be useful even when $\mathbf{A}$ is invertible. If $\mathbf{f}(s)$ is a constant vector $\mathbf{f}$, the general solution of $\dot{\mathbf{x}}(t) + \mathbf{B}\mathbf{x}(t) = \mathbf{f}(t)$ is given by

$$\mathbf{x}(t) = \left[e^{-\mathbf{B}t}\int_a^t e^{\mathbf{B}s}ds \right]\mathbf{f}. \tag{17}$$

If $\mathbf{B}^{-1}$ exists, then (17) is easily evaluated since $\int e^{\mathbf{B}s}ds = \mathbf{B}^{-1}e^{\mathbf{B}s} + \mathbf{G}$, $\mathbf{G} \in \mathbb{C}^{n\times n}$. However, if $\mathbf{B}$ is singular, then the evaluation of $\int e^{\mathbf{B}s}ds$ is more difficult. The next result shows how to do it using the Drazin inverse.

Theorem 9.2.4 *If* $\mathbf{B} \in \mathbb{C}^{n\times n}$ *and* $\mathrm{Ind}(\mathbf{B}) = k$, *then*

$$\int e^{\mathbf{B}s}ds = \mathbf{B}^{\mathrm{D}}e^{\mathbf{B}s} + (\mathbf{I} - \mathbf{B}\mathbf{B}^{\mathrm{D}})s\left[\mathbf{I} + \frac{\mathbf{B}s}{2!} + \frac{\mathbf{B}^2s^2}{3!} + \ldots + \frac{\mathbf{B}^{k-1}s^{k-1}}{k!}\right] + \mathbf{G},$$

$\mathbf{G} \in \mathbb{C}^{n\times n}$.

Proof Use the series expansion for $e^{\mathbf{B}s}$ to obtain

$$\frac{d}{ds}\left[\mathbf{B}^{\mathrm{D}}e^{\mathbf{B}s} + (\mathbf{I} - \mathbf{B}\mathbf{B}^{\mathrm{D}})s\sum_{i=0}^{k-1}\frac{\mathbf{B}^is^i}{(i+1)!} + \mathbf{G}\right] = e^{\mathbf{B}s}.$$

Corollary 9.2.1 *If* $\mathbf{f} \in \mathbb{C}^n$, *then*

$$\mathbf{x}(t) = \mathbf{B}^{\mathrm{D}}\mathbf{f} + t(\mathbf{I} - \mathbf{B}\mathbf{B}^{\mathrm{D}})\mathbf{f} - \frac{t^2(\mathbf{I} - \mathbf{B}\mathbf{B}^{\mathrm{D}})\mathbf{B}}{2!}\mathbf{f} + \frac{t^3(\mathbf{I} - \mathbf{B}\mathbf{B}^{\mathrm{D}})\mathbf{B}^2}{3!}\mathbf{f} + \ldots +$$

$\dfrac{(-1)^{k-1}t^k(\mathbf{I}-\mathbf{B}\mathbf{B}^{\mathrm{D}})\mathbf{B}^{k-1}}{k!}\mathbf{f}$ *is a solution of* $\dot{\mathbf{x}}(t)+\mathbf{B}\mathbf{x}(t)=\mathbf{f}$. (*Note, this is a polynomial in t.*)

Corollary 9.2.2 *For each t, let $C(t)$ denote the set of consistent initial vectors associated with t for $\mathbf{A}\dot{\mathbf{x}}(t)+\mathbf{B}\mathbf{x}(t)=\mathbf{f}(t)$ where $(\lambda\mathbf{A}+\mathbf{B})^{-1}$ exists for some λ and $\mathbf{f}(t)$ is k-times continuously differentiable. Then $\mathrm{d}(C(t),C(t_o))\to 0$ as $t\to t_o$, where $\mathrm{d}(C(t),C(t_o))=\sup_{y\in C(t)}\inf_{x\in C(t_o)}\|\mathbf{x}-\mathbf{y}\|_2$.*

The proof is left as an exercise. We also note in passing, the next theorem.

Theorem 9.2.5 *If $(\lambda\mathbf{A}+\mathbf{B})^{-1}$ exists for some λ, then*

$$\hat{\mathbf{A}}\hat{\mathbf{A}}^{\mathrm{D}}=\lim_{\lambda\to\infty}\frac{\hat{\mathbf{A}}_\lambda^{\mathrm{D}}}{\lambda},\quad \hat{\mathbf{A}}^{\mathrm{D}}\hat{\mathbf{B}}=\lim_{\lambda\to 0}\hat{\mathbf{A}}_\lambda^{\mathrm{D}}.$$

Proof Since $\mathbf{I}=\lambda\hat{\mathbf{A}}_\lambda+\hat{\mathbf{B}}_\lambda$ we obtain $\dfrac{\hat{\mathbf{A}}_\lambda^{\mathrm{D}}}{\lambda}=\hat{\mathbf{A}}\hat{\mathbf{A}}^{\mathrm{D}}+\dfrac{\hat{\mathbf{A}}^{\mathrm{D}}\hat{\mathbf{B}}}{\lambda}$. The first limit follows since $\hat{\mathbf{A}}^{\mathrm{D}}\hat{\mathbf{B}}$ is independent of λ. The second limit follows from $\hat{\mathbf{A}}^{\mathrm{D}}\hat{\mathbf{B}}=\hat{\mathbf{A}}_\lambda^{\mathrm{D}}-\lambda\hat{\mathbf{A}}\hat{\mathbf{A}}^{\mathrm{D}}$. ■

Example 9.2.2 Consider the homogeneous differential equation $\mathbf{A}\dot{\mathbf{x}}(t)+\mathbf{B}\mathbf{x}(t)=\mathbf{0}$ where

$$\mathbf{A}=\begin{bmatrix}1&0&-2\\-1&0&2\\2&3&2\end{bmatrix},\ \mathbf{B}=\begin{bmatrix}0&1&2\\-27&-22&-17\\18&14&10\end{bmatrix}.$$

Note that $\mathbf{A}$ and $\mathbf{B}$ are both singular and do not commute. Since $\mathbf{A}+\mathbf{B}$ turns out to be invertible we multiply on the left by $(\mathbf{A}+\mathbf{B})^{-1}$ to get $\hat{\mathbf{A}}\dot{\mathbf{x}}(t)+\hat{\mathbf{B}}\mathbf{x}(t)=\mathbf{0}$ where

$$\hat{\mathbf{A}}=(\mathbf{A}+\mathbf{B})^{-1}\mathbf{A}=\frac{1}{3}\begin{bmatrix}-3&-5&-4\\6&5&-2\\-3&2&10\end{bmatrix},\ \hat{\mathbf{B}}=\mathbf{I}-\hat{\mathbf{A}}=\frac{1}{3}\begin{bmatrix}6&5&4\\-6&-2&2\\3&-2&7\end{bmatrix}.$$

The eigenvalues of $\hat{\mathbf{A}}$ are 0, 1, 3 so that $\hat{\mathbf{A}}^{\mathrm{D}}$ may be computed by Theorem 7.5.2 to be

$$\hat{\mathbf{A}}^{\mathrm{D}}=\frac{1}{27}\begin{bmatrix}-27&-41&-28\\54&77&46\\-27&-34&-14\end{bmatrix}.$$

The consistency condition for initial conditions is thus

$$(\mathbf{I}-\hat{\mathbf{A}}\hat{\mathbf{A}}^{\mathrm{D}})\mathbf{x}(0)=\frac{1}{9}\begin{bmatrix}18&14&10\\-18&-14&-10\\9&7&5\end{bmatrix}\begin{bmatrix}x_1(0)\\x_2(0)\\x_3(0)\end{bmatrix}=\mathbf{0}.$$

There is only one independent equation involved,

$$9x_1(0) + 7x_2(0) + 5x_3(0) = 0. \tag{18}$$

Since $\sigma(-\hat{\mathbf{A}}^{\mathrm{D}}\hat{\mathbf{B}}) = \{0, 0, 2/3\}$, it is not difficult to compute the matrix exponential as

$$\mathbf{x}(t) = \mathrm{e}^{-\hat{\mathbf{A}}^{\mathrm{D}}\hat{\mathbf{B}}t}\mathbf{x}(0) = \frac{1}{18}\begin{bmatrix} 18 & 1-\mathrm{e}^{2t/3} & 2(1-\mathrm{e}^{2t/3}) \\ 0 & 26-8\mathrm{e}^{2t/3} & 16(1-\mathrm{e}^{2t/3}) \\ 0 & 13(\mathrm{e}^{2t/3}-1) & 26\mathrm{e}^{2t/3}-8 \end{bmatrix}\begin{bmatrix} x_1(0) \\ x_2(0) \\ x_3(0) \end{bmatrix}.$$

Equation (18) can be used to eliminate one of the $x_i(0)$.

For many applications it is desirable to be able to solve $\mathbf{A}\dot{\mathbf{x}} + \mathbf{B}\mathbf{x} = \mathbf{f}$ when $\mathbf{A}, \mathbf{B}$ are rectangular. We shall develop two important special cases. For each case, the general solution will be given. Derivation of the appropriate set of consistent initial conditions for the non-homogeneous equations is left to the reader. Another generalization of the results of this section may be found in Exercises 1–6.

We will first consider the case when $(\lambda\mathbf{A} + \mathbf{B})$ is one-to-one for some λ.

Theorem 9.2.6 *Suppose* $(\lambda\mathbf{A} + \mathbf{B})$ *is one-to-one. Then all solutions of* $\mathbf{A}\dot{\mathbf{x}} + \mathbf{B}\mathbf{x} = \mathbf{0}$ *are of the form*

$$\mathbf{x} = \mathrm{e}^{-\hat{\mathbf{A}}^{\mathrm{D}}\hat{\mathbf{B}}t}\mathbf{q} \textit{ where } \mathbf{q} \in R(\hat{\mathbf{A}}^{\mathrm{D}}\hat{\mathbf{A}})$$

and

$$[\mathbf{I} - (\lambda\mathbf{A} + \mathbf{B})(\lambda\mathbf{A} + \mathbf{B})^{\dagger}]\mathbf{A}\hat{\mathbf{A}}^{\mathrm{D}}\{\hat{\mathbf{A}}^{\mathrm{D}m}\hat{\mathbf{B}}^{m}\}\mathbf{q} = \mathbf{0} \text{ for } m = 0, 1, \ldots, n. \tag{19}$$

Here $\hat{\mathbf{A}} = (\lambda\mathbf{A} + \mathbf{B})^{\dagger}\mathbf{A}$, $\hat{\mathbf{B}} = (\lambda\mathbf{A} + \mathbf{B})^{\dagger}\mathbf{B}$.

Proof If $\mathbf{x}$ is a solution of $\mathbf{A}\dot{\mathbf{x}} + \mathbf{B}\mathbf{x} = \mathbf{0}$, then $\mathbf{x}$ is a solution of $\hat{\mathbf{A}}\dot{\mathbf{x}} + \hat{\mathbf{B}}\mathbf{x} = \mathbf{0}$. But $\hat{\mathbf{A}}\hat{\mathbf{B}} = \hat{\mathbf{B}}\hat{\mathbf{A}}$ and $\lambda\hat{\mathbf{A}} + \hat{\mathbf{B}} = \mathbf{I}$. Hence $\mathbf{x} = \mathrm{e}^{-\hat{\mathbf{A}}^{\mathrm{D}}\hat{\mathbf{B}}t}\hat{\mathbf{A}}^{\mathrm{D}}\hat{\mathbf{A}}\mathbf{q}$ by Theorem 3. Substituting back in gives $[-\mathbf{A}\hat{\mathbf{A}}^{\mathrm{D}}\hat{\mathbf{B}}\hat{\mathbf{A}}^{\mathrm{D}}\hat{\mathbf{A}} + \mathbf{B}\hat{\mathbf{A}}^{\mathrm{D}}\mathbf{A}]\mathrm{e}^{-\hat{\mathbf{A}}^{\mathrm{D}}\hat{\mathbf{B}}t}\mathbf{q} = \mathbf{0}$ for all t. Thus $[-\mathbf{A}\hat{\mathbf{A}}^{\mathrm{D}}\hat{\mathbf{B}} + \mathbf{B}\hat{\mathbf{A}}\hat{\mathbf{A}}^{\mathrm{D}}]\mathrm{e}^{-\hat{\mathbf{A}}^{\mathrm{D}}\hat{\mathbf{B}}t}\mathbf{q} = \mathbf{0}$ for all t, or equivalently, $[\mathbf{B}\hat{\mathbf{A}} - \mathbf{A}\hat{\mathbf{B}}]\hat{\mathbf{A}}^{\mathrm{D}}[\hat{\mathbf{A}}^{\mathrm{D}}\hat{\mathbf{B}}]^{m}\mathbf{q} = \mathbf{0}$ for $m = 0, 1, 2, \ldots$. But

$$\begin{aligned}\mathbf{A}\hat{\mathbf{B}} &= \mathbf{A}(\lambda\mathbf{A} + \mathbf{B})^{\dagger}\mathbf{B} = \mathbf{A}(\lambda\mathbf{A} + \mathbf{B})^{\dagger}(\lambda\mathbf{A} + \mathbf{B}) - \mathbf{A}(\lambda\mathbf{A} + \mathbf{B})^{\dagger}\lambda\mathbf{A} \\ &= \mathbf{A} - \lambda\mathbf{A}(\lambda\mathbf{A} + \mathbf{B})^{\dagger}\mathbf{A} \\ &= \mathbf{A} - (\lambda\mathbf{A} + \mathbf{B})(\lambda\mathbf{A} + \mathbf{B})^{\dagger}\mathbf{A} + \mathbf{B}(\lambda\mathbf{A} + \mathbf{B})^{\dagger}\mathbf{A} \\ &= [\mathbf{I} - (\lambda\mathbf{A} + \mathbf{B})(\lambda\mathbf{A} + \mathbf{B})^{\dagger}]\mathbf{A} + \mathbf{B}\hat{\mathbf{A}}. \quad \blacksquare\end{aligned}$$

Corollary 9.2.3 *If* $\lambda\mathbf{A} + \mathbf{B}$ *is one-to-one, and* $N(\bar{\lambda}\mathbf{A}^* + \mathbf{B}^*) = N(\mathbf{A}^*) \cap N(\mathbf{B}^*)$, *then all solutions of* $\mathbf{A}\dot{\mathbf{x}} + \mathbf{B}\mathbf{x} = \mathbf{0}$ *are of the form* $\mathbf{x} = \mathrm{e}^{-\hat{\mathbf{A}}^{\mathrm{D}}\hat{\mathbf{B}}t}\hat{\mathbf{A}}^{\mathrm{D}}\hat{\mathbf{A}}\mathbf{q}$ *where* $\mathbf{q}$ *is an arbitrary vector.*

Proof $R(\lambda\mathbf{A} + \mathbf{B})^{\perp} = N(\bar{\lambda}\mathbf{A}^* + \mathbf{B}^*) = N(\mathbf{A}^*) \cap N(\mathbf{B}^*)$. But $R(\mathbf{A}) \perp N(\mathbf{A}^*)$ so that $R(\mathbf{A}) \subseteq R(\lambda\mathbf{A} + \mathbf{B})^{\perp}$. Thus (19) holds for all $\mathbf{q} \in R(\mathbf{A}\mathbf{A}^{\mathrm{D}})$. $\blacksquare$

Example 9.2.3 Let $\mathbf{A} = \begin{bmatrix} 1 \\ 0 \end{bmatrix}$, $\mathbf{B} = \begin{bmatrix} 0 \\ 1 \end{bmatrix}$. Then $(\lambda\mathbf{A} + \mathbf{B})$ is one-to-one and $N(\lambda\mathbf{A} + \mathbf{B}) = N(\mathbf{A}) \cap N(\mathbf{B}) = \{\mathbf{0}\}$ for all λ. However, $N(\bar{\lambda}\mathbf{A}^* + \mathbf{B}^*) \neq N(\mathbf{A}^*) \cap N(\mathbf{B}^*)$ for all λ. $\mathbf{A}\dot{x} + \mathbf{B}x = \mathbf{0}$ has only $x = 0$ as a solution. Multiplying by $(\lambda\mathbf{A} + \mathbf{B})^\dagger = (|\lambda|^2 + 1)^{-1}[\lambda, 1]$ we get $\lambda(|\lambda|^2 + 1)^{-1}\dot{x} + (|\lambda|^2 + 1)^{-1}x = 0$ which has the non-zero solutions $x = e^{-\lambda^{-1}t}q$.

Theorem 9.2.7 *Suppose* $(\lambda\mathbf{A} + \mathbf{B})$ *is one-to-one and* $\mathbf{A}\dot{\mathbf{x}} + \mathbf{B}\mathbf{x} = \mathbf{f}$ *is consistent. Then all solutions of* $\mathbf{A}\dot{\mathbf{x}} + \mathbf{B}\mathbf{x} = \mathbf{f}$ *are of the form*

$$\mathbf{x} = e^{-\hat{\mathbf{A}}^D\hat{\mathbf{B}}t}\hat{\mathbf{A}}^D\hat{\mathbf{A}}\mathbf{q} + \hat{\mathbf{A}}^D e^{-\hat{\mathbf{A}}^D\hat{\mathbf{B}}t}\int_0^t e^{\hat{\mathbf{A}}^D\hat{\mathbf{B}}s}\hat{\mathbf{f}}(s)ds + (\mathbf{I} - \hat{\mathbf{A}}\hat{\mathbf{A}}^D)\sum_{n=0}^{k-1}(-1)^n(\hat{\mathbf{A}}\hat{\mathbf{B}}^D)^n\hat{\mathbf{B}}^D\hat{\mathbf{f}}^{(n)} \tag{20}$$

where $\hat{\mathbf{A}} = (\lambda\mathbf{A} + \mathbf{B})^\dagger\mathbf{A}$, $\hat{\mathbf{B}} = (\lambda\mathbf{A} + \mathbf{B})^\dagger\mathbf{B}$, $k = \mathrm{Ind}(\hat{\mathbf{A}})$, *and* $\hat{\mathbf{f}} = (\lambda\mathbf{A} + \mathbf{B})^\dagger\mathbf{f}$.

Proof If $\mathbf{x}$ solves $\mathbf{A}\dot{\mathbf{x}} + \mathbf{B}\mathbf{x} = \mathbf{f}$, then $\mathbf{x}$ solves $\hat{\mathbf{A}}\dot{\mathbf{x}} + \hat{\mathbf{B}}\mathbf{x} = \hat{\mathbf{f}}$ and $\lambda\hat{\mathbf{A}} + \hat{\mathbf{B}} = \mathbf{I}$. Thus (20) follows from Theorem 3. ■

Theorem 7 is not as completely satisfying as our other results since we have not stated precisely for which $\mathbf{f}$ is $\mathbf{A}\dot{\mathbf{x}} + \mathbf{B}\mathbf{x} = \mathbf{f}$ consistent when $\lambda\mathbf{A} + \mathbf{B}$ is one-to-one. While the general problem appears difficult, we do have the following.

Theorem 9.2.8 *Suppose* $\lambda\mathbf{A} + \mathbf{B}$ *is one-to-one and* $N(\lambda\mathbf{A}^* + \mathbf{B}^*) = N(\mathbf{A}^*) \cap N(\mathbf{B}^*)$. *Then* $\mathbf{A}\dot{\mathbf{x}} + \mathbf{B}\mathbf{x} = \mathbf{f}$ *is consistent if and only if* $(\mathbf{I} - (\lambda\mathbf{A} + \mathbf{B})(\lambda\mathbf{A} + \mathbf{B})^\dagger)\mathbf{f} = \mathbf{0}$.

Proof Suppose $\lambda\mathbf{A} + \mathbf{B}$ is one-to-one and $N(\bar{\lambda}\mathbf{A}^* + \mathbf{B}^*) = N(\mathbf{A}^*) \cap N(\mathbf{B}^*)$. Now $(\lambda\mathbf{A} + \mathbf{B})(\lambda\mathbf{A} + \mathbf{B})^\dagger$ is the identity on $R(\lambda\mathbf{A} + \mathbf{B}) = N(\bar{\lambda}\mathbf{A}^* + \mathbf{B}^*)^\perp = [N(\mathbf{A}^*) \cap N(\mathbf{B}^*)]^\perp \supseteq R(\mathbf{A}) \cup R(\mathbf{B})$. Thus $(\lambda\mathbf{A} + \mathbf{B})(\lambda\mathbf{A} + \mathbf{B})^\dagger\mathbf{A} = \mathbf{A}$ and $(\lambda\mathbf{A} + \mathbf{B})(\lambda\mathbf{A} + \mathbf{B})^\dagger\mathbf{B} = \mathbf{B}$. Hence for any $\mathbf{x}$, if we set $\mathbf{f} = \mathbf{A}\dot{\mathbf{x}} + \mathbf{B}\mathbf{x}$, we get $(\lambda\mathbf{A} + \mathbf{B})(\lambda\mathbf{A} + \mathbf{B})^\dagger\mathbf{f} = \mathbf{f}$. On the other hand, if $(\lambda\mathbf{A} + \mathbf{B})(\lambda\mathbf{A} + \mathbf{B})^\dagger\mathbf{f} = \mathbf{f}$, then $\mathbf{A}\dot{\mathbf{x}} + \mathbf{B}\mathbf{x} = \mathbf{f}$ is equivalent to $\hat{\mathbf{A}}\dot{\mathbf{x}} + \hat{\mathbf{B}}\mathbf{x} = \hat{\mathbf{f}}$. Since $\hat{\mathbf{A}}\dot{\mathbf{x}} + \hat{\mathbf{B}}\mathbf{x} = \hat{\mathbf{f}}$ is consistent, so is $\mathbf{A}\dot{\mathbf{x}} + \mathbf{B}\mathbf{x} = \mathbf{f}$. ■

The special cases when $\mathbf{A}$ or $\mathbf{B}$ is one-to-one are of some interest. $\mathbf{B}$ being one-to-one is the case of most interest for the applications of Section 5.

Theorem 9.2.9 *Suppose that* $\mathbf{A}$ *is one-to-one. Then* $\mathbf{A}\dot{\mathbf{x}} + \mathbf{B}\mathbf{x} = \mathbf{f}$ *is consistent if and only if* $\mathbf{f}$ *is of the form*

$$\mathbf{f} = \mathbf{A}\mathbf{A}^\dagger\mathbf{h} \oplus (\mathbf{I} - \mathbf{A}\mathbf{A}^\dagger)\mathbf{B}\mathbf{g} \tag{21}$$

where $\mathbf{h}$ *is an arbitrary function and*

$$\mathbf{g} = e^{-\mathbf{A}^\dagger\mathbf{B}t}\mathbf{q} + e^{-\mathbf{A}^\dagger\mathbf{B}t}\int_a^t e^{-\mathbf{A}^\dagger\mathbf{B}s}\mathbf{A}^\dagger\mathbf{h}(s)ds, \tag{22}$$

$\mathbf{q}$ *an arbitrary constant. Conversely, if* $\mathbf{f}$ *has the form* (21), *then* $\mathbf{g}$ *given in* (22) *is the general solution.*

Proof Suppose $\mathbf{A}$ is one-to-one. Then $\mathbf{A}\dot{\mathbf{x}} + \mathbf{B}\mathbf{x} = \mathbf{f}$ is equivalent to the pair of equations:

$$\dot{\mathbf{x}} + \mathbf{A}^{\dagger}\mathbf{B}\mathbf{x} = \mathbf{A}^{\dagger}\mathbf{f}, \text{ and} \tag{23}$$

$$(\mathbf{I} - \mathbf{A}\mathbf{A}^{\dagger})\mathbf{B}\mathbf{x} = (\mathbf{I} - \mathbf{A}\mathbf{A}^{\dagger})\mathbf{f}. \tag{24}$$

Now $\mathbf{A}\mathbf{A}^{\dagger}\mathbf{f}$ can be chosen arbitrarily, say $\mathbf{A}\mathbf{A}^{\dagger}\mathbf{h}$. Then (23) uniquely determines $\mathbf{x}$ giving (21). Substituting $\mathbf{x}$ into (24) gives $(\mathbf{I} - \mathbf{A}\mathbf{A}^{\dagger})\mathbf{f}$. ■

A similar result is possible if $\mathbf{B}$ is one-to-one.

Theorem 9.2.10 *Suppose that* $\mathbf{B}$ *is one-to-one. Then* $\mathbf{A}\dot{\mathbf{x}} + \mathbf{B}\mathbf{x} = \mathbf{f}$ *is consistent if and only if* $\mathbf{f}$ *is of the form*

$$\mathbf{f} = \mathbf{B}\mathbf{B}^{\dagger}\mathbf{h} + (\mathbf{I} - \mathbf{B}\mathbf{B}^{\dagger})\mathbf{A}\dot{\mathbf{g}} \tag{25}$$

where $\mathbf{h}$ *is arbitrary and*

$$\mathbf{g} = e^{-(\mathbf{B}^{\dagger}\mathbf{A})^{\mathrm{D}}t}(\mathbf{B}^{\dagger}\mathbf{A})^{\mathrm{D}}\mathbf{B}^{\dagger}\mathbf{A}\mathbf{q} + e^{-(\mathbf{B}^{\dagger}\mathbf{A})^{\mathrm{D}}t}(\mathbf{B}^{\dagger}\mathbf{A})^{\mathrm{D}}\int_0^t e^{(\mathbf{B}^{\dagger}\mathbf{A})^{\mathrm{D}}s}\mathbf{B}^{\dagger}\mathbf{h}(s)\,ds \tag{26}$$
$$+ [\mathbf{I} - (\mathbf{B}^{\dagger}\mathbf{A})^{\mathrm{D}}(\mathbf{B}^{\dagger}\mathbf{A})]\sum_{n=0}^{k-1}(-1)^n[\mathbf{B}^{\dagger}\mathbf{A}]^n\mathbf{B}^{\dagger}\mathbf{h}^{(n)},$$

$k = \mathrm{Ind}(\mathbf{B}^{\dagger}\mathbf{A})$, $\mathbf{q}$ *arbitrary. Conversely, if* $\mathbf{f}$ *has the form* (24), *then* $\mathbf{g}$ *in* (25) *is the general solution.*

Proof Suppose $\mathbf{B}$ is one-to-one. Then $\mathbf{A}\dot{\mathbf{x}} + \mathbf{B}\mathbf{x} = \mathbf{f}$ is equivalent to

$$\mathbf{B}^{\dagger}\mathbf{A}\dot{\mathbf{x}} + \mathbf{x} = \mathbf{B}^{\dagger}\mathbf{f}, \text{ and} \tag{27}$$

$$(\mathbf{I} - \mathbf{B}\mathbf{B}^{\dagger})\mathbf{A}\dot{\mathbf{x}} = (\mathbf{I} - \mathbf{B}\mathbf{B}^{\dagger})\mathbf{f}. \tag{28}$$

Again $\mathbf{B}\mathbf{B}^{\dagger}\mathbf{f}$ is arbitrary. From (27) $\mathbf{x}$ is determined uniquely in terms of $\mathbf{B}^{\dagger}\mathbf{f}$. Then $(\mathbf{I} - \mathbf{B}\mathbf{B}^{\dagger})\mathbf{f}$ must follow from (28). ■

We now turn to the case when $\lambda\mathbf{A} + \mathbf{B}$ is onto.

Let $\mathbf{A}, \mathbf{B}$ be $m \times n$ matrices. Let λ be such that $\lambda\mathbf{A} + \mathbf{B}$ is onto. Define $\mathbf{P} = (\lambda\mathbf{A} + \mathbf{B})^{\dagger}(\lambda\mathbf{A} + \mathbf{B})$. Then $\mathbf{A}\dot{\mathbf{x}} + \mathbf{B}\mathbf{x} = \mathbf{f}$ becomes

$$\mathbf{A}\mathbf{P}\dot{\mathbf{x}} + \mathbf{B}\mathbf{P}\mathbf{x} = \mathbf{f} - \mathbf{A}(\mathbf{I} - \mathbf{P})\dot{\mathbf{x}} - \mathbf{B}(\mathbf{I} - \mathbf{P})\mathbf{x}.$$

Or, equivalently,

$$\mathbf{A}(\lambda\mathbf{A} + \mathbf{B})^{\dagger}[(\lambda\mathbf{A} \dot{+} \mathbf{B})\mathbf{x}] + \mathbf{B}(\lambda\mathbf{A} + \mathbf{B})^{\dagger}[(\lambda\mathbf{A} + \mathbf{B})\mathbf{x}]$$
$$= \mathbf{f} - \mathbf{A}(\mathbf{I} - \mathbf{P})\dot{\mathbf{x}} - \mathbf{B}(\mathbf{I} - \mathbf{P})\mathbf{x}. \tag{29}$$

But $\lambda[\mathbf{A}(\lambda\mathbf{A} + \mathbf{B})^{\dagger}] + [\mathbf{B}(\lambda\mathbf{A} + \mathbf{B}^{\dagger})] = \mathbf{I}$. Thus (29) is, in terms of $(\lambda\mathbf{A} + \mathbf{B})\mathbf{x}$, a differential equation of the type already solved and hence has a solution for any choice of $(\mathbf{I} - \mathbf{P})\mathbf{x}$.

Theorem 9.2.11 *Suppose that* $\lambda\mathbf{A}+\mathbf{B}$ *is onto and* $\mathbf{f}$ *is n-times differentiable. Let* $\hat{\mathbf{A}}=\mathbf{A}(\lambda\mathbf{A}+\mathbf{B})^{\dagger}$, $\hat{\mathbf{B}}=\mathbf{B}(\lambda\mathbf{A}+\mathbf{B})^{\dagger}$. *Let* $\mathbf{g}=\mathbf{f}-\mathbf{A}[\mathbf{I}-(\lambda\mathbf{A}+\mathbf{B})^{\dagger}(\lambda\mathbf{A}+\mathbf{B})]\dot{\mathbf{h}}-\mathbf{B}[\mathbf{I}-(\lambda\mathbf{A}+\mathbf{B})^{\dagger}(\lambda\mathbf{A}+\mathbf{B})]\mathbf{h}$ *where* $\mathbf{h}$ *is an arbitrary* $(n+1)$*-times differentiable vector valued function. Then all solutions of* $\mathbf{A}\dot{\mathbf{x}}+\mathbf{B}\mathbf{x}=\mathbf{f}$ *are of the form*

$$\begin{aligned}\mathbf{x}=(\lambda\mathbf{A}+\mathbf{B})^{\dagger}\{&e^{-\hat{\mathbf{A}}^{D}\hat{\mathbf{B}}t}\hat{\mathbf{A}}\hat{\mathbf{A}}^{D}\mathbf{q}+\hat{\mathbf{A}}^{D}e^{-\hat{\mathbf{A}}^{D}\hat{\mathbf{B}}t}\int_0^t e^{\hat{\mathbf{A}}^{D}\hat{\mathbf{B}}s}\mathbf{g}(s)ds\\ &+(\mathbf{I}-\hat{\mathbf{A}}^{D}\hat{\mathbf{A}})\sum_{n=0}^{k-1}(-1)^n[\hat{\mathbf{A}}\hat{\mathbf{B}}^{D}]^n\hat{\mathbf{B}}^{D}\mathbf{g}^{(n)}\}+[\mathbf{I}-(\lambda\mathbf{A}+\mathbf{B})^{\dagger}\\ &\times(\lambda\mathbf{A}+\mathbf{B})]\mathbf{h},\end{aligned}$$

$\mathbf{q}$ *an arbitrary constant vector,* $k=\operatorname{Ind}(\hat{\mathbf{A}})$.

The formulas in Theorem 11 simplify considerably if $\mathbf{A}$ or $\mathbf{B}$ are onto. For the applications of Section 5, the case when $\mathbf{B}$ is onto is the more important.

Theorem 9.2.12 *Suppose that* $\mathbf{B}$ *is onto. Then all solutions of* $\mathbf{A}\dot{\mathbf{x}}+\mathbf{B}\mathbf{x}=\mathbf{f}$ *are of the form*

$$\begin{aligned}\mathbf{x}=\mathbf{B}^{\dagger}\{&e^{-\hat{\mathbf{C}}^{D}t}\hat{\mathbf{C}}\hat{\mathbf{C}}^{D}\mathbf{q}+\hat{\mathbf{C}}^{D}e^{-\hat{\mathbf{C}}^{D}t}\int_0^t e^{\hat{\mathbf{C}}^{D}s}\mathbf{g}(s)ds\\ &+(\mathbf{I}-\hat{\mathbf{C}}^{D}\hat{\mathbf{C}})\sum_{n=0}^{k-1}(-1)^n\hat{\mathbf{C}}^n\mathbf{g}^{(n)}+[\mathbf{I}-\mathbf{B}^{\dagger}\mathbf{B}]\mathbf{h},\end{aligned}$$

$\mathbf{h}$ *an arbitrary function,* $\mathbf{q}$ *an arbitary vector,* $\mathbf{g}=\mathbf{f}-\mathbf{A}[\mathbf{I}-\mathbf{B}^{\dagger}\mathbf{B}]\dot{\mathbf{h}}$, $\hat{\mathbf{C}}=\mathbf{A}\mathbf{B}^{\dagger}$, $k=\operatorname{Ind}(\mathbf{C})$.

Theorem 12 comes immediately from Theorem 11 by setting $\lambda=0$ and noting that $\hat{\mathbf{B}}=\mathbf{I}$.

Theorem 9.2.13 *Suppose that* $\mathbf{A}$ *is onto. Then all solutions of* $\mathbf{A}\dot{\mathbf{x}}+\mathbf{B}\mathbf{x}=\mathbf{f}$ *are of the form*

$$\mathbf{x}=\mathbf{A}^{\dagger}\left\{e^{-\mathbf{B}\mathbf{A}^{\dagger}t}\mathbf{q}+e^{-\mathbf{B}\mathbf{A}^{\dagger}t}\int_0^t e^{\mathbf{B}\mathbf{A}^{\dagger}s}\mathbf{g}(s)ds\right\}+[\mathbf{I}-\mathbf{A}^{\dagger}\mathbf{A}]\mathbf{h} \tag{30}$$

where $\mathbf{h}$ *is an arbitrary function and* $\mathbf{g}=\mathbf{f}-\mathbf{B}[\mathbf{I}-\mathbf{A}^{\dagger}\mathbf{A}]\mathbf{h}$.

Proof This one is easier to prove directly. Suppose $\mathbf{A}$ is onto and rewrite $\mathbf{A}\dot{\mathbf{x}}+\mathbf{B}\mathbf{x}=\mathbf{f}$ as $(\mathbf{A}\dot{\mathbf{x}})+\mathbf{B}\mathbf{A}^{\dagger}(\mathbf{A}\mathbf{x})=\mathbf{f}-\mathbf{B}[\mathbf{I}-\mathbf{A}^{\dagger}\mathbf{A}]\mathbf{x}$. Taking $[\mathbf{I}-\mathbf{A}^{\dagger}\mathbf{A}]\mathbf{x}$ arbitrary we can solve uniquely for $\mathbf{A}\mathbf{x}$, $\mathbf{A}^{\dagger}\mathbf{A}\mathbf{x}=\mathbf{x}$, to get (30). ■

3. Applications of the Drazin inverse to difference equations

The Drazin inverse also arises naturally in attempting to solve difference equations with singular coefficient matrices. To illustrate why the Drazin inverse works and other types of generalized inverses don't work in dealing with this type of difference equation, consider the following difference equation;

$$\mathbf{A}\mathbf{x}_{n+1} = \mathbf{x}_n, n \geq 0, \text{ where } \mathbf{A} \text{ is singular.}$$

At first glance, one might be tempted to introduce a (1)-inverse for **A**. However if one stops and thinks a moment, one can see that one must have that

$$\mathbf{x}_n = \mathbf{A}\mathbf{x}_{n+1} = \mathbf{A}^2\mathbf{x}_{n+2} = \ldots = \mathbf{A}^k\mathbf{x}_{n+k}$$

as well as $\mathbf{x}_{n+1} = \mathbf{A}\mathbf{x}_{n+2} = \mathbf{A}^2\mathbf{x}_{n+3} = \ldots = \mathbf{A}^k\mathbf{x}_{n+k+1}$, where $k = \text{Ind}(\mathbf{A})$. Thus, the problem could be stated as follows: given $\mathbf{x}_n \in R(\mathbf{A}^k)$, find a vector $\mathbf{x}_{n+1}$ such that $\mathbf{A}\mathbf{x}_{n+1} = \mathbf{x}_n$ and $\mathbf{x}_{n+1} \in R(\mathbf{A}^k)$.

By examining Definition 2.2, one can see that the above problem has a solution, the solution is unique, and is given by $\mathbf{x}_{n+1} = \mathbf{A}^{\mathrm{D}}\mathbf{x}_n$.

Not unexpectedly the solution of the difference equation proceeds much as for the differential equation.

Definition 9.3.1 *For* $\mathbf{A}, \mathbf{B} \in \mathbb{C}^{m \times m}, \mathbf{f}_n \in \mathbb{C}^m$, *the vector* $\mathbf{c} \in \mathbb{C}^m$ *is called a* consistent initial vector for the difference equation $\mathbf{A}\mathbf{x}_{n+1} = \mathbf{B}\mathbf{x}_n + \mathbf{f}_n$ *if the initial value problem* $\mathbf{A}\mathbf{x}_{n+1} = \mathbf{B}\mathbf{x}_n + \mathbf{f}_n, \mathbf{x}_0 = \mathbf{c}, n = 1, 2, \ldots$ *has a solution for* $\mathbf{x}_n$.

Definition 9.3.2 *The difference equation* $\mathbf{A}\mathbf{x}_{n+1} = \mathbf{B}\mathbf{x}_n + \mathbf{f}_n$ *is said to be* tractable *if the initial value problem* $\mathbf{A}\mathbf{x}_{n+1} = \mathbf{B}\mathbf{x}_n + \mathbf{f}_n, \mathbf{x}_0 = \mathbf{c}, n = 1, 2, \ldots$ *has a unique solution for each consistent initial vector* **c**.

Theorem 9.3.1 *The homogeneous difference equation*

$$\mathbf{A}\mathbf{x}_{n+1} = \mathbf{B}\mathbf{x}_n \qquad \mathbf{A}, \mathbf{B} \in \mathbb{C}^{m \times m}$$

is tractable if and only if there exists a scalar $\lambda \in \mathbb{C}$ *such that* $(\lambda\mathbf{A} + \mathbf{B})^{-1}$ *exists.*

Proof The proof follows the same lines as the proof of Theorem 2.1 except that $\mathbf{x}_{\lambda_i}(t) = e^{\lambda_i t}\mathbf{v}_{\lambda_i}$ is replaced with $\mathbf{x}_n^{(\lambda_i)} = \lambda_i^n \mathbf{v}_{\lambda_i}$. ∎

The difference analogue of Theorem 2.3 is as follows.

Theorem 9.3.2 *If the homogeneous equation*

$$\mathbf{A}\mathbf{x}_{n+1} = \mathbf{B}\mathbf{x}_n \tag{1}$$

is tractable, then the general solution is given by

$$\mathbf{x}_n = \begin{cases} \hat{\mathbf{A}}\hat{\mathbf{A}}^{\mathrm{D}}\mathbf{q} & \text{if } n = 0 \\ (\hat{\mathbf{A}}^{\mathrm{D}}\hat{\mathbf{B}})^n\mathbf{q} & \text{if } n = 1, 2, 3, \ldots \end{cases} \quad \mathbf{q} \in \mathbb{C}^m$$

where $\hat{\mathbf{A}} = (\lambda\mathbf{A} - \mathbf{B})^{-1}\mathbf{A}$ *and* $\hat{\mathbf{B}} = (\lambda\mathbf{A} - \mathbf{B})^{-1}\mathbf{B}$ *and* $\lambda \in \mathbb{C}$ *is such that* $(\lambda\mathbf{A} - \mathbf{B})^{-1}$ *exists. Furthermore,* $\mathbf{c} \in \mathbb{C}^m$ *is a consistent initial vector for* (1) *if and only if* $\mathbf{c} \in R(\hat{\mathbf{A}}^k)$, *where* $k = \text{Ind}(\hat{\mathbf{A}})$. *In this case the unique solution, subject to* $\mathbf{x}_0 = \mathbf{c}$, *is given by* $\mathbf{x}_n = (\hat{\mathbf{A}}^D\hat{\mathbf{B}})^n\mathbf{c}, n = 0, 1, 2, 3, \ldots$. *The inhomogeneous equation* $\mathbf{A}\mathbf{x}_{n+1} = \mathbf{B}\mathbf{x}_n + \mathbf{f}_n$ *is also tractable. Its general solution is, for* $n \geq 1$,

$$\mathbf{x}_n = (\hat{\mathbf{A}}^D\hat{\mathbf{B}})^n\hat{\mathbf{A}}\hat{\mathbf{A}}^D\mathbf{q} + \hat{\mathbf{A}}^D \sum_{i=0}^{n-1} (\hat{\mathbf{A}}^D\hat{\mathbf{B}})^{n-i-1}\hat{\mathbf{f}}_i - (\mathbf{I} - \hat{\mathbf{A}}\hat{\mathbf{A}}^D) \sum_{i=0}^{k-1} (\hat{\mathbf{A}}\hat{\mathbf{B}}^D)^i\hat{\mathbf{B}}^D\hat{\mathbf{f}}_{n+i}, \tag{2}$$

where $\hat{\mathbf{A}} = (\lambda\mathbf{A} - \mathbf{B})^{-1}\mathbf{A}, \hat{\mathbf{B}} = (\lambda\mathbf{A} - \mathbf{B})^{-1}\mathbf{B}, \hat{\mathbf{f}}_i = (\lambda\mathbf{A} - \mathbf{B})^{-1}\mathbf{f}_i, k = \text{Ind}(\hat{\mathbf{A}})$, *and* $\mathbf{q} \in \mathbb{C}^m$. *The solution* $\mathbf{x}_n$ *is independent of* λ. *Let* $\hat{\mathbf{w}} = -(\mathbf{I} - \hat{\mathbf{A}}\hat{\mathbf{A}}^D) \times \sum_{i=0}^{k-1} (\hat{\mathbf{A}}\hat{\mathbf{B}}^D)^i\hat{\mathbf{B}}^D\hat{\mathbf{f}}_i$. *The vector* $\mathbf{c}$ *is a consistent initial vector if and only if* $\mathbf{c}$ *lies in the flat* $\{\hat{\mathbf{w}} + R(\hat{\mathbf{A}}^k)\}$.

Proof Since (1) is tractable, multiplying by $(\lambda\mathbf{A} - \mathbf{B})^{-1}$ gives the equivalent equation $\hat{\mathbf{A}}\mathbf{x}_{n+1} = \hat{\mathbf{B}}\mathbf{x}_n$. After a similarity we get, as in the proof of Theorem 2.1,

$$\begin{bmatrix} \mathbf{C} & \mathbf{0} \\ \mathbf{0} & \mathbf{N} \end{bmatrix}\begin{bmatrix} \mathbf{x}_{n+1}^{(1)} \\ \mathbf{x}_{n+1}^{(2)} \end{bmatrix} = \begin{bmatrix} \mathbf{I} + \lambda\mathbf{C} & \\ \mathbf{0} & \mathbf{I} + \lambda\mathbf{N} \end{bmatrix}\begin{bmatrix} \mathbf{x}_n^{(1)} \\ \mathbf{x}_n^{(2)} \end{bmatrix}. \tag{3}$$

Thus $\mathbf{x}_n^{(2)} = (\mathbf{I} + \lambda\mathbf{N})^{-k}\mathbf{N}^k\mathbf{x}_{n+k}^{(2)} = \mathbf{0}, \mathbf{x}_n^{(1)} = \mathbf{C}^{-n}(\mathbf{I} + \lambda\mathbf{C})^n\mathbf{x}_0^{(1)}$, and the solution of the homogeneous equation follows. (2) may be verified directly as in the proof of Theorem 2.3. ■

It is interesting to note that the solution (2) for $\mathbf{x}_n$ depends not only on the $n + 1$ past vectors $\hat{\mathbf{f}}_0, \hat{\mathbf{f}}_1, \ldots, \hat{\mathbf{f}}_n$, but also on $k - 1$ 'future' vectors $\mathbf{f}_{n+1}, \hat{\mathbf{f}}_{n+2}, \ldots, \hat{\mathbf{f}}_{n+k-1}$. When $\mathbf{A}$ is non-singular, $\mathbf{x}_n = (\mathbf{A}^{-1}\mathbf{B})^n\mathbf{q} + \sum_{i=0}^{n-1} (\mathbf{A}^{-1}\mathbf{B})^{n-i-1}\mathbf{A}\mathbf{f}_i$ and $\mathbf{x}_n$ depends only on the past vectors $\mathbf{f}_0, \mathbf{f}_1, \ldots, \mathbf{f}_{n-1}$.

In many applications one has a difference equation holding for only a subset of the $\mathbf{x}_i$. The difference equation that is discussed in Section 5 is solved by the following theorem.

Theorem 9.3.3 *Suppose that* $\mathbf{A}, \mathbf{B}$ *are square matrices and there exists a scalar* λ *such that* $\lambda\mathbf{A} + \mathbf{B}$ *is non-singular. Set* $\hat{\mathbf{A}} = (\lambda\mathbf{A} + \mathbf{B})^{-1}\mathbf{A}$ *and* $\hat{\mathbf{B}} = (\lambda\mathbf{A} + \mathbf{B})^{-1}\mathbf{B}$. *Then all solutions of* $\mathbf{A}\mathbf{x}_{i+1} + \mathbf{B}\mathbf{x}_i = \mathbf{0}, i = 0, \ldots, \mathrm{N} - 1$ *are given by* $\mathbf{x}_i = (-\hat{\mathbf{A}}^D\hat{\mathbf{B}})^i\hat{\mathbf{A}}^D\hat{\mathbf{A}}\mathbf{x}_0 + (-\hat{\mathbf{B}}^D\hat{\mathbf{A}})^{\mathrm{N}-i}(\mathbf{I} - \hat{\mathbf{A}}^D\hat{\mathbf{A}})\mathbf{x}_\mathrm{N}$.

Proof Suppose that there exists a λ such that $\lambda\mathbf{A} + \mathbf{B}$ is non-singular. Taking a similarity we get as in (3)

$$\hat{\mathbf{A}} = \begin{bmatrix} \mathbf{A}_1 & \mathbf{0} \\ \mathbf{0} & \mathbf{M} \end{bmatrix}, \hat{\mathbf{B}} = \begin{bmatrix} \mathbf{B}_1 & \mathbf{0} \\ \mathbf{0} & \mathbf{B}_2 \end{bmatrix}, \mathbf{M}^k = \mathbf{0}, \mathbf{x}_i = \begin{bmatrix} \mathbf{w}_i \\ \mathbf{v}_i \end{bmatrix}$$

with $\mathbf{B}_1 = \mathbf{I} - \lambda\mathbf{A}_1, \mathbf{B}_2 = \mathbf{I} - \lambda\mathbf{M}$. Then the difference equation is equivalent to the decoupled equations

$$\mathbf{A}_1\mathbf{w}_{i+1} + \mathbf{B}_1\mathbf{w}_i = \mathbf{0}, \quad \mathbf{M}\mathbf{v}_{i+1} + \mathbf{B}_2\mathbf{v}_i = \mathbf{0}, i = 0, \ldots, \mathrm{N} - 1.$$

Thus $\mathbf{w}_i = (-\mathbf{A}_1^{-1}\mathbf{B}_1)^i\mathbf{w}_0$, and $\mathbf{v}_i = (-\mathbf{B}_2^{-1}\mathbf{M})^{\mathrm{N}-i}\mathbf{v}_{\mathrm{N}}$. ■

4. The Leslie population growth model and backward population projection

Suppose that a population is partitioned according to age groups. Given specific rates of fertility and mortality, along with an initial age distribution, the Leslie model provides the age distribution of the survivors and descendants of the initial population at successive, discrete points in time.

It is a standard demographic practice to consider only one sex at a time. We will consider only the female portion of a population. Select a unit of time (e.g. 5 years, or 1 year, or 10 s, or 0.5 μs, etc.). Let Δt denote one unit of time. Select an integer m such that $m(\Delta t)$ is the maximum age to be consider. Construct m disjoint age classes or age intervals;

$$\mathrm{A}_1 = (0, \Delta t], \mathrm{A}_2 = (\Delta t, 2\Delta t], \ldots, \mathrm{A}_m = ((m-1)\Delta t, m\Delta t].$$

Let t_0 denote an initial point in time and for some integer n let

$$t = n(\Delta t).$$

Let us agree to say a *female belongs* to A_k at time t if she is living at time t, and her age lies in A_k at time t. To define the survival and birth rates, let $p_k(t)$ be the probability that a female in A_k at time t will be in A_{k+1} at time $t + \Delta t$ (survival rates). Let $b_k(t)$ be the expected number of daughters produced in the time interval $[t, t + \Delta t)$, which are alive at time $t + \Delta t$, by a female in A_k (birth rates). Furthermore, let $n_k(t)$ be the expected number of females in A_k at time t. Finally, let $\mathbf{n}(t) = [n_1(t), \ldots, n_m(t)]^T$. For convenience, adopt the notation $\mathbf{n}(t_i) = \mathbf{n}(i)$, $p_k(t_i) = p_k(i)$, and $b_k(t_i) = b_k(i)$. Suppose we know the age distribution $\mathbf{n}(i)$ of our population at time t_i. From this together with the survival rates and birth rates, we can obtain the expected age distribution of the population at time t_{i+1} as

$$\begin{bmatrix} n_1(i+1) \\ n_2(i+1) \\ n_3(i+1) \\ \vdots \\ n_m(i+1) \end{bmatrix} = \begin{bmatrix} b_1(i) & b_2(i) & b_3(i) \ldots b_{m-1}(i) & b_m(i) \\ p_1(i) & 0 & 0 \quad \ldots \quad 0 & 0 \\ 0 & p_2(i) & 0 \quad \ldots \quad 0 & 0 \\ \vdots & & & \\ 0 & 0 & 0 \quad \ldots p_{m-1}(i) & 0 \end{bmatrix} \begin{bmatrix} n_1(i) \\ n_2(i) \\ n_3(i) \\ \vdots \\ n_m(i) \end{bmatrix} \tag{1}$$

or $\mathbf{n}(i+1) = \mathbf{T}(i)\mathbf{n}(i)$. The expression (1) is the *Leslie model*. Many times, the survival rates and birth rates are constant with respect to the time scale under consideration. Let us make this assumption and write

$$p_k(t) = p_k, \quad b_k(t) = b_k,$$

so that (1) becomes

$$\mathbf{n}(i+1) = \mathbf{T}\mathbf{n}(i) \tag{2}$$

We shall refer to $\mathbf{T}$ as the *Leslie matrix*. Suppose now that we are given an initial population distribution, $\mathbf{n}(0)$. It is easy to see that we can now project forward into time and produce the expected population at a future time, say $t = t_k$, by $\mathbf{n}(k) = \mathbf{T}\mathbf{n}(k-1) = \mathbf{T}^2\mathbf{n}(k-2) = \ldots = \mathbf{T}^k\mathbf{n}(0)$.

We wish to deal with the problem of projecting a population distribution backward in time in order to determine what kind of population distribution there had to exist in the past, in order to produce the present population distribution. Such a problem might arise, for example, in a situation where one has statistics giving the age distribution for population A at only the time t_i and other statistics giving the age distribution for population B at a different time, say t_{i+x}. If one wishes to make a comparison of the two populations at time t_i, then it is necessary to project population B backward in time.

Since $\mathbf{n}(i) = \mathbf{T}^{-1}\mathbf{n}(i+1)$ the problem of backward population projection is trivial in the case when the matrix $\mathbf{T}$ of (1) is non-singular. If $\mathbf{T}$ is singular, the problem is more interesting.

The Leslie matrix is very often singular. As a simple example, consider the population of human American females. Let $\Delta t = 5$ years and $m = 20$ so that the age classes are:

$A_1 = (0, 5], A_2 = (5, 10], \ldots, A_{19} = (90, 95], A_{20} = (95, 100]$. Almost everyone would agree that at least $b_{20} = 0$.

Suppose the Leslie matrix is given by

$$\mathbf{T} = \left[\begin{array}{cccccc|cccc} b_1 & b_2 & b_3 & \ldots & b_{m-k-1} & b_{m-k} & 0 & 0\ldots & 0 & 0 \\ p_1 & 0 & 0 & \ldots & 0 & 0 & 0 & 0\ldots & 0 & 0 \\ 0 & p_2 & 0 & \ldots & 0 & 0 & 0 & 0\ldots & 0 & 0 \\ \vdots & & & & & & & & & \\ 0 & 0 & 0 & \ldots & p_{m-k-1} & 0 & 0 & 0\ldots & 0 & 0 \\ \hline 0 & 0 & 0 & \ldots & 0 & p_{m-k} & 0 & 0\ldots & 0 & 0 \\ 0 & 0 & 0 & \ldots & 0 & 0 & p_{m-k+1} & 0\ldots & 0 & 0 \\ \vdots & & & & & & & \ddots & & \\ 0 & 0 & 0 & \ldots & 0 & 0 & 0 & 0\ldots & p_{m-1} & 0 \end{array}\right]_{m \times m} \tag{3}$$

$$= \left[\begin{array}{c|c} \mathbf{T}_{11} & 0 \\ \hline \mathbf{T}_{21} & \mathbf{T}_{22} \end{array}\right]$$

The complete statement that we can make concerning backward population projection is the following.

Theorem 9.4.1 For the Leslie population growth model whose matrix is given by (3), *and for an integer $x \geq 0$, let j be an integer such that $0 \leq j \leq k + x$. The future distributions $\mathbf{n}(k + x)$ determine the past distributions $\mathbf{n}(k + x - j)$ as follows.*

$$\mathbf{n}(k+x-j)=\begin{cases}(\mathbf{T}^{\mathrm{D}})^j\mathbf{n}(k+x) & \text{if } 0\le j\le x,\\ \begin{bmatrix}(\mathbf{M}^{\mathrm{D}})^j\mathbf{n}_1(k+x)\\ \hdashline \mathbf{v}\end{bmatrix} & \text{if } x<j\le k+x\end{cases} \tag{4}$$

where $\mathbf{v}\in\mathbb{R}^{j-x}$ *is arbitrary,* $\mathbf{n}_1(k+x)\in\mathbb{R}^{m-(j-x)}$ *is the vector of the first* $m-(j-x)$ *components of* $\mathbf{n}(k+x)$*, and* $\mathbf{M}$ *is the leading principal submatrix obtained from* $\mathbf{T}$ *by deleting the last* $j-x$ *rows and columns from* $\mathbf{T}$.

Proof To prove the top half of (4) note that since $0\le j\le x$, we have $\mathbf{n}(k+x-j)\in R(\mathbf{T}^k)$, $\mathbf{n}(k+x)\in R(\mathbf{T}^k)$ and $\mathbf{n}(k+x)=\mathbf{T}^j\mathbf{n}(k+x-j)$. Thus $\mathbf{n}(k+x-j)=(\mathbf{T}^j)^{\mathrm{D}}\mathbf{n}(k+x)=(\mathbf{T}^{\mathrm{D}})^j\mathbf{n}(k+x)$. To show the bottom half of (4), first write $\mathbf{T}$ as

$$\mathbf{T}=\left[\begin{array}{c|c}\mathbf{M} & \mathbf{0}\\ \hline \mathbf{R} & \mathbf{N}\end{array}\right],\ \text{where } \mathbf{M}\in\mathbb{R}^{(m-[j-x])\times(m-[j-x])},$$

so that, by Corollary 7.7.1, $\mathbf{T}^{\mathrm{D}}=\left[\begin{array}{c|c}\mathbf{M}^{\mathrm{D}} & \mathbf{0}\\ \hline \mathbf{X} & \mathbf{0}\end{array}\right]$.

Note that

$$\mathbf{n}(k)=(\mathbf{T}^{\mathrm{D}})^x\mathbf{n}(k+x)=\begin{bmatrix}(\mathbf{M}^{\mathrm{D}})^x\mathbf{n}_1(k+x)\\ \hdashline \mathbf{X}(\mathbf{M}^{\mathrm{D}})^{x-1}\mathbf{n}_1(k+x)\end{bmatrix}=\begin{bmatrix}\mathbf{n}_1(k)\\ \hdashline \mathbf{n}_2(k)\end{bmatrix}.$$

To complete the proof, it suffices to show

$$\mathbf{n}(k-[j-x])=\begin{bmatrix}(\mathbf{M}^{\mathrm{D}})^{j-x}\mathbf{n}_1(k)\\ \hdashline \mathbf{v}\end{bmatrix},\quad \mathbf{v}\ \text{arbitrary.} \tag{5}$$

To simplify our notation , let $l=[j-x]$.
Partition $\mathbf{M}$ as

$$\mathbf{M}=\left[\begin{array}{cccccc|cccc}
b_1 & b_2 & b_3 & \cdots & b_{m-k-1} & b_{m-k} & 0 & 0\cdots & 0 & 0\\
p_1 & 0 & 0 & \cdots & 0 & 0 & 0 & 0\cdots & 0 & 0\\
0 & p_2 & 0 & \cdots & 0 & 0 & 0 & 0\cdots & 0 & 0\\
\vdots & & & & & & & & & \\
0 & 0 & 0 & \cdots & p_{m-k-1} & 0 & 0 & 0\cdots & 0 & 0\\
\hline
0 & 0 & 0 & \cdots & 0 & p_{m-k} & 0 & 0\cdots & 0 & 0\\
0 & 0 & 0 & \cdots & 0 & 0 & p_{m-k+1} & 0\cdots & 0 & 0\\
\vdots & & & & & & & \ddots & & \\
0 & 0 & 0 & \cdots & 0 & 0 & 0 & 0\cdots & p_{m-l-1} & 0
\end{array}\right]_{(m-l)\times(m-l)} \tag{6}$$

Notice that $\operatorname{Ind}(\mathbf{M})=k-l$. For each positive integer, p, let

$$\mathbf{S}(p)=\sum_{i=0}^{p-1}\mathbf{N}^i\mathbf{R}\mathbf{M}^{p-1-i},\ \text{that is,}\ \begin{bmatrix}\mathbf{M} & \mathbf{0}\\ \mathbf{R} & \mathbf{N}\end{bmatrix}^p=\begin{bmatrix}\mathbf{M}^p & \mathbf{0}\\ \mathbf{S}(p) & \mathbf{N}^p\end{bmatrix}.$$

Given the distribution $\mathbf{n}(k)$, there had to exist some initial population, $\mathbf{n}(0)$, (not necessarily unique) and an intermediate population, $\mathbf{n}(k-l)$, which gave rise to $\mathbf{n}(k)$. Write $\mathbf{n}(0)$ and $\mathbf{n}(k-l)$ as

$$\mathbf{n}(0) = \begin{bmatrix} \mathbf{n}_1(0) \\ \hline \mathbf{n}_2(0) \end{bmatrix}, \mathbf{n}(k-l) = \begin{bmatrix} \mathbf{n}_1(k-l) \\ \hline \mathbf{n}_2(k-l) \end{bmatrix},$$

where $\mathbf{n}_2(0), \mathbf{n}_2(k-l) \in \mathbb{R}^l$. Now $\mathbf{T}^k\mathbf{n}(0) = \mathbf{n}(k)$ so that $\mathbf{n}_1(k) \in R(\mathbf{M}^k) = R(\mathbf{M}^{k-l})$. But $\mathbf{T}^{k-l}\mathbf{n}(0) = \mathbf{n}(k-l)$ so that $\mathbf{n}_1(k-j) \in R(\mathbf{M}^{k-l})$. Also note that $\mathbf{T}^l\mathbf{n}(k-l) = \mathbf{n}(k)$. Since $\mathbf{M}^l\mathbf{n}_1(k-l) = \mathbf{n}_1(k)$ and $\text{Ind}(\mathbf{M}) = k-l$ we have $\mathbf{n}_1(k-l)$ is uniquely determined by $\mathbf{n}_1(k)$ as $\mathbf{n}_1(k-l) = (\mathbf{M}^{\mathrm{D}})^l\mathbf{n}_1(k)$. To finish the proof it suffices to show that any vector $\mathbf{u}$ of the form (5) gives rise to the distribution $\mathbf{n}(k)$ after l intervals of time have elapsed.

Since $\mathbf{M}^l\mathbf{n}(k-l) = \mathbf{M}^l(\mathbf{M}^{\mathrm{D}})^l\mathbf{n}_1(k) = \mathbf{M}\mathbf{M}^{\mathrm{D}}\mathbf{n}_1(k) = \mathbf{n}_1(k)$, we have

$$\mathbf{T}^l\mathbf{u} = \begin{bmatrix} \mathbf{n}_1(k) \\ \hline \mathbf{S}(l)(\mathbf{M}^{\mathrm{D}})^l\mathbf{n}_1(k) \end{bmatrix}.$$

Thus it is only necessary to show that $\mathbf{S}(l)(\mathbf{M}^{\mathrm{D}})^l\mathbf{n}_1(k) = \mathbf{n}_2(k)$.

If $\mathbf{n}(0) = \begin{bmatrix} \mathbf{n}_1(0) \\ \hline \mathbf{n}_2(0) \end{bmatrix}$ is any initial distribution which gives rise to $\mathbf{n}(k)$, then $\mathbf{T}^k\mathbf{n}(0) = \mathbf{n}(k)$ implies $\mathbf{n}_1(k) = \mathbf{M}^k\mathbf{n}_1(0)$ and $\mathbf{n}_2(k) = \mathbf{S}(k)\mathbf{n}_1(0)$. Hence $\mathbf{S}(l)(\mathbf{M}^{\mathrm{D}})^l\mathbf{n}_1(k) = \mathbf{S}(l)(\mathbf{M}^{\mathrm{D}})^l\mathbf{M}^k\mathbf{n}_1(0) = \mathbf{S}(l)\mathbf{M}^{k-j}\mathbf{n}_1(0) = \mathbf{S}(k)\mathbf{n}_1(0) = \mathbf{n}_2(0)$. ■

5. Optimal control

In Section 2 we showed how to find solutions for linear systems of differential equations with singular coefficients provided that the solutions were uniquely determined by consistent initial conditions. In this section we shall apply those results to an optimal control problem. The problem presented will provide an interesting example of the type of differential equation studied in Section 2.

In general, an optimal control problem involves a process $\mathbf{x}$, which is regulated by a control $\mathbf{u}$. The problem is to choose a control $\mathbf{u}$ so as to cause $\mathbf{x}$ to have some type of desired behaviour and minimize a cost $J[\mathbf{x}, \mathbf{u}]$. The cost may, of course, take many forms. It may be time, total energy, or something else. The desired behaviour of the process may range from going to zero to hitting a moving 'target'. Finally, the process may depend on the control in a variety of ways, often non-linear. We shall present a particular problem and handle it in some detail. Of course, similar problems may also be analyzed using these techniques.

Let $\mathbf{A}, \mathbf{B}$ be $n \times n$ and $n \times m$ matrices respectively. All matrices and scalars are allowed to be complex though, of course, in many applications they are real. The usual inner product for complex (or real) vectors is denoted $(.,.)$. Let $\mathbf{Q}, \mathbf{H}$ be positive semi-definite $m \times m$ and $n \times n$ matrices. Finally, let $\mathbf{x}, \mathbf{u}$ denote vector valued functions of the real variable t. $\mathbf{x}$ is $n \times 1$ while $\mathbf{u}$ is $m \times 1$.

We consider the autonomous (time independent coefficients) control *process*

$$\dot{\mathbf{x}} = \mathbf{A}\mathbf{x} + \mathbf{B}\mathbf{u} \tag{1}$$

on the time interval $[t_0, t_1]$ with *quadratic cost functional*

$$J[\mathbf{x}, \mathbf{u}] = \frac{1}{2}\int_{t_0}^{t_1} (\mathbf{H}\mathbf{x}, \mathbf{x}) + (\mathbf{Q}\mathbf{u}, \mathbf{u})dt. \tag{2}$$

The dot in (1) denotes differentiation with respect to t.

If one has a fixed pair of vectors $\mathbf{x}_0, \mathbf{x}_1$ such that there exist controls $\mathbf{u}$ so that the process $\mathbf{x}$ is at $\mathbf{x}_0$ at a time t_0 and $\mathbf{x}_1$ at time t_1, then one can ask for a control that minimizes the cost (2) subject to the restraint that $\mathbf{x}(t_0) = \mathbf{x}_0, \mathbf{x}(t_1) = \mathbf{x}_1$.

Using the theory of Lagrange multipliers one gets the system of equations

$$\begin{aligned} &\dot{\boldsymbol{\lambda}} + \mathbf{A}^*\boldsymbol{\lambda} + \mathbf{H}\mathbf{x} = \mathbf{0} \\ &\dot{\mathbf{x}} - \mathbf{A}\mathbf{x} - \mathbf{B}\mathbf{u} = \mathbf{0} \\ &\mathbf{B}^*\boldsymbol{\lambda} + \mathbf{Q}\mathbf{u} = \mathbf{0} \end{aligned} \tag{3}$$

as necessary conditions for optimization in this sense [1].

If $\mathbf{Q}$ is invertible, then $\mathbf{u}$ can be eliminated from the second equation and the resulting system formed by the first two equations solved directly. We shall be most interested then in the case when $\mathbf{Q}$ is not invertible, though our results will include the case when $\mathbf{Q}$ is invertible.

The system (3) can be rewritten as

$$\begin{bmatrix} \mathbf{I} & \mathbf{0} & \mathbf{0} \\ \mathbf{0} & \mathbf{I} & \mathbf{0} \\ \mathbf{0} & \mathbf{0} & \mathbf{0} \end{bmatrix} \begin{bmatrix} \dot{\boldsymbol{\lambda}} \\ \mathbf{x} \\ \mathbf{u} \end{bmatrix} + \begin{bmatrix} \mathbf{A}^* & \mathbf{H} & \mathbf{0} \\ \mathbf{0} & -\mathbf{A} & -\mathbf{B} \\ \mathbf{B}^* & \mathbf{0} & \mathbf{Q} \end{bmatrix} \begin{bmatrix} \dot{\boldsymbol{\lambda}} \\ \mathbf{x} \\ \mathbf{u} \end{bmatrix} = \begin{bmatrix} \mathbf{0} \\ \mathbf{0} \\ \mathbf{0} \end{bmatrix}. \tag{4}$$

Note that (4) has leading coefficient singular.

We assume throughout that controls are continuous. All statements concerning optimality are made with respect to the control problem of this section and this linear manifold of controls.

Optimal control problems with singular matrices in the quadratic cost functional have received much attention. They occur naturally as a first order approximation to more general optimal control problems. [45] surveys the known results on one such problem with singular matrices in the cost.

The approach given here has the advantage that it leads to explicit solutions for the problem studied, as well as a procedure for solution. These explicit, closed form solutions, also simplify the proof and development of the mathematical theory for the problem studied.

We shall first show that if (3) has a solution satisfying the boundary conditions, then $\mathbf{u}$ must be an optimal control.

Theorem 9.5.1 *Suppose that* $\mathbf{x}, \mathbf{u}, \boldsymbol{\lambda}$ *is a solution of* (3) *and* $\mathbf{x}(t_0) = \mathbf{x}_0$, $\mathbf{x}(t_1) = \mathbf{x}_1$. *Then* $\mathbf{u}$ *is an optimal control.*

Proof To show that $J[\hat{\mathbf{x}}, \hat{\mathbf{u}}] \geq J[\mathbf{x}, \mathbf{u}]$ for all $\hat{\mathbf{x}}, \hat{\mathbf{u}}$ satisfying (1) and the boundary conditions it is clearly equivalent to show that

$$\phi(s) = J[s\mathbf{x} + (1-s)\hat{\mathbf{x}}, s\mathbf{u} + (1-s)\hat{\mathbf{u}}]$$

has a minimum at $s = 1$ for all $\hat{\mathbf{x}}, \hat{\mathbf{u}}$. Let $J_0 = \frac{1}{2}\int_{t_0}^{t} (\mathbf{Hx}, \hat{\mathbf{x}}) + (\mathbf{Qu}, \hat{\mathbf{u}})dt$, $\hat{J} = J[\hat{\mathbf{x}}, \hat{\mathbf{u}}]$, and $J = J[\mathbf{x}, \mathbf{u}]$. Then a direct calculation gives

$\phi(s) = s^2(J - 2J_0 + \hat{J}) + s(2J_0 - 2\hat{J}) + \hat{J}$. Since ϕ is quadratic in s it has a maximum or minimum at $s = 1$ if and only if $J_0 = J$, or equivalently,

$$\int_{t_0}^{t_1} (\mathbf{Hx}, \hat{\mathbf{x}}) + (\mathbf{Qu}, \hat{\mathbf{u}})dt = \int_{t_0}^{t_1} (\mathbf{Hx}, \mathbf{x}) + (\mathbf{Qu}, \mathbf{u})dt. \tag{5}$$

However, $\phi(s) \geq 0$ for all s so that if (5) holds there must be a minimum. Clearly (5) is equivalent to

$$\int_{t_0}^{t_1} (\mathbf{Hx}, \hat{\mathbf{x}} - \mathbf{x})dt = \int_{t_0}^{t_1} (\mathbf{Qu}, \mathbf{u} - \hat{\mathbf{u}})dt. \tag{6}$$

But

$$\int_{t_0}^{t_1} (\mathbf{Hx}, \hat{\mathbf{x}} - \mathbf{x})dt = \int_{t_0}^{t_1} (\mathbf{Hx}, \hat{\mathbf{x}}) - (\mathbf{Hx}, \mathbf{x})dt. \tag{7}$$

Now $(\mathbf{Hx}, \mathbf{x}) = (-\dot{\boldsymbol{\lambda}} - \mathbf{A}^*\boldsymbol{\lambda}, \mathbf{x}) = -(\dot{\boldsymbol{\lambda}}, \mathbf{x}) - (\mathbf{A}^*\boldsymbol{\lambda}, \mathbf{x})$. But $(\mathbf{A}^*\boldsymbol{\lambda}, \mathbf{x}) = (\boldsymbol{\lambda}, \mathbf{Ax}) = (\boldsymbol{\lambda}, \dot{\mathbf{x}} - \mathbf{Bu}) = (\boldsymbol{\lambda}, \dot{\mathbf{x}}) - (\boldsymbol{\lambda}, \mathbf{Bu}) = (\boldsymbol{\lambda}, \dot{\mathbf{x}}) - (\mathbf{B}^*\boldsymbol{\lambda}, \mathbf{u}) = (\boldsymbol{\lambda}, \dot{\mathbf{x}}) + (\mathbf{Qu}, \mathbf{u})$. Thus $(\mathbf{Hx}, \mathbf{x}) = (\dot{\boldsymbol{\lambda}}, \mathbf{x}) - (\boldsymbol{\lambda}, \dot{\mathbf{x}}) - (\mathbf{Qu}, \mathbf{u})$. Similarly, $(\mathbf{Hx}, \hat{\mathbf{x}}) = (\dot{\boldsymbol{\lambda}}, \hat{\mathbf{x}}) - (\boldsymbol{\lambda}, \dot{\hat{\mathbf{x}}}) - (\mathbf{Qu}, \hat{\mathbf{u}})$. Hence

$$\begin{aligned}\int_{t_0}^{t_1} (\mathbf{Hx}, \hat{\mathbf{x}} - \mathbf{x})dt &= \int_{t_0}^{t_1} (\dot{\boldsymbol{\lambda}}, \mathbf{x}) + (\boldsymbol{\lambda}, \dot{\mathbf{x}}) + (\mathbf{Qu}, \mathbf{u}) - (\dot{\boldsymbol{\lambda}}, \hat{\mathbf{x}}) - (\boldsymbol{\lambda}, \dot{\hat{\mathbf{x}}}) - (\mathbf{Qu}, \hat{\mathbf{u}})dt \\ &= (\boldsymbol{\lambda}, \mathbf{x})\Big|_{t_0}^{t_1} - (\boldsymbol{\lambda}, \hat{\mathbf{x}})\Big|_{t_0}^{t_1} + \int_{t_0}^{t_1} (\mathbf{Qu}, \mathbf{u}) - (\mathbf{Qu}, \hat{\mathbf{u}})dt \\ &= \int_{t_0}^{t_1} (\mathbf{Qu}, \mathbf{u} - \hat{\mathbf{u}})dt. \quad \blacksquare\end{aligned}$$

Note that Theorem 1 says that solutions of (3) satisfying the boundary conditions provide optimal controls even if the differential equation (3) has non-unique solutions for consistent initial conditions. Of course, in that case the optimal controls may not be unique.

As a useful by-product of the proof of Theorem 1 we have that

$$J[\mathbf{x}, \mathbf{u}] = -2(\boldsymbol{\lambda}, \mathbf{x})\Big|_{t_0}^{t_1}. \tag{8}$$

To simplify the solving of (4) rewrite it as

$$\mathbf{A}\dot{\mathbf{z}} + \mathbf{B}\mathbf{z} = \mathbf{0} \tag{9}$$

where $\mathbf{A} = \begin{bmatrix} \mathbf{I} & \mathbf{0} \\ \mathbf{0} & \mathbf{0} \end{bmatrix}$, $\mathbf{B} = \begin{bmatrix} \mathbf{B}_1 & \mathbf{B}_2 \\ \mathbf{B}_3 & \mathbf{B}_4 \end{bmatrix}$. Here $\mathbf{I}$ is $2n \times 2n$,

$$\mathbf{B}_1 = \begin{bmatrix} \mathbf{A}^* & \mathbf{H} \\ \mathbf{0} & -\mathbf{A} \end{bmatrix}, \mathbf{B}_2 = \begin{bmatrix} \mathbf{0} \\ -\mathbf{B} \end{bmatrix}, \mathbf{B}_3 = [\mathbf{B}^* \quad \mathbf{0}], \text{ and } \mathbf{B}_4 = \mathbf{Q}. \tag{10}$$

Clearly $(\mu + \mathbf{B}_1)^{-1}$ exists except for a finite number of μ. Define

$$\mathbf{Q}_\mu = \mathbf{B}_4 - \mathbf{B}_3(\mu + \mathbf{B}_1)^{-1}\mathbf{B}_2. \tag{11}$$

We now need the following easily verified result whose proof we omit.

Proposition 9.5.1 *$\mu\mathbf{A} + \mathbf{B}$ is invertible almost always if and only if $\mathbf{Q}_\mu$ is invertible almost always.*

Assume that $\mu, \mathbf{A}, \mathbf{B}$ are such that $\mu\mathbf{A} + \mathbf{B}, \mathbf{Q}_\mu, \mu + \mathbf{B}_1$ are invertible. Let $\hat{\mathbf{A}} = (\mu\mathbf{A} + \mathbf{B})^{-1}\mathbf{A}, \hat{\mathbf{B}} = (\mu\mathbf{A} + \mathbf{B})^{-1}\mathbf{B}$. Then $\mathbf{N}_\mu, \mathbf{M}_\mu, \mathbf{Z}_\mu$ are defined by

$$\hat{\mathbf{A}} = \begin{bmatrix} \mathbf{N}_\mu & \mathbf{0} \\ \mathbf{M}_\mu & \mathbf{0} \end{bmatrix}, \hat{\mathbf{B}} = \begin{bmatrix} \mathbf{Z}_\mu & \mathbf{0} \\ -\mu\mathbf{M}_\mu & \mathbf{I} \end{bmatrix} = \begin{bmatrix} \mathbf{I} - \mu\mathbf{N}_\mu & \mathbf{0} \\ -\mu\mathbf{M}_\mu & \mathbf{I} \end{bmatrix}.$$

Using Corollary 7.7.2 we get

$$\hat{\mathbf{A}}^{\mathrm{D}}\hat{\mathbf{A}} = \begin{bmatrix} \mathbf{N}_\mu^{\mathrm{D}} & \mathbf{0} \\ \mathbf{M}_\mu\mathbf{N}_\mu^{\mathrm{D}^2} & \mathbf{0} \end{bmatrix}\begin{bmatrix} \mathbf{N}_\mu & \mathbf{0} \\ \mathbf{M}_\mu & \mathbf{0} \end{bmatrix} = \begin{bmatrix} \mathbf{N}_\mu^{\mathrm{D}}\mathbf{N}_\mu & \mathbf{0} \\ \mathbf{M}_\mu\mathbf{N}_\mu^{\mathrm{D}} & \mathbf{0} \end{bmatrix},$$

$$\hat{\mathbf{A}}^{\mathrm{D}}\hat{\mathbf{B}} = \begin{bmatrix} \mathbf{N}_\mu^{\mathrm{D}} & \mathbf{0} \\ \mathbf{M}_\mu\mathbf{N}_\mu^{\mathrm{D}^2} & \mathbf{0} \end{bmatrix}\begin{bmatrix} \mathbf{Z}_\mu & \mathbf{0} \\ -\mu\mathbf{M}_\mu & \mathbf{I} \end{bmatrix} = \begin{bmatrix} \mathbf{N}_\mu^{\mathrm{D}}\mathbf{Z}_\mu & \mathbf{0} \\ \mathbf{M}_\mu\mathbf{N}_\mu^{\mathrm{D}^2}\mathbf{Z}_\mu & \mathbf{0} \end{bmatrix}.$$

To evaluate $\mathrm{e}^{-\hat{\mathbf{A}}^{\mathrm{D}}\hat{\mathbf{B}}t}$ note that for integers $r \geq 1$,

$$\begin{bmatrix} \mathbf{N}_\mu^{\mathrm{D}}\mathbf{Z}_\mu & \mathbf{0} \\ \mathbf{M}_\mu\mathbf{N}_\mu^{\mathrm{D}^2}\mathbf{Z}_\mu & \mathbf{0} \end{bmatrix}^r = \begin{bmatrix} [\mathbf{N}_\mu^{\mathrm{D}}\mathbf{Z}_\mu]^r & \mathbf{0} \\ \mathbf{M}_\mu\mathbf{N}_\mu^{\mathrm{D}^2}\mathbf{Z}_\mu[\mathbf{N}_\mu^{\mathrm{D}}\mathbf{Z}_\mu]^{r-1} & \mathbf{0} \end{bmatrix}.$$

Thus the power series expansion of the exponential gives

$$\mathrm{e}^{-\hat{\mathbf{A}}^{\mathrm{D}}\hat{\mathbf{B}}t} = \begin{bmatrix} \mathrm{e}^{-[\mathbf{N}_\mu^{\mathrm{D}}\mathbf{Z}_\mu]t} & \mathbf{0} \\ \mathbf{M}_\mu\mathbf{N}_\mu^{\mathrm{D}}\{\mathrm{e}^{-[\mathbf{N}_\mu^{\mathrm{D}}\mathbf{Z}_\mu]t} - \mathbf{I}\} & \mathbf{I} \end{bmatrix}.$$

Using Theorem 2.3 we see that the general solution of (9) is

$$\mathrm{e}^{-\hat{\mathbf{A}}^{\mathrm{D}}\hat{\mathbf{B}}t}\hat{\mathbf{A}}^{\mathrm{D}}\hat{\mathbf{A}} = \begin{bmatrix} \mathrm{e}^{-[\mathbf{N}_\mu^{\mathrm{D}}\mathbf{Z}_\mu]t} & \mathbf{0} \\ \mathbf{M}_\mu\mathbf{N}_\mu^{\mathrm{D}}\{\mathrm{e}^{-[\mathbf{N}_\mu^{\mathrm{D}}\mathbf{Z}_\mu]t} - \mathbf{I}\} & \mathbf{I} \end{bmatrix}\begin{bmatrix} \mathbf{N}_\mu^{\mathrm{D}}\mathbf{N}_\mu & \mathbf{0} \\ \mathbf{M}_\mu\mathbf{N}_\mu^{\mathrm{D}} & \mathbf{0} \end{bmatrix}$$

$$= \begin{bmatrix} \mathrm{e}^{-[\mathbf{N}_\mu^{\mathrm{D}}\mathbf{Z}_\mu]t}\mathbf{N}_\mu^{\mathrm{D}}\mathbf{N}_\mu & \mathbf{0} \\ \mathbf{M}_\mu\mathbf{N}_\mu^{\mathrm{D}}\mathrm{e}^{-[\mathbf{N}_\mu^{\mathrm{D}}\mathbf{Z}_\mu]t} & \mathbf{0} \end{bmatrix}.$$

From the original equation (4) we have that

$$\begin{bmatrix} \boldsymbol{\lambda} \\ \mathbf{x} \end{bmatrix} = \mathrm{e}^{-[\mathbf{N}_\mu^{\mathrm{D}}\mathbf{Z}_\mu](t-t_0)}\mathbf{N}_\mu^{\mathrm{D}}\mathbf{N}_\mu\begin{bmatrix} \boldsymbol{\lambda}_0 \\ \mathbf{x}_0 \end{bmatrix}, \text{ where } \boldsymbol{\lambda}_0 = \boldsymbol{\lambda}(t_0), \tag{12}$$

and

$$\mathbf{u} = \mathbf{M}_\mu \mathbf{N}_\mu^{\mathrm{D}} \begin{bmatrix} \boldsymbol{\lambda} \\ \mathbf{x} \end{bmatrix}. \tag{13}$$

Thus we have shown the following.

Theorem 9.5.2 *If $\mathbf{Q}_\mu$ is invertible, then the optimal control $\mathbf{u}$ is given in terms of $\mathbf{x}, \boldsymbol{\lambda}$ by (13) if an optimal control exists.*

While (13) gives $\begin{bmatrix} \boldsymbol{\lambda} \\ \mathbf{x} \end{bmatrix}$ explicitly, (13) does not give $\mathbf{u}$ directly in terms of $\mathbf{x}$. We now turn to this problem.

Let $\mathbf{E}(t) = e^{-[\mathbf{N}_\mu^{\mathrm{D}}\mathbf{Z}_\mu](t-t_0)}\mathbf{N}_\mu^{\mathrm{D}}\mathbf{N}_\mu = \begin{bmatrix} \mathbf{E}_1(t) & \mathbf{E}_2(t) \\ \mathbf{E}_3(t) & \mathbf{E}_4(t) \end{bmatrix}$ where the $\mathbf{E}_i(t)$, $i = 1, 2, 3, 4$ are all $n \times n$ matrices.

Suppose that (3) has a solution. Let $\boldsymbol{\lambda}(t_0) = \boldsymbol{\lambda}_0$. Then $\begin{bmatrix} \boldsymbol{\lambda}(t) \\ \mathbf{x}(t) \end{bmatrix} = \mathbf{E}(t)\begin{bmatrix} \boldsymbol{\lambda}_0 \\ \mathbf{x}_0 \end{bmatrix}$. Note that this is possible if and only if $\begin{bmatrix} \boldsymbol{\lambda}_0 \\ \mathbf{x}_0 \end{bmatrix}$ is in $R(\mathbf{N}_\mu^{\mathrm{D}}\mathbf{N}_\mu)$.
Now $\begin{bmatrix} \boldsymbol{\lambda}(t_1) \\ \mathbf{x}(t_1) \end{bmatrix} = \mathbf{E}(t_1)\begin{bmatrix} \boldsymbol{\lambda}_0 \\ \mathbf{x}_0 \end{bmatrix}$ or

$$\mathbf{x}_1 = \mathbf{E}_3(t_1)\boldsymbol{\lambda}_0 + \mathbf{E}_4(t_1)\mathbf{x}_0. \tag{14}$$

Once $\boldsymbol{\lambda}_0, \mathbf{x}_0$ are known, $\mathbf{x}, \mathbf{u}$ follow from (12) and (13).

On the other hand if (14) is viewed as defining $\mathbf{x}_1$, then from (12) $\mathbf{x}$ will go from $\mathbf{x}_0$ to $\mathbf{x}_1$. Thus we have established the following result.

Theorem 9.5.3 *Suppose that $\mathbf{Q}_\mu$ is invertible almost always. For a given $\mathbf{x}_0, \mathbf{x}_1$ there is an optimal control that takes $\mathbf{x}$ from $\mathbf{x}_0$ to $\mathbf{x}_1$ in the time interval $[t_0, t_1]$ if and only if the equation (14) has a solution $\boldsymbol{\lambda}_0$ such that $\begin{bmatrix} \boldsymbol{\lambda}_0 \\ \mathbf{x}_0 \end{bmatrix} \in R(\mathbf{N}_\mu^{\mathrm{D}}\mathbf{N}_\mu)$.*

It is possible, under our assumptions, for $\mathbf{x}$ to be able to go from $\mathbf{x}_0$ to $\mathbf{x}_1$ but not have an optimal control existing if $\mathbf{N}_\mu^{\mathrm{D}}\mathbf{N}_\mu \begin{bmatrix} \boldsymbol{\lambda}_0 \\ \mathbf{x}_0 \end{bmatrix} = \begin{bmatrix} \boldsymbol{\lambda}_0 \\ \mathbf{x}_0 \end{bmatrix}$, $\boldsymbol{\lambda}_0$ satisfying (14), is inconsistent in $\boldsymbol{\lambda}_0$. We shall give a simple example that illustrates this. It shall also serve to illustrate our method.

Example 9.5.1 Let $\mathbf{H} = \mathbf{I}, \mathbf{B} = \mathbf{I}, \mathbf{A} = \mathbf{0}, \mathbf{Q} = \begin{bmatrix} 1 & 0 \\ 0 & 0 \end{bmatrix}$ be two by two matrices. The process is then simply $\dot{\mathbf{x}} = \mathbf{u}$, and the cost is

$$\frac{1}{2}\int_{t_0}^{t_1} |x_1|^2 + |x_2|^2 + |u_1|^2 dt, \quad \mathbf{x} = [x_1, x_2]^{\mathrm{T}}, \quad \mathbf{u} = [u_1, u_2]^{\mathrm{T}}.$$

The system (4) becomes

$$\begin{bmatrix} \mathbf{I} & \mathbf{0} & \mathbf{0} \\ \mathbf{0} & \mathbf{I} & \mathbf{0} \\ \mathbf{0} & \mathbf{0} & \mathbf{0} \end{bmatrix} \begin{bmatrix} \dot{\boldsymbol{\lambda}} \\ \dot{\mathbf{x}} \\ \dot{\mathbf{u}} \end{bmatrix} + \begin{bmatrix} \mathbf{0} & \mathbf{I} & \mathbf{0} \\ \mathbf{0} & \mathbf{0} & -\mathbf{I} \\ \mathbf{I} & \mathbf{0} & \mathbf{Q} \end{bmatrix} \begin{bmatrix} \boldsymbol{\lambda} \\ \mathbf{x} \\ \mathbf{u} \end{bmatrix} = \begin{bmatrix} \mathbf{0} \\ \mathbf{0} \\ \mathbf{0} \end{bmatrix}. \tag{15}$$

Since $\mathbf{B}$ is invertible, we may take $\mu = \mathbf{0}$ in $(\mu\mathbf{A} + \mathbf{B})^{-1}$. Now

$$\mathbf{B}^{-1} = \begin{bmatrix} \mathbf{0} & \mathbf{I} & \mathbf{0} \\ \mathbf{0} & \mathbf{0} & -\mathbf{I} \\ \mathbf{I} & \mathbf{0} & -\mathbf{Q} \end{bmatrix}^{-1} = \begin{bmatrix} \mathbf{0} & \mathbf{Q} & \mathbf{I} \\ \mathbf{I} & \mathbf{0} & \mathbf{0} \\ \mathbf{0} & -\mathbf{I} & \mathbf{0} \end{bmatrix}$$

Multiplying (15) by $\mathbf{B}^{-1}$ gives

$$\begin{bmatrix} \mathbf{0} & \mathbf{Q} & \mathbf{0} \\ \mathbf{I} & \mathbf{0} & \mathbf{0} \\ \mathbf{0} & -\mathbf{I} & \mathbf{0} \end{bmatrix} \begin{bmatrix} \dot{\boldsymbol{\lambda}} \\ \dot{\mathbf{x}} \\ \dot{\mathbf{u}} \end{bmatrix} + \begin{bmatrix} \boldsymbol{\lambda} \\ \mathbf{x} \\ \mathbf{u} \end{bmatrix} = \mathbf{0}. \tag{16}$$

It is straightforward then, to get that the solutions to (15) are given by

$$\begin{bmatrix} \boldsymbol{\lambda} \\ \mathbf{x} \\ \mathbf{u} \end{bmatrix} = \exp\left(\begin{bmatrix} \mathbf{0} & \mathbf{Q} & \mathbf{0} \\ \mathbf{Q} & \mathbf{0} & \mathbf{0} \\ \mathbf{0} & -\mathbf{Q} & \mathbf{0} \end{bmatrix} (t - t_0) \right) \begin{bmatrix} \mathbf{Q} & \mathbf{0} & \mathbf{0} \\ \mathbf{0} & \mathbf{Q} & \mathbf{0} \\ -\mathbf{Q} & \mathbf{0} & \mathbf{0} \end{bmatrix} \begin{bmatrix} \boldsymbol{\lambda}_0 \\ \mathbf{x}_0 \\ \mathbf{u}_0 \end{bmatrix}. \tag{17}$$

It is clear that for any $\mathbf{x}_0, \mathbf{x}_1$ there exists a control $\mathbf{u}$ sending $\mathbf{x}_0$ to $\mathbf{x}_1$. But the $\mathbf{x}$ in (17) only takes on values of the form $\begin{bmatrix} c \\ 0 \end{bmatrix}$ for scalar c. Thus in order for an optimal control to exist, $\mathbf{x}_0, \mathbf{x}_1$ must be of the form $\begin{bmatrix} c_0 \\ 0 \end{bmatrix}$, $\begin{bmatrix} c_1 \\ 0 \end{bmatrix}$. A look at the power series for the exponential in (17) shows that

$$\begin{bmatrix} \boldsymbol{\lambda} \\ \mathbf{x} \\ \mathbf{u} \end{bmatrix} = \begin{bmatrix} \cosh(t-t_0)\mathbf{Q} + (\mathbf{I} - \mathbf{Q}) & -\sinh(t-t_0)\mathbf{Q} & \mathbf{0} \\ -\sinh(t-t_0)\mathbf{Q} & \cosh(t-t_0)\mathbf{Q} + (\mathbf{I} - \mathbf{Q}) & \mathbf{0} \\ -\cosh(t-t_0)\mathbf{Q} + \mathbf{Q} & -\sinh(t-t_0)\mathbf{Q} & \mathbf{I} \end{bmatrix}$$

$$\times \begin{bmatrix} \mathbf{Q} & \mathbf{0} & \mathbf{0} \\ \mathbf{0} & \mathbf{Q} & \mathbf{0} \\ -\mathbf{Q} & \mathbf{0} & \mathbf{0} \end{bmatrix} \begin{bmatrix} \boldsymbol{\lambda}_0 \\ \mathbf{x}_0 \\ \mathbf{u}_0 \end{bmatrix}$$

$$= \begin{bmatrix} \cosh(t-t_0)\mathbf{Q} & -\sinh(t-t_0)\mathbf{Q} & \mathbf{0} \\ -\sinh(t-t_0)\mathbf{Q} & \cosh(t-t_0)\mathbf{Q} & \mathbf{0} \\ -\cosh(t-t_0)\mathbf{Q} + \mathbf{Q} & -\sinh(t-t_0)\mathbf{Q} & \mathbf{0} \end{bmatrix} \begin{bmatrix} \boldsymbol{\lambda}_0 \\ \mathbf{x}_0 \\ \mathbf{u}_0 \end{bmatrix}.$$

If $\mathbf{x}_0 = \begin{bmatrix} c_0 \\ 0 \end{bmatrix}$ and $\mathbf{x}_1 = \begin{bmatrix} c_1 \\ 0 \end{bmatrix}$, we see that $t = t_0$ gives $\mathbf{u} = \mathbf{Q}\boldsymbol{\lambda}_0$. Since

$$\begin{bmatrix} \boldsymbol{\lambda}_0 \\ \mathbf{x}_0 \\ \mathbf{u}_0 \end{bmatrix} \in R\left(\begin{bmatrix} \mathbf{Q} & \mathbf{0} & \mathbf{0} \\ \mathbf{0} & \mathbf{Q} & \mathbf{0} \\ -\mathbf{Q} & \mathbf{0} & \mathbf{0} \end{bmatrix} \right)$$ we must have $\boldsymbol{\lambda}_0 = \begin{bmatrix} l_0 \\ 0 \end{bmatrix}$, and then $\mathbf{u}_0 = \begin{bmatrix} -l_0 \\ 0 \end{bmatrix}$.

Letting $t = t_1$ gives

$$c_1 = -\sinh(t_1 - t_0)l_0 + \cosh(t_1 - t_0)c_0. \tag{18}$$

Solving (18) for l_0, we have

$$\mathbf{u} = \begin{bmatrix} \cosh(t - t_0)\{c_1 - \cosh(t_1 - t_0)c_0\}/\sinh(t_1 - t_0) & -\sinh(t - t_0)c_0 \\ 0 & \end{bmatrix}$$

as the optimal control. $\mathbf{x}$ can also be easily solved for if desired.

We have arrived then at the following procedure for solving the original problem. Given $\mathbf{x}_0, \mathbf{x}_1$ determine whether it is possible to go from $\mathbf{x}_0$ to $\mathbf{x}_1$ with an optimal control by solving (if possible) (14) for $\boldsymbol{\lambda}_0$ such that $\begin{bmatrix} \boldsymbol{\lambda}_0 \\ \mathbf{x}_0 \end{bmatrix} \in R(\mathbf{N}_\mu^D \mathbf{N}_\mu)$. If $\boldsymbol{\lambda}_0$ is found, use the bottom half of (12) for $\mathbf{x}$ if $\mathbf{x}$ is needed. Use (12) and (3) to get the optimal control $\mathbf{u}$.

In working a given problem, it is sometimes simpler to solve (4) directly using the techniques used in deriving the formulas (12) and (13) as done in this example, rather than try to use the formulas directly.

At this point, an obvious question is 'What does $\mathbf{Q}_\mu$ being invertible mean?' That is, 'What is the physical significance of assuming the invertibility of $\mathbf{Q}_\mu$?' The answer itself is easily comprehended. The proof, however, requires some knowledge about Laplace transforms. The reader without an understanding of Laplace transforms and analytic functions is encouraged to read the statement of the theorems. From (10) and (11) we have

$$\mathbf{Q}_\mu = \mathbf{B}_4 - \mathbf{B}_3(\mu + \mathbf{B}_1)^{-1}\mathbf{B}_2 = \mathbf{Q} - [\mathbf{B}^* \ \ \mathbf{0}]\begin{bmatrix} \mu + \mathbf{A}^* & \mathbf{H} \\ \mathbf{0} & \mu - \mathbf{A} \end{bmatrix}\begin{bmatrix} \mathbf{0} \\ -\mathbf{B} \end{bmatrix}$$

$$= \mathbf{Q} - \mathbf{B}^*(\mu + \mathbf{A}^*)^{-1}\mathbf{H}(\mu - \mathbf{A})^{-1}\mathbf{B}. \tag{19}$$

Proposition 9.5.2 *If $\mathbf{Q}$ is invertible, then $\mathbf{Q}_\mu$ is almost always invertible.*

Proof $\lim_{\mu \to \infty} (\mu + \mathbf{A}^*)^{-1} = \mathbf{0}$, $\lim_{\mu \to \infty} (\mu - \mathbf{A})^{-1} = \mathbf{0}$ and the invertible $m \times m$ matrices form an open set. ■

If $\mathbf{Q}$ is invertible, then it is obvious from (4) that $\mathbf{u}$ can be solved for in terms of $\mathbf{x}, \boldsymbol{\lambda}$. Theorem 3 shows that this can happen even when $\mathbf{Q}$ is not invertible.

We note without proof the following proposition.

Proposition 9.5.3 *If $\mathbf{F}, \mathbf{G}$ are positive semi-definite $r \times r$ matrices, then $\mathbf{F} + \mathbf{G}$ is invertible if and only if $N(\mathbf{F}) \cap N(\mathbf{G}) = \{\mathbf{0}\}$.*

Of course, $\mathbf{Q}_\mu$ is invertible almost always for real μ if and only if it is almost always invertible for complex μ. Let $\mu = is$ where s is real. Then (19) becomes $\mathbf{Q}_\mu = \mathbf{Q} - \mathbf{B}^*(is + \mathbf{A}^*)^{-1}\mathbf{H}(is - \mathbf{A})^{-1}\mathbf{B} = \mathbf{Q} + \mathbf{B}^*(-is + \mathbf{A})^{-1*} \times \mathbf{H}(-is + \mathbf{A})^{-1}\mathbf{B}$. From Proposition 3 we have that $\mathbf{Q}_\mu$ is invertible almost always if and only if $\{\mathbf{0}\} = N(\mathbf{Q}) \cap N(\mathbf{B}^*(-is + \mathbf{A})^{-1*}\mathbf{H}(-is + \mathbf{A})^{-1}\mathbf{B}) = N(\mathbf{Q}) \cap N(\mathbf{H}^{1/2}(-is + \mathbf{A})^{-1}\mathbf{B}) = N(\mathbf{Q}) \cap N(\mathbf{H}(-is + \mathbf{A})^{-1}\mathbf{B})$ for almost all real s. Thus we have proven that:

Theorem 9.5.4 $\mathbf{Q}_\mu$ *is invertible for almost all* μ *if and only if* $N(\mathbf{Q}) \cap N(\mathbf{H}(-is + \mathbf{A})^{-1}\mathbf{B}) = \{\mathbf{0}\}$ *for almost every real* s.

We need a technical result on analytic (1)-inverses before proceeding.

Theorem 9.5.5 *Suppose that* $\mathbf{A}(\cdot)$ *is an* $m \times n$ *matrix valued function such that* $\mathbf{A}_{ij}(z)$ *is a fraction of polynomials for all* i *and* j. *Suppose also that* $N(\mathbf{A}(z))$ *is non-trivial for all* z *in the domain of* $\mathbf{A}(\cdot)$. *Then for any real number* $\omega > 0$, *there exists an* $m \times n$ *matrix valued function* $\mathbf{B}(\cdot)$ *such that*

(i) $\mathbf{B}_{ij}(z)$ *is a fraction of polynomials.*
(ii) $R(\mathbf{B}(z)) = N(\mathbf{A}(z))$ *for almost all* z.
(iii) *The poles of* $\mathbf{B}$ *are integral multiples of* ωi, $\omega > 0$, *are simple, and*
(iv) $\|\mathbf{B}(z)\| = O(1/|z|^3)$ *as* $|z| \to \infty$.

Proof Suppose that $\mathbf{A}(\cdot)$ is an $m \times n$ matrix valued function such that $\mathbf{A}_{ij}(z)$ is a fraction of polynomials for all i and j. Suppose also that $N(\mathbf{A}(z))$ is non-trivial for all z in the domain of $\mathbf{A}(\cdot)$. Let $\mathbf{X}$ be an $n \times m$ matrix of unknowns $\mathbf{X}_{ij}$. Then $\mathbf{AXA} = \mathbf{A}$ is a consistent linear system of at most mn equations in mn unknowns. Denote this new system by (s). Since the coefficients of s are fractions of polynomials, there exists a real number K such that all minors of (s) are identically zero, or identically non-zero, for $|z| \geq K$. Thus s can be solved by row operations (non-uniquely) to give a $\mathbf{F}(\cdot)$ such that for $|z| \geq K$; $\mathbf{AFA} = \mathbf{A}$, the entries of $\mathbf{F}(z)$ are fractions of polynomials in z, rank $(\mathbf{F}(z))$ is constant, and rank $(\mathbf{F}(z))$ is the maximum possible (dim $N(\mathbf{A}(z))$). Note that $(\mathbf{FA})_{ij}$ is a fraction of polynomials for all i and j. Let $z_1, \ldots, z_q$ be the poles of $\mathbf{FA}$. Let $r_1, \ldots, r_q$ denote their multiplicities. Let r_0 be such that $\|\mathbf{FA}\| = O(|z|^{r_0})$ as $|z| \to \infty$. Set $a = r_0 + r_1 + \ldots + r_q + 3$. Define

$$\mathbf{B}(z) = \left\{ \prod_{j=1}^{q} (z - z_j)^{r_j} \prod_{p=1}^{a} (z - ip\omega)^{-1} \right\} (\mathbf{I} - \mathbf{F}(z)\mathbf{A}(z)).$$

Then $\mathbf{B}$ clearly satisfies (i), (iii) and (iv). Since (ii) holds for $|z| \geq K$, it holds for almost all z by analytic continuation. ■

We can now prove the following.

Theorem 9.5.6 *The following are equivalent.*

(a) *There exists an* $\mathbf{x}_0, \mathbf{x}_1$ *for which optimal controls exists, but are not unique.*

(b) *There is a trajectory from zero to zero of zero cost with non-zero control.*

(c) $\mathbf{Q}_\mu$ *is not invertible for all* μ.

Proof Clearly (b) ⇒ (a) since $\mathrm{J}[\mathbf{0},\mathbf{0}] = 0$. To see that (a) ⇒ (b), let $(\mathbf{x},\mathbf{u})$, $(\hat{\mathbf{x}},\hat{\mathbf{u}})$ be two optimal solutions from $\mathbf{x}_0$ to $\mathbf{x}_1$. Then there exists $\boldsymbol{\lambda}, \hat{\boldsymbol{\lambda}}$ so that $(\boldsymbol{\lambda},\mathbf{x},\mathbf{u})$ and $(\hat{\boldsymbol{\lambda}},\hat{\mathbf{x}},\hat{\mathbf{u}})$ satisfy (3). Thus $(\boldsymbol{\lambda}-\hat{\boldsymbol{\lambda}},\mathbf{x}-\hat{\mathbf{x}},\mathbf{u}-\hat{\mathbf{u}})$ satisfies (3) and hence is optimal by Theorem 1. But $(\mathbf{x}-\hat{\mathbf{x}})(t_0) = (\mathbf{x}-\hat{\mathbf{x}})(t_1) = \mathbf{0}$ and $\mathbf{u}-\hat{\mathbf{u}}$ is not identically zero. That $\mathrm{J}[\mathbf{x}-\hat{\mathbf{x}},\mathbf{u}-\hat{\mathbf{u}}] = 0$ follows from (8).

Suppose now that (b) holds so that there exists $\mathbf{x},\mathbf{u}$ such that $\mathrm{J}[\mathbf{x},\mathbf{u}] = 0$, $\mathbf{x}(t_0) = \mathbf{0}, \mathbf{x}(t_1) = \mathbf{0}$, and $\mathbf{u}$ is non-zero. Since $\mathrm{J}[\mathbf{x},\mathbf{u}] = 0$ it is clear from (2) that $\mathbf{Hx} = \mathbf{0}$ and $\mathbf{Qu} = \mathbf{0}$. Extend $\mathbf{x},\mathbf{u}$ periodically to $[-\infty,\infty]$ and replace t by $t - t_0$. Call the new functions $\tilde{\mathbf{x}},\tilde{\mathbf{u}}$. Thus $\mathbf{H}\tilde{\mathbf{x}} = \mathbf{0}$, $\mathbf{Q}\tilde{\mathbf{u}} = \mathbf{0}$, and $\dot{\tilde{\mathbf{x}}} = \mathbf{A}\tilde{\mathbf{x}} + \mathbf{B}\tilde{\mathbf{u}}, t \neq n(t_1 - t_0), n = 0, \pm 1, \pm 2, \ldots$ Since $\tilde{\mathbf{u}}$ is bounded and sectionally continuous on finite intervals, $\tilde{\mathbf{x}}$ is continuous, and $\tilde{\mathbf{x}}$ is of exponential order, we can take Laplace transforms to get $\mathbf{H}L[\tilde{\mathbf{x}}] = \mathbf{0}$, $\mathbf{Q}L[\tilde{\mathbf{u}}] = \mathbf{0}$ and $L[\tilde{\mathbf{x}}] = (s-\mathbf{A})^{-1}\mathbf{B}L[\tilde{\mathbf{u}}]$. Thus $L[\tilde{\mathbf{u}}](s) \in N(\mathbf{Q}) \cap N(\mathbf{H}(s-\mathbf{A})^{-1}\mathbf{B})$ for all s in some right half plane. By Theorem 4, we have $\mathbf{Q}_\mu$ is not invertible for all μ.

Conversely, suppose that $\mathbf{Q}_\mu$ is not invertible for all μ. From the proof of Theorem 4 we have $N(\mathbf{Q}) \cap N(\mathbf{H}(\mu-\mathbf{A})^{-1}\mathbf{B}) = N(\mathbf{Q}_\mu)$ for $\mu = it$, t real. Thus $N(\mathbf{Q}) \cap N(\mathbf{H}(\mu-\mathbf{A})^{-1}\mathbf{B}) = N(\mathbf{Q}_\mu)$ for almost all μ. Now applying Theorem 5 to $\mathbf{Q}_\mu$ with $\omega = 2\pi/(t_1 - t_0)$ yields a $\mathbf{B}_\mu$ such that $\mathbf{Q}_\mu\mathbf{B}_\mu = \mathbf{0}$, and $\mathbf{B}_\mu$ satisfies (iii), (iv). But then $\mathbf{QB}_\mu = \mathbf{0}$, and $\mathbf{H}(\mu-\mathbf{A})^{-1}\mathbf{BB}_\mu = \mathbf{0}$. Let $\boldsymbol{\phi}$ be vector such that $\mathbf{B}_\mu\boldsymbol{\phi}$ is not identically zero. Denote $\mathbf{B}_\mu\boldsymbol{\phi}$ by $\boldsymbol{\phi}(\mu)$. Let $\tilde{\mathbf{x}}(s) = (s-\mathbf{A})^{-1}\mathbf{B}\boldsymbol{\phi}(s)$. Then we have that

$$\mathbf{H}\tilde{\mathbf{x}}(s) = \mathbf{0}, \mathbf{Q}\boldsymbol{\phi}(s) = \mathbf{0}, \text{ and } \tilde{\mathbf{x}}(s) = (s-\mathbf{A})^{-1}\mathbf{B}\boldsymbol{\phi}(s). \tag{20}$$

Let $\mathbf{x}$ be the inverse Laplace transform of $\tilde{\mathbf{x}}$, $\mathbf{u}$ the inverse Laplace transform of $\boldsymbol{\phi}$. From (20) and (iv) we have $\mathbf{H}\hat{\mathbf{x}} = \mathbf{0}$, $\mathbf{Q}\hat{\mathbf{u}} = \mathbf{0}$, $\dot{\mathbf{x}} = \mathbf{A}\hat{\mathbf{x}} + \mathbf{B}\hat{\mathbf{u}}$, $\hat{\mathbf{x}}(0) = \mathbf{0}$, and $\hat{\mathbf{u}}(0) = \mathbf{0}$. [29, p. 184]. Furthermore, $\hat{\mathbf{u}}$ is non-zero. Finally, since the poles of $\boldsymbol{\phi}(s)$ were simple and multiples of $2\pi i/(t_1 - t_0)$ we get that $\hat{\mathbf{x}},\hat{\mathbf{u}}$ are periodic with period $(t_1 - t_0)$ [29, p. 188]. Replace $\hat{\mathbf{x}},\hat{\mathbf{u}}$ by $\mathbf{x} = \hat{\mathbf{x}}(t+t_0), \mathbf{u} = \hat{\mathbf{u}}(t+t_0)$. Then $\mathbf{x}(t_0) = \mathbf{x}(t_1) = \mathbf{0}$, $\mathrm{J}[\mathbf{x},\mathbf{u}] = 0$, and $\dot{\mathbf{x}} = \mathbf{Ax} + \mathbf{Bu}$. Thus (c) ⇒ (b). ■

It is possible to have $\mathbf{Q}_\mu$ invertible almost always and still have non-zero optimal trajectories of zero cost. Of course, the control $\mathbf{u}$ must then be zero.

Example 9.5.2 Let $\mathbf{Q} = \mathbf{I}, \mathbf{A} = \mathbf{I}, \mathbf{B} = \mathbf{0}, \mathbf{H} = \mathbf{0}$ in (2) and (3). Then $\mathbf{Q}_\mu$ is invertible for large μ since $\mathbf{Q}$ is. Clearly $\mathbf{x} = \exp(\mathbf{A}(t-t_0))\mathbf{x}_0$ is a trajectory of zero cost from $\mathbf{x}_0$ to $\mathbf{x}_1 = \exp(\mathbf{A}(t_1-t_0))\mathbf{x}_0$. But $\mathbf{u} = \mathbf{0}$ and $\mathrm{J}[\mathbf{x},\mathbf{u}] = 0$. Note also if $\mathbf{x}_0 = \mathbf{0}$, then $\mathbf{x} \equiv \mathbf{0}$.

We will make no use of 'controllability', and hence will not define it. For the benefit of the reader familiar with the concept, note that the invertibility of $\mathbf{Q}_\mu$ is logically independent of the controllability of (2) since for any choice of $\mathbf{A},\mathbf{B}$, setting $\mathbf{Q} = \mathbf{I}$ makes $\mathbf{Q}_\mu$ invertible almost always, while setting $\mathbf{Q} = \mathbf{H} = \mathbf{0}$ makes $\mathbf{Q}_\mu \equiv \mathbf{0}$.

Note also that in Example 2, the pair $(\mathbf{A},\mathbf{B})$ was completely controllable and $\mathbf{Q}_\mu$ was invertible. However, optimal controls only existed for certain pairs $\mathbf{x}_0, \mathbf{x}_1$. Thus the assumption of controllability does not seem to simplify matters if $\mathbf{Q}, \mathbf{H}$ are allowed to be singular.

The method of this section can be applied, of course, to any problem which leads to a system of the form $\mathbf{A}\dot{\mathbf{z}} + \mathbf{B}\mathbf{z} = \mathbf{f}$. However, the special form of the $\mathbf{A}$ makes most of the calculations possible since it allowed us to use Theorem 7.8.1. Any problem which leads to a system with $\mathbf{A} = \begin{bmatrix} \mathbf{A}_1 & \mathbf{0} \\ \mathbf{A}_2 & \mathbf{0} \end{bmatrix}$ can be solved much as was (9), provided, of course, $\mu\mathbf{A} + \mathbf{B}$ is invertible for some μ. We shall now describe several such problems. The calculation of the solutions parallels those just done, so a description of the problem will suffice.

For example, suppose that the cost is given by $\int_{t_0}^{t_1} (\mathbf{Hx}, \mathbf{x}) + (\mathbf{Qu}, \mathbf{u}) + \langle \mathbf{x}, \mathbf{a} \rangle dt$ where $\mathbf{a}$ is a vector. Then the right hand side of (4) has $\boldsymbol{\alpha} = (\mathbf{a}^*, \mathbf{0}^*, \mathbf{0}^*)^*$ instead of the zero vector.

Theorem 2.3 can be used to solve this non-homogeneous system to get

$$\Omega = \hat{\mathbf{A}}^D e^{-\hat{\mathbf{A}}^D\hat{\mathbf{B}}t} \int_{t_0}^{t} e^{\hat{\mathbf{A}}^D\hat{\mathbf{B}}s}\hat{\mathbf{a}}\,ds + (\mathbf{I} - \hat{\mathbf{A}}\hat{\mathbf{A}}^D)\hat{\mathbf{B}}^D\hat{\boldsymbol{\alpha}} + e^{-\hat{\mathbf{A}}^D\hat{\mathbf{B}}t}\hat{\mathbf{A}}^D\hat{\mathbf{A}}\mathbf{q}.$$

The integral can be evaluated by using Theorem 2.4. For this problem, it is important to know whether or not the cost is positive.

Another variation on the same type of problem is process (1) with the cost functional $J[\mathbf{x}, \mathbf{u}] = \int_{t_0}^{t_1} \langle \mathbf{Hx}, \mathbf{x} \rangle + 2\langle \mathbf{u}, \mathbf{Cx} \rangle + \langle \mathbf{Qu}, \mathbf{u} \rangle dt$ where $\begin{bmatrix} \mathbf{H} & \mathbf{C}^* \\ \mathbf{C} & \mathbf{Q} \end{bmatrix}$ is positive semi-definite [1, pp. 461–463]. In this case the system to be solved is

$$\begin{bmatrix} \mathbf{I} & \mathbf{0} & \mathbf{0} \\ \mathbf{0} & \mathbf{I} & \mathbf{0} \\ \mathbf{0} & \mathbf{0} & \mathbf{0} \end{bmatrix} \begin{bmatrix} \dot{\boldsymbol{\lambda}} \\ \dot{\mathbf{x}} \\ \dot{\mathbf{u}} \end{bmatrix} + \begin{bmatrix} \mathbf{A}^* & \mathbf{H} & \mathbf{C}^* \\ \mathbf{0} & -\mathbf{A} & -\mathbf{B} \\ \mathbf{B}^* & \mathbf{C} & \mathbf{Q} \end{bmatrix} \begin{bmatrix} \boldsymbol{\lambda} \\ \mathbf{x} \\ \mathbf{u} \end{bmatrix} = \begin{bmatrix} \mathbf{0} \\ \mathbf{0} \\ \mathbf{0} \end{bmatrix}.$$

Solution proceeds almost exactly as when $\mathbf{C} = \mathbf{0}$, though $\mathbf{Q}_\mu$ has a slightly different form.

The analysis developed here can be also applied with little change to the following non-optimal control problem.

Given output $\mathbf{y}$, state vector $\mathbf{x}$, and process $\dot{\mathbf{x}} = \mathbf{Ax} + \mathbf{Bu}$, find a control $\mathbf{u}$ such that $\mathbf{y} = \mathbf{Cx} + \mathbf{Du}$. The appropriate system then is

$$\begin{bmatrix} \mathbf{I} & \mathbf{0} \\ \mathbf{0} & \mathbf{0} \end{bmatrix} \begin{bmatrix} \dot{\mathbf{x}} \\ \dot{\mathbf{u}} \end{bmatrix} + \begin{bmatrix} -\mathbf{A} & -\mathbf{B} \\ \mathbf{C} & \mathbf{D} \end{bmatrix} \begin{bmatrix} \mathbf{x} \\ \mathbf{u} \end{bmatrix} = \begin{bmatrix} \mathbf{0} \\ \mathbf{y} \end{bmatrix}. \tag{21}$$

If $\mathbf{y}$ and $\mathbf{u}$ are the same size vectors, then (21) is the non-homogeneous form of equation (9). It may be solved by our techniques under the assumption that $\mathbf{Q}_\mu = \mathbf{D} + \mathbf{C}(\mu - \mathbf{A})^{-1}\mathbf{B}$ is invertible.

One frequently does not want to have $\mathbf{D}$ a square matrix in (21). If $\mathbf{D}$ is not square, then (21) can often be solved using Theorems 7.10.16 and 7.10.20.

Discrete control systems arise both as discretized versions of continuous systems and as systems of independent interest. By using the results of Section 3 some discrete problems can be handled in much the same way as the continuous ones were handled. For example, consider the following:

Discrete control problem Let $\mathbf{A}, \mathbf{B}, \mathbf{Q}, \mathbf{H}$ be matrices of sizes $n \times n$, $n \times m$, $m \times m$, and $n \times n$ respectively. Assume that $\mathbf{Q}, \mathbf{H}$ are positive semi-definite. Let N be a fixed integer. Given the process

$$\mathbf{x}_{i+1} = \mathbf{A}\mathbf{x}_i + \mathbf{B}\mathbf{u}_i, i = 0, \ldots, N-1, \tag{22}$$

the cost

$$\mathbf{J}[\mathbf{x}, \mathbf{u}] = \frac{1}{2}\sum_{i=0}^{N} (\mathbf{H}\mathbf{x}_i, \mathbf{x}_i) + (\mathbf{Q}\mathbf{u}_i, \mathbf{u}_i), \tag{23}$$

and the initial position $\mathbf{x}_0$, find the control sequence which minimizes the cost (23). Here $\mathbf{x} = \{\mathbf{x}_i\}$, $\mathbf{u} = \{\mathbf{u}_i\}$.

Note that the terminal position is not specified whereas it was in the continuous problems considered earlier.

Theorem 9.5.7 *The Discrete Control Problem has a solution $\{\mathbf{x}_i\}, \{\mathbf{u}_i\}$, if and only if there exists $\{\boldsymbol{\lambda}_i\}$ such that the sequences $\{\mathbf{x}_i\}, \{\boldsymbol{\lambda}_i\}, \{\mathbf{u}_i\}$ satisfy*

$$\begin{bmatrix} \mathbf{I} & \mathbf{0} & \mathbf{0} \\ \mathbf{H} & -\mathbf{A}^* & \mathbf{0} \\ \mathbf{0} & \mathbf{0} & \mathbf{0} \end{bmatrix} \begin{bmatrix} \mathbf{x}_{i+1} \\ \boldsymbol{\lambda}_{i+1} \\ \mathbf{u}_{i+1} \end{bmatrix} + \begin{bmatrix} -\mathbf{A} & \mathbf{0} & -\mathbf{B} \\ \mathbf{0} & \mathbf{I} & \mathbf{0} \\ \mathbf{0} & -\mathbf{B}^* & \mathbf{Q} \end{bmatrix} \begin{bmatrix} \mathbf{x}_i \\ \boldsymbol{\lambda}_i \\ \mathbf{u}_i \end{bmatrix} = \begin{bmatrix} \mathbf{0} \\ \mathbf{0} \\ \mathbf{0} \end{bmatrix} \tag{24}$$

for $i = 0, 1, \ldots, N-1$, with $\mathbf{x}_0$ given and $\boldsymbol{\lambda}_N = \mathbf{0}, \mathbf{u}_N = \mathbf{0}$.

Proof Since $\mathbf{u}_N$ only appears in the cost and does not effect the $\{\mathbf{x}_i\}$, $\mathbf{u}_N$ may be taken to be any vector such that $\mathbf{Q}\mathbf{u}_N = \mathbf{0}$. Take $\mathbf{u}_N = \mathbf{0}$. To see the necessity of (24), consider $\mathrm{J}[\mathbf{x}, \mathbf{u}] + \sum_{i=0}^{N-1} (\boldsymbol{\lambda}_i, \mathbf{x}_{i+1} - \mathbf{A}\mathbf{x}_i - \mathbf{B}\mathbf{u}_i)$ and set $\boldsymbol{\lambda}_N = \mathbf{0}$. Then

$$\frac{\partial(\mathbf{x}_1 - \mathbf{A}\mathbf{x}_0 - \mathbf{B}\mathbf{u}_0, \ldots, \mathbf{x}_N - \mathbf{A}\mathbf{x}_{N-1} - \mathbf{B}\mathbf{u}_{N-1})}{\partial(\mathbf{x}_1, \ldots, \mathbf{x}_N)} = 1$$

where $(\mathbf{z}_1, \ldots, \mathbf{z}_N)$ is to be considered as a list of the n entries of $\mathbf{z}_1$ then the n entries of $\mathbf{z}_2$, etc. Thus one gets by the usual theory of Lagrange multipliers that

$$\begin{aligned} &\mathbf{H}\mathbf{x}_i - \mathbf{A}^*\boldsymbol{\lambda}_i + \boldsymbol{\lambda}_{i-1} = \mathbf{0}, \; i = 1, \ldots, N-1, \\ &\mathbf{H}\mathbf{x}_N + \boldsymbol{\lambda}_{N-1} = \mathbf{0}, \\ &\mathbf{Q}\mathbf{u}_i - \mathbf{B}^*\boldsymbol{\lambda}_i = \mathbf{0}, \qquad i = 0, \ldots, N-1 \\ &\mathbf{x}_{i+1} - \mathbf{A}\mathbf{x}_i - \mathbf{B}\mathbf{u}_i = \mathbf{0}, \quad i = 0, \ldots, N-1 \end{aligned} \tag{25}$$

is necessary. But (25) is equivalent to (24) since $\boldsymbol{\lambda}_N$ was taken equal to zero. On the other hand if $\{\mathbf{x}_i\}$, $\{\mathbf{u}_i\}$, $\{\boldsymbol{\lambda}_i\}$ satisfy (25), then one may show, almost exactly as for Theorem 1, that

$$\mathrm{J}[s\mathbf{x} + (1-s)\hat{\mathbf{x}}, s\mathbf{u} + (1-s)\hat{\mathbf{u}}],\ \hat{\mathbf{x}}_{i+1} = \mathbf{A}\hat{\mathbf{x}}_i + \mathbf{B}\hat{\mathbf{u}}_i,\ i = 0, \ldots, N-1,$$

has a minimum of $s = 1$. We omit the details. ■

For the control problem considered here, $\mathbf{x}_0$ can be arbitrary.

Theorem 9.5.8 Suppose that $\mathbf{Q} + \mathbf{B}^*(-\mu\mathbf{A}^* + \mathbf{I})^{-1}\mu\mathbf{H}(\mu - \mathbf{A})^{-1}\mathbf{B}$ *is invertible for some scalar* μ *(and hence for all but a finite number of* μ*). Then for every* $\mathbf{x}_0$ *there exists a solution to the Control Problem.*

Proof Given $\mathbf{x}_0$, $\mathbf{J}[\mathbf{x}, \mathbf{u}]$ defines a C^∞ function on $\mathbb{R}^{N\times m}$. Since J is bounded below, it suffices to show that $\mathrm{J}[\mathbf{x}, \mathbf{u}]$ goes to ∞ as $\sum_{i=0}^{N} \|\mathbf{u}_i\|^2$ does. If $\mathbf{Q}$ is invertible, this is clear. Suppose then that $\mathbf{Q}$ is singular and $\mathbf{Q} + \mathbf{B}^*(-\mu\mathbf{A}^* + \mathbf{I})^{-1}\mu\mathbf{H}(\mu - \mathbf{A})^{-1}\mathbf{B} = \mathbf{Q}_\mu$ is non-singular for almost all μ. Suppose for purposes of contradiction that there exists a sequence of control sequences $\{\mathbf{u}_{ir}\}$, $i = 0, \ldots, N-1$; $r = 0, \ldots,$ such that $\sum_{i=0}^{N-1} \|\mathbf{u}_{ir}\|^2 \to \infty$ but $\mathrm{J}[\mathbf{x}_r, \mathbf{u}_r]$ is bounded.

We shall show that, in fact, $\{\mathbf{u}_{ir}\}$ is bounded as $r \to \infty$. Since $\mathrm{J}[\mathbf{x}_r, \mathbf{u}_r]$ is bounded, one has $(\mathbf{Q}\mathbf{u}_{0r}, \mathbf{u}_{0r})$ is bounded. Hence $\mathbf{Q}^{1/2}\mathbf{u}_{0r}$ is bounded. But, $(\lambda\mathbf{H}\mathbf{x}_{1r}, \mathbf{x}_{1r}) = \|\mathbf{H}^{1/2}(\mathbf{A}\mathbf{x}_{0r} + \mathbf{B}\mathbf{u}_{0r})\|^2$ is also bounded. Then $\mathbf{H}^{1/2}\mathbf{B}\mathbf{u}_{0r}$ is bounded since $\mathbf{x}_{0r} = \mathbf{x}_0$ for all r. But $\mathbf{Q}_\mu$ is invertible for almost all μ, so that $\mathbf{u}_{0r}$ is bounded. Hence $\mathbf{x}_{1r}$ is bounded. Proceeding in this manner, one gets $\sum_{i=0}^{N-1} \|\mathbf{u}_{ir}\|^2$ is bounded. Thus J attains its minimum as desired. ■

We can now solve the Discrete Control Problem. Let $\mathbf{A}_1 = \begin{bmatrix} \mathbf{I} & \mathbf{0} \\ \mathbf{H} & -\mathbf{A} \end{bmatrix}$, $\mathbf{B}_1 = \begin{bmatrix} -\mathbf{A} & \mathbf{0} \\ \mathbf{0} & \mathbf{I} \end{bmatrix}$, $\mathbf{B}_2 = \begin{bmatrix} -\mathbf{B} \\ \mathbf{0} \end{bmatrix}$, $\mathbf{B}_3 = [\mathbf{0} - \mathbf{B}^*]$, $\mathbf{B}_4 = \mathbf{Q}$ and $\mathbf{z}_i = \begin{bmatrix} \mathbf{x}_i \\ \boldsymbol{\lambda}_i \end{bmatrix}$.

Then (24) becomes for $i = 0, \ldots, N-1$,

$$\begin{bmatrix} \mathbf{A}_1 & \mathbf{0} \\ \mathbf{0} & \mathbf{0} \end{bmatrix}\begin{bmatrix} \mathbf{z}_{i+1} \\ \mathbf{u}_{i+1} \end{bmatrix} + \begin{bmatrix} \mathbf{B}_1 & \mathbf{B}_2 \\ \mathbf{B}_3 & \mathbf{B}_4 \end{bmatrix}\begin{bmatrix} \mathbf{z}_i \\ \mathbf{u}_i \end{bmatrix} = \begin{bmatrix} \mathbf{0} \\ \mathbf{0} \end{bmatrix}. \tag{26}$$

Proposition 9.5.4 Let $\mathbf{Q}_\mu = \mathbf{B}_4 - \mathbf{B}_3\mathbf{H}_\mu^{-1}\mathbf{B}_2 = \mathbf{Q} + \mathbf{B}^*(-\mu\mathbf{A}^* + \mathbf{I})^{-1}\mathbf{H} \times (\mu - \mathbf{A})^{-1}\mathbf{B}$ *where* μ *is such that* $\mathbf{H}_\mu$ *and* $(\mu - \mathbf{A})$ *are invertible. Then*

$$\begin{bmatrix} \mu\mathbf{A}_1 + \mathbf{B}_1 & \mathbf{B}_2 \\ \mathbf{B}_3 & \mathbf{B}_4 \end{bmatrix} \tag{27}$$

is invertible if and only if $\mathbf{Q}_\mu$ *is invertible.*

It is assumed from here on that (27) and $\mu\mathbf{A}_1 + \mathbf{B}_1$ are invertible.

Multiply (26) by the inverse of (27) to get

$$\begin{bmatrix} \mathbf{N}_\mu & \mathbf{0} \\ \mathbf{M}_\mu & \mathbf{0} \end{bmatrix}\begin{bmatrix} \mathbf{z}_{i+1} \\ \mathbf{u}_{i+1} \end{bmatrix} + \begin{bmatrix} \mathbf{Z}_\mu & \mathbf{0} \\ -\mu\mathbf{M}_\mu & \mathbf{I} \end{bmatrix}\begin{bmatrix} \mathbf{z}_i \\ \mathbf{u}_i \end{bmatrix} = \begin{bmatrix} \mathbf{0} \\ \mathbf{0} \end{bmatrix}, \tag{28}$$

with $\mathbf{u}_N = \mathbf{0}$, $\boldsymbol{\lambda}_N = \mathbf{0}$, $\mathbf{x}_0$ given, $\mu\mathbf{N}_\mu + \mathbf{Z}_\mu = \mathbf{I}$. But

$$\begin{bmatrix} \mathbf{N}_\mu & \mathbf{0} \\ \mathbf{M}_\mu & \mathbf{0} \end{bmatrix}^{\mathrm{D}} = \begin{bmatrix} \mathbf{N}_\mu^{\mathrm{D}} & \mathbf{0} \\ \mathbf{M}_\mu\mathbf{N}_\mu^{\mathrm{D}^2} & \mathbf{0} \end{bmatrix}, \text{ and } \begin{bmatrix} \mathbf{Z}_\mu & \mathbf{0} \\ \mathbf{W}_\mu & \mathbf{I} \end{bmatrix}^{\mathrm{D}} = \begin{bmatrix} \mathbf{Z}_\mu^{\mathrm{D}} & \mathbf{0} \\ \mathbf{L}_\mu & \mathbf{0} \end{bmatrix}.$$

Here $\mathbf{L}_\mu = (\mathbf{W}_\mu + \mathbf{W}_\mu\mathbf{Z}_\mu + \ldots + \mathbf{W}_\mu\mathbf{Z}_u^{l-1})(\mathbf{I} - \mathbf{Z}_\mu^{\mathrm{D}}\mathbf{Z}_\mu) - \mathbf{W}_\mu\mathbf{Z}_\mu^{\mathrm{D}}$, $l = \mathrm{Ind}(\mathbf{Z}_\mu)$.

By Theorem 3.3, all solutions of (28) are given by

$$\begin{bmatrix} \mathbf{z}_i \\ \mathbf{u}_i \end{bmatrix} = \begin{bmatrix} -\mathbf{N}_\mu^{\mathrm{D}}\mathbf{Z}_\mu & \mathbf{0} \\ -\mathbf{M}_\mu\mathbf{N}_\mu^{\mathrm{D}^2}\mathbf{Z}_\mu & \mathbf{0} \end{bmatrix}^i \begin{bmatrix} \mathbf{N}_\mu^{\mathrm{D}}\mathbf{N}_\mu & \mathbf{0} \\ \mathbf{M}_\mu\mathbf{N}_\mu^{\mathrm{D}} & \mathbf{0} \end{bmatrix}\begin{bmatrix} \mathbf{z}_0 \\ \mathbf{u}_0 \end{bmatrix}$$

$$+ \begin{bmatrix} -\mathbf{Z}_\mu^{\mathrm{D}}\mathbf{N}_\mu & \mathbf{0} \\ -(\mathbf{L}_\mu\mathbf{N}_\mu + \mathbf{M}_\mu) & \mathbf{0} \end{bmatrix}^{N-i} \begin{bmatrix} \mathbf{I} - \mathbf{N}_\mu^{\mathrm{D}}\mathbf{N}_\mu & \mathbf{0} \\ -\mathbf{M}_\mu\mathbf{N}_\mu^{\mathrm{D}} & \mathbf{I} \end{bmatrix}\begin{bmatrix} \mathbf{z}_N \\ \mathbf{0} \end{bmatrix}. \tag{29}$$

A solution (29) will satisfy the boundary conditions if and only if

$$\mathbf{z}_0 = \mathbf{N}_\mu^{\mathrm{D}}\mathbf{N}_\mu\mathbf{z}_0 + (-\mathbf{Z}_\mu^{\mathrm{D}}\mathbf{N}_\mu)^N(\mathbf{I} - \mathbf{N}_\mu^{\mathrm{D}}\mathbf{N}_\mu)\mathbf{z}_N, \tag{30}$$

$$\mathbf{u}_0 = \mathbf{M}_\mu\mathbf{N}_\mu^{\mathrm{D}}\mathbf{z}_0 = (\mathbf{L}_\mu\mathbf{N}_\mu + \mathbf{M}_\mu)(-\mathbf{Z}_\mu^{\mathrm{D}}\mathbf{N}_\mu)^{N-1}(\mathbf{I} - \mathbf{N}_\mu^{\mathrm{D}}\mathbf{M}_\mu)\mathbf{z}_N, \tag{31}$$

$$\mathbf{z}_N = (-\mathbf{N}_\mu^{\mathrm{D}}\mathbf{Z}_\mu)^N\mathbf{N}_\mu^{\mathrm{D}}\mathbf{N}_\mu\mathbf{z}_0 + (\mathbf{I} - \mathbf{N}_\mu^{\mathrm{D}}\mathbf{N}_\mu)\mathbf{z}_N, \tag{32}$$

and

$$\mathbf{0} = \mathbf{M}_\mu(\mathbf{N}_\mu^{\mathrm{D}})^2\mathbf{Z}_\mu(-\mathbf{N}_\mu^{\mathrm{D}}\mathbf{Z}_\mu)^{N-1}\mathbf{N}_\mu^{\mathrm{D}}\mathbf{N}_\mu\mathbf{z}_0 - \mathbf{M}_\mu\mathbf{N}_\mu^{\mathrm{D}}\mathbf{z}_N. \tag{33}$$

Recall that $\mu\mathbf{N}_\mu + \mathbf{Z}_\mu = \mathbf{I}$. Thus $\mathbf{N}_\mu^{\mathrm{D}}$ and $\mathbf{Z}_\mu$ commute. Using (32), (33) becomes $-\mathbf{M}_\mu(-\mathbf{N}_\mu^{\mathrm{D}})^{N+1}\mathbf{Z}_\mu^N\mathbf{N}_\mu^{\mathrm{D}}\mathbf{N}_\mu\mathbf{z}_0 - \mathbf{M}_\mu\mathbf{N}_\mu^{\mathrm{D}}(-\mathbf{N}_\mu^{\mathrm{D}}\mathbf{Z}_\mu)^N\mathbf{N}_\mu^{\mathrm{D}}\mathbf{N}_\mu\mathbf{z}_0 = \mathbf{0}$. The preceding discussion is summarized in the following theorem.

Theorem 9.5.9 *Suppose that* $\mathbf{Q}_\mu$ *is invertible, that* $N > \mathrm{Ind}(\mathbf{N}_\mu)$, *and* $\mathbf{x}_0$ *is specified. Then* $\boldsymbol{\lambda}_0, \mathbf{x}_N$ *are obtained by solving*

$$(\mathbf{I} - \mathbf{N}_\mu^{\mathrm{D}}\mathbf{N}_\mu)\begin{bmatrix} \mathbf{x}_0 \\ \boldsymbol{\lambda}_0 \end{bmatrix} = \mathbf{0}, \text{ and } \mathbf{N}_\mu^{\mathrm{D}}\mathbf{N}_\mu\begin{bmatrix} \mathbf{x}_N \\ \mathbf{0} \end{bmatrix} = (-\mathbf{N}_\mu^{\mathrm{D}}\mathbf{Z}_\mu)^N\begin{bmatrix} \mathbf{x}_0 \\ \boldsymbol{\lambda}_0 \end{bmatrix}.$$

The control sequence is given by $\mathbf{u}_0 = \mathbf{M}_\mu\mathbf{N}_\mu^{\mathrm{D}}\begin{bmatrix} \mathbf{x}_0 \\ \boldsymbol{\lambda}_0 \end{bmatrix}$, *and for* $i > 0$,

$$\mathbf{u}_i = \mathbf{M}_\mu\mathbf{N}_\mu^{\mathrm{D}}\mathbf{Z}_\mu(-\mathbf{N}_\mu^{\mathrm{D}}\mathbf{Z}_\mu)^i\begin{bmatrix} \mathbf{x}_0 \\ \boldsymbol{\lambda}_0 \end{bmatrix} - (\mathbf{L}_\mu\mathbf{N}_\mu + \mathbf{M}_\mu)(-\mathbf{Z}_\mu^{\mathrm{D}}\mathbf{N}_\mu)^{N-i-1}$$

$$\times (\mathbf{I} - \mathbf{N}_\mu^{\mathrm{D}}\mathbf{N}_\mu)\begin{bmatrix} \mathbf{x}_N \\ \mathbf{0} \end{bmatrix}.$$

As mentioned earlier, one is probably better off to follow the steps in the proof of Theorem 9 rather than try to utilize the formulas.

If the process is not completely controllable, and $\mathbf{x}_0$ is a point that cannot be steered to the origin, then $\mathbf{x}_N$ will be unequal to zero. A very simple example is obtained by taking $\mathbf{A} = \mathbf{I}, \mathbf{B} = \mathbf{0}$, and $\mathbf{Q}$ invertible. Then $\mathbf{Q}_\mu$ is invertible. In this case, of course, one would get $\mathbf{x}_i = \mathbf{x}_0$, and $\mathbf{u}_i = \mathbf{0}$ for all i.

It is possible to have $\mathbf{Q}$ not invertible, the process not completely controllable and $\mathbf{Q}_\mu$ still be invertible and our results apply. One may take $\mathbf{A} = \mathbf{0}, \mathbf{H} = \begin{bmatrix} 1 & 0 \\ 0 & 0 \end{bmatrix}, \mathbf{B} = \begin{bmatrix} 1 & 0 \\ 0 & 0 \end{bmatrix}$ and $\mathbf{Q} = \begin{bmatrix} 0 & 0 \\ 0 & 1 \end{bmatrix}$ as an example.

In applications it frequently happens that $\mathbf{Q}$ is invertible. Unlike the continuous control problem, the discrete problem can still give rise to a singular difference equation when $\mathbf{Q}$ is invertible.

If $\mathbf{Q}$ in the Discrete Control Problem is non-singular, then $\mathbf{u}_i = \mathbf{Q}^{-1}\mathbf{B}^*\boldsymbol{\lambda}_i$ for $i = 0, 1, \ldots, N-1$ and (24) becomes

$$\tilde{\mathbf{A}}\mathbf{z}_{i+1} + \tilde{\mathbf{B}}\mathbf{z}_i = \begin{bmatrix} \mathbf{I} & \mathbf{0} \\ \mathbf{H} & -\mathbf{A}^* \end{bmatrix}\begin{bmatrix} \mathbf{x}_{i+1} \\ \boldsymbol{\lambda}_{i+1} \end{bmatrix} + \begin{bmatrix} -\mathbf{A} & -\mathbf{B}\mathbf{Q}^{-1}\mathbf{B}^* \\ \mathbf{0} & \mathbf{I} \end{bmatrix}\begin{bmatrix} \mathbf{x}_i \\ \boldsymbol{\lambda}_i \end{bmatrix}$$
$$= \begin{bmatrix} \mathbf{0} \\ \mathbf{0} \end{bmatrix}, \tag{34}$$

for $i = 0, 1, \ldots, N-1$, and $\boldsymbol{\lambda}_N = \mathbf{0}$.

$\tilde{\mathbf{A}}$ is invertible if and only if $\mathbf{A}$ is. However, there always exists a μ such that $\mu\tilde{\mathbf{A}} + \tilde{\mathbf{B}}$ is invertible so that Theorem 3.3 can always be applied. The difference equation (34) has the advantage that one can work with matrices that are $2n \times 2n$ instead of $(2n+m) \times (2n+m)$.

While $N \geq \operatorname{Ind}(\hat{\mathbf{A}})$ was assumed in the statement of the theorems, the assumption is not really necessary. If $N < \operatorname{Ind}(\hat{\mathbf{A}})$, one may still use Theorem 3.3 to solve (24). Note that Theorem 5.7 holds even if $\mathbf{Q}_\mu$ is a singular for all μ.

6. Functions of a matrix

If $f(\lambda)$ is an analytic function defined on a neighbourhood of the eigenvalues of the $n \times n$ matrix $\mathbf{A}$, then there exists a family of projections $\mathbf{P}_i, i = 1, \ldots, r$, such that $\mathbf{P}_i\mathbf{P}_j = \mathbf{0}$ if $i \neq j$, $\sum_i \mathbf{P}_i = \mathbf{I}$, and

$$f(\mathbf{A}) = \sum_{i=1}^{r} \sum_{m=0}^{k_i - 1} \frac{f^{(m)}(\lambda_i)}{m!}(\mathbf{A} - \lambda_i)^m \mathbf{P}_i \tag{1}$$

where $k_i = \operatorname{Ind}(\mathbf{A} - \lambda_i)$. (Equivalently, k_i is the multiplicity of λ_i as a root of the minimal polynomial of $\mathbf{A}$.) Formula (1) has been known for some time, see [49] for example. The purpose of this short section is to observe that the $\mathbf{P}_i$ may be explicitly written in terms of the Drazin inverse.

Theorem 9.6.1 *Suppose that* $\mathbf{A} \in \mathbb{C}^{n \times n}$ *has r distinct eigenvalues* $\{\lambda_1, \ldots, \lambda_r\}$, *that* $\operatorname{Ind}(\mathbf{A} - \lambda_i) = k_i$, *and that* $f(\lambda)$ *is analytic on a neighbourhood of the* λ_i.

Then

$$f(\mathbf{A}) = \sum_{i=1}^{r} \sum_{m=0}^{k_i - 1} \frac{f^{(m)}(\lambda_i)}{m!} (\mathbf{A} - \lambda_i)^m (\mathbf{I} - (\mathbf{A} - \lambda_i)^{\mathrm{D}} (\mathbf{A} - \lambda_i)). \tag{2}$$

Proof Let $\mathbf{Q}_i = (\mathbf{I} - (\mathbf{A} - \lambda_i)^{\mathrm{D}}(\mathbf{A} - \lambda_i))$. From [49] it suffices to show that

$$\mathbf{A}\mathbf{Q}_i = \mathbf{Q}_i\mathbf{A},\ (\mathbf{A} - \lambda_i)\mathbf{Q}_i \text{ is nilpotent, } \sum_{i=1}^{r} \mathbf{Q}_i = \mathbf{I}, \text{ and } \mathbf{Q}_i^2 = \mathbf{Q}_i \neq \mathbf{0}. \tag{3}$$

Suppose the Jordan form of $\mathbf{A}$ is given by $\mathbf{T}\mathbf{A}\mathbf{T}^{-1} = \mathrm{Diag}\{\mathbf{J}_1, \ldots, \mathbf{J}_r\}$ where $\mathbf{J}_i = \lambda_i\mathbf{I} + \mathbf{N}_i$ with $\mathbf{N}_i$ nilpotent. Then $\mathbf{T}\mathbf{Q}_i\mathbf{T}^{-1} = \mathrm{Diag}\{\mathbf{Q}_{i1}, \ldots, \mathbf{Q}_{ir}\}$ with $\mathbf{Q}_{il} = \mathbf{0}$ for $l \neq i$, $\mathbf{Q}_{il} = \mathbf{I}$ for $i = l$. Thus (3) holds. ■

The $\mathbf{Q}_i$ are often referred to as the *idempotent component matrices* of $\mathbf{A}$.

Corollary 9.6.1 *Let* $\mathbf{A}$ *be as in Theorem* 1. *Then*

$$\mathrm{e}^{\mathbf{A}t} = \sum_{i=1}^{r} \left[\sum_{m=0}^{k_i - 1} \frac{t^m}{m!} (\mathbf{A} - \lambda_i)^m (\mathbf{I} - (\mathbf{A} - \lambda_i)^{\mathrm{D}} (\mathbf{A} - \lambda_i)) \right] \mathrm{e}^{\lambda_i t}. \tag{4}$$

Proof Let $\mathrm{f}(\lambda) = \mathrm{e}^{\lambda}$, $\mathbf{B} = t\mathbf{A}$ in (2). Observe that the eigenvalues of $t\mathbf{A}$ are $\{t\lambda_1, \ldots, t\lambda_r\}$ and for $t > 0 (t\mathbf{A} - t\lambda_i)^{\mathrm{D}}(t\mathbf{A} - t\lambda_i) = (\mathbf{A} - \lambda_i)^{\mathrm{D}}(\mathbf{A} - \lambda_i)$. ■

Using Corollary 9.6.1, one may get the following version of Theorem 2.3.

Theorem 9.6.2 *Suppose there exists a c such that* $c\mathbf{A} + \mathbf{B}$ *is invertible. Let* $\hat{\mathbf{A}} = (c\mathbf{A} + \mathbf{B})^{-1}\mathbf{A}$, $\hat{\mathbf{B}} = (c\mathbf{A} + \mathbf{B})^{-1}\mathbf{B}$, *and* $\mathbf{C}_\lambda = \lambda\hat{\mathbf{A}} + \hat{\mathbf{B}}$ *for all* λ. *Suppose* $\mathbf{C}_\lambda$ *is not invertible for some* λ. *Let* $\{\lambda_1, \ldots, \lambda_r\}$ *be the* λ *for which* $\lambda\mathbf{A} + \mathbf{B}$ *is not invertible. Then all solutions of* $\mathbf{A}\dot{\mathbf{x}} + \mathbf{B}\mathbf{x} = \mathbf{0}$ *are of the form*

$$\mathbf{x} = \sum_{i=1}^{r} \sum_{m=0}^{k_i - 1} (-1)^m \frac{t^m}{m!} \hat{\mathbf{A}}^{\mathrm{D}m} \mathbf{C}_{\lambda_i} \mathrm{e}^{\lambda_i t} (\mathbf{I} - \mathbf{C}_{\lambda_i}^{\mathrm{D}} \mathbf{C}_{\lambda_i}) \mathbf{q}, \tag{5}$$

or equivalently,

$$\sum_{i=1}^{r} \sum_{m=0}^{k_i - 1} (-1)^m \frac{t^m}{m!} (\lambda_i - c + \hat{\mathbf{A}}^{\mathrm{D}})^m \mathrm{e}^{\lambda_i t} (\mathbf{I} - [(\lambda_i - c)\hat{\mathbf{A}} + \mathbf{I}]^{\mathrm{D}} [(\lambda_i - c)\hat{\mathbf{A}} + \mathbf{I}]) \mathbf{q} \tag{6}$$

where $k_i = \mathrm{Ind}\,\mathbf{C}_{\lambda_i}$, $\mathbf{q}$ *an arbitrary vector in* $\mathbb{C}^n$.

Proof The general solution of $\mathbf{A}\dot{\mathbf{x}} + \mathbf{B}\mathbf{x} = \mathbf{0}$ is $\mathbf{x} = \mathrm{e}^{-\hat{\mathbf{A}}^{\mathrm{D}}\hat{\mathbf{B}}t}\hat{\mathbf{A}}^{\mathrm{D}}\hat{\mathbf{A}}\mathbf{q}$, $\mathbf{q}$ an arbitrary vector by Theorem 10.13. Since $c\hat{\mathbf{A}} + \hat{\mathbf{B}} = \mathbf{I}$, we have

$$\mathbf{x} = \mathrm{e}^{-\hat{\mathbf{A}}^{\mathrm{D}}(\mathbf{I} - c\hat{\mathbf{A}})t}\hat{\mathbf{A}}^{\mathrm{D}}\hat{\mathbf{A}}\mathbf{q} = \mathrm{e}^{(-\hat{\mathbf{A}}^{\mathrm{D}} + c\mathbf{I})t}\hat{\mathbf{A}}^{\mathrm{D}}\hat{\mathbf{A}}\mathbf{q}. \tag{7}$$

Also if $\lambda_i\mathbf{A} + \mathbf{B}$ is not invertible, then $\lambda_i\hat{\mathbf{A}} + \hat{\mathbf{B}}$ is not. Hence $\lambda_i\hat{\mathbf{A}} + \mathbf{I} - c\hat{\mathbf{A}}$ is not. Thus $\lambda_i\mathbf{A} + \mathbf{B}$ is not invertible if and only if $-(\lambda_i - c)^{-1}$ is an eigenvalue of $\hat{\mathbf{A}}$. But then $(-\lambda_i + c)$ is an eigenvalue of $\hat{\mathbf{A}}^{\mathrm{D}}$. Thus $\lambda_i\mathbf{A} + \mathbf{B}$ is not invertible if and only if λ_i is an eigenvalue of $c\mathbf{I} - \hat{\mathbf{A}}^{\mathrm{D}}$. Both (4) and (6) now follow from (4), (7) and a little algebra. ■

Note that if the k_i in Theorem 1 are unknown or hard to compute, one may use n in their place.

It is interesting to note that while $\lambda\hat{\mathbf{A}} + \hat{\mathbf{B}}$ and $\lambda\mathbf{A} + \mathbf{B}$ have the same eigenvalues (λ for which $\det(\lambda\mathbf{A} + \mathbf{B}) = 0$), it is the algebraic multiplicity of the eigenvalue in the pencil $\lambda\hat{\mathbf{A}} + \hat{\mathbf{B}}$ that is important and not the multiplicity in $\lambda\mathbf{A} + \mathbf{B}$. In some sense, $\hat{\mathbf{A}}\dot{\mathbf{x}} + \hat{\mathbf{B}}\mathbf{x} = \mathbf{0}$ is a more natural equation to consider than $\mathbf{A}\dot{\mathbf{x}} + \mathbf{B}\mathbf{x} = \mathbf{0}$.

It is possible to get formulas like (5), (6) using inverses other than the Drazin. However, they tend to often be less satisfying, either because they apply to more restrictive cases, introduce extraneous solutions, or are more cumbersome. For example, one may prove the following corollary.

Corollary 9.6.2 *Let* $\mathbf{A}, \mathbf{B}$ *be as in Theorem* 1. *Then all solutions of* $\mathbf{A}\dot{\mathbf{x}} + \mathbf{B}\mathbf{x} = \mathbf{0}$ *are of the form*

$$\mathbf{x} = \sum_{i=1}^{r} \sum_{m=0}^{k_i - 1} (-1)^m \frac{t^m}{m!} (\hat{\mathbf{A}}^\dagger \mathbf{C}_{\lambda_i})^m e^{\lambda_i t} (\mathbf{I} - (\mathbf{C}_{\lambda_i}^{k_i})^\dagger \mathbf{C}_{\lambda_i}^{k_i}) \mathbf{q}_i \tag{8}$$

for some $\mathbf{q}_1, \ldots, \mathbf{q}_r$.

The proof of Corollary 2 is left to the exercises.

Formula (8) has several disadvantages in comparison to (5) or (6). First $\hat{\mathbf{A}}^\dagger$ and $\mathbf{C}_{\lambda_i}$ do not necessarily commute. Secondly, in (5) or (6) one has $\mathbf{q} = \mathbf{x}(0)$ while in (8) one needs to find $\mathbf{q}_k, \ldots, \mathbf{q}_r$ such that $\sum_{i=1}^{r} (\mathbf{I} - (\mathbf{C}_{\lambda_i}^{k_i})^\dagger \mathbf{C}_{\lambda_i}^{k_i}) \mathbf{q}_i = \mathbf{q}$, which may be a non-trivial task.

7. Weak Drazin inverses

The preceding sections have given several applications of the Drazin inverse.

It can, however, be difficult to compute the Drazin inverse. One way to lessen this latter problem is to look for a generalized inverse that would play much the same role for $\mathbf{A}^{\mathrm{D}}$ as the (1)-inverses play for $\mathbf{A}^\dagger$. One would not expect such an inverse to be unique. It should, at least in some cases of interest, be easier to compute. It should also be usable as a replacement for $\mathbf{A}^{\mathrm{D}}$ in many of the applications. Finally, it should have additional applications of its own.

Consider the difference equation

$$\mathbf{A}\mathbf{x}_{m+1} = \mathbf{x}_m, m \geq 0, \mathbf{A} \in \mathbb{C}^{n \times n}. \tag{1}$$

From Section 3 we know that all solutions of (1) are of the form $\mathbf{x}_m = (\mathbf{A}^{\mathrm{D}})^m \mathbf{A}^{\mathrm{D}} \mathbf{A} \mathbf{q}$, $\mathbf{q} \in \mathbb{C}^n$. It is the fact that the Drazin inverse solves (1) that helps explain its applications to differential equations in Section 2. We shall define an inverse so that it solves (1) when (1) is consistent. Note that in (1), we have $\mathbf{x}_m = \mathbf{A}^l \mathbf{x}_{m+l}$ for $l \geq 0$. Thus if our inverse is to always solve (1) it must send $R(\mathbf{A}^k)$, $k = \mathrm{Ind}(\mathbf{A})$, onto itself and have its restriction to $R(\mathbf{A}^k)$ the same as the inverse of $\mathbf{A}$ restricted to $R(\mathbf{A}^k)$. That is, it provides the unique solution to $\mathbf{Ax} = \mathbf{b}, \mathbf{x} \in R(\mathbf{A}^k)$, when $\mathbf{b} \in R(\mathbf{A}^k)$.

Definition 9.7.1 *Suppose that* $\mathbf{A} \in \mathbb{C}^{n \times n}$ *and* $k = \mathrm{Ind}(\mathbf{A})$. *Then* $\mathbf{B}$ *is a* weak Drazin inverse, *denoted* $\mathbf{A}^{\mathrm{d}}$, *if*

(d) $\mathbf{BA}^{k+1} = \mathbf{A}^k$.

$\mathbf{B}$ *is called a* projective weak Drazin inverse *of* $\mathbf{A}$ *if* $\mathbf{B}$ *satisfies* (d) *and*

(p) $R(\mathbf{BA}) = R(\mathbf{AA}^{\mathrm{D}})$.

$\mathbf{B}$ *is called a* commuting weak Drazin inverse of $\mathbf{A}$ *if* $\mathbf{B}$ *satisfies* (d) *and*

(c) $\mathbf{AB} = \mathbf{BA}$.

$\mathbf{B}$ *is called a* minimal rank weak Drazin inverse of $\mathbf{A}$ *if* $\mathbf{B}$ *satisfies* (d) *and*

(m) $\mathit{rank}(\mathbf{B}) = \mathit{rank}(\mathbf{A}^{\mathrm{D}})$.

Definition 9.7.2 *An* $(i_1, \ldots, i_m)$*-inverse of* $\mathbf{A}$ *is a matrix* $\mathbf{B}$ *satisfying the properties listed in the m-tuple. Here* $i_l \in \{1, 2, 3, 4, \mathrm{d}, \mathrm{m}, \mathrm{c}, \mathrm{p}\}$. *The integers* 1, 2, 3, 4 *represent the usual defining relations of the Moore–Penrose inverse. Properties d, m, c, p are as in Definition* 1.

We shall only be concerned with properties $\{1, 2, \mathrm{m}, \mathrm{d}, \mathrm{c}, \mathrm{p}\}$. Note that they are all invariant under a simultaneous similarity of $\mathbf{A}$ and $\mathbf{B}$. Also note that one could define a right weak (d)-inverse by $\mathbf{A}^{k+1}\mathbf{B} = \mathbf{A}^k$, and get a theory analogous to that developed here.

Theorem 9.7.1 *Suppose that* $\mathbf{A} \in \mathbb{C}^{n \times n}, k = \mathrm{Ind}\,\mathbf{A}$. *Suppose* $\mathbf{T} \in \mathbb{C}^{n \times n}$ *is a non-singular matrix such that*

$$\mathbf{TAT}^{-1} = \begin{bmatrix} \mathbf{C} & \mathbf{0} \\ \mathbf{0} & \mathbf{N} \end{bmatrix}, \quad \mathbf{C} \text{ non-singular, } \mathbf{N}^k = \mathbf{0}. \tag{2}$$

Then $\mathbf{B}$ *is a* (d)*-inverse of* $\mathbf{A}$ *if and only if*

$$\mathbf{TBT}^{-1} = \begin{bmatrix} \mathbf{C}^{-1} & \mathbf{X} \\ \mathbf{0} & \mathbf{Y} \end{bmatrix}, \quad \mathbf{X}, \mathbf{Y} \textit{ arbitrary}. \tag{3}$$

$\mathbf{B}$ *is an* (m, d)*-inverse for* $\mathbf{A}$ *if and only if*

$$\mathbf{TBT}^{-1} = \begin{bmatrix} \mathbf{C}^{-1} & \mathbf{X} \\ \mathbf{0} & \mathbf{0} \end{bmatrix}, \quad \mathbf{X} \textit{ arbitrary}. \tag{4}$$

$\mathbf{B}$ *is a* (p, d)*-inverse of* $\mathbf{A}$ *if and only if*

$$\mathbf{TBT}^{-1} = \begin{bmatrix} \mathbf{C}^{-1} & \mathbf{X} \\ \mathbf{0} & \mathbf{Y} \end{bmatrix}, \quad \mathbf{X} \textit{ arbitrary}, \ \mathbf{YN} = \mathbf{0}. \tag{5}$$

$\mathbf{B}$ *is a* (c, d)*-inverse of* $\mathbf{A}$ *if and only if*

$$\mathbf{TBT}^{-1} = \begin{bmatrix} \mathbf{C}^{-1} & \mathbf{0} \\ \mathbf{0} & \mathbf{Y} \end{bmatrix}, \quad \mathbf{YN} = \mathbf{NY}. \tag{6}$$

B *is a* (1, d)-*inverse of* **A** *if and only if*

$$\mathbf{TBT}^{-1} = \begin{bmatrix} \mathbf{C}^{-1} & \mathbf{X} \\ \mathbf{0} & \mathbf{N}^{-} \end{bmatrix}, \quad \mathbf{XN} = \mathbf{0},\ \mathbf{N}^{-} \text{ a (1)-}\textit{inverse of } \mathbf{N}. \tag{7}$$

B *is a* (2, d)-*inverse of* **A** *if and only if*

$$\mathbf{TBT}^{-1} = \begin{bmatrix} \mathbf{C}^{-1} & \mathbf{X} \\ \mathbf{0} & \mathbf{Y} \end{bmatrix}, \quad \mathbf{YNY} = \mathbf{Y}, \mathbf{XNY} = \mathbf{0}. \tag{8}$$

If $\mathbf{TAT}^{-1}$ *is nilpotent, then* (3)–(8) *are to be interpreted as the* (2, 2)-*block in the matrix. If* **A** *is invertible, then all reduce to* $\mathbf{A}^{-1}$.

Proof Let **A** be written as in (2). That each of (3)–(8) are the required types of inverses is a straight-forward verification. Suppose then that **B** is a (d)-inverse of **A**. The case when **A** is nilpotent or invertible is trivial, so assume that **A** is neither nilpotent nor invertible. Since **B** leaves $R(\mathbf{A}^k)$ invariant, we have $\mathbf{TBT}^{-1} = \begin{bmatrix} \mathbf{Z} & \mathbf{X} \\ \mathbf{0} & \mathbf{Y} \end{bmatrix}$ for some **Z**, **X**, **Y**. Substituting into (d) gives only $\mathbf{ZC}^{k+1} = \mathbf{C}^k$. Hence $\mathbf{Z} = \mathbf{C}^{-1}$ and (3) follows. (4) is clear. Assume now that **B** satisfies (3). If **B** is a (p, d)-inverse then

$$R\left(\begin{bmatrix} \mathbf{I} & \mathbf{XN} \\ \mathbf{0} & \mathbf{YN} \end{bmatrix}\right) = R\left(\begin{bmatrix} \mathbf{I} & \mathbf{0} \\ \mathbf{0} & \mathbf{0} \end{bmatrix}\right)$$

Thus (5) follows. If **B** is a (c, d)-inverse of **A**, then

$$\begin{bmatrix} \mathbf{I} & \mathbf{CX} \\ \mathbf{0} & \mathbf{NY} \end{bmatrix} = \begin{bmatrix} \mathbf{I} & \mathbf{XN} \\ \mathbf{0} & \mathbf{YN} \end{bmatrix}.$$

But then $\mathbf{C}^k\mathbf{X} = \mathbf{XN}^k = \mathbf{0}$ and (6) follows. Similarly, (7) and (8) follow from (3) and the definition of properties $\{1, 2\}$. ■

Note that any number $l \geq \text{Ind}(\mathbf{A})$ can be used in place of k in (d). In general, $\mathbf{A}^k\mathbf{A}^{\mathrm{d}}\mathbf{A} \neq \mathbf{A}^k$ and $\mathbf{A}^{k+1}\mathbf{A}^{\mathrm{d}} \neq \mathbf{A}^k$. Although $\mathbf{A}^{\mathrm{d}}\mathbf{A}$ and $\mathbf{A}\mathbf{A}^{\mathrm{d}}$ are not always projections, both are the identity on $R(\mathbf{A}^k)$. From (3), (4), and (6);

Corollary 9.7.1 $\mathbf{A}^{\mathrm{D}}$ *is the unique* (*p*, *c*, *d*)-*inverse of* **A**. $\mathbf{A}^{\mathrm{D}}$ *is also a* (2, *p*, *c*, *d*)-*inverse and is the unique* (2, *c*, *d*)-*inverse of* **A** *by definition.*

Corollary 9.7.2 *Suppose that* $\text{Ind}(\mathbf{A}) = 1$. *Then*

(i) **B** *is a* (1, *d*)-*inverse of* **A** *if and only if* **B** *is a* (*d*)-*inverse, and*
(ii) **B** *is a* (2, *d*)-*inverse of* **A** *if and only if* **B** *is an* (*m*, *d*)-*inverse.*

Corollary 9.7.3 *Suppose that* $\text{Ind}(\mathbf{A}) \geq 2$. *Then there are no* (1, *c*, *d*)-*inverses or* (1, *p*, *d*)-*inverses.*

Proof Suppose that $\text{Ind}(\mathbf{A}) \geq 2$ and **B** is a (1, c, d)-inverse of **A**. Then by (3), (6), (7) we have $\mathbf{X} = \mathbf{0}$, $\mathbf{NYN} = \mathbf{N}$, and $\mathbf{NY} = \mathbf{YN}$. But then $\mathbf{N}^{k-1} = \mathbf{0}$

which is a contradiction. If $\mathbf{B}$ is a (1, p, d)-inverse we have by (3), (5), (7) that $\mathbf{X} = \mathbf{0}$, $\mathbf{Y} = \mathbf{0}$, and $\mathbf{NON} = \mathbf{N}$ which is a contradiction. ■

Most of the (d)-inverses are not spectral in the sense of [40] since no assumptions have been placed on $N(\mathbf{A})$, $N(\mathbf{A}^d)$. However;

Corollary 9.7.4 *The operation of taking (m, d)-inverses has the spectral mapping property. That is, λ is a non-zero eigenvalue for $\mathbf{A}$ if and only if $1/\lambda$ is a non-zero eigenvalue for the (m, d)-inverse $\mathbf{B}$. Furthermore, the eigenspaces for λ and $1/\lambda$ are the same. Both $\mathbf{A}$ and $\mathbf{B}$ either have a zero eigenvalue or are invertible. The zero eigenspaces need not be the same.*

Note that if λ is a non-zero eigenvalue of $\mathbf{A}$, then $1/\lambda$ is an eigenvalue of any (d)-inverse of $\mathbf{A}$.

Corollary 9.7.5 *If $\mathbf{B}_1, \ldots, \mathbf{B}_r$ are (d)-inverses of $\mathbf{A}$, then $\mathbf{B}_1\mathbf{B}_2\cdots\mathbf{B}_r$ is a (d)-inverse of $\mathbf{A}^r$. In particular, $(\mathbf{A}^d)^m$ is a (d)-inverse of $\mathbf{A}^m$.*

Corollary 5 is not true for (1)-inverses. For $\mathbf{B} = \begin{bmatrix} 1 & -1 \\ 1 & -1 \end{bmatrix}$ is a (1, 2)-inverse of $\mathbf{A} = \begin{bmatrix} 1 & 0 \\ 0 & 0 \end{bmatrix}$, but $\mathbf{B}^2 = \mathbf{0}$ and hence $\mathbf{B}^2$ is not a (1)-inverse of $\mathbf{A}^2 = \mathbf{A}$. This is not surprising, for $(\mathbf{A}^\dagger)^2$ may not be even a (1)-inverse of $\mathbf{A}^2$.

Theorem 9.7.2 *Suppose that $\mathbf{A} \in \mathbb{C}^{n \times n}$, $\operatorname{Ind}(\mathbf{A}) = k$. Then*

(i) $\{\mathbf{A}^D + \mathbf{Z}(\mathbf{I} - \mathbf{A}^D\mathbf{A}) \mid \mathbf{Z} \in \mathbb{C}^{n \times n}\}$ *is the set of all (d)-inverses of* $\mathbf{A}$,
(ii) $\{\mathbf{A}^D + \mathbf{A}^D\mathbf{A}\mathbf{Z}(\mathbf{I} - \mathbf{A}^D\mathbf{A} \mid \mathbf{Z} \in \mathbb{C}^{n \times n}\}$ *is the set of all (m, d)-inverses of* $\mathbf{A}$,
(iii) $\{\mathbf{A}^D + \mathbf{Z}(\mathbf{I} - \mathbf{A}^D\mathbf{A}) \mid \mathbf{Z}\mathbf{A} = \mathbf{A}\mathbf{Z}\}$ *is the set of all (c, d)-inverses of* $\mathbf{A}$,

and

(iv) $\{\mathbf{A}^D + (\mathbf{I} - \mathbf{A}^D\mathbf{A})[\mathbf{A}(\mathbf{I} - \mathbf{A}^D\mathbf{A})]^- \mid (\mathbf{I} - \mathbf{A}^D\mathbf{A})[\mathbf{A}(\mathbf{I} - \mathbf{A}^D\mathbf{A})]^- \mathbf{A}(\mathbf{I} - \mathbf{A}^D\mathbf{A}) = \mathbf{0}\}$ *is the set of all (1, d)-inverses of* $\mathbf{A}$.

Proof (i)–(iv) follow from Theorem 1. We have omitted the (p, d)- and (2, d)-inverses since they are about as appealing as (iv). ■

Just as it is possible to calculate $\mathbf{A}^\dagger$ given an $\mathbf{A}^-$, one may calculate $\mathbf{A}^D$ from any $\mathbf{A}^d$.

Corollary 9.7.6 *If $k = \operatorname{Ind}(\mathbf{A})$, then $\mathbf{A}^D = (\mathbf{A}^d)^{l+1}\mathbf{A}^l$ for any $l \geq k$.*

The next two results are the weak Drazin equivalents of Theorem 7.8.1.

Theorem 9.7.3 *Suppose that $\mathbf{A} \in \mathbb{C}^{n \times n}$ and $\mathbf{A} = \begin{bmatrix} \mathbf{C} & \mathbf{D} \\ \mathbf{0} & \mathbf{E} \end{bmatrix}$ where $\mathbf{C}$ is invertible. Then all (d)-inverses of $\mathbf{A}$ are given by*

$$\mathbf{A}^d = \begin{bmatrix} \mathbf{C}^{-1} & -\mathbf{C}^{-1}\mathbf{D}\mathbf{E}^d + \mathbf{Z}(\mathbf{I} - \mathbf{E}^d\mathbf{E}) \\ \mathbf{0} & \mathbf{E}^d \end{bmatrix}; \tag{9}$$

$\mathbf{E}^{\mathrm{d}}$ *any (d)-inverse of* $\mathbf{E}$, $\mathbf{E}^{\tilde{\mathrm{d}}}$ *an (m, d)-inverse of* $\mathbf{E}$, $\mathbf{Z}$ *an arbitrary matrix of the correct size.*

Proof Suppose $\mathbf{A} = \begin{bmatrix} \mathbf{C} & \mathbf{D} \\ \mathbf{0} & \mathbf{E} \end{bmatrix}$ with $\mathbf{C}$ invertible. Let $k = \operatorname{Ind}(\mathbf{A}) = \operatorname{Ind}(\mathbf{E})$. Then $\mathbf{A}^k = \begin{bmatrix} \mathbf{C}^k & \Theta \\ \mathbf{0} & \mathbf{E}^k \end{bmatrix}$, where Θ is some matrix. Now the range of $\begin{bmatrix} \mathbf{I} & \mathbf{0} \\ \mathbf{0} & \mathbf{0} \end{bmatrix}$ is in $R(\mathbf{A}^k)$. Hence $\mathbf{A}^{\mathrm{D}}$ and any $\mathbf{A}^{\mathrm{d}}$ agree on it. Thus

$$\mathbf{A}^{\mathrm{d}} = \begin{bmatrix} \mathbf{C}^{-1} & \mathbf{X}_1 \\ \mathbf{0} & \mathbf{X}_2 \end{bmatrix}. \tag{10}$$

Now suppose (10) is a (d)-inverse of $\mathbf{A}$. Then $\mathbf{A}\mathbf{A}^{\mathrm{d}}\mathbf{A}^k = \mathbf{A}^k$. Hence

$$\begin{bmatrix} \mathbf{C} & \mathbf{D} \\ \mathbf{0} & \mathbf{E} \end{bmatrix}\begin{bmatrix} \mathbf{C}^{-1} & \mathbf{X}_1 \\ \mathbf{0} & \mathbf{X}_2 \end{bmatrix}\begin{bmatrix} \mathbf{C}^k & \Theta \\ \mathbf{0} & \mathbf{E}^k \end{bmatrix} = \begin{bmatrix} \mathbf{C}^k & \Theta \\ \mathbf{0} & \mathbf{E}^k \end{bmatrix}.$$

Thus $\begin{bmatrix} \mathbf{I} & \mathbf{C}\mathbf{X}_1 + \mathbf{D}\mathbf{X}_2 \\ \mathbf{0} & \mathbf{E}\mathbf{X}_2 \end{bmatrix}\begin{bmatrix} \mathbf{C}^k & \Theta \\ \mathbf{0} & \mathbf{E}^k \end{bmatrix} = \begin{bmatrix} \mathbf{C}^k & \Theta \\ \mathbf{0} & \mathbf{E}^k \end{bmatrix}$, or

$$\Theta + (\mathbf{C}\mathbf{X}_1 + \mathbf{D}\mathbf{X}_2)\mathbf{E}^k = \Theta, \tag{11}$$

$$\mathbf{E}\mathbf{X}_2\mathbf{E}^k = \mathbf{E}^k. \tag{12}$$

If $\mathbf{A}^{\mathrm{d}}\mathbf{A}^{k+1} = \mathbf{A}^k$ is to hold, one must have $\mathbf{X}_2$ a (d)-inverse of $\mathbf{E}$. Let $\mathbf{X}_2 = \mathbf{E}^{\mathrm{d}}$ for some (d)-inverse of $\mathbf{E}$. Then (12) holds. Now (11) becomes $\mathbf{X}_1\mathbf{E}^k = -\mathbf{C}^{-1}\mathbf{D}\mathbf{E}^{\mathrm{d}}\mathbf{E}^k$. Let $\mathbf{E}^{\tilde{\mathrm{d}}}$ be an (m, d)-inverse of $\mathbf{E}$. Then $\mathbf{E}^{\tilde{\mathrm{d}}}\mathbf{E}$ is a projection onto $R(\mathbf{E}^k)$. Hence $\mathbf{X}_1$ must be of the form $-\mathbf{C}^{-1}\mathbf{D}\mathbf{E}^{\mathrm{d}} + \mathbf{Z}(\mathbf{I} - \mathbf{E}^{\tilde{\mathrm{d}}}\mathbf{E})$ and (9) follows. To see that (9) defines a (d)-inverse of $\mathbf{A}$ is a direct computation.

It should be pointed out that while $\mathbf{B}\mathbf{A}^{k+1} = \mathbf{A}^k$ implies $\mathbf{A}\mathbf{B}\mathbf{A}^k = \mathbf{A}^k$, the two conditions are not equivalent.

Corollary 9.7.7 *Suppose there exists an invertible* $\mathbf{T}$ *such that*

$$\mathbf{T}\mathbf{A}\mathbf{T}^{-1} = \begin{bmatrix} \mathbf{C} & \mathbf{X} \\ \mathbf{0} & \mathbf{N} \end{bmatrix}, \tag{13}$$

with $\mathbf{C}$ *invertible and* $\mathbf{N}$ *nilpotent. Then* $\mathbf{T}^{-1}\begin{bmatrix} \mathbf{C}^{-1} & \mathbf{0} \\ \mathbf{0} & \mathbf{0} \end{bmatrix}\mathbf{T}$ *is an (m, d)-inverse for* $\mathbf{A}$.

If one wanted $\mathbf{A}^{\mathrm{D}}$ from (13) it would be given by the more complicated expression $\mathbf{T}\mathbf{A}^{\mathrm{D}}\mathbf{T}^{-1} = \begin{bmatrix} \mathbf{C}^{-1} & \tilde{\mathbf{X}} \\ \mathbf{0} & \mathbf{0} \end{bmatrix}$, $\tilde{\mathbf{X}} = \mathbf{C}^{-2}\left(\sum_{n=0}^{k-1} \mathbf{C}^{-k}\mathbf{X}\mathbf{N}^k\right)$.

Although for block triangular matrices it is easier to compute a weak Drazin than a Drazin inverse, in practice one frequently does not have a block triangular matrix to begin with. We now give two results which are the weak Drazin analogues of Algorithm 7.6.1.

Theorem 9.7.4 *Suppose that* $\mathbf{A} \in \mathbb{C}^{n \times n}$ *and that* $p(x) = x^l(c_0 + \ldots + c_r x^r)$,

$c_0 \neq 0$, *is the characteristic (or minimal) polynomial of* $\mathbf{A}$. *Then*

$$\mathbf{A}^{\mathrm{d}} = -\frac{1}{c_0}(c_1\mathbf{I} + \ldots + c_r\mathbf{A}^{r-1}) \tag{14}$$

is a (c, d)*-inverse of* $\mathbf{A}$. *If* (14) *is not invertible, then* $\mathbf{A}^{\mathrm{d}} + (\mathbf{I} - \mathbf{A}^{\mathrm{d}}\mathbf{A})$ *is an invertible* (c, d)*-inverse of* $\mathbf{A}$.

Proof Since $p(\mathbf{A}) = \mathbf{0}$, we have $(c_0\mathbf{I} + \ldots + c_r\mathbf{A}^r)\mathbf{A}^l = \mathbf{0}$. Hence $(c_1\mathbf{I} + \ldots + c_r\mathbf{A}^{r-1})\mathbf{A}^{l+1} = -c_0\mathbf{A}^l$. Since $\operatorname{Ind}(\mathbf{A}) \leq l$, we have that (14) is a (d)-inverse. It is commuting since it is a polynomial in $\mathbf{A}$. Now let $\mathbf{A}$ be as in (2). Then since $\mathbf{A}^{\mathrm{d}}$ is a (c, d)-inverse it is in the form (6). But then $\mathbf{Y} = -\frac{1}{c_0}(c_1\mathbf{I} + c_2\mathbf{N} + \ldots + c_r\mathbf{N}^{r-1})$. If $c_1 \neq 0$, then $\mathbf{Y}$ is invertible since $\mathbf{N}$ is nilpotent and we are done. Suppose that $c_1 = 0$, then

$$\mathbf{A}^{\mathrm{d}} + (\mathbf{I} - \mathbf{A}^{\mathrm{d}}\mathbf{A}) = \begin{bmatrix} \mathbf{C}^{-1} & \mathbf{0} \\ \mathbf{0} & \mathbf{I} - \mathbf{YN} + \mathbf{Y} \end{bmatrix}$$ and $\mathbf{I} - \mathbf{YN} + \mathbf{Y}$ is

invertible since $\mathbf{Y} - \mathbf{YN}$ is nilpotent. That $\mathbf{A}^{\mathrm{d}} + (\mathbf{I} - \mathbf{A}^{\mathrm{d}}\mathbf{A})$ is a (c, d)-inverse follows from the fact that $\mathbf{A}^{\mathrm{d}}$ is. ■

Note that Theorem 4 requires no information on eigenvalues or their multiplicities to calculate a (c, d)-inverse. If $\mathbf{A}$ has rational entries, (14) would provide an exact answer if exact arithmetic were used.

Theorem 4 suggests that a variant of the Souriau–Frame algorithm could be used to compute (c, d)-inverses. In fact, the algorithm goes through almost unaltered.

Theorem 9.7.5 *Suppose that* $\mathbf{A} \in \mathbb{C}^{n \times n}$. *Let* $\mathbf{B}_0 = \mathbf{I}$. *For* $j = 1, 2, \ldots, n$, *let* $p_j = \frac{1}{j}\operatorname{Tr}(\mathbf{AB}_{j-1})$ *and* $\mathbf{B}_j = \mathbf{AB}_{j-1} - p_j\mathbf{I}$. *If* $p_s \neq \mathbf{0}$, *but* $p_{s+1} = p_{s+2} = \ldots = p_n = 0$, *then*

$$\mathbf{A}^{\mathrm{d}} = \frac{1}{p_s}\mathbf{B}_{s-1} \tag{15}$$

is a (c, d)*-inverse. In fact,* (14) *and* (15) *are the same matrix.*

Proof Let $k = \operatorname{Ind}(\mathbf{A})$. Observe that $\mathbf{B}_j = \mathbf{A}^j - p_1\mathbf{A}^{j-1} - p_2\mathbf{A}^{j-2} - \ldots - p_j\mathbf{I}$. If r is the smallest integer such that $\mathbf{B}_r = 0$ and s is the largest such that $p_s \neq 0$, then $\operatorname{Ind}(\mathbf{A}) = r - s$. Since $\mathbf{B}_r = \mathbf{0}$, we have $\mathbf{A}^r = p_1\mathbf{A}^{r-1} - \ldots - p_{s-1}\mathbf{A}^{r-s+1} - p_s\mathbf{A}^{r-s} = \mathbf{0}$. Hence,

$$\mathbf{A}^{r-s} = \frac{1}{p_s}(\mathbf{A}^r - p_1\mathbf{A}^{r-1} - \ldots - p_{s-1}\mathbf{A}^{r-s+1}) =$$

$\frac{1}{p_s}(\mathbf{A}^{s-1} - p_1\mathbf{A}^{s-2} - \ldots - p_{s-1}\mathbf{I})\mathbf{A}^{r-s+1}$. That is, $\mathbf{A}^k = \left(\frac{1}{p_s}\mathbf{B}_{s-1}\right)\mathbf{A}^{k+1}$ as desired. ■

Lemma 9.7.1 *Suppose that* $\mathbf{A}, \mathbf{B} \in \mathbb{C}^{n \times n}$ *and* $\mathbf{AB} = \mathbf{BA}$. *Let* $\mathbf{A}^{\mathrm{d}}$ *be any (d)-inverse of* $\mathbf{A}$. *Then* $\mathbf{A}^{\mathrm{d}}\mathbf{BAA}^{\mathrm{D}} = \mathbf{BA}^{\mathrm{d}}\mathbf{AA}^{\mathrm{D}} = \mathbf{BA}^{\mathrm{D}}\mathbf{AA}^{\mathrm{D}} = \mathbf{A}^{\mathrm{D}}\mathbf{BAA}^{\mathrm{D}}$.

Proof If $\mathbf{AB} = \mathbf{BA}$, then $\mathbf{A}^{\mathrm{D}}\mathbf{B} = \mathbf{BA}^{\mathrm{D}}$. Also if $\mathbf{A}$ is given by (2), then $\mathbf{TBT}^{-1} = \begin{bmatrix} \mathbf{B}_1 & \mathbf{0} \\ \mathbf{0} & \mathbf{B}_2 \end{bmatrix}$ with $\mathbf{B}_1\mathbf{C} = \mathbf{CB}_1$. Lemma 1 now follows from Theorem 1. ∎

As an immediate consequence of Lemma 1, one may use a (d)-inverse in many of the applications of the Drazin inverse. For example, see the next theorem.

Theorem 9.7.6 *Suppose that* $\mathbf{A}, \mathbf{B} \in \mathbb{C}^{n \times n}$. *Suppose that* $\mathbf{A}\dot{\mathbf{x}} + \mathbf{Bx} = \mathbf{0}$ *has unique solutions for consistent initial conditions, that is, there is a scalar c such that* $(c\mathbf{A} + \mathbf{B})$ *is invertible. Let* $\hat{\mathbf{A}} = (c\mathbf{A} + \mathbf{B})^{-1}\mathbf{A}$, $\hat{\mathbf{B}} = (c\mathbf{A} + \mathbf{B})^{-1}\mathbf{B}$. *Let* $k = \mathrm{Ind}(\hat{\mathbf{A}})$, *if* $\mathbf{A}\dot{\mathbf{x}} + \mathbf{Bx} = \mathbf{0}$, $\mathbf{x}(0) = \mathbf{q}$, *is consistent, then the solution is* $\mathbf{x} = e^{-\hat{\mathbf{A}}^{\mathrm{d}}\hat{\mathbf{B}}t}\mathbf{q}$. *If* $\hat{\mathbf{A}}^{\mathrm{d}}$ *is an (m, d)-inverse of* $\hat{\mathbf{A}}$, *then all solutions of* $\mathbf{A}\dot{\mathbf{x}} + \mathbf{Bx} = \mathbf{0}$ *are of the form* $x = e^{-\hat{\mathbf{A}}^{\mathrm{d}}\hat{\mathbf{B}}t}\hat{\mathbf{A}}\hat{\mathbf{A}}^{\mathrm{d}}\mathbf{q}$, $\mathbf{q} \in \mathbb{C}^n$, *and the space of consistent initial conditions is* $R(\hat{\mathbf{A}}\hat{\mathbf{A}}^{\mathrm{d}}) = R(\hat{\mathbf{A}}^{\mathrm{d}}\hat{\mathbf{A}})$.

Note in Theorem 6 that $\hat{\mathbf{A}}\hat{\mathbf{A}}^{\mathrm{d}}$ need not equal $\hat{\mathbf{A}}^{\mathrm{d}}\hat{\mathbf{A}}$ even if $\hat{\mathbf{A}}^{\mathrm{d}}$ is an (m, d)-inverse of $\mathbf{A}$.

Weak Drazin inverses can also be used in the theory of Markov chains. For example, the next result follows from the results of Chapter 8.

Theorem 9.7.8 *If* $\mathbf{T}$ *is the transition matrix of an m-state ergodic chain and if* $\mathbf{A} = \mathbf{I} - \mathbf{T}$, *then the rows of* $\mathbf{I} - \mathbf{A}^{\mathrm{d}}\mathbf{A}$ *are all equal to the unique fixed probability vector* $\mathbf{w}^*$ *of* $\mathbf{T}$ *for any (d)-inverse of* $\mathbf{A}$.

8. Exercises

Exercises 1–6 provide a generalization of some of the results in Section 2. Proofs may be found in [18].

1. Suppose that $\mathbf{A}, \mathbf{B}$ are $m \times n$ matrices. Let $(\cdot)^{\circ}$ denote a (2)-inverse. Show that the following are equivalent:

   (i) $(\lambda\mathbf{A} + \mathbf{B})^{\circ}\mathbf{A}$, $(\lambda\mathbf{A} + \mathbf{B})^{\circ}\mathbf{B}$ commute.
   (ii) $(\lambda\mathbf{A} + \mathbf{B})(\lambda\mathbf{A} + \mathbf{B})^{\circ}\mathbf{A}[\mathbf{I} - (\lambda\mathbf{A} + \mathbf{B})^{\circ}(\lambda\mathbf{A} + \mathbf{B})] = \mathbf{0}$.
   (iii) $(\lambda\mathbf{A} + \mathbf{B})(\lambda\mathbf{A} + \mathbf{B})^{\circ}\mathbf{B}[\mathbf{I} - (\lambda\mathbf{A} + \mathbf{B})^{\circ}(\lambda\mathbf{A} + \mathbf{B})] = \mathbf{0}$.

2. Prove that if $\mathbf{A}, \mathbf{B}$ are hermitian, then $(\lambda\mathbf{A} + \mathbf{B})^{\dagger}\mathbf{A}$, $(\lambda\mathbf{A} + \mathbf{B})^{\dagger}\mathbf{B}$ commute if and only if there exists a $\tilde{\lambda}$ such that $N(\tilde{\lambda}\mathbf{A} + \mathbf{B}) = N(\mathbf{A}) \cap N(\mathbf{B})$. Furthermore, if $\tilde{\lambda}$ exists, then $(\tilde{\lambda}\mathbf{A} + \mathbf{B})^{\dagger}\mathbf{A}$, $(\tilde{\lambda}\mathbf{A} + \mathbf{B})^{\dagger}\mathbf{B}$ commute.
3. Prove that if $\mathbf{A}, \mathbf{B} \in \mathbb{C}^{n \times n}$ are such that one is EP and the other is positive semi-definite, then there exists λ such that $\lambda\mathbf{A} + \mathbf{B}$ is invertible if and only if $N(\mathbf{A}) \cap N(\mathbf{B}) = \{\mathbf{0}\}$.
4. Prove that if $\mathbf{A}, \mathbf{B} \in \mathbb{C}^{n \times n}$ are such that one is EP and one is positive semi-definite, then there exists λ such that $N(\lambda\mathbf{A} + \mathbf{B}) = N(\mathbf{A}) \cap N(\mathbf{B})$.

5. Suppose that $\mathbf{A}, \mathbf{B} \in \mathbb{C}^{n \times n}$ are such that $N(\mathbf{A}) \cap N(\mathbf{B})$ reduces both $\mathbf{A}$ and $\mathbf{B}$. Suppose also that there exists a λ such that $N(\lambda\mathbf{A} + \mathbf{B}) = N(\mathbf{A}) \cap N(\mathbf{B})$. Prove that when $\mathbf{A}\dot{\mathbf{x}} + \mathbf{B}\mathbf{x} = \mathbf{f}$, $\mathbf{f}$ n-times continuously differentiable is consistent if and only if $\mathbf{f}(t) \in R(\lambda\mathbf{A} + \mathbf{B})$ for all t, that is, $(\lambda\mathbf{A} + \mathbf{B}) \times (\lambda\mathbf{A} + \mathbf{B})^{\dagger}\mathbf{f} = \mathbf{f}$. And that if it is consistent, then all solutions are of the form

$$\mathbf{x} = \hat{\mathbf{A}}^{\mathrm{D}} e^{-\hat{\mathbf{A}}^{\mathrm{D}}\hat{\mathbf{B}}t} \int_0^t e^{\hat{\mathbf{A}}^{\mathrm{D}}\hat{\mathbf{B}}s}\hat{\mathbf{f}}(s)\,ds$$
$$+ [(\lambda\mathbf{A} + \mathbf{B})^{\mathrm{D}}(\lambda\mathbf{A} + \mathbf{B}) - \hat{\mathbf{A}}\hat{\mathbf{A}}^{\mathrm{D}}] \sum_{m=0}^{k-1} (-1)^m [\hat{\mathbf{A}}\hat{\mathbf{B}}^{\mathrm{D}}]^m \hat{\mathbf{B}}^{\mathrm{D}}\hat{\mathbf{f}}^{(m)}$$
$$+ e^{-\hat{\mathbf{A}}^{\mathrm{D}}\hat{\mathbf{B}}t}\hat{\mathbf{A}}^{\mathrm{D}}\hat{\mathbf{A}}\mathbf{q} + [\mathbf{I} - (\lambda\mathbf{A} + \mathbf{B})^{\mathrm{D}}(\lambda\mathbf{A} + \mathbf{B})]\mathbf{g}$$

where $\hat{\mathbf{A}} = (\lambda\mathbf{A} + \mathbf{B})^{\mathrm{D}}\mathbf{A}$, $\hat{\mathbf{B}} = (\lambda\mathbf{A} + \mathbf{B})^{\mathrm{D}}\mathbf{B}$, $\hat{\mathbf{f}} = (\lambda\mathbf{A} + \mathbf{B})^{\mathrm{D}}\mathbf{f}$, $\mathbf{q}$ is an arbitary vector, $\mathbf{g}$ an arbitrary vector valued function, and $k = \operatorname{Ind}(\hat{\mathbf{A}})$.

6. Prove that if $\mathbf{A}, \mathbf{B}$ are EP and one is positive semi-definite, λ as in Exercise 4, then all solutions of $\mathbf{A}\dot{\mathbf{x}} + \mathbf{B}\mathbf{x} = \mathbf{f}$ are in the form given in Exercise 5.
7. Derive formula (8) in Corollary 9.6.2.
8. Derive an expression for the consistent set of initial conditions for $\mathbf{A}\dot{\mathbf{x}} + \mathbf{B}\mathbf{x} = \mathbf{f}$ when $\mathbf{f}$ is n-times differentiable and $\lambda\mathbf{A} + \mathbf{B}$ is onto.
9. Verify that Corollary 9.7.6 is true.
10. Fill in the details in the proof of Theorem 9.5.7.
11. If $\mathbf{A} \in \mathbb{C}^{n \times n}$ and $k = \operatorname{Ind}(\mathbf{A})$, define a right weak Drazin by $\mathbf{A}^{k+1}\mathbf{B} = \mathbf{A}^k$. Develop the right equivalent of Theorems 1, 2, 3 and their corollaries.
12. Solve $\mathbf{A}\dot{\mathbf{x}}(t) + \mathbf{B}\mathbf{x}(t) = \mathbf{b}$, $\mathbf{A}, \mathbf{B}$ as in Example 2.2, $\mathbf{b} = [1\,2\,0]^*$.

Answer:
$$x_1(t) = -\tfrac{1}{18}e^{(2/3)t}(x_2(0) + 2x_3(0)) - \tfrac{13}{18}x_2(0) - \tfrac{4}{9}x_3(0) - \tfrac{1}{9} - t$$
$$x_2(t) = -\tfrac{4}{9}e^{(2/3)t}(x_2(0) + 2x_3(0)) - \tfrac{13}{9}x_2(0) + \tfrac{8}{9}x_3(0) + 2t$$
$$x_3(t) = \tfrac{13}{18}e^{(2/3)t}(x_2(0) + 2x_3(0)) - \tfrac{13}{18}x_2(0) - \tfrac{4}{3}x_3(0) - t$$

13. Let $\mathbf{T}$ be a matrix of the form (1). Assume each $p_k > 0$ (If any $p_k = 0$, then there would never be anyone in the age interval $\mathbf{A}_{k+1}$. Thus, we agree to truncate the matrix $\mathbf{T}$ just before the first zero survival probability.) The matrix $\mathbf{T}$ is non-singular if and only if b_m (the last birth rate) is non-zero. Show that the characteristic equation for $\mathbf{T}$ is

$$0 = x^m - b_1x^{m-1} - p_1b_2x^{m-2} - p_1p_2b_3x^{m-3} - \dots$$
$$- (p_1p_2 \cdots p_{m-2}b_{m-1})x - (p_1p_2 \cdots p_{m-1}b_m).$$

10
Continuity of the generalized inverse

1. Introduction

Consider the following statement:
(A) If $\{\mathbf{A}_j\} \subseteq \mathbb{C}^{m \times m}$ is a sequence of matrices and $\mathbf{A}_j$ converges to an invertible matrix $\mathbf{A}$, then for large enough j, $\mathbf{A}_j$ is invertible and $\mathbf{A}_j^{-1}$ converges to $\mathbf{A}^{-1}$.

In addition to its obvious theoretical interest, statement (A) has practical computational content.

First, if we have a sequence of 'nice' matrices $\{\mathbf{A}_j\}$ which gets close to $\mathbf{A}$, it tells us that $\mathbf{A}_j^{-1}$ gets close to $\mathbf{A}^{-1}$. Thus approximation methods might be of use in computing $\mathbf{A}^{-1}$.

Secondly, statement (A) gives us information on how sensitive the inverse of $\mathbf{A}$ is to 'errors' in determining $\mathbf{A}$. It tells us that if our error in determining $\mathbf{A}$ was 'small', then the error resulting in $\mathbf{A}^{-1}$ due to the error in $\mathbf{A}$ will also be 'small'.

This chapter will determine to what extent statement (A) is true for the Moore–Penrose and Drazin generalized inverses. But first, we must discuss what we mean by 'near', 'small', and 'converges to'.

2. Matrix norms

In linear algebra the most common way of telling when things are close is by the use of norms. Norms are to vectors what absolute value is to numbers.

Definition 10.2.1. A function ρ sending a vector space V into the positive reals is called a norm *if for all $\mathbf{u}, \mathbf{v} \in V$ and $\alpha \in \mathbb{C}$;*

(i) $\rho(\mathbf{u}) = 0$ iff $\mathbf{u} = \mathbf{0}$.
(ii) $\rho(\alpha\mathbf{u}) = |\alpha|\rho(\mathbf{u})$, and
(iii) $\rho(\mathbf{u}+\mathbf{v}) \leq \rho(\mathbf{u}) + \rho(\mathbf{v})$ *(triangle inequality)*.

We will usually denote $\rho(\mathbf{u})$ by $\|\mathbf{u}\|$.

There are many different norms that can be put on $\mathbb{C}^n$. If $\mathbf{u}\in\mathbb{C}^n$ and $\mathbf{u}$ has coordinates $(u_1,\ldots,u_n)$, then the *sup norm* of $\mathbf{u}$ is given by

$$\|\mathbf{u}\|_\infty = \sup_{1\le i\le n}\{|u_i|\}.$$

The *p-norm* of $\mathbf{u}$ is given by

$$\|\mathbf{u}\|_p = \left(\sum_{i=1}^{\infty}|u_i|^p\right)^{1/p} \quad \text{for } p\ge 1.$$

The function $\|\mathbf{u}\|_p$ is not a norm for $0\le p<1$ since it fails to satisfy (iii). The norm $\|\mathbf{u}\|_2$ is the ordinary Euclidean norm of $\mathbf{u}$, that is, $\|\mathbf{u}\|_2$ is the geometric length of $\mathbf{u}$.

We are using the term norm a little loosely. To be precise we would have to say that $\|\cdot\|_p$ is actually a family of norms, one for each $\mathbb{C}^n$. However, to avoid unenlightening verbage we shall continue to talk of the norm $\|\cdot\|_p$, the norm $\|\cdot\|_\infty$, etc.

The $m\times n$ matrices, $\mathbb{C}^{m\times n}$, are isomorphic, as a vector space, to $\mathbb{C}^{mn}$. Thus $\mathbb{C}^{m\times n}$ can be equipped with any of the above norms. However, in working out estimates it is extremely helpful if $\|\mathbf{AB}\|\le\|\mathbf{A}\|\,\|\mathbf{B}\|$ whenever $\mathbf{AB}$ is defined. A norm for which $\|\mathbf{AB}\|\le\|\mathbf{A}\|\,\|\mathbf{B}\|$ for $\mathbf{A}\in\mathbb{C}^{m\times n}, \mathbf{B}\in\mathbb{C}^{n\times r}$ is called a *matrix norm.* Not all norms are matrix norms.

Example 10.2.1 If $\mathbf{A}=[a_{ij}]$, then define $\|\mathbf{A}\| = \max_{i,j}|a_{ij}|$. This is just the $\|\cdot\|_\infty$ of $\mathbb{C}^{mn}$ applied to $\mathbb{C}^{m\times n}$. Now let $\mathbf{A}=\begin{bmatrix}1 & 1\\ 0 & 1\end{bmatrix}$ and $\mathbf{B}=\begin{bmatrix}0 & 1\\ 0 & 1\end{bmatrix}$. Then $\|\mathbf{A}\|=\|\mathbf{B}\|=1$, but $\|\mathbf{AB}\|=2$. Thus this norm is not a matrix norm.

There is a standard way to develop a matrix norm from a vector norm. Suppose that $\mathbf{A}\in\mathbb{C}^{m\times n}$ and $\|\cdot\|_s$ is a norm on $\mathbb{C}^r$ for all r. Define $\|\mathbf{A}\|_{os}$ by $\|\mathbf{A}\|_{os}=\sup\{\|\mathbf{Au}\|_s:\mathbf{u}\in\mathbb{C}^n, \|\mathbf{u}\|_s=1\}$. It is possible to generalize this definition by using a different norm in $\mathbb{C}^m$ to 'measure' $\mathbf{Au}$ than the one used in $\mathbb{C}^n$ to 'measure' $\mathbf{u}$. However, in working problems it is usually easier to use a fixed vector norm such as $\|\cdot\|_p$ or $\|\cdot\|_\infty$. We will not need the more general definition.

There is another formulation of $\|\cdot\|_{os}$.

Proposition 10.2.1 *Suppose that* $\mathbf{A}\in\mathbb{C}^{m\times n}$ *and* $\|\cdot\|_s$ *is a norm on* $\mathbb{C}^m$ *and* $\mathbb{C}^n$. *Thus* $\|\mathbf{A}\|_{os}=\inf\{K\in\mathbb{R}:\|\mathbf{Au}\|_s\le K\|\mathbf{u}\|_s$ *for every* $\mathbf{u}\in\mathbb{C}^n\}$.

The proof of Proposition 1 is left to the exercises.

If $\mathbf{A}$ is thought of as a linear transformation from $\mathbb{C}^n$ to $\mathbb{C}^m$, where $\mathbb{C}^n$ and $\mathbb{C}^m$ are equipped with the norm $\|\cdot\|_s$, then $\|\mathbf{A}\|_{os}$ is the norm that is usually used. $\|\mathbf{A}\|_{os}$ is also referred to as the *operator norm* of $\mathbf{A}$ relative to $\|\cdot\|_s$, hence the subscript os.

Conversely, if $\|\cdot\|$ is a matrix norm, then by identifying $\mathbb{C}^r$ and $\mathbb{C}^{r\times 1}$ it induces a norm on $\mathbb{C}^r$, say $\|\cdot\|_s$. Since $\|\cdot\|$ is a matrix norm we have

$$\|\mathbf{Au}\|_s\le\|\mathbf{A}\|\,\|\mathbf{u}\|_s. \tag{1}$$

If (1) occurs for a pair of norms $\|\cdot\|$ on $\mathbb{C}^{m\times n}$ and $\|\cdot\|_s$ on $\mathbb{C}^n$ and $\mathbb{C}^m$, we say that $\|\cdot\|$ is *consistent* with the vector norm $\|\cdot\|_s$. By Proposition 1 we know that $\|\cdot\|$ is consistent with $\|\cdot\|_s$ if and only if $\|\mathbf{A}\| \geq \|\mathbf{A}\|_{os}$.

We pause briefly to recall the definition of a limit.

Definition 10.2.2 *Suppose that* $\{\mathbf{A}_j\}$ *is a sequence of m × n matrices and* $\|\cdot\|$ *is a norm on* $\mathbb{C}^{m\times n}$. *Then* $\mathbf{A}_j$ converges *to* $\mathbf{A}$, (*written* $\mathbf{A}_j \to \mathbf{A}$ or $\lim_{j\to\infty} \mathbf{A}_j = \mathbf{A}$) *if for every real number* $\varepsilon > 0$, *there exists a real number* j_ε *such that if* $j \geq j_\varepsilon$, *then* $\|\mathbf{A}_j - \mathbf{A}\| < \varepsilon$.

Notice that the definition of convergence seems to depend on the norm used. However, in finite dimensional spaces there is no difficulty.

Theorem 10.2.1 *Suppose that* $\|\cdot\|_s$ *and* $\|\cdot\|_t$ *are two norms on a finite dimensional vector space V. Then* $\|\cdot\|_s$ *and* $\|\cdot\|_t$ *are* equivalent. *That is, there exist constants, k, l such that* $k\|\mathbf{u}\|_s \leq \|\mathbf{u}\|_t \leq l\|\mathbf{u}\|_s$ *for all* $\mathbf{u}\in V$.

Theorem 1 is a standard result in most introductory courses in linear algebra. A proof may be found, for example, in [49]. Theorem 1 tells us that if $\mathbf{A}_j \to \mathbf{A}$ with regard to one norm, then $\mathbf{A}_j \to \mathbf{A}$ with regard to any norm.

It is worth noting that $\mathbf{A}_j \to \mathbf{A}$ if and only if the entries of $\mathbf{A}_j$ converge to the corresponding entries of $\mathbf{A}$.

To further develop this circle of ideas, and for future reference, let us see what form Theorem 1 takes for the norms we have been looking at. Recall that an inequality is called *sharp* if it cannot be improved by multiplying one side by a scalar.

Theorem 10.2.2 *Suppose that* $\mathbf{u}\in\mathbb{C}^n$ *and that p, q are two real numbers greater or equal to one. Then*

(i) $\|\mathbf{u}\|_\infty \leq \|\mathbf{u}\|_p \leq n^{1/p}\|\mathbf{u}\|_\infty$,

(ii) $n^{-1/p}\|\mathbf{u}\|_p \leq \|\mathbf{u}\|_\infty \leq \|\mathbf{u}\|_p$,

(iii) $n^{-1/p}\|\mathbf{u}\|_p \leq \|\mathbf{u}\|_q \leq \|\mathbf{u}\|_p$, if $p < q$, *and*

(iv) $n^{-1/p}\|\mathbf{u}\|_p \leq \|\mathbf{u}\|_q \leq n^{-1/q}\|\mathbf{u}\|_p$ for all $p, q \geq 1$.

Furthermore (i) *and* (ii) *are sharp.*

Proof (iv) follows from (i) and (ii) while (ii) is merely a rewriting of (i). To prove (i) assume $\mathbf{u}\in\mathbb{C}^n$ and note that

$$\|\mathbf{u}\|_\infty = \max_{1\leq i\leq n} |u_i| = (\max|u_i|^p)^{1/p} \leq \left(\sum_{i=1}^n |u_i|^p\right)^{1/p} = \|\mathbf{u}\|_p$$

$$\leq \left(\sum_{i=1}^n (\max|u_i|)^p\right)^{1/p} \leq (\max|u_i|)\left(\sum_{i=1}^n 1^p\right)^{1/p} = n^{1/p}\|\mathbf{u}\|_\infty.$$

If $\mathbf{u} = \mathbf{e}_1$, then $\|\mathbf{u}\|_\infty = \|\mathbf{u}\|_p = 1$ so $\|\mathbf{u}\|_\infty \leq \|\mathbf{u}\|_p$ is sharp. Of $\mathbf{u}$ consists of n ones, then $\|\mathbf{u}\|_\infty = 1$ while $\|\mathbf{u}\|_p = n^{1/p}$. Thus $\|\mathbf{u}\|_p \leq n^{1/p}\|\mathbf{u}\|_\infty$ is sharp.

Let us now consider (iii). To see that $n^{-1/p}\|\mathbf{u}\|_p \le \|\mathbf{u}\|_q$ observe that from (i) and (ii) we have $n^{-1/p}\|\mathbf{u}\|_p \le \|\mathbf{u}\|_\infty \le \|\mathbf{u}\|_q$. The statement that $\|\mathbf{u}\|_q \le \|\mathbf{u}\|_p$ if $1 \le p < q$ is known as *Jensen's inequality*. The vector $\mathbf{e}_1$ shows that Jensen's inequality is sharp. ■

We now turn to the problem of determining the matrix norms $\|\cdot\|_{op}$ and $\|\cdot\|_{o\infty}$. We begin with $\|\cdot\|_{o\infty}$. Suppose that $\mathbf{A} = [a_{ij}] \in \mathbb{C}^{m\times n}$ and $\mathbf{u} = [u_i] \in \mathbb{C}^n$. We shall find a K for which $\|\mathbf{Au}\|_\infty \le \mathrm{K}\|\mathbf{u}\|_\infty$ and show that K is the best possible. Now

$$\begin{aligned}\|\mathbf{Au}\|_\infty &= \sup_i\left\{\left|\sum_{j=1}^n a_{ij}u_j\right|\right\} \le \sup_i\left\{\sum_{j=1}^n |a_{ij}|\,|u_j|\right\}\\ &\le \sup_i\left\{\left(\sup_j |u_j|\right)\sum_{j=1}^n |a_{ij}|\right\}\\ &= \sup_i\left\{\sum_{j=1}^n |a_{ij}|\right\}\|\mathbf{u}\|_\infty. \end{aligned} \tag{2}$$

Thus $$\|\mathbf{A}\|_{o\infty} \le \sup_i\left\{\sum_{j=1}^n |a_{ij}|\right\}. \tag{3}$$

To get equality in (3) we need a vector $\mathbf{u}$ such that $\|\mathbf{u}\|_\infty = 1$ and equality occurs in (2). To get equality in the second inequality of (2) we can take $|u_j| = 1$, $1 \le j \le n$. The first inequality of (2) will be an equality if $a_{ij}u_j = |a_{ij}|\,|u_j|$ for that row of $\mathbf{A}$ for which $\sum_{j=1}^n |a_{ij}|$ is maximum. Denote this row by k. Then define $\mathbf{u}$ as follows,

$$\mathbf{u}_j = \begin{cases} \dfrac{a_{kj}}{|a_{kj}|} & \text{if } a_{kj} \ne 0\\ 1 & \text{if } a_{kj} = 0.\end{cases}$$

Then $\|\mathbf{A}\|_{o\infty}\|\mathbf{u}\| \ge \|\mathbf{Au}\| = \sup_i\left\{\sum_{j=1}^n |a_{ij}|\right\}\|\mathbf{u}\|_\infty$. This and (3) gives us the following.

Proposition 10.2.2 *If* $\mathbf{A} \in \mathbb{C}^{m\times n}$ *and* $\mathbf{A} = [a_{ij}]$, *then* $\|\mathbf{A}\|_{o\infty} = \sup_i\left\{\sum_{j=1}^n |a_{ij}|\right\}$.

The $\|\cdot\|_{op}$ norms do not have a common formula. However, in the special case $p = 1$ or $p = \infty$, $\|\cdot\|_{op}$ is reasonably calculable. Since $p = 1, 2, \infty$ are the most common of the p-norms used this is not as bad as it sounds. We leave the derivation of the $\|\cdot\|_{o1}$ norm to the exercises.

Proposition 10.2.3 *If* $\mathbf{A} \in \mathbb{C}^{m\times n}$, *then* $\|\mathbf{A}\|_{o1} = \max_j\left\{\sum_{i=1}^n |a_{ij}|\right\}$.

The Euclidean norm $\|\cdot\|_2$ is of much interest. Suppose that $\mathbf{A}\in\mathbb{C}^{m\times n}$ and $\mathbf{u}\in\mathbb{C}^n$, then

$$\|\mathbf{Au}\|_2^2 = (\mathbf{Au},\mathbf{Au}) = (\mathbf{A^*Au},\mathbf{u}). \tag{4}$$

Since $\mathbf{A^*A}$ is self-adjoint, every vector $\mathbf{u}\in\mathbb{C}^n$ can be written as $\mathbf{u} = \mathbf{v}_1\oplus\ldots\oplus\mathbf{v}_r$ where $\mathbf{v}_1,\ldots,\mathbf{v}_r$ satisfy $(\mathbf{A^*A})\mathbf{v}_1 = \lambda_i\mathbf{v}_i$ for a real number λ_i. Let λ_o be the largest eigenvalue of $\mathbf{A^*A}$ with associated eigenvector $\mathbf{u}_o$. Then from (4), $\|\mathbf{Au}\|_2^2\le\lambda_o\|\mathbf{u}\|_2^2$ and $\|\mathbf{Au}_o\|_2^2=\lambda_o\|\mathbf{u}_o\|_2^2$. Thus we have Proposition 10.2.4.

Proposition 10.2.4 *If* $\mathbf{A}\in\mathbb{C}^{m\times n}$, *then* $\|\mathbf{A}\|_{o2}=\lambda^{1/2}$ *where* λ *is the largest eigenvalue of* $\mathbf{A^*A}$. *That is,* $\|\mathbf{A}\|_{o2}$ *is the largest singular value of* $\mathbf{A}$.

In addition to $\|\cdot\|_{o\infty}$ and $\|\cdot\|_{op}$ there are other matrix norms which can be used either to estimate $\|\cdot\|_{o\infty}$ and $\|\cdot\|_{op}$ or in their own right. Two common ones are given in the exercises.

3. Matrix norms and invertibility

In order to understand the convergence properties of the generalized inverse we shall briefly review the situation for invertible matrices. Throughout this section $\|\cdot\|$ will denote any matrix norm on $\mathbb{C}^{n\times n}$ such that $\|\mathbf{I}\|=1$. The assumption that $\|\mathbf{I}\|=1$ does not rule out any of the matrix norms discussed in Section 2 and simplifies our formulas.

Proposition 10.3.1 *Suppose that* $\mathbf{A}\in\mathbb{C}^{n\times n}$ *and* $\|\cdot\|$ *is a matrix norm on* $\mathbb{C}^{n\times n}$ *which is consistent with a vector norm* $\|\cdot\|_s$. *Then for every eigenvalue* λ *of* $\mathbf{A}$, $|\lambda|\le\|\mathbf{A}\|$.

The proof of Proposition 1 is straightforward and is left to the exercises. Note that the vector norm of $\mathbb{C}^n$ does not appear explicitly in Proposition 1.

We will prove statement (A) of the introduction and develop some norm estimates for $\mathbf{A}^{-1}-\mathbf{B}^{-1}$ in terms of $\|\mathbf{A}-\mathbf{B}\|$. The next proposition is basic.

Proposition 10.3.2 *If* $\|\mathbf{A}\|<1$, *then* $(\mathbf{I}-\mathbf{A})$ *is invertible and*

$$\|(\mathbf{I}-\mathbf{A})^{-1}\|\le\frac{1}{1-\|\mathbf{A}\|}.$$

Proof Let $\mathbf{B}=\sum_{n=0}^{\infty}\mathbf{A}^n$. The series converges absolutely since $\|\mathbf{A}\|<1$. Let $\mathbf{S}_k=\sum_{n=0}^{k}\mathbf{A}^n$. Then $(\mathbf{I}-\mathbf{A})\mathbf{S}_k=\mathbf{I}-\mathbf{A}^{n+1}\to\mathbf{I}$ while $(\mathbf{I}-\mathbf{A})\mathbf{S}_k\to(\mathbf{I}-\mathbf{A})\mathbf{B}$. Thus $\mathbf{B}=(\mathbf{I}-\mathbf{A})^{-1}$. The estimate for $\|\mathbf{B}\|$ follows from the series representation for $\mathbf{B}$, $\|\mathbf{B}\|=\left\|\sum_{n=0}^{\infty}\mathbf{A}^n\right\|\le\sum_{n=0}^{\infty}\|\mathbf{A}\|^n=(1-\|\mathbf{A}\|)^{-1}$. ■

Corollary 10.3.1 *If* $\mathbf{A}\in\mathbb{C}^{n\times n}$ *and* $\|\mathbf{I}-\mathbf{A}\|<1$ *then* $\mathbf{A}$ *is invertible and*

$$\|\mathbf{A}^{-1}\|\leq\frac{1}{1-\|\mathbf{I}-\mathbf{A}\|}$$

Proof If $\|\mathbf{I}-\mathbf{A}\|<1$, then $\mathbf{A}=\mathbf{I}-(\mathbf{I}-\mathbf{A})$ and the result follows from Proposition 2. ■

The next two results are also basic to this section. We begin by establishing that if $\mathbf{A}$ is invertible and $\mathbf{B}$ is close to $\mathbf{A}$, then $\mathbf{B}$ is invertible. Alternatively we can show that if $\mathbf{B}$ is small then $\mathbf{A}+\mathbf{B}$ is invertible. We choose the latter. Now $(\mathbf{A}+\mathbf{B})=\mathbf{A}(\mathbf{I}+\mathbf{A}^{-1}\mathbf{B})$ where we are assuming that $\mathbf{A}$ is invertible. By Proposition 2, $(\mathbf{I}+\mathbf{A}^{-1}\mathbf{B})$ will be invertible if $\|\mathbf{A}^{-1}\mathbf{B}\|<1$. A sufficient condition for this is clearly $\|\mathbf{B}\|<1/\|\mathbf{A}^{-1}\|$ since then $\|\mathbf{A}^{-1}\mathbf{B}\|\leq\|\mathbf{A}^{-1}\|\,\|\mathbf{B}\|<\|\mathbf{A}^{-1}\|/\|\mathbf{A}^{-1}\|=1$. Now assume $\|\mathbf{B}\|<1/\|\mathbf{A}^{-1}\|$. Let us estimate $\|\mathbf{A}^{-1}-(\mathbf{A}+\mathbf{B})^{-1}\|$. The idea is to remove invertible factors so that we can use Propositions 1, 2 and Corollary 1. Now

$$\begin{aligned}\mathbf{A}^{-1}-(\mathbf{A}+\mathbf{B})^{-1}&=[\mathbf{A}^{-1}(\mathbf{A}+\mathbf{B})-\mathbf{I}](\mathbf{A}+\mathbf{B})^{-1}\\&=[\mathbf{I}+\mathbf{A}^{-1}\mathbf{B}-\mathbf{I}](\mathbf{A}+\mathbf{B})^{-1}\\&=\mathbf{A}^{-1}\mathbf{B}(\mathbf{A}+\mathbf{B})^{-1}=\mathbf{A}^{-1}\mathbf{B}[\mathbf{A}(\mathbf{I}+\mathbf{A}^{-1}\mathbf{B})]^{-1}\\&=\mathbf{A}^{-1}\mathbf{B}[\mathbf{I}+\mathbf{A}^{-1}\mathbf{B}]^{-1}\mathbf{A}^{-1}.\end{aligned}$$

Thus $\|\mathbf{A}^{-1}-(\mathbf{A}+\mathbf{B})^{-1}\|\leq\|\mathbf{B}\|\,\|\mathbf{A}^{-1}\|^2\|(\mathbf{I}+\mathbf{A}^{-1}\mathbf{B})^{-1}\|\leq\dfrac{\|\mathbf{A}^{-1}\|^2\|\mathbf{B}\|}{1-\|\mathbf{A}^{-1}\mathbf{B}\|}$ $\leq\dfrac{\|\mathbf{A}^{-1}\|^2\|\mathbf{B}\|}{1-\|\mathbf{A}^{-1}\|\,\|\mathbf{B}\|}$. The second inequality follows from Proposition 2 while the third follows from $\|\mathbf{A}^{-1}\mathbf{B}\|\leq\|\mathbf{A}^{-1}\|\,\|\mathbf{B}\|<1$. We have proven the following result.

Theorem 10.3.1 *Suppose that* $\mathbf{A}\in\mathbb{C}^{n\times n}$ *is invertible. If* $\|\mathbf{B}\|<1/\|\mathbf{A}^{-1}\|$, *then* $(\mathbf{A}+\mathbf{B})$ *is invertible. Furthermore*

$$\|\mathbf{A}^{-1}-(\mathbf{A}+\mathbf{B})^{-1}\|\leq\frac{\|\mathbf{A}^{-1}\|^2\|\mathbf{B}\|}{1-\|\mathbf{A}^{-1}\|\,\|\mathbf{B}\|}. \tag{2}$$

Corollary 10.3.2 *If* $\mathbf{A}\in\mathbb{C}^{m\times m}$ *is invertible and* $\{\mathbf{A}_j\}\subseteq\mathbb{C}^{m\times n}$ *such that* $\mathbf{A}_j\to\mathbf{A}$, *then for large enough j,* $\mathbf{A}_j^{-1}$ *exists and* $\mathbf{A}_j^{-1}\to\mathbf{A}^{-1}$.

Proof Let $\mathbf{B}_j=\mathbf{A}_j-\mathbf{A}$ and apply Theorem 1. ■

Example 10.3.1 Let $\mathbf{A}=\begin{bmatrix}1/2 & -1/2\\ -1/2 & 1/2\end{bmatrix}$ so that $\mathbf{I}-\mathbf{A}=\begin{bmatrix}1/2 & 1/2\\ 1/2 & 1/2\end{bmatrix}$. Then $\|\mathbf{A}\|_\infty=1/2<1$. But $\mathbf{I}-\mathbf{A}$ is not invertible. Thus Proposition 2 depends on the fact $\|\cdot\|$ was a matrix norm.

The estimates given in Proposition 2, Corollary 1, and Theorem 1 are all sharp as can be seen from the scalar case.

4. Continuity of the Moore—Penrose generalized inverse

We now wish to see to what extent the results of Section 3 can be extended to the generalized inverse. The first thing to notice is that statement (A) of Section 1 is not true for the generalized inverse.

Example 10.4.1 Let $\mathbf{A}_j = \begin{bmatrix} 1 & 0 \\ 0 & 1/j \end{bmatrix}$ and $\mathbf{A} = \begin{bmatrix} 1 & 0 \\ 0 & 0 \end{bmatrix}$. Then $\mathbf{A}_j^\dagger = \begin{bmatrix} 1 & 0 \\ 0 & j \end{bmatrix}$ while $\mathbf{A}^\dagger = \begin{bmatrix} 1 & 0 \\ 0 & 0 \end{bmatrix}$. Thus $\mathbf{A}_j \to \mathbf{A}$ but $\mathbf{A}_j$ does not converge to anything, much less to $\mathbf{A}^\dagger$.

The problem that immediately presents itself is to determine necessary and sufficient conditions so that $\mathbf{A}_j \to \mathbf{A}$ implies $\mathbf{A}_j^\dagger \to \mathbf{A}^\dagger$. Surprisingly it has a nice answer.

In order to try and get some feeling for what is happening let us return to Example 1. Rather than talking of $\mathbf{A}_j \to \mathbf{A}$ we will talk of $\mathbf{A} + \mathbf{E}_j$ where $\mathbf{E}_j \to \mathbf{0}$.

Let $\mathbf{E}_j = \begin{bmatrix} a_j & b_j \\ c_j & d_j \end{bmatrix}$ where for simplicity we assume that a_j, b_j, c_j, d_j are all real. We will also assume that $\mathbf{E}_j \to \mathbf{0}$ so that $a_j \to 0$, $b_j \to 0$, $c_j \to 0$, and $d_j \to 0$. We wish to investigate when does $(\mathbf{A} + \mathbf{E}_j)^\dagger \to \mathbf{A}^\dagger$ where $\mathbf{A} = \begin{bmatrix} 1 & 0 \\ 0 & 0 \end{bmatrix}$. Since $a_j \to 0$ we might as well assume that $|a_j| < 1$ for all j.
Then

$$\mathbf{A} + \mathbf{E}_j = \begin{bmatrix} a_j + 1 & b_j \\ c_j & d_j \end{bmatrix}$$

has rank at least one. That is, $\operatorname{rank}(\mathbf{A} + \mathbf{E}_j) \geq \operatorname{rank}(\mathbf{A})$. We shall find out later that this is typical. There are two cases to consider for a particular j.

Case I $\mathbf{A} + \mathbf{E}_j$ is invertible. In this case

$$(\mathbf{A} + \mathbf{E}_j)^\dagger = (\mathbf{A} + \mathbf{E}_j)^{-1} = \frac{1}{(a_j + 1)d_j - b_j c_j} \begin{bmatrix} d_j & -b_j \\ -c_j & a_j + 1 \end{bmatrix}. \tag{1}$$

Case II $\mathbf{A} + \mathbf{E}_j$ is singular. In this case $\operatorname{rank}(\mathbf{A} + \mathbf{E}_j) = 1$ so there exists α_j such that

$$\mathbf{A} + \mathbf{E}_j = \begin{bmatrix} a_j + 1 & b_j \\ c_j & d_j \end{bmatrix} = \begin{bmatrix} (a_j + 1) & \alpha_j(a_j + 1) \\ c_j & \alpha_j c_j \end{bmatrix} = \begin{bmatrix} a_j + 1 \\ c_j \end{bmatrix} [1 \quad \alpha_j].$$

Thus

$$(\mathbf{A} + \mathbf{E}_j)^\dagger = \frac{1}{(1 + \alpha_j^2)((1 + a_j)^2 + c_j^2)} \begin{bmatrix} a_j + 1 & c_j \\ b_j & d_j \end{bmatrix}. \tag{2}$$

Now suppose that $\mathbf{A} + \mathbf{E}_j$ has rank 1 for j greater than some j_0. Then we are in Case II for $j \geq j_0$. Now $\alpha_j = b_j/(a_j + 1)$ so that $\alpha_j \to 0$. Since $a_j \to 0, b_j \to 0, c_j \to 0, \alpha_j \to 0$, and $d_j \to 0$ we get from (2) that $(\mathbf{A} + \mathbf{E}_j)^\dagger \to \mathbf{A}^\dagger$ as desired.

Suppose however that $\mathbf{A} + \mathbf{E}_j$ does not have rank 1 for all j greater than some j_0. Then there exists a subsequence $\mathbf{E}_m$ such that $\operatorname{rank}(\mathbf{A} + \mathbf{E}_m) = 2$ and $\mathbf{E}_m \to 0$ for all integers m in the subsequence. Thus $a_m \to 0, d_m \to 0$, $b_m \to 0$, and $c_m \to 0$. But the (2, 2) entry of $(\mathbf{A} + \mathbf{E}_m)^\dagger$ is

$$\frac{a_m + 1}{(a_m + 1)d_m - b_m c_m} \tag{3}$$

which does not converge to anything much less zero.

We have then that for our particular example, the following.

(B) Given $\mathbf{E}_j \to \mathbf{0}$ we have $(\mathbf{A} + \mathbf{E}_j)^\dagger \to \mathbf{A}^\dagger$ if and only if there is a j_0 such that $\operatorname{rank}(\mathbf{A} + \mathbf{E}_j) = \operatorname{rank}(\mathbf{A})$ for $j \geq j_0$.

Statement (B) turns out to be valid for any $\mathbf{A} \in \mathbb{C}^{m \times n}$, $\{\mathbf{E}_j\} \subseteq \mathbb{C}^{m \times n}$ such that $\mathbf{E}_j \to \mathbf{0}$. The proof will proceed much as in the special case. After some preliminary results we shall consider different cases involving $\operatorname{rank}(\mathbf{A})$ and $\operatorname{rank}(\mathbf{A} + \mathbf{E}_j)$.

To begin, recall the following fact.

Fact 10.4.1 If $\mathbf{A} \in \mathbb{C}^{n \times n}$ is invertible. $\|\cdot\|$ is a matrix norm on $\mathbb{C}^{n \times n}$ and $\mathbb{C}^{n \times 1}$, then $\|\mathbf{Ax}\| \geq \|\mathbf{x}\| \, \|\mathbf{A}^{-1}\|^{-1}$.

For all $\mathbf{x} \in \mathbb{C}^n$ we will also need the following.

Fact 10.4.2 If $\|\cdot\|$ is any norm on a vector space V, then $\|\mathbf{u} + \mathbf{v}\| \geq \|\mathbf{u}\| - \|\mathbf{v}\|$ for all $\mathbf{u}, \mathbf{v} \in V$.

The generalized inverse version of Fact 1 takes the following form.

Proposition 10.4.1 *Suppose that* $\mathbf{A} \in \mathbb{C}^{m \times n}$ *and* $\|\cdot\|$ *is a matrix norm on* $\mathbb{C}^{p \times q}$, $p \geq 1, q \geq 1$. *Then* $\|\mathbf{Ax}\| \geq \|\mathbf{x}\| / \|\mathbf{A}^\dagger\|$ *for* $\mathbf{x} \in R(\mathbf{A}^*)$.

Proof Suppose $\mathbf{x} \in R(\mathbf{A}^*)$. Then $\|\mathbf{x}\| = \|\mathbf{A}^\dagger \mathbf{A} \mathbf{u}\| \leq \|\mathbf{A}^\dagger\| \, \|\mathbf{Ax}\|$ so that $\|\mathbf{Ax}\| \geq \|\mathbf{x}\| / \|\mathbf{A}^\dagger\|$ as desired. ■

In the discussion following Example 1 we noted that in that example $\operatorname{rank}(\mathbf{A} + \mathbf{E}_j) \geq \operatorname{rank}(\mathbf{A})$. This is typical as the next proposition shows.

Proposition 10.4.2 *Suppose that* $\mathbf{A} \in \mathbb{C}^{m \times n}$ *and* $\|\cdot\|$ *is a matrix norm on* $\mathbb{C}^{p \times q}$, $p \geq 1, q \geq 1$. *If* $\mathbf{E} \in \mathbb{C}^{m \times n}$ *and* $\|\mathbf{E}\| < 1/\|\mathbf{A}^\dagger\|$, *then* $\operatorname{rank}(\mathbf{A} + \mathbf{E}) \geq \operatorname{rank}(\mathbf{A})$.

Proof Suppose that $\operatorname{rank}(\mathbf{A}) = r$ and $\|\mathbf{E}\| < 1/\|\mathbf{A}^\dagger\|$. Let $\{\mathbf{u}_1, \ldots, \mathbf{u}_r\}$ be a basis for $R(\mathbf{A}^*)$. It suffices to show that $\{(\mathbf{A} + \mathbf{E})\mathbf{u}_1, \ldots, (\mathbf{A} + \mathbf{E})\mathbf{u}_r\}$ is a linearly independent subset of $R(\mathbf{A} + \mathbf{E})$. Suppose that $\mathbf{0} = \sum_{i=1}^{r} \alpha_i (\mathbf{A} + \mathbf{E})\mathbf{u}_i$;

$\alpha_i \in \mathbb{C}$. Then if $\mathbf{x} = \sum_{i=1}^{r} \alpha_i \mathbf{u}_i \neq \mathbf{0}$ we get that

$$\begin{aligned} 0 = \left\| \sum_{i=1}^{r} \alpha_i (\mathbf{A} + \mathbf{E})\mathbf{u}_i \right\| &= \left\| (\mathbf{A} + \mathbf{E}) \sum_{i=1}^{r} \alpha_i \mathbf{u}_i \right\| = \| \mathbf{Ax} + \mathbf{Ex} \| \\ &\geq \| \mathbf{Ax} \| - \| \mathbf{Ex} \| \text{ (by Fact 2)} \geq \| \mathbf{Ax} \| - \| \mathbf{E} \| \, \| \mathbf{x} \| \\ &> \| \mathbf{Ax} \| - \| \mathbf{x} \| / \| \mathbf{A}^\dagger \| \geq \| \mathbf{x} \| / \| \mathbf{A}^\dagger \| - \| \mathbf{x} \| / \| \mathbf{A}^\dagger \| \end{aligned}$$

(by Proposition 1) $= 0$.
But $0 > 0$ is a contradiction so that $\mathbf{x} = \mathbf{0}$ and hence $\alpha_1 = \alpha_2 = \ldots = \alpha_r = 0$. Thus $\operatorname{rank}(\mathbf{A} + \mathbf{E}) \geq \operatorname{rank}(\mathbf{A})$. ■

Proposition 2 has several useful corollaries.

Corollary 10.4.1 *If* $\mathbf{A}, \mathbf{B} \in \mathbb{C}^{m \times n}$ *and*

$$\| \mathbf{A} - \mathbf{B} \| < \frac{1}{\max\{ \| \mathbf{A}^\dagger \|, \| \mathbf{B}^\dagger \| \}}, \tag{1}$$

then $\operatorname{rank}(\mathbf{A}) = \operatorname{rank}(\mathbf{B})$.

Proof Let $\mathbf{E} = \mathbf{A} - \mathbf{B}$ and notice that (1) implies that $\| \mathbf{A} - \mathbf{B} \| < 1/\| \mathbf{A}^\dagger \|$ and $\| \mathbf{A} - \mathbf{B} \| < 1/\| \mathbf{B}^\dagger \|$. Thus $\operatorname{rank}(\mathbf{A}) = \operatorname{rank}(\mathbf{B} + \mathbf{E}) \geq \operatorname{rank}(\mathbf{B})$ and $\operatorname{rank}(\mathbf{B}) = \operatorname{rank}(\mathbf{A} - \mathbf{E}) \geq \operatorname{rank}(\mathbf{A})$ so that $\operatorname{rank}(\mathbf{A}) = \operatorname{rank}(\mathbf{B})$.

As a special case of Corollary 1 we have the next corollary.

Corollary 10.4.2 *If* $\mathbf{P}, \mathbf{Q}$ *are orthogonal projectors in* $\mathbb{C}^n$ *and* $\| \mathbf{P} - \mathbf{Q} \|_{o2} < 1$, *then* $\operatorname{rank}(\mathbf{P}) = \operatorname{rank}(\mathbf{Q})$.

The proof of half of statement (B) is now immediate.

Lemma 10.4.1 *Suppose that* $\mathbf{E}_j \to \mathbf{0}$ *and* $(\mathbf{A} + \mathbf{E}_j)^\dagger \to \mathbf{A}^\dagger$ *where* $\mathbf{A} \in \mathbb{C}^{m \times n}$, $\{\mathbf{E}_j\} \subseteq \mathbb{C}^{m \times n}$. *Then there exists a* j_o *such that* $\operatorname{rank}(\mathbf{A} + \mathbf{E}_j) = \operatorname{rank}(\mathbf{A})$ *if* $j \geq j_o$

Proof Suppose that $\mathbf{E}_j \to \mathbf{0}$ and $(\mathbf{A} + \mathbf{E}_j)^\dagger \to \mathbf{A}^\dagger$. But $(\mathbf{A} + \mathbf{E}_j) \to \mathbf{A}$. Thus

$$\mathbf{P}_{R(\mathbf{A}+\mathbf{E}_j)} = (\mathbf{A} + \mathbf{E}_j)(\mathbf{A} + \mathbf{E}_j)^\dagger \to \mathbf{AA}^\dagger = \mathbf{P}_{R(\mathbf{A})}.$$

But the limit can be taken with respect to any norm by Theorem 2.1. Thus by Definition 2.2 there exists a j_o such that if $j \geq j_o$, then $\| \mathbf{P}_{R(\mathbf{A}+\mathbf{E}_j)} - \mathbf{P}_{R(\mathbf{A})} \|_{o2} < 1$. That $\operatorname{rank}(\mathbf{A} + \mathbf{E}_j) = \operatorname{rank}(\mathbf{A})$ for $j \geq j_o$ now follows from Corollary 2. ■

The rest of this section will be divided into two parts. The first will be devoted to a proof of statement (B). The proof will be somewhat qualitative in nature. The second part will consist of a quantitative discussion of the same ideas.

Theorem 10.4.1 *Suppose that* $\mathbf{A} \in \mathbb{C}^{m \times n}$ *and* $\{\mathbf{E}_j\} \subseteq \mathbb{C}^{m \times n}$ *such that* $\mathbf{E}_j \to \mathbf{0}$. *Then* $(\mathbf{A} + \mathbf{E}_j)^\dagger \to \mathbf{A}^\dagger$ *if and only if* $\operatorname{rank}(\mathbf{A} + \mathbf{E}_j) = \operatorname{rank}(\mathbf{A})$ *for j greater than some fixed* j_o.

Proof Lemma 1 takes care of the only if part. Suppose that $\mathbf{E}_j \to \mathbf{0}$ and $\operatorname{rank}(\mathbf{A}+\mathbf{E}_j) = \operatorname{rank}\mathbf{A} = r$ for $j \geq j_o$. Now there exists unitary matrices $\mathbf{U} \in \mathbb{C}^{m\times m}$, $\mathbf{V} \in \mathbb{C}^{n\times n}$ and invertible matrix $\mathbf{B} \in \mathbb{C}^{r\times r}$ such that

$$\mathbf{C} = \mathbf{UAV} = \begin{bmatrix} \mathbf{B} & \mathbf{0} \\ \mathbf{0} & \mathbf{0} \end{bmatrix}. \tag{2}$$

Let

$$\mathbf{F}_j = \mathbf{UE}_j\mathbf{V} = \begin{bmatrix} \mathbf{F}_{11}(j) & \mathbf{F}_{12}(j) \\ \mathbf{F}_{21}(j) & \mathbf{F}_{22}(j) \end{bmatrix}. \tag{3}$$

Notice that $\operatorname{rank}(\mathbf{C}+\mathbf{F}_j) = \operatorname{rank}(\mathbf{A}+\mathbf{E}_j)$ and $\operatorname{rank}(\mathbf{C}) = \operatorname{rank}(\mathbf{A})$. Furthermore since $\mathbf{C}^\dagger = \mathbf{V}^*\mathbf{A}^\dagger\mathbf{U}^*$ and $(\mathbf{C}+\mathbf{F}_j)^\dagger = [\mathbf{U}(\mathbf{A}+\mathbf{E}_j)\mathbf{V}]^\dagger = \mathbf{V}^*(\mathbf{A}+\mathbf{E}_j)^\dagger\mathbf{U}^*$, $(\mathbf{C}+\mathbf{F}_j)^\dagger \to \mathbf{C}^\dagger$ if and only if $(\mathbf{A}+\mathbf{E}_j)^\dagger \to \mathbf{A}^\dagger$. For notational convenience we will omit the j in (3). We wish to get a formula for $(\mathbf{C}+\mathbf{F})^\dagger$. Let $\|\cdot\| = \|\cdot\|_{o2}$. Since $\mathbf{F} \to \mathbf{0}$ we may assume $\|\mathbf{F}\| < \|\mathbf{B}^{-1}\|^{-1} = \|\mathbf{A}^\dagger\|^{-1}$. But $\|\mathbf{F}\| \geq \sup\{\|\mathbf{F}_{11}\|, \|\mathbf{F}_{12}\|, \|\mathbf{F}_{21}\|, \|\mathbf{F}_{22}\|\}$. Thus $\operatorname{rank}(\mathbf{B}+\mathbf{F}_{11}) = \operatorname{rank}(\mathbf{B})$ by Proposition 2 and the fact that $\mathbf{B}$ is of full rank. Thus

$$\operatorname{rank}\begin{bmatrix} \mathbf{B}+\mathbf{F}_{11} & \mathbf{F}_{12} \\ \mathbf{F}_{21} & \mathbf{F}_{22} \end{bmatrix} = \operatorname{rank}(\mathbf{B}).$$

By Lemma 3.3.1 $\mathbf{C}+\mathbf{F}$ can be written as

$$\begin{bmatrix} \mathbf{B}+\mathbf{F}_{11} & \mathbf{F}_{12} \\ \mathbf{F}_{21} & \mathbf{F}_{21}(\mathbf{B}+\mathbf{F}_{11})^{-1}\mathbf{F}_{12} \end{bmatrix} = \left[\begin{array}{c} \mathbf{I} \\ \hline \mathbf{F}_{21}(\mathbf{B}+\mathbf{F}_{11})^{-1} \end{array}\right] \times \mathbf{B}[\mathbf{I} \mid (\mathbf{B}+\mathbf{F}_{11})^{-1}\mathbf{F}_{12}].$$

Thus

$$(\mathbf{C}+\mathbf{F})^\dagger = \left[\begin{array}{c} \mathbf{I} \\ \hline \mathbf{F}_{12}^*(\mathbf{B}^*+\mathbf{F}_{11}^*)^{-1} \end{array}\right] \mathbf{X}_1\mathbf{B}^{-1}\mathbf{X}_2[\mathbf{I} \mid (\mathbf{B}_1^*+\mathbf{F}_{11}^*)^{-1}\mathbf{F}_{21}^*] \tag{4}$$

where $\mathbf{X}_1 = [\mathbf{I} + (\mathbf{B}+\mathbf{F}_{11})^{-1}\mathbf{F}_{12}\mathbf{F}_{12}^*(\mathbf{B}^*+\mathbf{F}_{11}^*)^{-1}]^{-1}$ and $\mathbf{X}_2 = [\mathbf{I} + (\mathbf{B}^*+\mathbf{F}_{11}^*)^{-1}\mathbf{F}_{21}^*\mathbf{F}_{21}(\mathbf{B}+\mathbf{F}_{11})^{-1}]^{-1}$, by Theorem 3.3.4. Now $\mathbf{F}_{11}$, $\mathbf{F}_{12}, \mathbf{F}_{21}, \mathbf{F}_{22} \to \mathbf{0}$ since $\mathbf{F} \to \mathbf{0}$. By Corollary 2, $(\mathbf{B}+\mathbf{F}_{11})^{-1} \to \mathbf{B}^{-1}$. But then $(\mathbf{B}+\mathbf{F}_{11})^{-1}\mathbf{F}_{12}\mathbf{F}_{12}^*(\mathbf{B}^*+\mathbf{F}_{11})^{-1} \to \mathbf{0}$. Similarly $(\mathbf{B}^*+\mathbf{F}_{11}^*)^{-1}\mathbf{F}_{21}^*\mathbf{F}_{21} \times (\mathbf{B}+\mathbf{F}_{11})^{-1} \to \mathbf{0}$. Thus $\mathbf{X}_1 \to \mathbf{I}$, $\mathbf{X}_2 \to \mathbf{I}$ as $\mathbf{F} \to \mathbf{0}$. So

$$\left[\begin{array}{c} \mathbf{I} \\ \hline \mathbf{F}_{12}^*(\mathbf{B}^*+\mathbf{F}_{11}^*)^{-1} \end{array}\right] \to \begin{bmatrix} \mathbf{I} \\ \mathbf{0} \end{bmatrix} \text{ and } [\mathbf{I} \mid (\mathbf{B}+\mathbf{F}_{11})^{-1}\mathbf{F}_{12}] \to [\mathbf{I} \mid \mathbf{0}] \text{ as } \mathbf{F} \to \mathbf{0}.$$

Thus

$$(\mathbf{C}+\mathbf{F})^\dagger \to \begin{bmatrix} \mathbf{I} \\ \mathbf{0} \end{bmatrix} \mathbf{IB}^{-1}\mathbf{I}[\mathbf{I} \mid \mathbf{0}] = \begin{bmatrix} \mathbf{B}^{-1} & \mathbf{0} \\ \mathbf{0} & \mathbf{0} \end{bmatrix} = \mathbf{C}^\dagger$$

as $\mathbf{F} \to \mathbf{0}$ and the proof is complete. ■

We assumed that $\|\cdot\| = \|\cdot\|_{o2}$ so that we could assert that $\|\mathbf{F}_j\| = \|\mathbf{E}_j\|$ and $\|\mathbf{A}\| = \|\mathbf{C}\|$.

Theorem 1 can be proved without using Theorem 3.3.4 in several ways. Some are modifications of our argument here. Others, such as the original one due to Penrose, are completely different.

In example 1 we had $\operatorname{rank}(\mathbf{A}_j) > \operatorname{rank}(\mathbf{A})$ and $\|\mathbf{A}_j^\dagger\| = j = 1/\|\mathbf{A} - \mathbf{A}_j\|$. This behaviour is typical as the next result shows.

Theorem 10.4.2 *Suppose that* $\mathbf{A}, \mathbf{E} \in \mathbb{C}^{m \times n}$ *and that* $\operatorname{rank}(\mathbf{A}+\mathbf{E}) > \operatorname{rank}(\mathbf{A})$. *Then* $\|(\mathbf{A}+\mathbf{E})^\dagger\| \geq 1/\|\mathbf{E}\|$ *for any operator norm* $\|\cdot\|$.

Proof Suppose that $\mathbf{A}, \mathbf{E} \in \mathbb{C}^{m \times n}$ and that $\operatorname{rank}(\mathbf{A}+\mathbf{E}) > \operatorname{rank} \mathbf{A}$. Then $\dim N(\mathbf{A}) > \dim N(\mathbf{A}+\mathbf{E})$. Hence there is a vector $\mathbf{u} \in N(\mathbf{A})$ of norm 1 such that $\mathbf{u} \in N(\mathbf{A}+\mathbf{E})^\perp$. The proof of this last fact is left to the exercises. Now $\mathbf{u} \in R(\mathbf{A}^* + \mathbf{E}^*)$. Hence $(\mathbf{A}+\mathbf{E})^\dagger(\mathbf{A}+\mathbf{E})\mathbf{u} = \mathbf{u}$ or $(\mathbf{A}+\mathbf{E})^\dagger\mathbf{E}\mathbf{u} = \mathbf{u}$. But then $1 \leq \|(\mathbf{A}+\mathbf{E})^\dagger\mathbf{E}\| \leq \|(\mathbf{A}+\mathbf{E})^\dagger\|\,\|\mathbf{E}\|$ and $\|(\mathbf{A}+\mathbf{E})^\dagger\| \geq 1/\|\mathbf{E}\|$ as desired. ■

Example 1 shows that the inequality of Theorem 2 is sharp.

An obvious consequence of Theorem 2 is that if $\mathbf{E}_j \to \mathbf{0}$ but $\operatorname{rank}(\mathbf{A}+\mathbf{E}_j) > \operatorname{rank}(\mathbf{A})$, then not only does $(\mathbf{A}+\mathbf{E}_j)^\dagger \nrightarrow \mathbf{A}^\dagger$ but $(\mathbf{A}+\mathbf{E}_j)^\dagger$ is not even bounded in norm.

Theorem 1 is theoretically satisfying in the sense that it completely characterizes when $(\mathbf{A}+\mathbf{E}_j)^\dagger \to \mathbf{A}^\dagger$. However, in some situations it is important to be able to estimate the difference (error) between $(\mathbf{A}+\mathbf{E}_j)^\dagger$ and $\mathbf{A}^\dagger$. The rest of this section is devoted to estimating $\|(\mathbf{A}+\mathbf{E}_j)^\dagger - \mathbf{A}^\dagger\|$. In proving Theorem 1 we used unitary matrices. Unitary matrices are especially good when working with the Euclidean operator norm $\|\cdot\|_{o2}$ for if $\mathbf{U}^* = \mathbf{U}^{-1}$, $\mathbf{U} \in \mathbb{C}^{m \times n}$, and $\mathbf{A} \in \mathbb{C}^{n \times m}$, then $\|\mathbf{UA}\|_{o2} = \|\mathbf{A}\|_{o2}$.

We will use the Euclidean operator norm since it allows us to use the simplified block terms of Theorem 1. We shall also use the notation of the proof of Theorem 1. Let

$$\Theta_1 = \begin{bmatrix} \mathbf{X}_1 \\ \mathbf{F}_{12}^*(\mathbf{B}^* + \mathbf{F}_{11}^*)^{-1}\mathbf{X}_1 \end{bmatrix}, \Theta_2 = [\mathbf{X}_2 \mid \mathbf{X}_2(\mathbf{B}^* + \mathbf{F}_{11}^*)^{-1}\mathbf{F}_{21}^*].$$

Also set $\Psi_1 = \begin{bmatrix} \mathbf{I} \\ \mathbf{0} \end{bmatrix}$ and $\Psi_2 = [\mathbf{I} \quad \mathbf{0}]$. Then $(\mathbf{C}+\mathbf{F})^\dagger = \Theta_1\mathbf{B}^{-1}\Theta_2$ while $\mathbf{C}^\dagger = \Psi_1\mathbf{B}^{-1}\Psi_2$.

Thus $(\mathbf{C}+\mathbf{F})^\dagger - \mathbf{C}^\dagger = \Theta_1\mathbf{B}^{-1}\Theta_2 - \Psi_1\mathbf{B}^{-1}\Psi_2 = \Theta_1\mathbf{B}^{-1}\Theta_2 - \Theta_1\mathbf{B}^{-1}\Psi_2 + \Theta_1\mathbf{B}^{-1}\Psi_2 - \Psi_1\mathbf{B}^{-1}\Psi_2 = \Theta_1\mathbf{B}^{-1}(\Theta_2 - \Psi_2) + (\Theta_1 - \Psi_1) \times \mathbf{B}^{-1}\Psi_2$. Taking the norm of both sides gives

$$\|(\mathbf{C}+\mathbf{F})^\dagger - \mathbf{C}^\dagger\| \leq \|\Theta_1\|\,\|\mathbf{B}^{-1}\|\,\|\Theta_2 - \Psi_2\| + \|\Theta_1 - \Psi_1\| \times \|\mathbf{B}^{-1}\|\,\|\Psi_2\|. \tag{5}$$

Certain factors in (5) are obvious. $\|\mathbf{B}^{-1}\| = \|\mathbf{C}^\dagger\| = \|\mathbf{A}^\dagger\|$ and $\|\Psi_2\| = 1$. In order to estimate the rest we will use the following lemma.

Lemma 10.4.2 *If* $\mathbf{A} \in \mathbb{C}^{m \times n}$, $\mathbf{B} \in \mathbb{C}^{r \times n}$, *and* $\|\cdot\|$ *denotes the Euclidean operator, norm, then* $\left\| \begin{bmatrix} \mathbf{A} \\ \mathbf{B} \end{bmatrix} \right\| \leq (\|\mathbf{A}\|^2 + \|\mathbf{B}\|^2)^{1/2}$.

Proof of Lemma Suppose $\mathbf{u} \in \mathbb{C}^n$, $\|\mathbf{u}\|_2 = 1$. Then

$$\left\| \begin{bmatrix} \mathbf{A} \\ \mathbf{B} \end{bmatrix} \mathbf{u} \right\|_2^2 = \left\| \begin{bmatrix} \mathbf{Au} \\ \mathbf{Bu} \end{bmatrix} \right\|_2^2 = \|\mathbf{Au}\|_2^2 + \|\mathbf{Bu}\|_2^2 \leq \|\mathbf{A}\|^2 + \|\mathbf{B}\|^2.$$

Thus $\left\| \begin{bmatrix} \mathbf{A} \\ \mathbf{B} \end{bmatrix} \right\|^2 \leq \|\mathbf{A}\|^2 + \|\mathbf{B}\|^2$ as desired. Notice that since $\|\mathbf{D}^*\| = \|\mathbf{D}\|$ for the Euclidean operator norm we also have $\|[\mathbf{A}, \mathbf{B}]\|^2 \leq \|\mathbf{A}\|^2 + \|\mathbf{B}\|^2$. ■

We now begin to estimate the various terms in Θ_1 and Θ_2. To simplify matters we assume that

$$\|\mathbf{B}^{-1}\| \, \|\mathbf{F}_{11}\| < 1, \ \|(\mathbf{B} + \mathbf{F}_{11})^{-1}\| \, \|\mathbf{F}_{12}\| < 1, \text{ and}$$
$$\|(\mathbf{B} + \mathbf{F}_{11})^{-1}\| \, \|\mathbf{F}_{21}\| < 1. \tag{6}$$

These assumptions will be discussed more fully later. Now

$$\begin{aligned} \|\mathbf{F}_{12}^*(\mathbf{B}^* + \mathbf{F}_{11}^*)\| &\leq \|\mathbf{F}_{12}\| \, \|(\mathbf{B} + \mathbf{F}_{11})^{-1}\| \\ &= \|\mathbf{F}_{12}\| \, \|(\mathbf{I} + \mathbf{B}^{-1}\mathbf{F}_{11})^{-1}\mathbf{B}^{-1}\| \\ &\leq \|\mathbf{F}_{12}\| \, \|(\mathbf{I} + \mathbf{B}^{-1}\mathbf{F}_{11})^{-1}\| \, \|\mathbf{B}^{-1}\| \\ &\leq \|\mathbf{F}_{12}\| \frac{1}{1 - \|\mathbf{B}^{-1}\mathbf{F}_{11}\|} \|\mathbf{B}^{-1}\| \end{aligned} \tag{7}$$

by Proposition 3.2 and assumptions (6). Let $\alpha = \|\mathbf{B}^{-1}\|(1 - \|\mathbf{B}^{-1}\| \times \|\mathbf{F}_{11}\|)^{-1}$. Then (7) becomes

$$\|\mathbf{F}_{12}^*(\mathbf{B}^* + \mathbf{F}_{11}^*)^{-1}\| \leq \alpha \|\mathbf{F}_{12}\|. \tag{8}$$

Similarly,

$$\|(\mathbf{B}^* + \mathbf{F}_{11}^*)^{-1}\mathbf{F}_{21}^*\| \leq \alpha \|\mathbf{F}_{21}\|. \tag{9}$$

Observe that for any $\mathbf{D} \in \mathbb{C}^{m \times n}$ that $\|(\mathbf{I} + \mathbf{DD}^*)\mathbf{u}\| \geq \|\mathbf{u}\|$ for all $\mathbf{u} \in \mathbb{C}^m$. Thus $\|(\mathbf{I} + \mathbf{DD}^*)^{-1}\| \leq 1$. Hence

$$\|\mathbf{X}_1\| \leq 1, \text{ and } \|\mathbf{X}_2\| \leq 1. \tag{10}$$

By Lemma 2 we can now conclude from (8), (9), (10) that

$$\|\Theta_1\|^2 \leq 1 + \alpha^2 \|\mathbf{F}_{12}\|^2. \tag{11}$$

Now

$$\Theta_1 - \Psi_1 = \begin{bmatrix} \mathbf{X}_1 - \mathbf{I} \\ \mathbf{F}_{12}^*(\mathbf{B}^* + \mathbf{F}_{11}^*)^{-1}\mathbf{X}_1 \end{bmatrix} \tag{12}$$

and

$$\Theta_2 - \Psi_2 = [\mathbf{X}_2 - \mathbf{I} \,|\, \mathbf{X}_2(\mathbf{B}^* + \mathbf{F}_{11}^*)^{-1}\mathbf{F}_{21}^*]. \tag{13}$$

By Theorem 3.1 we calculate that

$$\|\mathbf{I}-\mathbf{X}_1\| \leq \frac{\|(\mathbf{B}+\mathbf{F}_{11})^{-1}\mathbf{F}_{12}\mathbf{F}_{12}^*(\mathbf{B}^*+\mathbf{F}_{11}^*)^{-1}\|}{1-\|(\mathbf{B}+\mathbf{F}_{11})^{-1}\mathbf{F}_{12}\mathbf{F}_{12}^*(\mathbf{B}^*+\mathbf{F}_{11}^*)^{-1}\|} = \frac{\|(\mathbf{B}+\mathbf{F}_{11})^{-1}\mathbf{F}_{12}\|^2}{1-\|(\mathbf{B}+\mathbf{F}_{11})^{-1}\mathbf{F}_{12}\|^2}.$$

Or

$$\|\mathbf{I}-\mathbf{X}_1\| \leq \frac{\alpha^2\|\mathbf{F}_{12}\|^2}{1-\alpha^2\|\mathbf{F}_{12}\|^2} \tag{14}$$

by (8). Similarly,

$$\|\mathbf{I}-\mathbf{X}_2\| \leq \frac{\alpha^2\|\mathbf{F}_{21}\|^2}{1-\alpha^2\|\mathbf{F}_{21}\|^2}. \tag{15}$$

Combining (8), (10), (12) and (14) with Lemma 2 gives

$$\|\Theta_1-\Psi_1\|^2 \leq \left(\frac{\alpha^2\|\mathbf{F}_{12}\|^2}{1-\alpha^2\|\mathbf{F}_{12}\|^2}\right)^2 + \|\mathbf{F}_{12}\|^2\alpha^2. \tag{16}$$

In the same manner we get that

$$\|\Theta_2-\Psi_2\|^2 \leq \left(\frac{\alpha^2\|\mathbf{F}_{21}\|^2}{1-\alpha^2\|\mathbf{F}_{21}\|^2}\right)^2 + \|\mathbf{F}_{21}\|^2\alpha^2. \tag{17}$$

Substituting (11), (16) and (17) into (5) gives us that

$$\begin{aligned}\|(\mathbf{C}+\mathbf{F})^\dagger-\mathbf{C}^\dagger\| &\leq (1+\alpha^2\|\mathbf{F}_{12}\|^2)^{1/2}\|\mathbf{C}^\dagger\|\,\|\mathbf{F}_{21}\|\alpha \\ &\quad\times\left(\frac{\alpha^2\|\mathbf{F}_{21}\|^2}{(1-\alpha^2\|\mathbf{F}_{21}\|^2)^2}+1\right)^{1/2} \\ &\quad+\alpha\|\mathbf{F}_{12}\|\left(\frac{\alpha^2\|\mathbf{F}_{12}\|^2}{(1-\alpha^2\|\mathbf{F}_{21}\|^2)^2}+1\right)^{1/2}\|\mathbf{C}^\dagger\|.\end{aligned} \tag{18}$$

It follows from (2) and (3) that if we define

$$\begin{aligned}&\mathbf{E}_{11}=\mathbf{P}_{R(\mathbf{A})}\mathbf{E}\mathbf{P}_{R(\mathbf{A}^*)}, \quad \mathbf{E}_{12}=\mathbf{P}_{R(\mathbf{A})}\mathbf{E}\mathbf{P}_{N(\mathbf{A})}, \\ &\mathbf{E}_{21}=\mathbf{P}_{N(\mathbf{A}^*)}\mathbf{E}\mathbf{P}_{R(\mathbf{A}^*)}, \quad \mathbf{E}_{22}=\mathbf{P}_{N(\mathbf{A}^*)}\mathbf{E}\mathbf{P}_{N(\mathbf{A})},\end{aligned} \tag{19}$$

then $\|\mathbf{E}_{ij}\| = \|\mathbf{F}_{ij}\|$ for $1 \leq i,j \leq 2$.

Theorem 10.4.3 *Suppose that* $\mathbf{A},\mathbf{E}\in\mathbb{C}^{m\times n}$ *and* $\mathrm{rank}(\mathbf{A}+\mathbf{E})=\mathrm{rank}(\mathbf{A})$. *Suppose further that* $\|\mathbf{A}^\dagger\|\,\|\mathbf{E}\|<1/2$ *where* $\|\cdot\|=\|\cdot\|_{o2}$. *Define* $\mathbf{E}_{ij}$ *as in* (19). *Let* $\alpha=\|\mathbf{A}^\dagger\|/(1-\|\mathbf{A}^\dagger\|\,\|\mathbf{E}_{11}\|)$. *Then*

$$\begin{aligned}\|(\mathbf{A}+\mathbf{E})^\dagger-\mathbf{A}^\dagger\| &\leq \alpha\|\mathbf{E}_{21}\|\,\|\mathbf{A}^\dagger\|(1+\alpha^2\|\mathbf{E}_{12}\|^2)^{1/2} \\ &\quad\times\left(\frac{\alpha^2\|\mathbf{E}_{21}\|^2}{(1-\alpha^2\|\mathbf{E}_{21}\|^2)^2}+1\right)^{1/2} \\ &\quad+\alpha\|\mathbf{E}_{12}\|\,\|\mathbf{A}^\dagger\|\left(\frac{\alpha^2\|\mathbf{E}_{12}\|^2}{(1-\alpha^2\|\mathbf{E}_{21}\|^2)^2}+1\right)^{1/2}.\end{aligned} \tag{20}$$

Proof We need only show that $\|\mathbf{A}^\dagger\|\,\|\mathbf{E}\| < 1/2$ implies conditions (6) are satisfied. Notice that $\|\mathbf{A}^\dagger\| = \|\mathbf{B}^{-1}\|$ and that $\|\mathbf{E}_{ij}\| \leq \|\mathbf{E}\|$ for $1 \leq i, j \leq 2$. Thus $\|\mathbf{B}^{-1}\|\,\|\mathbf{E}_{11}\| \leq \|\mathbf{A}^\dagger\|\,\|\mathbf{E}\| < 1/2 < 1$ so the first condition is satisfied. To see that the second and third conditions are satisfied we calculate that

$$\|(\mathbf{B}+\mathbf{F}_{11})^{-1}\|\,\|\mathbf{F}_{ij}\| \leq \frac{\|\mathbf{B}^{-1}\|}{1-\|\mathbf{B}^{-1}\|\,\|\mathbf{F}_{11}\|}\|\mathbf{F}_{ij}\| < \frac{1/2}{1-1/2} = 1$$

as desired. ■

There are several nice features to (20). The first is that $\|\mathbf{A}^\dagger\|$ factors out of the right hand side. Thus the 'percentage error' $100\|(\mathbf{A}+\mathbf{E})^\dagger - \mathbf{A}^\dagger\|/\|\mathbf{A}^\dagger\|$ is easy to estimate. The second is that it is probably undesirable in many cases to actually compute $\|\mathbf{E}_{ij}\|$. However, if $\|\mathbf{E}_{ij}\|$ is replaced by any number K, $\|\mathbf{E}_{ij}\| \leq K < 1/\alpha^2$, then an estimate can be obtained. Since $\|\mathbf{E}\|_{o2} \geq \|\mathbf{E}_{ij}\|_{o2}$, one could use some of the more easy to compute norms in Section 2 to estimate $\|\mathbf{E}\|_{o2}$ and hence to estimate $\|\mathbf{E}_{ij}\|_{o2}$. In particular, we get

$$\|(\mathbf{A}+\mathbf{E})^\dagger - \mathbf{A}^\dagger\| \leq \alpha\|\mathbf{A}^\dagger\|\,\|\mathbf{E}\|(1+(1+\alpha^2\|\mathbf{E}\|^2)^{1/2}) \times\left(1+\frac{\alpha^2\|\mathbf{E}\|^2}{(1-\alpha^2\|\mathbf{E}\|^2)^2}\right)^{1/2} \tag{21}$$

for any $\|\mathbf{E}\|_{o2} \leq \|\mathbf{E}\| < 1/\alpha^2$.

Example 10.4.2 Let $\mathbf{A} = \begin{bmatrix} 10 & 0 & 0 \\ 0 & 11 & 0 \\ 0 & 0 & 0 \end{bmatrix}$ and $\mathbf{B} = \begin{bmatrix} 10 & 1 & 1 \\ 0 & 11 & 0 \\ 0 & 0 & 0 \end{bmatrix}$. We wish to estimate $\|\mathbf{B}^\dagger - \mathbf{A}^\dagger\|$ where $\|\cdot\| = \|\cdot\|_{o2}$. Let $\mathbf{E} = \begin{bmatrix} 0 & 1 & 1 \\ 0 & 0 & 0 \\ 0 & 0 & 0 \end{bmatrix}$ so that $\mathbf{B} = \mathbf{A} + \mathbf{E}$. Observe that $\|\mathbf{A}^\dagger\| = 0.1$. Also $\|\mathbf{E}\| \leq \|\mathbf{E}\|_2 = \sqrt{2}$. Thus $\|\mathbf{A}^\dagger\|\,\|\mathbf{E}\| \leq \sqrt{2}/10 < 1/2$ and Theorem 3 can be used. Now $\alpha \leq 0.1(1-61)\sqrt{2})^{-1} \doteqdot 0.11647$. Substituting into (21) gives $\|\mathbf{B}^\dagger - \mathbf{A}^\dagger\| \leq 0.03364$. Note that $\|\cdot\|_\infty \leq \|\cdot\|_{o2}$. Thus one conclusion of our estimate is that

$$\mathbf{B}^\dagger = \begin{bmatrix} 1/(10+\varepsilon) & \varepsilon & \varepsilon \\ \varepsilon & 1/11+\varepsilon & \varepsilon \\ \varepsilon & \varepsilon & \varepsilon \end{bmatrix}$$

where ε denotes a term < 0.03364 in absolute value. The exact value for $\mathbf{B}^\dagger$ is

$$\mathbf{B}^\dagger = \begin{bmatrix} 1/10 - 1/1010 & -10/1111 & 0 \\ 0 & 1/11 & 0 \\ 1/101 & -1/1111 & 0 \end{bmatrix}.$$

For some purposes (21) is sufficient. However, in a particular problem one might want a better estimate such as (20). Other estimates exist. We will give one due to Stewart. Its proof is left to the (starred) exercises.

Theorem 10.4.4 *Suppose* $\mathbf{A}, \mathbf{E} \in \mathbb{C}^{m \times n}$. *Define the* $\mathbf{E}_{ij}$ *as in* (19). *Suppose further that* $\|\mathbf{A}^\dagger\| \, \|\mathbf{E}_{11}\| < 1$ *and that* $rank(\mathbf{A}+\mathbf{E}) = rank(\mathbf{A})$. *Let* $\kappa = \|\mathbf{A}\| \, \|\mathbf{A}^\dagger\|$ *where* $\|\cdot\| = \|\cdot\|_{o2}$. *Then*

$$\|(\mathbf{A}+\mathbf{E})^\dagger - \mathbf{A}^\dagger\| \le \|\mathbf{A}^\dagger\| \left(\beta_{11} + \gamma \sum_{(i,j) \ne (1,1)} \frac{\beta_{ij}^2}{1+\beta_{ij}^2} \right)^{1/2},$$

where $\gamma = \left(1 - \dfrac{\kappa \|\mathbf{E}_{11}\|}{\|\mathbf{A}\|}\right)^{-1}$ *and* $\beta_{ij} = \gamma\kappa \|\mathbf{E}_{ij}\| / \|\mathbf{A}\|$.

The number $\kappa = \|\mathbf{A}\| \, \|\mathbf{A}^\dagger\|$ is called the *condition number* of $\mathbf{A}$. κ measures the amount of distortion of the unit ball of $\mathbb{C}^n$ caused by the linear transformation induced by $\mathbf{A}$. Theorem 3.1 may be written using κ.

Since $\mathbf{P}_{R(\mathbf{A})} = \mathbf{A}\mathbf{A}^\dagger$ we have $\kappa \ge 1$ if $\mathbf{A} \ne \mathbf{0}$. Let λ_l and λ_s denote the largest and smallest non-zero singular values of $\mathbf{A}$. (See page 6 for the definition of singular values.) It follows from the singular value decomposition, Theorem 0.2.2, that $\|\mathbf{A}\| = \lambda_l$ and $\|\mathbf{A}^\dagger\| = 1/\lambda_s$. Thus $\kappa = \lambda_l/\lambda_s$.

It is sometimes useful to have an estimate for $\|\mathbf{A}^\dagger - \mathbf{B}^\dagger\|$ even if $\text{rank}(\mathbf{A}) \ne \text{rank}(\mathbf{B})$. The following Proposition is helpful.

Proposition 10.4.3 *Suppose that* $\mathbf{A}, \mathbf{B} \in \mathbb{C}^{m \times n}$. *Then*

$$\mathbf{A}^\dagger - \mathbf{B}^\dagger = \mathbf{B}^\dagger(\mathbf{B}-\mathbf{A})\mathbf{A}^\dagger + (\mathbf{I} - \mathbf{B}^\dagger\mathbf{B})(\mathbf{A}^* - \mathbf{B}^*)\mathbf{A}^{\dagger *}\mathbf{A}^\dagger + \mathbf{B}^\dagger\mathbf{B}^{\dagger *}(\mathbf{A}^* - \mathbf{B}^*) \times (\mathbf{I} - \mathbf{A}\mathbf{A}^\dagger).$$

The proof of Proposition 3 is straightforward and is left to the exercises. From Proposition 3 we get quickly that

Theorem 10.4.5 *Suppose that* $\mathbf{A}, \mathbf{B} \in \mathbb{C}^{m \times n}$. *If* $\|\cdot\| = \|\cdot\|_{o2}$, *then* $\|\mathbf{A}^\dagger - \mathbf{B}^\dagger\| \le 3 \max\{\|\mathbf{A}^\dagger\|^2, \|\mathbf{B}^\dagger\|^2\} \, \|\mathbf{A} - \mathbf{B}\|$.

Proof From Proposition 3 we have that

$$\begin{aligned}\|\mathbf{A}^\dagger - \mathbf{B}^\dagger\| &\le \|\mathbf{A}-\mathbf{B}\| (\|\mathbf{B}^\dagger\| \, \|\mathbf{A}^\dagger\| + \|\mathbf{A}^\dagger\|^2 + \|\mathbf{B}^\dagger\|^2) \\ &\le \|\mathbf{A}-\mathbf{B}\| \, 3 \max\{\|\mathbf{A}^\dagger\|^2, \|\mathbf{B}^\dagger\|^2\}. \quad \blacksquare\end{aligned}$$

The 3 in Theorem 5 can be replaced by $(1+\sqrt{5})/2$. See [89].

5. Matrix valued functions

Theorem 10.4.1 has another formulation which is of interest. Let $\mathbf{A}(t)$ be an $m \times n$ *matrix valued function* for t in some interval $[a, b]$. That is,

$$\mathbf{A}(t) = \begin{bmatrix} a_{11}(t) & \dots & a_{1n}(t) \\ \vdots & & \vdots \\ a_{m1}(t) & \dots & a_{mn}(t) \end{bmatrix}, \quad a \le t \le b,$$

where $a_{ij}(t)$ is a complex valued function and $a_{ij}(t)$ is defined on $[a,b]$. If $a_{ij}(t)$ is continuous for $1 \leq i \leq m$, $1 \leq j \leq n$, then $\mathbf{A}$ is called continuous. This is equivalent to saying $\lim_{t \to t_0} \|\mathbf{A}(t) - \mathbf{A}(t_0)\| = \mathbf{0}$, for $a \leq t_0 \leq b$. If $\mathbf{A}(t)$ is invertible for $t \in [a,b]$, then $[\mathbf{A}(t)]^{-1}$ defines a matrix function for $t \in [a,b]$. We denote this function by $\mathbf{A}^{-1}(t)$. It is immediate from Theorem 3.1 that:

Proposition 10.5.1 *If $\mathbf{A}(t)$ is a continuous $n \times n$ matrix valued function on $[a,b]$ such that $\mathbf{A}(t)$ is invertible for $a \leq t \leq b$, then $\mathbf{A}^{-1}(t) = [\mathbf{A}(t)]^{-1}$ is a continuous matrix valued function on $[a,b]$.*

Proposition 1 may also be proved by Cramer's rule.

If $\mathbf{A}(t)$ is an $m \times n$ matrix valued function we define the $n \times m$ matrix valued function $\mathbf{A}^{\dagger}(\cdot)$ by $\mathbf{A}^{\dagger}(t) = [\mathbf{A}(t)]^{\dagger}$.

Theorem 4.1 gives us the following extension of Proposition 1.

Theorem 10.5.1 *Suppose that $\mathbf{A}(\cdot)$ is a continuous $m \times n$ matrix valued function defined on $[a,b]$. Then $\mathbf{A}^{\dagger}(t)$ is continuous on $[a,b]$ if and only if rank$(\mathbf{A}(t))$ is constant on $[a,b]$.*

Define the *rank function* $r(t)$ by $r(t) = \operatorname{rank}(\mathbf{A}(t))$. The discontinuities of $\mathbf{A}^{\dagger}(t)$ occur when $\mathbf{A}(t)$ changes rank. That is, at the discontinuities of $r(t)$. We wish to establish 'how many' discontinuities $\mathbf{A}^{\dagger}(t)$ may have if $\mathbf{A}(t)$ is continuous. The discussion requires a certain familiarity with the concepts of open and closed sets such as is found in a standard first course in real analysis.

Example 10.5.1 Let $f(t) = t\sin(\pi/t)$ if $0 < t \leq 1$ and $f(0) = 0$. Let $\mathbf{A}(t) = \begin{bmatrix} f(t) & 0 \\ 0 & 1 \end{bmatrix}$. Then $r(t) = 2$ if $t \neq \frac{1}{n}$ or $t \neq 0$, while $r\left(\frac{1}{n}\right) = r(0) = 1$.

Let $\mathscr{S} = \{t_0 \mid \mathbf{A}^{\dagger}$ is not continuous at $t_0\}$. Then $\mathscr{S} = \left\{\frac{1}{n} : n \text{ an integer}\right\} \bigcup \{0\}$.

Notice that the set $\mathscr{S}$ of Example 1 has an infinite number of points in it. However, it is *closed* (contains all its limit points) and has no interior (contains no open sets). This behaviour is typical.

Theorem 10.5.2 *Suppose that $\mathbf{A}(\cdot)$ is a continuous $m \times n$ matrix valued function defined on $[a,b]$. Let $\mathscr{S} = \{t_0 \in [a,b] \mid \mathbf{A}^{\dagger}(t)$ is not continuous at $t_0\}$. Then $\mathscr{S}$ is closed and has no interior. Thus there exists a collection of open intervals $\{(a_i, b_i)\}$ such that $\mathbf{A}^{\dagger}(\cdot)$ is continuous on each (a_i, b_i) and the closure of $\bigcup_i (a_i, b_i)$ is all of $[a,b]$.*

Proof Suppose that $\mathbf{A}(\cdot)$ is a continuous $m \times n$ matrix valued function defined on $[a,b]$. Let $\mathscr{S} = \{t_0 \mid r(t)$ is not continuous at $t_0\}$. Since $r(t)$ is not continuous only when it is not constant this $\mathscr{S}$ is the same as the $\mathscr{S}$ of Theorem 2. If the determinant of a *fixed* submatrix of $\mathbf{A}(t)$ is taken we get a continuous function of t. Let $\phi_i(t)$ be the sum of the absolute values of the determinants of all $i \times i$ submatrices of $\mathbf{A}(t)$. (For convenience suppose $n \leq m$.) Then ϕ_i, $1 \leq i \leq n$, are continuous functions on $[a, b]$ and if $\phi_i(t_o) = 0$, then $\phi_j(t_o) = 0$ for $j \geq i$. Notice that $r(t) = \sup\{i \mid \phi_i(t) \neq 0\}$.

Let $\mathscr{S}_i = \{t \mid \phi_i(t) = 0\}$. Notice that $\mathscr{S}_i \subseteq \mathscr{S}_{i+1}$, and $\mathscr{S}_i$ is closed since ϕ_i is continuous. Let $\partial\mathscr{S}_i$ denote the *boundary* of $\mathscr{S}_i$. For a closed set $\mathscr{S}_i$, $\partial\mathscr{S}_i$ is those points in $\mathscr{S}_i$ which are not interior points (loosely speaking the edge of $\mathscr{S}_i$). For the $\mathscr{S}$ of Example 1 we have $\partial\mathscr{S} = \mathscr{S}$. Now $\partial\mathscr{S}_i$ is a closed set with no interior. Hence $\bigcup_{i=1}^{n} \partial\mathscr{S}_i$ is a closed set with no interior.

We shall show that $\mathscr{S} = \bigcup_{i=1}^{n} \partial\mathscr{S}_i$.

Suppose that $r(t)$ is not continuous at t_o, that is, $t_o \in \mathscr{S}$. Let $r = r(t_o)$. By the continuity of $\mathbf{A}(t)$ and Proposition 4.2 we have $r(t) \geq r(t_o)$ for t near t_o. Since $r(t)$ is not continuous at t_o there exists a sequence $t_j \to t_o$ such that $r(t_j) \geq r(t_o) + 1$. But $\phi_{r+1}(t_o) = 0$ and $\phi_{r+1}(t_j) \neq 0$. Thus $t_o \in \partial\mathscr{S}_{r+1}$. Hence $\mathscr{S} \subseteq \bigcup_{i=1}^{n} \partial\mathscr{S}_i$.

To show equality suppose that $t_o \in \bigcup_{i=1}^{n} \partial\mathscr{S}_i$. Let r be the smallest integer such that $t_o \in \partial\mathscr{S}_r$. Let t_j be a sequence such that $t_j \to t_o$ but $t_j \notin \mathscr{S}_r$. Then $r(t_o) = r$ and $r(t_j) \geq r + 1$. Thus $r(t)$ is not continuous at t_o and $t_o \in \mathscr{S}$. Hence $\mathscr{S} = \bigcup_{i=1}^{n} \partial\mathscr{S}_i$. ■

It should be noted that set of discontinuities of $\mathbf{A}^\dagger(t)$ can still be very complicated for there exist closed sets with no interior which are uncountable.

Example 10.5.2 Let $\mathscr{S}$ be *any* closed subset of $[0, 1]$ which has no interior. Define $f(t) = \inf\{|t - s| : s \in \mathscr{S}\}$. Then f is continuous and $\mathscr{S} = \{t \in [0, 1] : f(t) = 0\}$. Let $\mathrm{A}(t) = [f(t)]$ so that $\mathrm{A}^\dagger(t) = [(f(t))^\dagger]$. Then the set of discontinuities of $\mathrm{A}^\dagger(t)$ is $\mathscr{S}$.

It is also useful to be able to differentiate matrix valued functions. If $\mathbf{A}(t)$ is an $m \times n$ matrix valued function on $[a, b]$, we define $\mathbf{dA}(t_o)$ by

$$\mathbf{dA}(t_o) = \lim_{t \to t_o} [\mathbf{A}(t) - \mathbf{A}(t_o)]/(t - t_o)$$

provided the limit exists (any matrix norm may be used). This is equivalent to saying that $\mathbf{dA}(t_o) = [a'_{ij}(t_o)]$, where $a'_{ij}(t_o) = \frac{da_{ij}}{dt}(t_o)$. $\mathbf{dA}(t_o)$ is called the derivative of $\mathbf{A}$ at t_o.

If $\mathbf{A}^\dagger(t)$ is to be differentiable at t_o, it must be continuous at t_o, hence of constant rank in some open interval containing t_o. Provided that this happens the differentiability of $\mathbf{A}(t)$ at t_o implies the differentiability of $\mathbf{A}^\dagger(t)$ at t_o.

Theorem 10.5.3 *Suppose that $\mathbf{A}(t)$ is an $m \times n$ differentiable matrix valued function defined on $[a,b]$. If rank$(\mathbf{A}(t))$ is constant, then $\mathbf{A}^\dagger(t)$ is differentiable on $[a,b]$ and*

$$d\mathbf{A}^\dagger = -\mathbf{A}^\dagger(d\mathbf{A})\mathbf{A}^\dagger + \mathbf{P}_{R(\mathbf{A}^*)^\perp}(d\mathbf{A}^*)\mathbf{A}^{\dagger *}\mathbf{A}^\dagger + \mathbf{A}^\dagger\mathbf{A}^{\dagger *}(d\mathbf{A}^*)\mathbf{P}_{R(\mathbf{A})}$$

Proof Suppose that $\mathbf{A}(t)$ is differentiable on $[a,b]$ and that rank$(\mathbf{A}(t))$ is constant. Then $\lim_{t\to t_o}[\mathbf{A}(t)-\mathbf{A}(t_o)]/(t-t_o) = d\mathbf{A}(t_o)$, $\lim_{t\to t_o}\mathbf{A}^\dagger(t) = \mathbf{A}^\dagger(t_o)$. Since we are differentiating with respect to a real variable we have $[d(\mathbf{A}^*)](t_o) = [(d\mathbf{A})(t_o)]^*$, that is $(d\mathbf{A})^* = d(\mathbf{A}^*)$, where $\mathbf{A}^*(t) = [\mathbf{A}(t)]^*$. Thus the symbol $d\mathbf{A}^*$ is well defined. By Proposition 4.3 we have

$$\begin{aligned}(\mathbf{A}^\dagger(t)-\mathbf{A}^\dagger(t_o))/(t-t_o) &= -\mathbf{A}^\dagger(t_o)\{[\mathbf{A}(t)-\mathbf{A}(t_o)]/(t-t_o)\}\mathbf{A}^\dagger(t_o)\\ &\quad + (\mathbf{I}-\mathbf{A}^\dagger(t_o)\mathbf{A}(t_o))\{[\mathbf{A}^*(t)-\mathbf{A}^*(t_o)]/(t-t_o)\}\mathbf{A}^\dagger(t)^*\mathbf{A}^\dagger(t)\\ &\quad + \mathbf{A}(t_o)^\dagger\mathbf{A}(t_o)^{\dagger *}\{[\mathbf{A}^*(t)-\mathbf{A}^*(t_o)]/(t-t_o)\}(\mathbf{I}-\mathbf{A}(t)\mathbf{A}(t)^\dagger).\end{aligned}$$

Taking the limit as $t \to t_o$ of both sides gives the formula for $d\mathbf{A}^\dagger$ and the differentiability of $\mathbf{A}^\dagger$. ■

It is important to notice that in the proof of Theorem 3 we used the fact that t was a real variable. Theorem 3, as stated, is not valid for t a complex variable. Let us see why.

For the remainder of this section suppose that z is a complex variable and $\mathbf{A}(z)$ is an analytic, $m \times n$ matrix valued function defined on a connected open set Ω. That is, $a_{ij}(z)$ is analytic for $z \in \Omega$, $1 \le i \le m$, $1 \le j \le n$. If $\mathbf{A}^\dagger(z)$ were also analytic, then so would be $\mathbf{A}(z)\mathbf{A}^\dagger(z)$ and $\mathbf{A}^\dagger(z)\mathbf{A}(z)$. But $\|\mathbf{A}(z)\mathbf{A}^\dagger(z)\|_{o2}$ and $\|\mathbf{A}(z)\mathbf{A}^\dagger(z)\|_{o2}$ are identically one on Ω. Thus $\mathbf{A}(z)\mathbf{A}^\dagger(z)$ and $\mathbf{A}^\dagger(z)\mathbf{A}(z)$ are identically constant. (A vector-valued version of the maximum modulus theorem is used to prove this last assertion.) Thus $R(\mathbf{A}(z))$ and $N(\mathbf{A}(z))$ are independent of z if both $\mathbf{A}(z)$ and $\mathbf{A}^\dagger(z)$ are analytic.

Suppose now that $R(\mathbf{A}(z))$ and $N(\mathbf{A}(z))$ are independent of z. Then there exist constant unitary matrices $\mathbf{U}, \mathbf{V}$ such that

$$\mathbf{A}(z) = \mathbf{U}\begin{bmatrix}\mathbf{A}_1(z) & \mathbf{0}\\ \mathbf{0} & \mathbf{0}\end{bmatrix}\mathbf{V}, \quad \mathbf{A}^\dagger(z) = \mathbf{V}^*\begin{bmatrix}\mathbf{A}_1^{-1}(z) & \mathbf{0}\\ \mathbf{0} & \mathbf{0}\end{bmatrix}\mathbf{U}^*$$

where $\mathbf{A}_1(z)$ is analytic and invertible. Of course, some of the zero submatrices in this decomposition of $\mathbf{A}(z)$ may not be present. But then, $\mathbf{A}^\dagger(z)$ is analytic since $\mathbf{A}_1^{-1}(z)$ is.

Theorem 10.5.4 *If $\mathbf{A}(z)$ is an analytic, $m \times n$ matrix valued function defined on a connected open set Ω, then $\mathbf{A}^\dagger(z)$ is analytic on Ω if and only if $R(\mathbf{A}(z))$ and $N(\mathbf{A}(z))$ are independent of z.*

The assumption that $R(\mathbf{A}(z))$ and $N(\mathbf{A}(z))$ are independent of z may be too restrictive for some applications. If one is willing to use an inverse other than the Moore–Penrose inverse, then the situation is somewhat more flexible. It is possible to pick a (1, 2)-inverse $\mathbf{A}_2^-(z)$ so that $\mathbf{A}_2^-(z)$ and $\mathbf{A}(z)$ are both meromorphic on the same domain. (Recall that a function is called meromorphic on a domain if it is analytic except at isolated points z_i called poles where $\mathbf{A}(z)$ satisfies $\lim_{z \to z_i} (z - z_i)^m \mathbf{A}(z) = \mathbf{0}$ for some m.) An example will help illustrate the general situation.

Example 10.5.3 Let $\mathbf{A}(z) = \begin{bmatrix} z & 0 & 1 \\ 0 & z & -1 \\ 0 & z & -1 \end{bmatrix}$ and

$\mathbf{A}_2^-(z) = \begin{bmatrix} z^\dagger & 0 & 0 \\ 0 & 0 & 0 \\ 0 & -1/2 & -1/2 \end{bmatrix}$. Notice that $N(\mathbf{A}(z))$ is not independent of z, but $\mathbf{A}_2^-(z)$ is analytic except at $z = 0$ where $\mathbf{A}(z)$ has a rank change. The projectors

$$\mathbf{A}_2^-(z)\mathbf{A}(z) = \begin{bmatrix} 1 & 0 & 0 \\ 0 & 0 & 0 \\ 0 & -z & 1 \end{bmatrix} \text{ and } \mathbf{A}(z)\mathbf{A}_2^-(z) = \begin{bmatrix} 1 & 0 & 0 \\ 0 & 1/2 & 1/2 \\ 0 & 1/2 & 1/2 \end{bmatrix}$$

are both analytic for all z, including zero, but $\mathbf{A}_2^-(z)\mathbf{A}(z)$ is of non-constant norm.

This behaviour is typical.

Theorem 10.5.5 *Let $\mathbf{A}(z)$ be a meromorphic $m \times n$ matrix valued function defined on a neighbourhood of z_0. Then there is a $n \times m$ matrix valued function $\mathbf{A}_2^-(z)$ defined on a neighbourhood of z_0 such that:*

(i) *$\mathbf{A}_2^-(z)$ is a (1, 2)-inverse for $\mathbf{A}(z)$ for each $z \neq z_0$.*
(ii) *$\mathbf{A}_2^-(z)$ is analytic on a deleted neighbourhood of z_0.*
(iii) *$\mathbf{A}_2^-(z)\mathbf{A}(z)$ and $\mathbf{A}(z)\mathbf{A}_2^-(z)$ are analytic on a neighbourhood of z_0.*

The proof of Theorem 5 may be found in [6]. Theorem 5 is a local result in that it talks about behaviour near a point. It is possible to get global versions.

Theorem 4 and 5 could be useful in a variety of settings. For example, equations of the form $\mathbf{A}(z)\mathbf{x} = \mathbf{0}$ occur in the general eigenvalue problem, the study of vibrating mechanical systems, and some damping problems. If the ideas of Chapter 5 are applied to circuits with capacitors and inductors, then the impedance matrix is a meromorphic matrix valued function since it involves polynomials in w and $1/w$. In particular, Theorem 5 might be useful when using the characterization of the impedance matrix of a shorted n-port network given in Proposition 5.3.1.

6. Non-linear least squares problems: an example

Theorem 5.2 can be useful in *non-linear least squares problems*. Some non-linear problems are very similar to those discussed in Chapter 2, Sections 6 and 7. In Section 2.7 we discussed fitting a function of the form

$$y = \beta_1 g_1(x) + \beta_2 g_2(x) + \ldots + \beta_k g_k(x)$$

to a set of data. There the $g_i(x)$ were known functions and the β_i were parameters to be estimated.

A somewhat more general problem is to fit a function of the form

$$y = \beta_1 g_1(x, \boldsymbol{\alpha}) + \beta_2 g_2(x, \boldsymbol{\alpha}) + \ldots + \beta_k g_k(x, \boldsymbol{\alpha}) \tag{1}$$

to a set of data. Here the $g_i(x, \boldsymbol{\alpha})$ are known functions while the β_i, and $\boldsymbol{\alpha} = (\alpha_1, \ldots, \alpha_r)$ are parameters. The g_i may be vector valued. Equations of the form (1) appear widely.

For example,

$$\begin{aligned} y &= \beta_1 \cos(\alpha_1 x) + \beta_2 \sin(\alpha_2 x), \quad \text{and} \\ y &= \beta_1 e^{\alpha_1 t} + \beta_2 t e^{\alpha_2 t} + \beta_3 t^2 e^{\alpha_3 t} \end{aligned} \tag{2}$$

are both in the form of (1). Functions of the form of (2) are common throughout the sciences. Any process that can be described by a linear differential equation (with reasonably good coefficients) will have solutions of the form (1) where k will be the order of the equation. Sometimes it is clear from the problem what the $g_i(x, \boldsymbol{\alpha})$ look like.

A particular example is the iron Mössbauer spectrum with two sites of different electric field gradient and one single line. Here (1) takes the form:

$$\begin{aligned} y = \beta_1 + \beta_2 t + \beta_3 t^2 - \beta_4 &\left[\frac{\alpha_3^2}{\alpha_3^2 + (\alpha_1 + 0.5\alpha_2 - t)^2} \right. \\ &\left. + \frac{\alpha_3^2}{\alpha_3^2 + (\alpha_1 - 0.5\alpha_2 - t)^2} \right] \\ - \beta_5 &\left[\frac{\alpha_6^2}{\alpha_6^2 + (\alpha_4 + 0.5\alpha_2 - t)^2} \right. \\ &\left. + \frac{\alpha_6^2}{\alpha_6^2 + (\alpha_1 - 0.5\alpha_2 - t)^2} \right] \\ - \beta_6 &\left[\frac{\alpha_8^2}{\alpha_8^2 + (\alpha_7 - t)^2} \right]. \end{aligned} \tag{3}$$

We are not going into the full theory of fitting (1) to a set of data since it would lead us too far astray. We will however work out an example that illustrates the basic ideas and difficulties.

Example 10.6.1 Suppose that we wish to fit the equation

$$y = \beta_0 + \beta_1 e^{\alpha t} \tag{4}$$

to the data points (t_i, y_i) which are $(0, 0)$, $(1, 1)$, and $(2, 3)$. Here we assume α is a real parameter. Thus we wish to minimize the error in

$$\begin{aligned} 0 &= \beta_0 + \beta_1 + e_1 \\ 1 &= \beta_0 + \beta_1 \mathrm{e}^{\alpha} + e_2 \\ 3 &= \beta_0 + \beta_1 \mathrm{e}^{2\alpha} + e_3 . \end{aligned}$$

This can be written as

$$\mathbf{y} = \boldsymbol{\Phi}(\alpha)\boldsymbol{\beta} + \mathbf{e} \text{ where} \tag{5}$$

$$\mathbf{y} = \begin{bmatrix} y_1 \\ y_2 \\ y_3 \end{bmatrix} = \begin{bmatrix} 0 \\ 1 \\ 3 \end{bmatrix}, \boldsymbol{\Phi}(\alpha) = \begin{bmatrix} 1 & 1 \\ 1 & \mathrm{e}^{\alpha} \\ 1 & \mathrm{e}^{2\alpha} \end{bmatrix}, \text{ and } \mathbf{e} = \begin{bmatrix} e_1 \\ e_2 \\ e_3 \end{bmatrix}.$$

We will minimize $\mathbf{e}$ with respect to the Euclidean norm. For any value of α, we minimize $\|\mathbf{e}\|$ by choosing $\boldsymbol{\beta} = \boldsymbol{\Phi}(\alpha)^{\dagger}\mathbf{y}$. Substituting $\boldsymbol{\beta}$ into (5) gives that $\mathbf{e} = \mathbf{y} - \boldsymbol{\Phi}(\alpha)\boldsymbol{\beta} = \mathbf{y} - \boldsymbol{\Phi}(\alpha)\boldsymbol{\Phi}(\alpha)^{\dagger}\mathbf{y} = (\mathbf{I} - \boldsymbol{\Phi}(\alpha)\boldsymbol{\Phi}(\alpha)^{\dagger})\mathbf{y}$. To minimize $\|\mathbf{e}\|$, we must minimize $\|(\mathbf{I} - \boldsymbol{\Phi}(\alpha)\boldsymbol{\Phi}(\alpha)^{\dagger})\mathbf{y}\|$, or equivalently, maximize $\|\boldsymbol{\Phi}(\alpha)\boldsymbol{\Phi}(\alpha)^{\dagger}\mathbf{y}\| = \mathbf{y}^*\boldsymbol{\Phi}(\alpha)\boldsymbol{\Phi}(\alpha)^{\dagger}\mathbf{y}$. It is clear from the data, that $\alpha = 0$ gives a poor fit. Assume then that $\mathrm{e}^{\alpha} \neq 1$. Then

$$\boldsymbol{\Phi}(\alpha)^{\dagger} = \frac{1}{2 - 2\mathrm{e}^{\alpha} - 2\mathrm{e}^{3}\alpha + 2\mathrm{e}^{4\alpha}} \times \begin{bmatrix} -\mathrm{e}^{\alpha} + \mathrm{e}^{4\alpha} & 1 - \mathrm{e}^{\alpha} - \mathrm{e}^{3\alpha} - \mathrm{e}^{4\alpha} & 1 - \mathrm{e}^{3\alpha} \\ 2 - \mathrm{e}^{\alpha} - \mathrm{e}^{2\alpha} & -1 + 2\mathrm{e}^{\alpha} - \mathrm{e}^{2\alpha} & -1 - \mathrm{e}^{\alpha} + 2\mathrm{e}^{2\alpha} \end{bmatrix}$$

A direct calculation gives

$$\mathbf{y}^*\boldsymbol{\Phi}(\alpha)\boldsymbol{\Phi}(\alpha)^{\dagger}\mathbf{y} = \frac{16 - 8\mathrm{e}^{\alpha} - 13\mathrm{e}^{2\alpha} - 14\mathrm{e}^{3\alpha} + 19\mathrm{e}^{4\alpha}}{2 - 2\mathrm{e}^{\alpha} - 2\mathrm{e}^{3\alpha} + 2\mathrm{e}^{4\alpha}}.$$

Differentiate this and set equal to zero to locate potential maxima. The result is $8 - 26\mathrm{e}^{\alpha} + 19\mathrm{e}^{2\alpha} + 24\mathrm{e}^{3\alpha} - 46\mathrm{e}^{4\alpha} + 26\mathrm{e}^{5\alpha} - 5\mathrm{e}^{6\alpha} = 0$. This is a sixth degree polynomial in e^{α}. Its roots are -0.8, 1, and 2. The root -0.8 is out since $\mathrm{e}^{\alpha} \geq 0$, and 1 is ruled out by assumption. We shall discuss the one root later. That leaves $\mathrm{e}^{\alpha} = 2$ or $\alpha = \ln 2$. Then $\boldsymbol{\beta}^* = [-1, 1]$ so that $y = -1 + \mathrm{e}^{t \ln 2}$ or $y = -1 + 2^t$ is the curve of form (4) which bests fits our data. Note that, in fact, we have an exact fit.

In Example 1 we discarded two roots -0.8 and 1. Clearly the -0.8 was extraneous. Where did the 1 come from? Consider the problem of finding extrema of f/g where f, g are two differentiable functions. Proceeding formally we get $f'/g - fg'/g^2 = 0$ or $gf' - fg' = 0$. If we now notice that $g(x) = x^2$ we would conclude that $x = 0$ was a potential extremum. But f/g is possibly not even defined for $x = 0$. This is exactly what happened in Example 1. The root $\mathrm{e}^{\alpha} = 1$, corresponded to when $\boldsymbol{\Phi}(\alpha)$ was not of full rank and the term $2 - 2\mathrm{e}^{\alpha} - 2\mathrm{e}^{3\alpha} + 2\mathrm{e}^{4\alpha}$ of $\boldsymbol{\Phi}(\alpha)^{\dagger}$ was zero.

There need not always exist a best fit.

Example 10.6.2 If the process of Example 1 is used to try to fit the equation

$$y = \beta_0 + \beta_1 e^{\alpha t} \tag{7}$$

to the data (0, 0), (1, 1) and (2, 2) then, as in Example 1, we get a polynomial but the only positive root of this one is 1. It is not too difficult to find values of $\alpha \neq 0, \beta_0$, and β_1 which give a better fit than $\alpha = 0$ ever does. What has happened here is that there is no best fit. By picking α close enough to zero and correctly choosing the β_0, β_1 one can fit (7) with as small an error as desired. But an exact fit is impossible since the data is colinear and (7) is strictly concave up or constant.

Aside from these problems this least squares technique is, in general, considerably more complicated. Even if $\boldsymbol{\Phi}(\alpha)^\dagger$ is of full rank the formulas for $\boldsymbol{\Phi}(\alpha)^\dagger$ can be very complicated. Secondly, if there are several α's, then $\mathbf{y}^*\boldsymbol{\Phi}(\alpha)\boldsymbol{\Phi}(\alpha)^\dagger\mathbf{y}$ will be a function of several variables which makes maximization more difficult. Finally, even when the maximization of $\mathbf{y}^*\boldsymbol{\Phi}(\alpha)\boldsymbol{\Phi}(\alpha)^\dagger\mathbf{y}$ reduces to finding the roots of a polynomial as our example did, the polynomial may be of large degree and require numerical methods to find its roots.

The reader interested in the numerical methods necessary for solving such problems is referred to the paper of Golub and Pereyra [37].

Note that in working our example we used the differentiability of $\boldsymbol{\Phi}(\alpha)\boldsymbol{\Phi}(\alpha)^\dagger$ and not $\boldsymbol{\Phi}(\alpha)^\dagger$. Where Theorem 5.2 comes in explicitly is in the theoretical development of the general technique.

7. Other inverses

It should be pointed out that it does not make sense to try and duplicate the results on the continuity of $\mathbf{A}^\dagger$ for (i, j, k)-inverses. The reason is obvious, for $\mathbf{A}^-(t)$ (or any other (i, j, k)-inverse) is not a well defined function. Thus if $\mathbf{A}_j \to \mathbf{A}$ and $\operatorname{rank}(\mathbf{A}_j) = \operatorname{rank}(\mathbf{A})$ one could still have $\mathbf{A}_j^-$ fail to even converge.

Example 10.7.1 Let $\mathbf{A} = \begin{bmatrix} 1 & 0 \\ 0 & 0 \end{bmatrix}$, $\mathbf{A}_j = \begin{bmatrix} 1 + 1/j & 0 \\ 0 & 0 \end{bmatrix}$ and

$\mathbf{A}_j^- = \begin{bmatrix} \dfrac{j}{j+1} & (-1)^j \\ 0 & 0 \end{bmatrix}$. Then $\mathbf{A}_j \to \mathbf{A}$, $\operatorname{rank}(\mathbf{A}_j) = \operatorname{rank}(\mathbf{A})$ but $\{\mathbf{A}_j^-\}$ does not even converge.

If $\mathbf{A}_j \to \mathbf{A}$ and $\mathbf{A}_j^\phi$ is a uniquely defined matrix for each j, then it is possible to discuss the convergence of the sequence $\{\mathbf{A}_j^\phi\}$. This would be necessary, for example, in discussing iterative algorithms or error bounds for particular methods of calculating inverses. See Chapter 12 for some such results.

For uniquely defined inverses, such as the Drazin, it is possible to consider continuity. The conditions under which the Drazin inverse is continuous are similar to but not identical to those under which the Moore–Penrose inverse is continuous.

Example 10.7.2 Let $\mathbf{A}_j = \begin{bmatrix} 1 & 0 & 0 \\ 0 & 0 & 1/j \\ 0 & 0 & 0 \end{bmatrix}$ and $\mathbf{A} = \begin{bmatrix} 1 & 0 & 0 \\ 0 & 0 & 0 \\ 0 & 0 & 0 \end{bmatrix}$

Then $\mathbf{A}_j \to \mathbf{A}$, $\mathbf{A}_j^{\mathrm{D}} \to \mathbf{A}^{\mathrm{D}}$, but $\operatorname{rank}(\mathbf{A}_j) > \operatorname{rank}(\mathbf{A})$ and $\operatorname{Ind}(\mathbf{A}_j) > \operatorname{Ind}(\mathbf{A})$.

Example 10.7.3 Let $\mathbf{A}_j = \begin{bmatrix} 1/j & 1 & 0 & 0 \\ 0 & 0 & 0 & 0 \\ 0 & 0 & 0 & 1 \\ 0 & 0 & 0 & 0 \end{bmatrix}$, $\mathbf{A} = \begin{bmatrix} 0 & 1 & 0 & 0 \\ 0 & 0 & 0 & 0 \\ 0 & 0 & 0 & 1 \\ 0 & 0 & 0 & 0 \end{bmatrix}$

Then $\mathbf{A}_j^{\mathrm{D}} = \begin{bmatrix} j & j^2 & 0 & 0 \\ 0 & 0 & 0 & 0 \\ 0 & 0 & 0 & 0 \\ 0 & 0 & 0 & 0 \end{bmatrix}$ while $\mathbf{A}^{\mathrm{D}} = \mathbf{0}$. Thus $\mathbf{A}_j \to \mathbf{A}$, $\operatorname{rank}(\mathbf{A}_j) = \operatorname{rank}(\mathbf{A})$, $\operatorname{Ind}(\mathbf{A}_j) = \operatorname{Ind}(\mathbf{A})$, but $\mathbf{A}_j^{\mathrm{D}} \not\to \mathbf{A}^{\mathrm{D}}$.

Recall that the core-rank of $\mathbf{A} \in \mathbb{C}^{m \times m}$ is the rank of $\mathbf{A}^k$ where $k = \operatorname{Ind}(\mathbf{A})$. Notice in Example 2 that core-rank $\mathbf{A}_j$ = core-rank $\mathbf{A}$. This turns out to be the key.

Theorem 10.7.1 *Suppose that $\mathbf{A}_j, \mathbf{A} \in \mathbb{C}^{m \times m}$ and that $\mathbf{A}_j \to \mathbf{A}$. Then $\mathbf{A}_j^{\mathrm{D}} \to \mathbf{A}^{\mathrm{D}}$ if and only if there exists an j_0 such that core-rank$(\mathbf{A}_j)$ = core-rank$(\mathbf{A})$ for $j \geq j_0$.*

Before proving Theorem 1 we need two preparatory results. The first is a generalization of Corollary 4.2.

Proposition 10.7.1 *Suppose that $\mathbf{P}_j, \mathbf{P} \in \mathbb{C}^{m \times m}$ are projectors (not necessarily orthogonal). Suppose further that $\mathbf{P}_j \to \mathbf{P}$. Then there is a j_0 such that rank$(\mathbf{P}_j)$ = rank$(\mathbf{P})$* for $j \geq j_0$.

Proof Suppose that $\mathbf{P}_j \to \mathbf{P}$ where $\mathbf{P}_j^2 = \mathbf{P}_j$ and $\mathbf{P}^2 = \mathbf{P}$. Since $\mathbf{P}_j \to \mathbf{P}$ we have $\operatorname{rank}(\mathbf{P}_j) \geq \operatorname{rank}(\mathbf{P})$ for large enough j.

Let $\mathbf{E}_j = \mathbf{P}_j - \mathbf{P}$ so that $\mathbf{P}_j = \mathbf{P} + \mathbf{E}_j$, $\mathbf{E}_j \to \mathbf{0}$. Suppose that there does not exist a j_0 such that $\operatorname{rank}(\mathbf{P}_j) = \operatorname{rank}(\mathbf{P})$ for $j \geq j_0$. Then there exists a subsequence $\mathbf{P}_{j_k}$ such that $\operatorname{rank}(\mathbf{P}_{j_k}) > \operatorname{rank}(\mathbf{P})$. That is, $\dim R(\mathbf{P}_{j_k}) > \dim R(\mathbf{P})$. But $N(\mathbf{P})$ is complementary to $R(\mathbf{P})$. Thus for each j_k there is a vector $\mathbf{u}_{j_k}$ such that $\mathbf{u}_{j_k} \in R(\mathbf{P}_{j_k})$, $\mathbf{u}_{j_k} \in N(\mathbf{P})$, and $\mathbf{u}_{j_k} \neq \mathbf{0}$. But then $\mathbf{u}_{j_k} = \mathbf{P}_{j_k}\mathbf{u}_{j_k} = (\mathbf{P} + \mathbf{E}_{j_k})\mathbf{u}_{j_k} = \mathbf{P}\mathbf{u}_{j_k} + \mathbf{E}_{j_k}\mathbf{u}_{j_k} = \mathbf{E}_{j_k}\mathbf{u}_{j_k}$. Let $\|\cdot\|$ be an operator norm. Then $\|\mathbf{E}_{j_k}\| \geq 1$ for all j_k. But $\|\mathbf{E}_{j_k}\| \to \mathbf{0}$ and we have a contradiction. Thus the required j_0 does exist. ■

We next prove a special case of Theorem 1.

Proposition 10.7.2 *Suppose that $\mathbf{A}_j, \mathbf{A} \in \mathbb{C}^{m \times m}$ and $\mathbf{A}_j \to \mathbf{A}$. Suppose* *further that* $\mathrm{Ind}(\mathbf{A}_j) = \mathrm{Ind}(\mathbf{A})$ *and core-rank*$(\mathbf{A}_j) =$ *core-rank*$(\mathbf{A})$ *for* j *greater than some fixed* j_0. *Then* $\mathbf{A}_j^{\mathrm{D}} \to \mathbf{A}^{\mathrm{D}}$.

Proof Suppose that $\mathbf{A}_j \to \mathbf{A}$, $\mathrm{Ind}(\mathbf{A}_j) = \mathrm{Ind}(\mathbf{A}) = k$ and core-rank $(\mathbf{A}_j) =$ core-rank $(\mathbf{A})$. From Chapter 7 we know $\mathbf{A}_j^{\mathrm{D}} = \mathbf{A}_j^k(\mathbf{A}^{2k+1})^\dagger \mathbf{A}_j^k$ while $\mathbf{A}^{\mathrm{D}} = \mathbf{A}^k(\mathbf{A}^{2k+1})^\dagger \mathbf{A}^k$. But rank $(\mathbf{A}_j^{2k+1}) =$ core-rank $(\mathbf{A}_j) =$ core-rank $(\mathbf{A}) =$ rank $(\mathbf{A}^{2k+1})$. Thus $(\mathbf{A}_j^{2k+1})^\dagger \to (\mathbf{A}^{2k+1})^\dagger$ by Theorem 4.1 Proposition 2 now follows. ■

We are now ready to prove Theorem 1.

Proof of Theorem 1 Suppose that $\mathbf{A}_j, \mathbf{A}$ are $m \times m$ matrices and $\mathbf{A}_j \to \mathbf{A}$. We will first prove the only if part of Theorem 1. Suppose that $\mathbf{A}_j^{\mathrm{D}} \to \mathbf{A}^{\mathrm{D}}$. Then $\mathbf{A}_j\mathbf{A}_j^{\mathrm{D}} \to \mathbf{A}\mathbf{A}^{\mathrm{D}}$. But $\mathbf{A}_j\mathbf{A}_j^{\mathrm{D}}$ is a projector onto $R(\mathbf{A}_j^{\mathrm{Ind}(\mathbf{A}_j)})$ and $\mathbf{A}\mathbf{A}^{\mathrm{D}}$ is a projector onto $R(\mathbf{A}^{\mathrm{Ind}(\mathbf{A})})$. Thus rank $(\mathbf{A}_j\mathbf{A}_j^{\mathrm{D}}) =$ core-rank $(\mathbf{A}_j)$ and rank $(\mathbf{A}\mathbf{A}^{\mathrm{D}}) =$ core-rank $(\mathbf{A})$. That core-rank $(\mathbf{A}_j) =$ core-rank $(\mathbf{A})$ for large j now follows from Proposition 1 and the fact that $\mathbf{A}_j\mathbf{A}_j^{\mathrm{D}} \to \mathbf{A}\mathbf{A}^{\mathrm{D}}$.

To prove the if part of Theorem 1 assume that core-rank $(\mathbf{A}_j) =$ core-rank $(\mathbf{A})$.

Let $\mathbf{A}_j = \mathbf{C}_j + \mathbf{N}_j$ and $\mathbf{A} = \mathbf{C} + \mathbf{N}$ be the core-nilpotent decompositions of $\mathbf{A}_j$ and $\mathbf{A}$ discussed in Chapter 7. Now $\mathbf{A}_j^l \to \mathbf{A}^l$ for all integers l. Pick $l \geq \sup\{\mathrm{Ind}(\mathbf{A}_j), \mathrm{Ind}(\mathbf{A})\}$. Then $\mathbf{A}_j^l = \mathbf{C}_j^l$ and $\mathbf{A}^l = \mathbf{C}^l$. Hence $\mathbf{C}_j^l \to \mathbf{C}^l$. Now $\mathrm{rank}(\mathbf{C}_j^l) = \mathrm{rank}(\mathbf{C}_j) =$ core-rank $(\mathbf{A}_j) =$ core-rank $(\mathbf{A}) = \mathrm{rank}(\mathbf{C}) = \mathrm{rank}(\mathbf{C}^l)$. This implies that $\mathrm{Ind}(\mathbf{C}_j)$ and $\mathrm{Ind}(\mathbf{C})$ are either both zero or both one. Thus $(\mathbf{C}_j^l)^{\mathrm{D}} \to (\mathbf{C}^l)^{\mathrm{D}}$ by Proposition 2. We may assume the indices are one else $\mathbf{A}$ is invertible and we are done. But $(\mathbf{C}_j^l)^{\mathrm{D}} = (\mathbf{C}_j^{\mathrm{D}})^l$ and $(\mathbf{C}^l)^{\mathrm{D}} = (\mathbf{C}^{\mathrm{D}})^l$. Hence $\mathbf{A}_j^{\mathrm{D}} = \mathbf{C}_j^{\mathrm{D}} = (\mathbf{C}_j^{\mathrm{D}})^{l+1}\mathbf{C}_j^l = (\mathbf{C}_j^{l+1})^{\mathrm{D}}\mathbf{C}_j^l$ converges to $(\mathbf{C}^{l+1})^{\mathrm{D}}\mathbf{C}^l = (\mathbf{C}^{\mathrm{D}})^{l+1}\mathbf{C}^l = \mathbf{C}^{\mathrm{D}} = \mathbf{A}^{\mathrm{D}}$. ■

We would like to conclude this section by examining the continuity of the index. In working with the rank it was helpful to observe that if $\mathbf{A}_j \to \mathbf{A}$, then rank $(\mathbf{A}_j) \geq$ rank $(\mathbf{A})$ for large enough j. This is not true for the index.

Example 10.7.4 Let $\mathbf{A}_j = \begin{bmatrix} 1/j & 1 \\ 0 & 1/j \end{bmatrix}$, $\mathbf{A} = \begin{bmatrix} 0 & 1 \\ 0 & 0 \end{bmatrix}$. Then $\mathbf{A}_j \to \mathbf{A}$ but $\mathrm{Ind}(\mathbf{A}_j) = 0$ while $\mathrm{Ind}(\mathbf{A}) = 2$.

Notice that $\mathbf{A}_j^{\mathrm{D}} \nrightarrow \mathbf{A}^{\mathrm{D}}$ in Example 4.

Proposition 10.7.3 Suppose that $\mathbf{A}_j, \mathbf{A} \in \mathbb{C}^{m \times m}$. *If* $\mathbf{A}_j \to \mathbf{A}$ *and* $\mathbf{A}_j^{\mathrm{D}} \to \mathbf{A}^{\mathrm{D}}$, *then there exists a* j_0 *such that* $\mathrm{Ind}(\mathbf{A}) \leq \mathrm{Ind}(\mathbf{A}_j)$ *for* $j \geq j_0$.

Proof Suppose $\mathbf{A}_j \to \mathbf{A}$ and $\mathbf{A}_j^{\mathrm{D}} \to \mathbf{A}^{\mathrm{D}}$. Let j_0 be such that $\mathrm{Ind}(\mathbf{A}_j)$ does not take on any of its finite number of values a finite number of times for $j \geq j_0$. Let $l = \inf\{\mathrm{Ind}(\mathbf{A}_j) : j \geq j_0\}$ and let j_k be the subsequence such that $\mathrm{Ind}(\mathbf{A}_{j_k}) = l$. Let $\mathbf{N}_j, \mathbf{N}$ be the nilpotent parts of $\mathbf{A}_j, \mathbf{A}$ respectively. Then $\mathbf{0} = (\mathbf{N}_{j_k})^l = (\mathbf{A}_{j_k})^l(\mathbf{I} - (\mathbf{A}_{j_k})^{\mathrm{D}}(\mathbf{A}_{j_k})) \to \mathbf{A}^l(\mathbf{I} - \mathbf{A}^{\mathrm{D}}\mathbf{A}) = \mathbf{N}^l$. Hence $\mathrm{Ind}(\mathbf{N}) \leq l$. ■

8. Exercises

1. Prove that $\|\cdot\|_p$, $p \geq 1$ defines a norm on $\mathbb{C}^n$.
2. Prove that for $\mathbf{A} \in \mathbb{C}^{m \times n}$ and norm $\|\cdot\|_s$ on $\mathbb{C}^n$ and $\mathbb{C}^m$ that

$$\begin{aligned}&\sup\{\|\mathbf{Au}\|_s : \mathbf{u} \in \mathbb{C}^n, \|\mathbf{u}\|_s = 1\}\\ &\quad = \inf\{\mathrm{K} : \|\mathbf{Au}\|_s \leq \mathrm{K}\|\mathbf{u}\|_s \text{ for every } \mathbf{u} \in \mathbb{C}^n\}.\end{aligned}$$

3. Suppose that $\{\mathbf{A}_k\}$ is a sequence of $m \times n$ matrices. Let $\mathbf{A}_k = [a_{ij}(k)]$. Suppose that $\mathbf{A} = [a_{ij}] \in \mathbb{C}^{m \times n}$. Prove that $\mathbf{A}_k \to \mathbf{A}$ if and only if $\lim_{k \to \infty} a_{ij}(k) = a_{ij}$ for $1 \leq i \leq m$, $1 \leq j \leq n$.
4. Show that if $\mathbf{A} \in \mathbb{C}^{m \times n}$, then $\|\mathbf{A}\|_{o1} = \max_j \left\{ \sum_{i=1}^{n} |a_{ij}| \right\}$.
5. If $\mathbf{A} \in \mathbb{C}^{m \times n}$, then let $\|\mathbf{A}\|_\alpha = n \max_{i,j} |a_{ij}| = n\|\mathbf{A}\|_\infty$. Since $\mathbb{C}^{m \times n}$ is isomorphic to $\mathbb{C}^{mn}$ we can give $\mathbb{C}^{m \times n}$ the $\|\cdot\|_2$ norm. $\|\mathbf{A}\|_2 = \left(\sum_{i,j} |a_{ij}|^2\right)^{1/2}$.
   (a) Prove that $\|\cdot\|_\alpha$ is a matrix norm in the following sense. If $\mathbf{A}, \mathbf{B} \in \mathbb{C}^{n \times n}$, then $\|\mathbf{AB}\|_\alpha \leq \|\mathbf{A}\|_\alpha \|\mathbf{B}\|_\alpha$.
   (b) Prove that $\|\cdot\|_2$ is a matrix norm. (This will probably require the following inequality:

$$\sum_{i=1}^{n} |x_i y_i| \leq \left(\sum_{i=1}^{n} |x_i|^2\right)^{1/2} \left(\sum_{i=1}^{n} |y_i|^2\right)^{1/2}$$

   which goes by the name of *Cauchy's inequality*.)
   Suppose that $\mathbf{A} \in \mathbb{C}^{n \times m}$. Show that
   (c) $\frac{1}{n}\|\mathbf{A}\|_\alpha \leq \|\mathbf{A}\|_{o\infty} \leq \|\mathbf{A}\|_\alpha$,
   (d) $\frac{1}{n}\|\mathbf{A}\|_\alpha \leq \|\mathbf{A}\|_{o1} \leq \|\mathbf{A}\|_\alpha$, and
   (e) $\frac{1}{n}\|\mathbf{A}\|_\alpha \leq \|\mathbf{A}\|_{o2} \leq \|\mathbf{A}\|_\alpha$.
   (f) $n^{-1/2}\|\mathbf{A}\|_2 \leq \|\mathbf{A}\|_{o2} \leq \|\mathbf{A}\|_2$
   (g) $n^{-1/2}\|\mathbf{A}\|_2 \leq \|\mathbf{A}\|_{o\infty} \leq n^{1/2}\|\mathbf{A}\|_2$
   (h) $n^{-1/2}\|\mathbf{A}\|_2 \leq \|\mathbf{A}\|_{o1} \leq n^{1/2}\|\mathbf{A}\|_{o2}$
   (i) $n^{-1/2}\|\mathbf{A}\|_{o2} \leq \|\mathbf{A}\|_{o\infty} \leq n^{1/2}\|\mathbf{A}\|_{o2}$
   (j) $n^{-1/2}\|\mathbf{A}\|_{o2} \leq \|\mathbf{A}\|_{o1} \leq n^{1/2}\|\mathbf{A}\|_{o2}$
   (k) $n^{-1}\|\mathbf{A}\|_{o\infty} \leq \|\mathbf{A}\|_{o1} \leq n\|\mathbf{A}\|_{o\infty}$.
   (l) In parts (f)–(k) determine the correct inequalities if $\mathbf{A} \in \mathbb{C}^{m \times n}$ rather than $\mathbb{C}^{n \times n}$.
6. Show that the inequalities (c)–(k) of Exercise 5 are all sharp.
7. Prove Proposition 10.3.1.

8. Let $\mathbf{I}$ be the identity matrix on $\mathbb{C}^{n\times n}$. Show that $\|\mathbf{I}\| \geq 1$ for any matrix norm $\|\cdot\|$ on $\mathbb{C}^{n\times n}$.
9. Let $\|\cdot\|$ be any matrix norm $\|\mathbf{I}\| > 1$ is permitted. If $\|\mathbf{A}\| < 1$, prove that $(\mathbf{I}-\mathbf{A})$ is invertible and estimate $\|(\mathbf{I}-\mathbf{A})^{-1}\|$.
10. Prove Fact 10.4.1.
11. Prove Fact 10.4.2.
12. Give an example to show that Proposition 10.4.2 is no longer true if $\|\mathbf{E}\| < 1/\|\mathbf{A}^\dagger\|$ is weakened to $\|\mathbf{E}\| \leq 1/\|\mathbf{A}^\dagger\|$.
13. Suppose M, N are subspaces of $\mathbb{C}^n$, $\|\cdot\|$ a norm on $\mathbb{C}^n$. Prove that if $\dim M > \dim N$, and K is a complementary subspace to N then there exists a $\mathbf{u} \in M \cap K$ such that $\|\mathbf{u}\| = 1$.

*14. Prove Theorem 10.4.4. (Hint: first prove for the case that $\mathbf{A} \in \mathbb{C}^{m\times n}$ has rank n.)

15. Prove Proposition 10.4.3.

9. References and further reading

A more complete discussion of matrix norms may be found in [33] and [37]. In [37] norms are discussed in terms of their unit balls $\{\mathbf{u} : \|\mathbf{u}\| \leq 1\}$. The relationship between various norms is studied in [33]. In particular, Exercise 5 is from [33].

Theorem 4.1 was given in Penrose's first paper on generalized inverses [67]. His proof was based on the characteristic function of $\mathbf{A}^*\mathbf{A}$ and appears as an aside on the bottom of page 408. In a follow up paper [68] he discussed approximating $\mathbf{A}^\dagger$.

The development given here is similar to that of Stewart [88]. A more restrictive treatment that applies some of the ideas to error estimation is [9]. An infinite dimensional treatment is given in [63]. Another treatment that is restricted to hermitian matrices is [76].

The paper by Robertson and Rosenberg [75] deals with matrix valued measures. In particular, they prove matrix versions of the Hahn–Jordan decomposition, the Radon–Nikodym theorem, and the Lebesgue decomposition. Their results use generalized inverses. As a lemma they establish that if $\mathbf{A}(t)$ is a *measurable* $m \times n$ matrix valued function (that is, $a_{ij}(\cdot)$ is measurable for $1 \leq i \leq m$, $1 \leq j \leq n$) and if

$$\mathbf{B}(t) = [\mathbf{A}(t)]^\dagger,$$

then $\mathbf{B}(t)$ is a measurable $n \times m$ matrix valued function.

Proposition 4.3 and Theorem 5.3 are from [37]. There is a nice bibliography at the end of [37] which includes a reference which explains formula (3).

11
Linear programming

1. Introduction and basic theory

This chapter will discuss how the theory of linear programming relates to the theory of generalized inverses. The chapter is not designed to teach the reader the full theory or applications of linear programming. We ignore, for example, the simplex method.

We will begin by describing the basic linear programming problem. Then several basic theorems will be presented. Selected proofs will be given to give the reader an idea of some of the techniques involved. We conclude by showing how the generalized inverse can occur in working with linear programming problems. Hopefully by the end of this chapter, the reader will have a good idea of the part that the theory of the generalized inverse can play in the theory of linear programming.

To begin with, we should probably point out that the name 'linear programming' can be somewhat misleading. It is not concerned with computer programming as such, though computational algorithms play an important part. Rather the theory concerns maximizing and minimizing linear expressions with respect to linear constraints and linear inequalities. These problems arise, for example, in allocation of resources and transportation problems. The 'program' is thus more of a 'schedule' or 'allocation scheme'.

In order to motivate the formulation of the general mathematical problem, let us consider the following simplified situation. A manufacturing company makes three products P_1, P_2 and P_3. Each product uses the inputs of electricity, labour, iron and copper. Suppose the amounts used are given by Table 4.1.

The numbers are in terms of amount of input required (in some appropriate units) per unit output of product. For example, each unit of P_1 uses 4 units of electricity. The problem is to maximize the profit where the profit per unit of product is 1 for P_1, 2 for P_2, and 1 for P_3. We assume that all available product can be sold. However, there are certain constraints. We suppose that there are only 20 units of labour, 10 units of

Table 4.1

	P_1	P_2	P_3
Elect.	4	8	4
Labour	3	3	1
Iron	1	1	0
Copper	1	2	2

iron, and 5 units of copper available each week. The problem may be formalized mathematically as follows. Let x_1, x_2, x_3 denote the quantities of products P_1, P_2, P_3 to be produced in a week.

Problem A Maximize $x_1 + 2x_2 + x_3$ subject to the constraints

$$\begin{aligned} 3x_1 + 3x_2 + x_3 &\le 20, \\ x_1 + x_2 &\le 10, \\ x_1 + 2x_2 + 2x_3 &\le 5, \end{aligned}$$

and $x_1 \ge 0, x_2 \ge 0, x_3 \ge 0$.

This can be put into a more standard form as follows. Let

$$\begin{aligned} x_4 &= 20 - 3x_1 - 3x_2 - x_3, \\ x_5 &= 10 - x_1 - x_2, \text{ and} \\ x_6 &= 5 - x_1 - 2x_2 - 2x_3. \end{aligned}$$

Then Problem A becomes Problem B.

Problem B Maximize $x_1 + 2x_2 + x_3$ subject to

$$\begin{aligned} 3x_1 + 3x_2 + x_3 + x_4 &= 20, \\ x_1 + x_2 + x_3 &= 10, \\ x_1 + 2x_2 + 2x_3 + x_6 &= 5, \end{aligned}$$

and $x_1 \ge 0, x_2 \ge 0, x_3 \ge 0, x_4 \ge 0, x_5 \ge 0, x_6 \ge 0$.

The variables x_4, x_5, x_6 represent unused available input and are called *slack variables*. The general linear programming problem can be formulated using Problem A as a model. One might be tempted to define it as maximization of a linear function subject to any combination of inequalities and equalities. However, by multiplying inequalities by the appropriate sign we may get all the equalities in the same direction.

If $\mathbf{x} = (x_1, \ldots, x_n) \in \mathbb{R}^n$ and $\mathbf{y} = (y_1, \ldots, y_n) \in \mathbb{R}^n$ and if $x_i \le y_i$ for all i, then we write $\mathbf{x} \le \mathbf{y}$. This notation will simplify our calculations. Let $\mathbb{R}^n_+ = \{\mathbf{x} \in \mathbb{R}^n : \mathbf{x} \ge \mathbf{0}\}$.

Since linear programming problems are usually done with real values we will work with them. However, most of the theory is valid for $\mathbb{C}^n$ if $\mathbf{x} \ge \mathbf{y}$ is interpreted to mean $\operatorname{Re}(\mathbf{x} - \mathbf{y}) \ge \mathbf{0}$ and we maximize the real valued functional, $\operatorname{Re}(\mathbf{x}, \mathbf{c})$. Here $\operatorname{Re} \mathbf{z}$ is the vector whose ith entry is the real part of the ith entry of $\mathbf{z}$.

Definition 11.1.1 *Suppose that* $\mathbf{A} \in \mathbb{R}^{m \times n}$, $\mathbf{b} \in \mathbb{R}^m$, *and* $\mathbf{c} \in \mathbb{R}^n$. *The* general linear programming problem *is to maximize* $(\mathbf{x}, \mathbf{c})$ *subject to the constraints* $\mathbf{Ax} \leq \mathbf{b}$ *and* $\mathbf{x} \geq \mathbf{0}$.

Note that by maximizing $(\mathbf{x}, -\mathbf{c})$ we can obtain the minimum of $(\mathbf{x}, \mathbf{c})$.

Definition 11.1.2 *A linear programming problem is in standard form if it is of the form* $\mathbf{Ax} = \mathbf{b}$, $\mathbf{x} \geq \mathbf{0}$, $\mathbf{A} \in \mathbb{R}^{m \times n}$, $\mathbf{b} \in \mathbb{R}^m$.

Note that the general linear programming problem can be put in standard form by using slack variables. Problem B was the standard form of Problem A. Problem B has the advantage that one is working with equality constraints. Problem A will in general have a smaller matrix **A**. Thus while Problem B might be easier to work with theoretically, Problem A might be simpler since it involves smaller sized matrices and vectors.

Going along with each linear programming problem is another one called its dual.

Definition 11.1.2 *The* dual *of the linear programming problem: maximize* $(\mathbf{c}, \mathbf{x})$ *subject to* $\mathbf{Ax} \leq \mathbf{b}$, $\mathbf{x} \geq \mathbf{0}$ is, *minimize* $(\mathbf{b}, \mathbf{y})$ *subject to* $\mathbf{A}^*\mathbf{y} \geq \mathbf{c}$, $\mathbf{y} \geq \mathbf{0}$.

The dual of Problem A would be Problem C.

Problem C Minimize $20y_1 + 10y_2 + 5y_3$ subject to

$$\begin{aligned} 3y_1 + y_2 + y_3 &\geq 1 \\ 3y_1 + y_2 + 2y_3 &\geq 2 \\ y_1 + 2y_3 &\geq 1 \end{aligned} \tag{1}$$

and $y_1 \geq 0, y_2 \geq 0, y_3 \geq 0$.

This dual problem has an interpretation related to that of Problem A. Problem A amounts to maximizing total net revenue, while Problem C consists of minimizing the total 'accounting value' of the inputs. The y_i's are the values to be assigned to the inputs. They are sometimes called 'shadow prices.' The equations (1) say that the values given to inputs must not be less than the contribution of the inputs to net revenue. We shall see shortly that Problem A and Problem C are equivalent in an appropriate sense.

We now turn to the mathematical treatment of linear programming problems. For either the general problem or its dual, a vector is called feasible if it satisfies the constraints. A feasible vector which maximizes the functional in the general problem (or minimizes the functional in the dual problem) will be called *optimal* for the general (or dual) problem.

It should be noted that the dual of the general problem and the dual of the problem in standard form are equivalent. That is, they have the same feasible vectors, optimal vectors, and minimal value of the functional.

Proposition 11.1.1 *The general problem and its standard form have equivalent duals.*

It will be more convenient for us to work with the standard form.

The first problem is to determine when feasible solutions of $\mathbf{Ax} = \mathbf{b}$, $\mathbf{x} \geq \mathbf{0}$ exist, that is, when the constraints are consistent.

Proposition 11.1.2 *The constraints* $\mathbf{Ax} = \mathbf{b}, \mathbf{x} \geq \mathbf{0}$, *are consistent if and only if both* $\mathbf{b} \in R(\mathbf{A})$ *and* $\mathbf{A}^{-}\mathbf{b} \in N(\mathbf{A}) + \mathbb{R}^n_+$.

Proof Solutions of $\mathbf{Ax} = \mathbf{b}$ exist, of course, if and only if $\mathbf{b} \in R(\mathbf{A})$. If $\mathbf{b} \in R(\mathbf{A})$, then the set of all solutions to $\mathbf{Ax} = \mathbf{b}$ is $\mathbf{A}^{-}\mathbf{b} + N(\mathbf{A})$ for any (1)-inverse $\mathbf{A}^{-}$. Thus the set of constraints it consistent if and only if $(\mathbf{A}^{-}\mathbf{b} + N(\mathbf{A})) \cap \mathbb{R}^n_+$ is non-empty. This happens if and only if $\mathbf{A}^{-}\mathbf{b} \in N(\mathbf{A}) + \mathbb{R}^n_+$. ■

From Proposition 2 and a lot more work we can get a characterization of consistent constraints due to Farkas. A proof using generalized inverse notation may be found in [11].

Theorem 11.1.1 *(Farkas). Suppose that* $\mathbf{A} \in \mathbb{R}^{m \times n}, \mathbf{b} \in \mathbb{R}^m$. *Then the following are equivalent:*

(i) $\mathbf{Ax} = \mathbf{b}, \mathbf{x} \geq \mathbf{0}$ *is consistent,*
(ii) $\mathbf{A}^*\mathbf{y} \geq \mathbf{0}$ *implies that* $(\mathbf{b}, \mathbf{y}) \geq \mathbf{0}$.

A linear programming problem and its dual are closely related. We summarize this relationship in the following fundamental theorem. The theorem is standard. A proof may be found in Simonnard [84], for example.

Theorem 11.1.2 *A linear programming problem and its dual either both have optimal solutions or neither does. If they both do, the maximum value of the original problem equals the minimum value of the dual. This common value is called the* optimal value *of the problem.*

This theorem has several consequences. First of all, it says that if either the original or the dual has no feasible vector, then the other cannot achieve a maximum (or minimum) even if it has a feasible vector.

Suppose that $\mathbf{x}_0$ is a feasible vector for a linear programming problem while $\mathbf{y}_0$ is feasible for its dual. Then $\mathbf{Ax}_0 \leq \mathbf{b}$ and $\mathbf{A}^*\mathbf{y}_0 \geq \mathbf{c}$.
But then

$$(\mathbf{b}, \mathbf{y}_0) \geq \mathbf{y}_0^*\mathbf{Ax}_0 \geq \mathbf{c}^*\mathbf{x}_0 = (\mathbf{c}, \mathbf{x}_0). \tag{2}$$

According to Theorem 2, if $\mathbf{y}, \mathbf{x}$ are optimal feasible vectors, then $(\mathbf{b}, \mathbf{y}) = (\mathbf{c}, \mathbf{x})$. This provides a way of testing two feasible vectors to see if they are both optimal solutions. Notice that (2) also says that if the original problem has a feasible vector $\mathbf{x}_0$, then $(\mathbf{b}, \mathbf{y})$ is bounded below. There would then seem to be hope for a minimum. Similarly, if $\mathbf{y}_0$ is a feasible vector for the dual, we have $(\mathbf{c}, \mathbf{x})$ is bounded above by $(\mathbf{b}, \mathbf{y}_0)$. In fact, the following is true.

Theorem 11.1.3 *If $f(\mathbf{x}) = (\mathbf{c}, \mathbf{x})$ is bounded above (below) on $\{\mathbf{x}: \mathbf{Ax} = \mathbf{b}, \mathbf{x} \geq \mathbf{0}\}$, then f attains its maximum (minimum) on $\{\mathbf{x}: \mathbf{Ax} = \mathbf{b}, \mathbf{x} \geq \mathbf{0}\}$.*

It should be noted that Theorem 3 is not true if $\{\mathbf{x}: \mathbf{Ax} = \mathbf{b}, \mathbf{x} \geq \mathbf{0}\}$ is replaced by an arbitrary convex set.

In light of our observation that feasible vectors of the general problem and its dual form bounds for $(\mathbf{x}, \mathbf{c})$ and $(\mathbf{x}, \mathbf{b})$, the next result follows from Theorem 3.

Theorem 11.1.4 *A necessary and sufficient condition for one of the two problems (and hence both) to have optimal vectors is that they both have feasible vectors.*

Under certain conditions it is easy to show $(\mathbf{x}, \mathbf{c})$ is bounded above. As pointed out earlier, any vector satisfying the constraints is of the form $\mathbf{x}_0 = \mathbf{A}^-\mathbf{b} + (\mathbf{I} - \mathbf{A}^-\mathbf{A})\mathbf{x}_0$. Thus $\sup_{\mathbf{x}_0}(\mathbf{x}_0, \mathbf{c}) = (\mathbf{A}^-\mathbf{b}, \mathbf{c}) + \sup_{\mathbf{x}_0}((\mathbf{I} - \mathbf{A}^-\mathbf{A})\mathbf{x}_0, \mathbf{c})$, where $\mathbf{x}_0$ ranges over all feasible vectors. One way to get that the supremum on the right exists is to have $(\mathbf{I} - \mathbf{A}^-\mathbf{A})^*\mathbf{c} = \mathbf{0}$. For then, $((\mathbf{I} - \mathbf{A}^-\mathbf{A})\mathbf{x}_0, \mathbf{c}) = 0$ and the supremum is $(\mathbf{A}^-\mathbf{b}, \mathbf{c})$. If $(\mathbf{I} - \mathbf{A}^-\mathbf{A})^*\mathbf{c} = \mathbf{0}$, then $\mathbf{c}^*(\mathbf{I} - \mathbf{A}^-\mathbf{A}) = \mathbf{0}$. Thus we have:

Proposition 11.1.3 *A sufficient condition that the function $f(\mathbf{x}) = (\mathbf{c}, \mathbf{x})$ be bounded above is that $\mathbf{c}^*\mathbf{A}^-\mathbf{A} = \mathbf{c}^*$.*

Proposition 3 is a very special case. If $\mathbf{c}^*\mathbf{A}^-\mathbf{A} = \mathbf{c}^*$, then $f(\mathbf{x}) = (\mathbf{A}^-\mathbf{b}, \mathbf{c})$ for all feasible $\mathbf{x}$. Thus every feasible vector is optimal.

Another, and more reasonable, way to guarantee that $f(\mathbf{x}) = (\mathbf{c}, \mathbf{x})$ attains a maximum is to have that $(\mathbf{A}^-\mathbf{b} + N(\mathbf{A})) \cap \mathbb{R}^n_+$, the set of feasible vectors, is a closed, bounded set. That it is closed is clear since $\mathbf{A}^-\mathbf{b} + N(\mathbf{A})$ and $\mathbb{R}^n_+$ are closed. (Closed being used here in the sense of having all of its limit points.)

One way that $(\mathbf{A}^-\mathbf{b} + N(\mathbf{A})) \cap \mathbb{R}^n_+$ could be unbounded would be the existence of an $\mathbf{h}_0 \in N(\mathbf{A}), \mathbf{h}_0 \neq \mathbf{0}$, such that $\mathbf{h}_0 \geq \mathbf{0}$. In this case if $\mathbf{h} \in N(\mathbf{A})$ were such that $\mathbf{A}^-\mathbf{b} + \mathbf{h} \geq \mathbf{0}$, we would have $\mathbf{A}^-\mathbf{b} + \mathbf{h} + \lambda\mathbf{h}_0 \geq \mathbf{0}$ for $\lambda \geq 0$.

In fact, the existence of non-zero $\mathbf{h}_0 \in N(\mathbf{A}) \cap \mathbb{R}^n_+$ turns out to be necessary and sufficient if feasible solutions exist.

Proposition 11.1.4 *Suppose that the set $\{\mathbf{A}^-\mathbf{b} + N(\mathbf{A})\} \cap \mathbb{R}^n_+$ is non-empty. Then it is unbounded if and only if there exists a non-zero $\mathbf{h}_0 \in N(\mathbf{A})$ such that $\mathbf{h}_0 \geq \mathbf{0}$.*

Proof We need only show the only if part. Suppose that $\{\mathbf{A}^-\mathbf{b} + N(\mathbf{A})\} \cap \mathbb{R}^n_+$ is unbounded. Let $\{\mathbf{A}^-\mathbf{b} + \mathbf{h}_m\}$ be an unbounded sequence made up of vectors in $(\mathbf{A}^-\mathbf{b} + N(\mathbf{A})) \cap \mathbb{R}^n_+$. Then $\{\mathbf{h}_m\}$ is an unbounded sequence. Let $\mathbf{k}_m = \mathbf{h}_m / \|\mathbf{h}_m\|$. Since $\{\mathbf{k}_m\}$ is a bounded sequence in $\mathbb{R}^n$, it has a convergent subsequence which we will denote by $\mathbf{k}_l$. Let $\mathbf{k}_0$ denote the limit of this subsequence. That is $\mathbf{k}_l \to \mathbf{k}_0$. Since $\mathbf{k}_l \in N(\mathbf{A})$ and $N(\mathbf{A})$ is closed we have $\mathbf{k}_0 \in N(\mathbf{A})$. Note that $(\mathbf{A}^-\mathbf{b} + \mathbf{h}_l)/\|\mathbf{h}_l\| \to \mathbf{0} + \mathbf{k}_0 = \mathbf{k}_0$. Thus $\mathbf{k}_0 \in \mathbb{R}^n_+$ since $\mathbf{A}^-\mathbf{b} + \mathbf{h}_l \in \mathbb{R}^n_+$, and $\mathbf{k}_0$ is the required vector. ■

2. Pyle's reformulation

To illustrate one way that the generalized inverse can be used in linear programming problems we will discuss a method developed by Pyle to reformulate a linear programming problem into a non-negative fixed point problem. The ways in which the generalized inverse are used in this section are typical of many applications of the theory of generalized inverses to to linear programming. Another application is discussed in the final section.

We will not give all the proofs but rather outline Pyle's argument. The proofs are assigned as exercises or may be found in his paper.

Consider the following problem and its dual:

(P1) Maximize $(\mathbf{x}, \mathbf{c})$ where $\mathbf{A}\mathbf{x} = \mathbf{b}, \mathbf{x} \geq \mathbf{0}$.
(D1) Minimize $(\mathbf{y}, \mathbf{b})$ where $\mathbf{A}^*\mathbf{y} \geq \mathbf{c}, \mathbf{y} \geq \mathbf{0}$.

Here $\mathbf{A} \in \mathbb{R}^{m \times n}, \mathbf{c} \in \mathbb{R}^m, \mathbf{b} \in \mathbb{R}^n$.

The first step is to reformulate (P1) and (D1) as problems in which every feasible solution of the new problems is an optimal solution of (P1) and (D1). This requires being able to express optimality of vectors as an algebraic condition. This algebraic condition will then be added to both of our original problems to get one large problem. The derivation of the algebraic condition requires us to first reformulate (P1) and (D1).

If $\mathbf{x}$ satisfies the constraints of (P1), we know that $\mathbf{x} = \mathbf{A}^\dagger\mathbf{b} + (\mathbf{I} - \mathbf{A}^\dagger\mathbf{A})\mathbf{x}$. Thus $(\mathbf{x}, \mathbf{c}) = (\mathbf{A}^\dagger\mathbf{b}, \mathbf{c}) + ((\mathbf{I} - \mathbf{A}^\dagger\mathbf{A})\mathbf{x}, \mathbf{c}) = (\mathbf{A}^\dagger\mathbf{b}, \mathbf{c}) + (\mathbf{x}, (\mathbf{I} - \mathbf{A}^\dagger\mathbf{A})\mathbf{c})$. (P1) is thus equivalent to

(P2) Maximize $(\mathbf{x}, (\mathbf{I} - \mathbf{A}^\dagger\mathbf{A})\mathbf{c})$ where $\mathbf{A}\mathbf{x} = \mathbf{b}$ (or $\mathbf{A}^\dagger\mathbf{A}\mathbf{x} = \mathbf{A}^\dagger\mathbf{b}$), $\quad \mathbf{x} \geq \mathbf{0}$.

This is obviously equivalent to

(P3) Minimize $(\mathbf{x}, -(\mathbf{I} - \mathbf{A}^\dagger\mathbf{A})\mathbf{c})$ where $\mathbf{A}^\dagger\mathbf{A}\mathbf{x} = \mathbf{A}^\dagger\mathbf{b}, \quad \mathbf{x} \geq \mathbf{0}$.

Now we shall rewrite (D1). The dual of (P2) is

(D2) Minimize $(\mathbf{y}, \mathbf{A}^\dagger\mathbf{b})$ where $\mathbf{A}^\dagger\mathbf{A}\mathbf{y} \geq (\mathbf{I} - \mathbf{A}^\dagger\mathbf{A})\mathbf{c}$.

To remove the inequality in (D2) observe that if $\mathbf{y}$ is a feasible solution of (D2). then $\mathbf{A}^\dagger\mathbf{A}\mathbf{y} \geq (\mathbf{I} - \mathbf{A}^\dagger\mathbf{A})\mathbf{c}$. Thus $\mathbf{z} = -(\mathbf{I} - \mathbf{A}^\dagger\mathbf{A})\mathbf{c} + \mathbf{A}^\dagger\mathbf{A}\mathbf{y} \geq \mathbf{0}$. Thus $\mathbf{z}$ is a feasible solution of $(\mathbf{I} - \mathbf{A}^\dagger\mathbf{A})\mathbf{z} = -(\mathbf{I} - \mathbf{A}^\dagger\mathbf{A})\mathbf{c}, \quad \mathbf{z} \geq \mathbf{0}$.

Conversely, if $(\mathbf{I} - \mathbf{A}^\dagger\mathbf{A})\mathbf{z} = -(\mathbf{I} - \mathbf{A}^\dagger\mathbf{A})\mathbf{c}, \quad \mathbf{z} \geq \mathbf{0}$. Then $\mathbf{z} = -(\mathbf{I} - \mathbf{A}^\dagger\mathbf{A})\mathbf{c} + (\mathbf{A}^\dagger\mathbf{A})\mathbf{y}$ and $\mathbf{A}^\dagger\mathbf{A}\mathbf{y} \geq \mathbf{0}$. Furthermore $(\mathbf{y}, \mathbf{A}^\dagger\mathbf{b}) = (\mathbf{z}, \mathbf{A}^\dagger\mathbf{b})$ if $\mathbf{z}, \mathbf{y}$ are so related. Thus there is a many to one correspondence between feasible solutions of (D2) and feasible solutions of

(D3) Minimize $(\mathbf{z}, \mathbf{A}^\dagger\mathbf{b})$, where $(\mathbf{I} - \mathbf{A}^\dagger\mathbf{A})\mathbf{z} = -(\mathbf{I} - \mathbf{A}^\dagger\mathbf{A})\mathbf{c}, \quad \mathbf{z} \geq \mathbf{0}$.

The next theorem provides the basis for the algebraic condition we are looking for.

Theorem 11.2.1 *Suppose that* $\mathbf{x}, \mathbf{z}$ *are feasible vectors for* (P3) *and* (D3). *Then* $(\mathbf{x}, \mathbf{z}) = 0$ *if and only if* $\mathbf{x}$ *is optimal for* (P3) *and* $\mathbf{z}$ *is optimal for* (D3).

It will simplify some of our later calculations, and reduce the size of the matrices involved, if we use partial isometries to replace the projectors appearing in the constraints of (P3) and (D3). Let $q = \dim N(\mathbf{A})$. $\mathbf{E}^*$ will be an isometry from $\mathbb{R}^{n-q}$ onto $R(\mathbf{A}^*)$ while $\mathbf{F}^*$ will be an isometry from $\mathbb{R}^q$ onto $N(\mathbf{A})$. Then $\mathbf{F} = [\mathbf{a}_1, \ldots, \mathbf{a}_q]^*$ and $\mathbf{E} = [\mathbf{a}_{q+1}, \ldots, \mathbf{a}_n]^*$, where $\{\mathbf{a}_1, \ldots, \mathbf{a}_q\}$ is an orthonormal basis for $N(\mathbf{A})$ and $\{\mathbf{a}_{q+1}, \ldots, \mathbf{a}_n\}$ is an orthonormal basis for $R(\mathbf{A}^*)$. The only restraint that we put on the choice of basis is that $\mathbf{a}_1 = (\mathbf{I} - \mathbf{A}^\dagger\mathbf{A})\mathbf{c}/\|(\mathbf{I} - \mathbf{A}^\dagger\mathbf{A})\mathbf{c}\|$, and $\mathbf{a}_{q+1} = \mathbf{A}^\dagger\mathbf{b}/\|\mathbf{A}^\dagger\mathbf{b}\|$. We rule out the possibility that $(\mathbf{I} - \mathbf{A}^\dagger\mathbf{A})\mathbf{c} = \mathbf{0}$, for then all feasible solutions to the original problem are optimal by Proposition 1.3. If $\mathbf{A}^\dagger\mathbf{b} = \mathbf{0}$, then for consistent constraints we have $\mathbf{b} = \mathbf{0}$ and $(\mathbf{x}, \mathbf{c})$ is either zero or unbounded on the set of feasible vectors. So we assume $\mathbf{A}^\dagger\mathbf{b} \neq \mathbf{0}$. Since $\mathbf{E}^*, \mathbf{F}^*$ are isometries, we have that $\mathbf{E}\mathbf{E}^*\mathbf{E} = \mathbf{E}$ and $\mathbf{F}\mathbf{F}^*\mathbf{F} = \mathbf{F}$. Note that $\mathbf{A}^\dagger\mathbf{A} = \mathbf{E}^*\mathbf{E}$. Thus $\mathbf{E}\mathbf{A}^\dagger\mathbf{A} = \mathbf{E}$. Similarly $(\mathbf{I} - \mathbf{A}^\dagger\mathbf{A}) = \mathbf{F}^*\mathbf{F}$. Let $\mathbf{c}_o = (\mathbf{I} - \mathbf{A}^\dagger\mathbf{A})\mathbf{c}$. Problems (P3) and (D3) may now be written as

(P4) Minimize $(\mathbf{x}, -\mathbf{c}_o)$, where $\mathbf{E}\mathbf{x} = \mathbf{E}\mathbf{A}^\dagger\mathbf{b}$, $\mathbf{x} \geq \mathbf{0}$

and

(D4) Minimize $(\mathbf{y}, \mathbf{A}^\dagger\mathbf{b})$, where $\mathbf{F}\mathbf{y} = \mathbf{F}(-\mathbf{c}_o)$, $\mathbf{y} \geq \mathbf{0}$.

Notice that $\mathbf{x} = \mathbf{A}^\dagger\mathbf{b}$ is a solution of $\mathbf{E}\mathbf{x} = \mathbf{E}\mathbf{A}^\dagger\mathbf{b}$ and $N(\mathbf{E}) = R(\mathbf{A}^*)^\perp = N(\mathbf{A})$. Thus if $\mathbf{x}$ is a feasible vector for (P4), then $\mathbf{x}$ must be of the form

$$\mathbf{x} = \mathbf{A}^\dagger\mathbf{b} + \sum_{i=1}^{q} \alpha_i \mathbf{a}_i. \tag{1}$$

Similarily, if $\mathbf{y}$ is a feasible vector for (D4), then

$$\mathbf{y} = -\mathbf{c}_o + \sum_{i=q+1}^{n} \alpha_i \mathbf{a}_i. \tag{2}$$

Now if $\mathbf{x}, \mathbf{y}$ are feasible for (P4) and (D4), then they are feasible for (P3) and (D3). They will both be optimal if and only if $(\mathbf{x}, \mathbf{y}) = 0$. Substituting (1) and (2) into $(\mathbf{x}, \mathbf{y}) = 0$ gives

$$\begin{aligned} 0 &= \left(\mathbf{A}^\dagger\mathbf{b} + \sum_{i=1}^{q} \alpha_i \mathbf{a}_i, -\mathbf{c}_o + \sum_{i=q+1}^{n} \alpha_i \mathbf{a}_i\right) \\ &= (\mathbf{A}^\dagger\mathbf{b}, -\mathbf{c}_o) - \sum_{i=1}^{q} \alpha_i (\mathbf{a}_i, \mathbf{c}_o) + \sum_{i=q+1}^{n} \alpha_i (\mathbf{A}^\dagger\mathbf{b}, \mathbf{a}_i). \end{aligned}$$

But $\mathbf{c}_o = \|\mathbf{c}_o\|\mathbf{a}_1$ and $\mathbf{a}_{q+1} = \mathbf{A}^\dagger\mathbf{b}/\|\mathbf{A}^\dagger\mathbf{b}\|$. Thus $0 = -\alpha_1(\mathbf{a}_1, \mathbf{c}_o) + \alpha_{q+1}(\mathbf{A}^\dagger\mathbf{b}, \mathbf{A}^\dagger\mathbf{b})/\|\mathbf{A}^\dagger\mathbf{b}\|$, or

$$\alpha_1 = \alpha_{q+1}\|\mathbf{A}^\dagger\mathbf{b}\|/\|\mathbf{c}_o\|. \tag{3}$$

But

$$\begin{aligned} \alpha_1 &= (\mathbf{x}, \mathbf{a}_1) = \mathbf{c}_o^*\mathbf{x}/\|\mathbf{c}_o\|, \text{ and} \\ \alpha_{q+1} &= (\mathbf{y}, \mathbf{a}_{q+1}) = (\mathbf{A}^\dagger\mathbf{b})^*\mathbf{y}/\|\mathbf{A}^\dagger\mathbf{b}\|. \end{aligned} \tag{4}$$

Equations (3), (4) and the equality constraints of (P4), (D4) may be expressed in one set of constraints as

$$\begin{bmatrix} \mathbf{E} & \mathbf{0} & 0 \\ \mathbf{0} & \mathbf{F} & 0 \\ \mathbf{0} & \mathbf{a}_{q+1}^* & -\|\mathbf{c}_o\|/\|\mathbf{A}^\dagger\mathbf{b}\| \\ \mathbf{a}_1^* & \mathbf{0} & -1 \end{bmatrix} \begin{bmatrix} \mathbf{x} \\ \mathbf{y} \\ \alpha_1 \end{bmatrix} = \begin{bmatrix} \mathbf{E}\mathbf{A}^\dagger\mathbf{b} \\ -\mathbf{E}\mathbf{c}_o \\ 0 \\ 0 \end{bmatrix} \tag{5}$$

We have then that:

Theorem 11.2.2 *Any solution of* (5) *which is non-negative in its first 2n components provides optimal solutions of* (P4) and (D4).

It is easy to modify (5) so that the desired solutions have all components non-negative. Let $\alpha_1 = \beta_1 - \beta_2$ where $\beta_1 \geq 0$ and $\beta_2 \geq 0$. Then (5) becomes

$$\begin{bmatrix} \mathbf{E} & \mathbf{0} & 0 & 0 \\ \mathbf{0} & \mathbf{F} & 0 & 0 \\ \mathbf{0} & \mathbf{a}_{q+1}^* & -\|\mathbf{c}_o\|/\|\mathbf{A}^\dagger\mathbf{b}\| & \|\mathbf{c}_o\|/\|\mathbf{A}^\dagger\mathbf{b}\| \\ \mathbf{a}_1^* & \mathbf{0} & 1 & -1 \end{bmatrix} \begin{bmatrix} \mathbf{x} \\ \mathbf{y} \\ \beta_1 \\ \beta_2 \end{bmatrix} = \begin{bmatrix} \mathbf{E}\mathbf{A}^\dagger\mathbf{b} \\ -\mathbf{E}\mathbf{c}_o \\ 0 \\ 0 \end{bmatrix} \tag{6}$$

Theorem 11.2.3 *Any solution of* (6) *which is non-negative in all of its components provides optimal solutions of* (P4) *and* (D4).

Let $\mathbf{B}$ be the coefficient matrix of (6). Then Theorem 3 says that solutions of

(P5) $\mathbf{Bz} = \mathbf{d}, \quad \mathbf{z} \geq \mathbf{0}.$

provide optimal solutions of (P4) and (D4). Problem (P5) can be rewritten so as to get

(P6) $\mathbf{Px} = \mathbf{x}, \quad \mathbf{x} \geq \mathbf{0}, \quad \mathbf{P}$ an orthogonal projector.

To see what $\mathbf{P}$ has to be, observe that we are asking for $R(\mathbf{P}) \cap \mathbb{R}_+^n$ and $\{\mathbf{z} : \mathbf{Bz} = \mathbf{d}\} \cap \mathbb{R}_+^n$ to be the same. A reasonable way to try and guarantee this is to require that $\{\mathbf{z} : \mathbf{Bz} = \mathbf{d}\} \subseteq R(\mathbf{P})$. If $\mathbf{Bz} = \mathbf{d}$, then $\mathbf{z} = \mathbf{B}^\dagger\mathbf{d} \oplus (\mathbf{I} - \mathbf{B}^\dagger\mathbf{B})\mathbf{z}$. $\mathbf{P}$ would thus be the sum of $(\mathbf{I} - \mathbf{B}^\dagger\mathbf{B})$ and a projector onto the subspace spanned by $\mathbf{B}^\dagger\mathbf{d}$. Let $\mathbf{P} = (\mathbf{I} - \mathbf{B}^\dagger\mathbf{B}) + (\mathbf{B}^\dagger\mathbf{d})(\mathbf{B}^\dagger\mathbf{d})^* / \|\mathbf{B}^\dagger\mathbf{d}\|^2$. Then $\mathbf{P}$ is a hermitian projector and $R(\mathbf{P}) \supseteq \{\mathbf{z} : \mathbf{Bz} = \mathbf{d}\}$.

Let us examine the relationship between solutions of (P5) and (P6). Suppose that $\mathbf{z}$ is a solution of (P5). Then $\mathbf{z} \geq \mathbf{0}$ and $\mathbf{z} = \mathbf{B}^\dagger\mathbf{d} + (\mathbf{I} - \mathbf{B}^\dagger\mathbf{B})\mathbf{z} \in R(\mathbf{P})$. Thus $\mathbf{z}$ is a solution of (P6). Suppose that $\mathbf{x}$ is a solution of (P6). Then $\mathbf{x} \geq \mathbf{0}$ and $\mathbf{Px} = \mathbf{x}$. This implies that

$$\mathbf{B}^\dagger\mathbf{Bx} = \frac{(\mathbf{B}^\dagger\mathbf{d})(\mathbf{B}^\dagger\mathbf{d})^*}{\|\mathbf{B}^\dagger\mathbf{d}\|^2}\mathbf{x} = \frac{(\mathbf{B}^\dagger\mathbf{d}, \mathbf{x})}{\|\mathbf{B}^\dagger\mathbf{d}\|^2}\mathbf{B}^\dagger\mathbf{d}.$$

Multiplying both sides by $\mathbf{B}$ and using the fact that $\mathbf{Bx} = \mathbf{d}$ is consistent, we get that $\mathbf{Bx} = (\mathbf{B}^\dagger\mathbf{d}, \mathbf{x})\mathbf{d}/\|\mathbf{B}^\dagger\mathbf{d}\|^2$. Let

$$\mathbf{z} = \frac{\|\mathbf{B}^\dagger\mathbf{d}\|^2}{(\mathbf{B}^\dagger\mathbf{d}, \mathbf{x})}\mathbf{x}, \text{ if } (\mathbf{B}^\dagger\mathbf{d}, \mathbf{x}) \neq 0. \tag{7}$$

Then $\mathbf{Bz} = \mathbf{d}$. If $(\mathbf{B}^\dagger\mathbf{d}, \mathbf{x}) > 0$, then $\mathbf{z}$ will be a solution of (P5) since we assumed $\mathbf{x} \geq \mathbf{0}$. If $(\mathbf{B}^\dagger\mathbf{d}, \mathbf{x}) = 0$, then $\mathbf{x} \in R(\mathbf{I} - \mathbf{B}^\dagger\mathbf{B}) = N(\mathbf{B})$ and $\mathbf{x} \geq \mathbf{0}$. We would then have $(\mathbf{B}^\dagger\mathbf{d} + N(\mathbf{B})) \cap \mathbb{R}^n_+$ is unbounded. This in turn would imply that in our original problems the constraints defined an unbounded set. Suppose that $(\mathbf{B}^\dagger\mathbf{d}, \mathbf{x}) < 0$. Then $\mathbf{z} \leq \mathbf{0}$ so that $\mathbf{z} = \mathbf{B}^\dagger\mathbf{d} \oplus (\mathbf{I} - \mathbf{B}^\dagger\mathbf{B})\mathbf{z} \leq \mathbf{0}$ and there is an $\mathbf{h} \in N(\mathbf{A})$ such that $\mathbf{B}^\dagger\mathbf{d} \oplus \mathbf{h} \geq \mathbf{0}$ since (P5) is assumed consistent. But then $\mathbf{h} - (\mathbf{I} - \mathbf{B}^\dagger\mathbf{B})\mathbf{z} = (\mathbf{B}^\dagger\mathbf{d} + \mathbf{h}) - (\mathbf{B}^\dagger\mathbf{d} + (\mathbf{I} - \mathbf{B}^\dagger\mathbf{B})\mathbf{z}) \geq \mathbf{0}$. Again we have $N(\mathbf{B}) \cap \mathbb{R}^n_+$ is non-empty and $(\mathbf{B}^\dagger\mathbf{d} + N(\mathbf{B})) \cap \mathbb{R}^n_+$ would be unbounded. Summarizing these observations, we have the next theorem.

Theorem 11.2.4 *If in problems* (P1) *and* (D1) *the constraints define a bounded subset of* $\mathbb{R}^n$, *then solutions of* (P6) *provide solutions of* (P5) *by means of equation* (7). *Thus solutions of* (P6) *will provide optimal solutions of* (P1) *and* (D1).

It is worth noting that $\mathbf{B}^\dagger$, and hence $\mathbf{P}$, are fairly easy to compute. One way is to observe that $\mathbf{B}$ is of full row rank and hence $\mathbf{B}^\dagger = \mathbf{B}^*(\mathbf{BB}^*)^{-1}$. Because of the particular entries involved, the matrix $(\mathbf{BB}^*)$ is easy to invert. An even easier way to find $\mathbf{B}^\dagger$ is to observe that by slightly modifying the last two rows of $\mathbf{B}$ we can get a matrix $\mathbf{C}$ such that $\mathbf{Bz} = \mathbf{d}$ and $\mathbf{Cz} = \mathbf{d}$ have the same solutions but $\mathbf{C}$ is a partial isometry. Then $\mathbf{C}$ is used in place of $\mathbf{B}$. From Chapter 2 we know that $\mathbf{C}^\dagger = \mathbf{C}^*$.

3. Exercises

1. For a set $\mathscr{S} \subseteq \mathbb{R}^n$, define $\mathscr{S}^\circ = \{\mathbf{z} \in \mathbb{R}^n : (\mathbf{z}, \mathbf{s}) \geq 0 \text{ for all } \mathbf{s} \in \mathscr{S}\}$. Prove that for $\mathscr{S}, \mathscr{T}$ any sets in $\mathbb{R}^n$:
   (a) $\mathscr{S} \subseteq \mathscr{S}^{\circ\circ}$
   (b) $\mathscr{S} \subseteq \mathscr{T}$ implies $\mathscr{T}^\circ \subset \mathscr{S}^\circ$
   (c) $\mathscr{S} \subseteq (\mathscr{S}^\circ)^\circ = \mathscr{S}^{\circ\circ}$
   (d) $\mathscr{S}^\circ = \mathscr{S}^{\circ\circ\circ}$
   (e) $\mathscr{S}^\circ = (\text{cl}\mathscr{S})^\circ$ $\text{cl}\mathscr{S}$ is the closure of $\mathscr{S}$ in the Euclidean norm on $\mathbb{R}^n$.
   (f) $\mathscr{S}^\circ + \mathscr{T}^\circ \subseteq (\mathscr{S} + \mathscr{T})^\circ$
2. Verify that if $M \subseteq \mathbb{R}^n$ is a subspace, then $M^\circ = M^\perp$.
3. Verify that $(\mathbb{R}^n_+)^\circ = \mathbb{R}^n_+$.
4. Prove that $\mathbb{R}^n_+ + N(\mathbf{A})$ is closed. (Hint: It is possible to define an $n \times p$ matrix $\mathbf{C}$ such that $\mathbf{C}\mathbb{R}^p_+ = \mathbb{R}^n_+ + N(\mathbf{A})$ where $p = n + \dim N(\mathbf{A})$. In other words, $\mathbb{R}^n_+ + N(\mathbf{A})$ is a *polyhedral cone*. Using $\mathbf{C}$, $\mathbf{C}^\dagger$, and the fact that a closed convex set has an element of minimal norm, one can show that $\mathbb{R}^n_+ + N(\mathbf{A})$ is closed. Details may be found in [11, p. 380].)
5. Prove Theorem 11.2.1.

6. Verify that if $\mathbf{B} \in \mathbb{R}^{m \times n}, \mathbf{d} \in \mathbb{R}^m$, then $\mathbf{P} = (\mathbf{I} - \mathbf{B}^\dagger \mathbf{B}) + \frac{1}{\|\mathbf{B}^\dagger \mathbf{d}\|^2} (\mathbf{B}^\dagger \mathbf{d})(\mathbf{B}^\dagger \mathbf{d})^*$ is an orthogonal projector.
7. Let $\mathbf{B}$ be the coefficient matrix in equation (6), Section 2. Verify that $\mathbf{B}$ is of full row rank and calculate $\mathbf{B}^\dagger$ by using $\mathbf{B}^\dagger = \mathbf{B}^*(\mathbf{B}\mathbf{B}^*)^{-1}$.
8. Let $\mathbf{B}$ be the coefficient matrix in equation (6), Section 2. Modify the last two rows of $\mathbf{B}$ to get a new matrix $\mathbf{C}$ such that $\mathbf{Cz} = \mathbf{d}$ and $\mathbf{Bz} = \mathbf{d}$ have the same solutions and $\mathbf{C}$ is a partial isometry.
9. Prove Proposition 11.1.1.

4. References and further reading

There are many books on linear programming. We list three [48], [84], [85] in the references. They are the ones we have found most useful. [48] is the most technical and [85] the least. The exposition in [84] seemed well written.

The generalized inverse is also discussed in [70] by Pyle and Cline. They are concerned with gradient projection methods. That is, they approach optimal vectors by moving through the set of feasible vectors. This is in contrast to the simplex method which goes from vertex to vertex around the edge of the convex set of feasible vectors. (If there is an optimal vector, there must be one at a vertex.)

A proof of Farkas's theorem can be found in Ben-Israel [11]. The complex case and a more thorough discussion of polars and cones is also given in [11].

12
Computational concerns

1. Introduction

This chapter will consider the problem of computing generalized inverses. We will not go into a detailed analysis of the different methods. Books exist on just computation of least squares problems. Rather we shall discuss some of the common methods and when they would be most useful. The bibliography at the end of this book will have references that go into a more detailed analysis.

Our procedure will be as follows. We shall first discuss some of the difficulty with calculating generalized inverses. We will talk mainly about the Moore–Penrose inverse but the difficulties apply to all. Then we shall consider the problem of computing $\mathbf{A}^{\dagger}$. A particular algorithm involving the singular value decomposition will be developed in some detail. Then a section on $\mathbf{A}^{-}$, and finally a section on $\mathbf{A}^{\mathrm{D}}$.

The first thing to note is that we are talking about calculating, say $\mathbf{A}^{\dagger}$, and not necessarily about solving $\mathbf{Ax} = \mathbf{b}$ in the least squares sense. One has the same distinction in working with invertible matrices $\mathbf{A}$. If one wishes to solve $\mathbf{Ax} = \mathbf{b}$, the quickest way is not to calculate $\mathbf{A}^{-1}$, and then $\mathbf{A}^{-1}\mathbf{b}$. It takes of order n^3 operations to calculate $\mathbf{A}^{-1}$ and another n^2 to form $\mathbf{A}^{-1}\mathbf{b}$. The direct solution of $\mathbf{Ax} = \mathbf{b}$ by Gaussian elimination can be done in $n^3/3$ operations. (Here operations are multiplications or divisions.) Similarly, it takes more time to calculate $\mathbf{A}^{\dagger}$ and then $\mathbf{A}^{\dagger}\mathbf{b}$, than to directly calculate $\mathbf{A}^{\dagger}\mathbf{b}$.

The algorithms we shall discuss fall into three broad groups. The first is full rank factorizations and singular value decompositions. The second is iterative. The third we shall loosely describe as 'other'. It consists of various special ways of calculating generalized inverses. These methods are usually of most use for small matrices, and will tend to be mainly for the Drazin inverse.

The reader interested in actual programs, numerical experiments, and error analysis is referred to the references of the last section.

It will be assumed that the calculations are not being done entirely in exact arithmetic. Thus there is some number $\varepsilon_o > 0$, which depends on the equipment used, such that numbers less than ε_o are considered zero. Several algorithms suitable for hand calculation or exact arithmetic on small matrices have been given earlier. For small matrices, those methods are sometimes preferable to the more complicated ones we shall now discuss. This chapter is primarily interested in computer calculation for 'large' matrices.

Throughout this chapter $\|\cdot\|$ denotes a matrix norm as described in Chapter 11. For invertible **A**, we define the condition number of **A** with respect to the norm $\|\cdot\|$ as $\kappa(\mathbf{A}) = \|\mathbf{A}\|\,\|\mathbf{A}^{-1}\|$. We frequently write κ instead of $\kappa(\mathbf{A})$. If **A** is singular, then $\kappa(\mathbf{A}) = \|\mathbf{A}\|\,\|\mathbf{A}^\dagger\|$.

2. Calculation of $\mathbf{A}^\dagger$

This section will be concerned with computing $\mathbf{A}^\dagger$. The first difficulty is that this is not, as stated, a well-posed problem. If **A** is a matrix which is not of full column or row rank then it is possible to change the rank of **A** by an arbitrarily small perturbation. Using the notation of Chapter 12 we have:

Unpleasant fact Suppose $\mathbf{A} \in \mathbb{C}^{m \times n}$ is of neither full column nor full row rank. Then for any real number K, and any $\varepsilon > 0$, there exists a matrix **E**, $\|\mathbf{E}\| < \varepsilon$, such that $\|(\mathbf{A}+\mathbf{E})^\dagger - \mathbf{A}^\dagger\| \geq \mathrm{K}$.

Proof Let $\mathbf{u} \in N(\mathbf{A})$ and $\mathbf{v} \in R(\mathbf{A})^\perp$ be vectors of norm one. Let $\mathbf{E} = \varepsilon_o \mathbf{v}\mathbf{u}^*$ where $\varepsilon_o = \min\{1/\mathrm{K}, \varepsilon\}$. Then $\|\mathbf{E}\| = \varepsilon_o$ and $\operatorname{rank}(\mathbf{A}+\mathbf{E}) = \operatorname{rank}(\mathbf{A}) + 1$. Hence $\|(\mathbf{A}+\mathbf{E})^\dagger - \mathbf{A}^\dagger\| \geq \mathrm{K}$ by Theorem 10.4.2.

If **A** is determined experimentally, or is entered in decimal notation on a computer, or there is round off, then it is not obvious whether it makes sense to talk about computing $\mathbf{A}^\dagger$ unless **A** is of full rank. One method of posing the problem is by using the singular value decomposition (Theorem 0.2.2). Let **X** be an $m \times n$ matrix. By Theorem 0.2.2 there exist unitary matrices **U**, **V** so that

$$\mathbf{X} = \mathbf{U}\begin{bmatrix} \Sigma(\mathbf{X}) & \mathbf{0} \\ \mathbf{0} & \mathbf{0} \end{bmatrix}\mathbf{V} \tag{1}$$

where $\Sigma(\mathbf{X}) = \operatorname{Diag}[\sigma_1, \ldots, \sigma_r]$. Then

$$\mathbf{X}^\dagger = \mathbf{V}^*\begin{bmatrix} \Sigma(\mathbf{X})^\dagger & \mathbf{0} \\ \mathbf{0} & \mathbf{0} \end{bmatrix}\mathbf{U}^*. \tag{2}$$

Now let

$$\mathbf{X}_\varepsilon = \mathbf{U}\begin{bmatrix} \Sigma(\mathbf{X})_\varepsilon & \mathbf{0} \\ \mathbf{0} & \mathbf{0} \end{bmatrix}\mathrm{V}$$

where $\Sigma(\mathbf{X})_\varepsilon = \operatorname{Diag}[\tilde{\sigma}_1, \ldots, \tilde{\sigma}_r]$; $\tilde{\sigma}_i = \sigma_i$ if $\sigma_i \geq \varepsilon$, $\tilde{\sigma}_i = 0$ if $\sigma_i < \varepsilon$. Here ε

depends on the computing equipment available and the desired accuracy. Now take $\mathbf{A} \in \mathbb{C}^{m \times n}$. The entries of $\Sigma(\mathbf{A})$ are the eigenvalues of $(\mathbf{A}^*\mathbf{A})^{1/2}$. Since eigenvalues vary continuously with the matrix's entries, one has that $\Sigma(\mathbf{A}+\mathbf{E})_\varepsilon$ and $\Sigma(\mathbf{A})_\varepsilon$ have the same rank if $\|\mathbf{E}\|$ is small enough. Thus

$$\lim_{\|\mathbf{E}\| \to 0} \|(\mathbf{A}_\varepsilon)^\dagger - ((\mathbf{A}+\mathbf{E})_\varepsilon)^\dagger\| = \mathbf{0} \tag{3}$$

by Theorem 10.4.1. Thus the problem of computing $(\mathbf{A}_\varepsilon)^\dagger$ is well-posed if (2) is used as the definition of the Moore–Penrose inverse.

It should be noted that the rate of convergence of (3) depends on $\mathbf{A}$ and $\mathbf{E}$ and not just the condition number of $\mathbf{A}$.

Example 12.2.1 Let $\mathbf{A} = \begin{bmatrix} 1 & 0 \\ 0 & 0 \end{bmatrix}$, $\mathbf{E}_\alpha = \begin{bmatrix} 0 & 0 \\ 0 & \alpha \end{bmatrix}$. Then $\|\mathbf{A}_\varepsilon^\dagger - (\mathbf{A}+\mathbf{E}_\alpha)_\varepsilon^\dagger\| = 1/\alpha$ if $\alpha > \varepsilon$, it equals 0 if $\alpha < \varepsilon$. For this $\mathbf{A}$, we have a large error if $\varepsilon < \|\mathbf{E}_\alpha\| < 2\varepsilon$ and no error for $\|\mathbf{E}_\alpha\| < \varepsilon$. Note $\kappa(\mathbf{A}) = 1$ for $\|\ \|_2$.

There is another way to view the calculation of $\mathbf{A}^\dagger$ as a well-posed problem. Fix $\mathbf{A} \in \mathbb{C}^{m \times n}$ and consider the equations

$$\mathbf{AXA} - \mathbf{A} = \mathbf{E}_1, \tag{4}$$

$$\mathbf{XAX} - \mathbf{X} = \mathbf{E}_2, \tag{5}$$

$$\mathbf{AX} - \mathbf{X}^*\mathbf{A}^* = \mathbf{E}_3, \tag{6}$$

and

$$\mathbf{XA} - \mathbf{A}^*\mathbf{X}^* = \mathbf{E}_4. \tag{7}$$

In terms of our previous discussion, $\mathbf{X}$ may be thought of as the computed estimate and the $\mathbf{E}_i$ as error terms. We shall now solve (4)–(7) for $\mathbf{X}$ in terms of $\mathbf{A}$, $\mathbf{A}^\dagger$ and the $\mathbf{E}_i$. Equation (7) gives $\mathbf{AXA} - \mathbf{AA}^*\mathbf{X}^* = \mathbf{AE}_4$, or by (4), $\mathbf{E}_1 + \mathbf{A} - \mathbf{AA}^*\mathbf{X}^* = \mathbf{AE}_4$. Thus $\mathbf{XAA}^* = \mathbf{A}^* + \mathbf{E}_1^* - \mathbf{E}_4^*\mathbf{A}^*$. Multiplying on the right by $\mathbf{A}^{*\dagger}\mathbf{A}^\dagger$ gives

$$\mathbf{XAA}^\dagger = [\mathbf{A}^* + \mathbf{E}_1^* - \mathbf{E}_4^*\mathbf{A}^*]\mathbf{A}^{*\dagger}\mathbf{A}^\dagger. \tag{8}$$

Similarly from (6) we get $\mathbf{AXA} - \mathbf{X}^*\mathbf{A}^*\mathbf{A} = \mathbf{E}_3\mathbf{A}$ and hence

$$\mathbf{A}^\dagger\mathbf{AX} = \mathbf{A}^\dagger\mathbf{A}^{*\dagger}[\mathbf{A}^* + \mathbf{E}_1^* - \mathbf{A}^*\mathbf{E}_3^*]. \tag{9}$$

Substituting (8) and (9) into (5) gives

$$\begin{aligned}
\mathbf{E}_2 + \mathbf{X} = \mathbf{XAX} &= \mathbf{XAA}^\dagger\mathbf{AA}^\dagger\mathbf{AX} \\
&= [\mathbf{A}^* + \mathbf{E}_1^* - \mathbf{E}_4^*\mathbf{A}^*]\mathbf{A}^{*\dagger}\mathbf{A}^\dagger\mathbf{AA}^\dagger\mathbf{A}^{*\dagger}[\mathbf{A}^* + \mathbf{E}_1^* - \mathbf{A}^*\mathbf{E}_3^*] \\
&= [\mathbf{A}^\dagger + \mathbf{E}_1^*\mathbf{A}^{*\dagger}\mathbf{A}^\dagger - \mathbf{E}_4^*\mathbf{A}^\dagger]\mathbf{AA}^\dagger\mathbf{A}^{*\dagger}[\mathbf{A}^* + \mathbf{E}_1^* - \mathbf{A}^*\mathbf{E}_3^*] \\
&= [\mathbf{I} + \mathbf{E}_1^*\mathbf{A}^{*\dagger} - \mathbf{E}_4^*]\mathbf{A}^\dagger\mathbf{A}^{*\dagger}[\mathbf{A}^* + \mathbf{E}_1^* - \mathbf{A}^*\mathbf{E}_3^*] \\
&= [\mathbf{I} + \mathbf{E}_1^*\mathbf{A}^{*\dagger} - \mathbf{E}_4^*][\mathbf{A}^\dagger + \mathbf{A}^\dagger\mathbf{A}^{*\dagger}\mathbf{E}_1^* - \mathbf{A}^\dagger\mathbf{E}_3^*] \\
&= \mathbf{A}^\dagger + [\mathbf{E}_1^*\mathbf{A}^{*\dagger} - \mathbf{E}_4^*]\mathbf{A}^\dagger + \mathbf{A}^\dagger[\mathbf{A}^{*\dagger}\mathbf{E}_1^* - \mathbf{E}_3^*] \\
&\quad + [\mathbf{E}_1^*\mathbf{A}^{*\dagger} - \mathbf{E}_4^*]\mathbf{A}^\dagger[\mathbf{A}^{*\dagger}\mathbf{E}_1^* - \mathbf{E}_3^*].
\end{aligned} \tag{10}$$

As an immediate consequence of (10) we have the following results.

Theorem 12.2.1 *If* $\{\mathbf{X}_r\}$ *is a sequence of* $n \times m$ *matrices such that the sequences* $\{\mathbf{X}_r\mathbf{A}\mathbf{X}_r - \mathbf{X}_r\}$, $\{\mathbf{A}\mathbf{X}_r - \mathbf{X}_r^*\mathbf{A}^*\}$, $\{\mathbf{X}_r\mathbf{A} - \mathbf{A}^*\mathbf{X}_r^*\}$ *and* $\{\mathbf{A}\mathbf{X}_r\mathbf{A} - \mathbf{A}\}$ *all converge to zero, then* $\{\mathbf{X}_r\}$ *converges to* $\mathbf{A}^\dagger$.

Theorem 12.2.2 *Let* $\mathbf{A} \in \mathbb{C}^{m\times n}$, $\mathbf{X} \in \mathbb{C}^{n\times m}$ *and* $\|\cdot\|$ *denote a matrix norm such that* $\|\mathbf{B}^*\| = \|\mathbf{B}\|$. *Define* $\mathbf{E}_i$, $i = 1,2,3,4$ *by* (4)–(7). *Then*

$$\begin{aligned}\|\mathbf{X} - \mathbf{A}^\dagger\| \le \|\mathbf{E}_2\| &+ \|\mathbf{A}^\dagger\|\{\|\mathbf{E}_1\| + \|\mathbf{E}_3\| + \|\mathbf{E}_3\|\,\|\mathbf{E}_4\|\} \\ &+ \|\mathbf{A}^\dagger\|^2\{2\|\mathbf{E}_1\| + \|\mathbf{E}_1\|\,\|\mathbf{E}_3\| + \|\mathbf{E}_4\|\,\|\mathbf{E}_1\|\} \\ &+ \|\mathbf{A}^\dagger\|^3\|\mathbf{E}_1\|^2\end{aligned} \tag{1}$$

Note that estimate (11) in Theorem 2 seems to suggest that minimizing $\|\mathbf{E}_1\|$ is important if $\mathbf{E}$ has large condition number.

Theorems 1 and 2 show that if $\|\mathbf{A}\|\,\|\mathbf{A}^\dagger\|\,\varepsilon_0$ and $\|\mathbf{A}^\dagger\|^3\varepsilon_0$ are small, then calculating $\mathbf{A}^\dagger$ is well-posed in the sense that if $\mathbf{X}$ comes close to satisfying the defining conditions of the Moore–Penrose inverse, then $\mathbf{X}$ must be close to $\mathbf{A}^\dagger$.

Algorithms exist for calculating the singular value decompositions that are stable. That is, error accumulates no faster than the condition number would seem to warrant. Perhaps the best known one is due to Golub and Reinsch [39]. Their method consists of using Givens rotations and Householder transformations to reduce $\mathbf{A}$ to the form $\begin{bmatrix}\mathbf{J}\\ \mathbf{0}\end{bmatrix}$ where $\mathbf{J}$ is diagonal.

This method is stable primarily because the Householder matrices and Givens rotations are unitary. Thus their application does not increase the norm of the error matrix. This algorithm is discussed in more detail in the next section.

One might think that the full-rank case would always be easier to work with. Assume $\mathbf{A} \in \mathbb{C}^{m\times n}$ is full column rank. Then from Theorem 1.3.2

$$\mathbf{A}^\dagger = (\mathbf{A}^*\mathbf{A})^{-1}\mathbf{A}^*. \tag{12}$$

However, taking $\mathbf{A}^*\mathbf{A}$ is to be avoided unless one knows something about the singular values of $\mathbf{A}$. Let $\mathbf{A} = \mathbf{U}\Sigma\mathbf{V}$ where $\Sigma = \mathrm{Diag}\{\sigma_1, \dots, \sigma_n\}$. Then $\mathbf{A}^*\mathbf{A} = \mathbf{V}^*\Sigma^2\mathbf{V}$ where $\Sigma^2 = \mathrm{Diag}\{\sigma_1^2, \dots, \sigma_n^2\}$. If σ_i is small, then σ_i^2 might be negligible. For example, if round off is 10^{-11}, a singular value of 10^{-6} would be lost.

Example 12.2.2 Let $\mathbf{A} = \begin{bmatrix}1 & 1\\ \beta & 0\\ 0 & \beta\end{bmatrix}$. Then $\mathbf{A}^*\mathbf{A} = \begin{bmatrix}1+\beta^2 & 1\\ 1 & 1+\beta^2\end{bmatrix}$

If β is small, we could have β^2 negligible and $(\mathbf{A}^*\mathbf{A})_\varepsilon = \begin{bmatrix}1 & 1\\ 1 & 1\end{bmatrix}$.

In any event, for small β there is a loss of information.

These same comments apply to computing $\mathbf{A}^\dagger$ by writing $\mathbf{A} = \mathbf{BC}$ where $\mathbf{BC}$ is a full rank factorization. Then $\mathbf{A}^\dagger = \mathbf{C}^\dagger\mathbf{B}^\dagger$ by Theorem 1.3.2. It is possible with a little effort to still utilize a full rank factorization and avoid the possible ill-conditioning caused by taking $\mathbf{C}^\dagger = \mathbf{C}^*(\mathbf{CC}^*)^{-1}$ or $\mathbf{B}^\dagger = (\mathbf{B}^*\mathbf{B})^{-1}\mathbf{B}^*$ if only one of $\mathbf{CC}^*$ or $\mathbf{B}^*\mathbf{B}$ is ill-conditioned. If, however, *both* $\mathbf{CC}^*$ and $\mathbf{B}^*\mathbf{B}$ are ill-conditioned then the method will not help much. The interested reader is referred to [35].

If (12) is to be used, then double precision arithmetic is recommended. In fact, some numerical analysts recommend always using double precision. This will be especially true when working with the Drazin inverse as will be explained later.

Another way of calculating the Moore–Penrose inverse is by an iterative scheme. Iterative methods exist for finding the inverse of a non-singular matrix. Their principal use is in taking a computed inverse and yielding a more accurate one. However, the algorithms for iteratively calculating the Moore–Penrose inverse are not generally self-correcting.

One of the more commonly discussed iterative methods is the one due to Ben-Israel [8]. One form of it is the following. Its proof and some related results are left to the exercises.

Theorem 12.2.3 *Let λ_1 denote the largest eigenvalue of $\mathbf{AA}^*$, $\mathbf{A} \in \mathbb{C}^{m \times n}$. If $0 < \alpha < 2/\lambda_1$, set $\mathbf{X}_0 = \alpha\mathbf{A}^*$ and define $\mathbf{X}_{r+1} = \mathbf{X}_r(2\mathbf{I} - \mathbf{AX}_r)$. Then $\mathbf{X}_r \to \mathbf{A}^\dagger$ as $r \to \infty$.*

Stewart has shown that the iterative scheme of Theorem 3 takes at least $2\log_2[\kappa(\mathbf{A}_\varepsilon)]$ iterations to produce an approximation to $(\mathbf{A}_\varepsilon)^\dagger$. Thus it can be slow. The method does turn out to be stable. Error does not accumulate more rapidly than would be expected. Each iteration requires $2mn^2$ multiplications. This method and its variants are not competitive for large matrices with the singular value decomposition. Stewart has also shown that if the entries of $\mathbf{A}$ are large so that the round-off error is large in magnitude, then errors can accumulate rapidly so a careful analysis of rounding error is needed.

Proper splittings have also been recently proposed as a method of calculating either (1)-inverses or the Moore–Penrose inverse [14]. These methods require first the splitting and then an iteration. Details may be found in Exercises 11–16 and in [14].

At this point it should be pointed out that while the singular value decomposition is a widely used method for computing $\mathbf{A}^\dagger$ there are two schools of thought. Let us use the phrase 'elimination method' to designate those methods of computing $\mathbf{A}^\dagger$ which are based on some variation of Gaussian elimination. Algorithms 1.3.1, 1.3.2, 3.3.1 and 3.3.2 are of this type.

To begin with there is no such thing as a universal algorithm. Given any non-trivial algorithm there is usually an example for which it works poorly. Secondly, if an algorithm is to be able to handle less well conditioned problems it frequently must be made more sophisticated.

On this basis there are those who argue that elimination methods have their place. Elimination methods usually are much simpler with substantially lower operation counts. Consequently, they are quicker to run with less accumulation of machine error.

It would be unfortunate if after reading this chapter, a reader computed $\mathbf{A}^\dagger$ for a 4×4 matrix with integer entries by using the singular value decomposition. By the time he had it entered on a machine, he could have obtained the answer at this desk, exactly if need be, by an elimination method and drunk a second cup of coffee. For well conditioned matrices of moderate to small size there are strong arguments for using an elimination method.

In this same vein we might point out that in our computing experiments with $\mathbf{A}^{\mathrm{D}}$ we tried powering $\mathbf{A}$ by using QR factorization. While this is a stable method we discovered, not suprisingly, that for 6×6 or 8×8 matrices we got more error than if we just naively multiplied them out. The reason was the substantial increase in the operation count and the loss of information. When $\mathbf{A}$ was entered exactly (had integer entries for example), some error was introduced in the factorization whereas direct multiplication often produced the exact answer.

If one suspects that a matrix is severely ill-conditioned and/or very high accuracy is needed, then the best idea of all may be to compute in exact (residue) arithmetic. While this may substantially increase the computational effort needed, it will provide $\mathbf{A}^\dagger$ exactly if $\mathbf{A}$ has rational entries. Elimination methods are preferable with residue arithmetic since they have lower operation counts and produce exact answers.

The basic idea is to rescale the original matrix so that it has integral entries. Then compute in modular arithmetic. If the numbers involved are large, then multiple modulus arithmetic may be necessary. A good discussion of three elimination methods and associated algorithms in modular arithmetic may be found in [14]. See also [87]. Residue arithmetic can be further studied in [90], [92]. See also the other references of [14]. Exact computation may also be done in p-adic number systems (see [14] for references).

3. Computation of the singular value decomposition

This section will hopefully develop the Golub–Reinsch algorithm in sufficient detail so that the interested reader will understand its basic structure. We shall first briefly discuss Householder transformations and Givens rotations. Then we shall give an outline of the algorithm. Finally, we shall discuss each step in greater detail. We shall not discuss the actual organization of such a program or worry about storage. The interested reader is referred to [52, Chapter 18] or [15], [36], [38], [39].

Recall from Chapter 10 that $\|\mathbf{v}\|_2 = (\mathbf{v}^*\mathbf{v})^{1/2}$.

Definition 12.3.1 *If* $\mathbf{u} \in \mathbb{C}^{n \times n}$, $\mathbf{u}^*\mathbf{u} = 1$, *then* $\mathbf{H} = \mathbf{I} - 2\mathbf{u}\mathbf{u}^*$, *is called a Householder transformation.*

It is easy to see that $\mathbf{H}^* = \mathbf{H} = \mathbf{H}^{-1}$ and hence $\mathbf{H}$ is unitary.

Proposition 12.3.1 *Given a vector* $\mathbf{v} \in \mathbb{C}^n$, *let* $\mathbf{u} = \mathbf{v} + \sigma\|\mathbf{v}\|_2\mathbf{e}_1$ *where*

$\sigma = +1$ *if* $v_1 \geq 0$, $\sigma = -1$ *if* $v_1 < 0$. *Let* $\mathbf{H} = \mathbf{I} - 2\mathbf{u}\mathbf{u}^*/\mathbf{u}^*\mathbf{u}$. *Then* $\mathbf{H}\mathbf{v} = -\sigma\|\mathbf{v}\|_2\mathbf{e}_1$.

Definition 12.3.2 *A Givens rotation is a matrix* $\mathbf{G} = [\mathbf{g}_1, \ldots, \mathbf{g}_n]$, $\mathbf{g}_i \in \mathbb{C}^n$ *such that* $\mathbf{g}_i = \mathbf{e}_i$ *except for two values of* $i; i_1 < i_2$. *For these two columns* $g_{i_1i_1} = c$, $g_{i_2i_1} = -s$, $g_{i_1i_2} = s$, $g_{i_2i_2} = c$ *with* $|c|^2 + |s|^2 = 1$. *All other entries in* $\mathbf{g}_{i_1}, \mathbf{g}_{i_2}$ *are zero.*

If $i_2 = i_1 + 1$, *then* $\mathbf{G}$ *is of the form*

$$\begin{bmatrix} \mathbf{I} & \mathbf{0} & \mathbf{0} \\ \mathbf{0} & \begin{bmatrix} c & s \\ -s & c \end{bmatrix} & \mathbf{0} \\ \mathbf{0} & \mathbf{0} & \mathbf{I} \end{bmatrix} \text{ with } |c|^2 + |s|^2 = 1.$$

A Givens rotation is unitary and is used to introduce zeros.

Proposition 12.3.2 *Given the vector* $\mathbf{v} \in \mathbb{C}^n$ *and indices* $1 \leq i_1 < i_2 \leq n$, *set* $c = v_{i_1}/(|v_{i_1}|^2 + |v_{i_2}|^2)^{1/2}$, $s = v_{i_2}/(|v_{i_1}|^2 + |v_{i_2}|^2)^{1/2}$. *Then the Givens rotation* $\mathbf{G}$ *defined by* i_1, i_2, c, s *is such that* $(\mathbf{G}\mathbf{v})_i = v_i$ *if* $i \neq i_1$ *or* i_2, $(\mathbf{G}\mathbf{v})_{i_1} = (|v_{i_1}|^2 + |v_{i_2}|^2)^{1/2}$, $(\mathbf{G}\mathbf{v})_{i_2} = 0$.

The operation performed by $\mathbf{G}$ in Proposition 2 is often referred to as zeroing the i_2-entry and placing it in the i_1-position. Note that only i_1, i_2 are affected at all. Note that $(n-1)$ Givens rotation could be used on a vector $\mathbf{v}$ to get a multiple of $\mathbf{e}_1$ but one Householder transformation will do the job.

We now present the algorithm.

Algorithm 12.3.1 Given $\mathbf{A} \in \mathbb{C}^{m \times n}$. Assume $m \geq n$. If $m \leq n$, work with $\mathbf{A}^*$.

(I) Perform at most $2n - 1$ Householder transformations $\mathbf{Q}_i, \mathbf{H}_i$ to $\mathbf{A}$ to get

$$\mathbf{Q}^{\mathrm{T}}\mathbf{A}\mathbf{H} = \mathbf{Q}_n(\ldots(\mathbf{Q}_1\mathbf{A})\mathbf{H}_2 \ldots \mathbf{H}_n) = \begin{bmatrix} \mathbf{B} \\ \mathbf{0} \end{bmatrix}, \mathbf{B} \in \mathbb{C}^{n \times n}$$

where

$$\mathbf{B} = \begin{bmatrix} q_1 & e_2 & & \bigcirc \\ & q_2 & \ddots & \\ & & \ddots & e_n \\ \bigcirc & & & q_n \end{bmatrix} \tag{1}$$

Note $\mathbf{B}$ is bidiagonal.

(II) If some $e_i = 0$, then $\mathbf{B} = \begin{bmatrix} \mathbf{B}_1 & \mathbf{0} \\ \mathbf{0} & \mathbf{B}_2 \end{bmatrix}$ and the singular value decompositions of $\mathbf{B}_1, \mathbf{B}_2$ may be computed separately.

(III) If all $e_i \neq 0$ but some $q_k = 0$, pre-multiplication by $(n-k)$ Given s rotation $\mathbf{G}_i$ gives

$$\mathbf{G}_n \cdots \mathbf{G}_{k+1}\mathbf{B} = \tilde{\mathbf{B}} = \left[\begin{array}{cccc|ccc} q_1 & e_2 & & & & & \\ & q_2 & \ddots & \bigcirc & & \bigcirc & \\ & & q_{k-1} & e_k & & & \\ & \bigcirc & & 0 & & & \\ \hline & & & & \tilde{q}_{k-1} & \tilde{e}_{k+1} & \bigcirc \\ & & \bigcirc & & & \tilde{q}_{k+2} & \tilde{e}_n \\ & & & & \bigcirc & & \tilde{q}_n \end{array}\right]$$

$$= \begin{bmatrix} \tilde{\mathbf{B}}_1 & \mathbf{0} \\ \mathbf{0} & \tilde{\mathbf{B}}_2 \end{bmatrix}.$$

Application of Givens rotations on the right to $\tilde{\mathbf{B}}_1$ produces a matrix with zero last column.

(IV) Steps II, III have reduced the problem to computing the singular value decomposition of $\mathbf{B}$ in the form (1) where all $q_i \neq 0$, all the $e_i \neq 0$. Say $\mathbf{B}$ is $n \times n$. Now set $\mathbf{B}_1 = \mathbf{B}$, $\mathbf{B}_{k+1} = \mathbf{U}_k^*\mathbf{B}_k\mathbf{V}_k$ where $\mathbf{U}_k, \mathbf{V}_k$ are orthogonal and to be defined. The $\mathbf{U}_k$, $\mathbf{V}_k$ are chosen so that $\mathbf{B}_k$ is upper bidiagonal for all k and the $(n-1, n)$-entry of $\mathbf{B}_k$ goes to zero as $k \to \infty$. Thus after a finite number of steps

$$\mathbf{B}_k = \begin{bmatrix} \hat{q}_1 & \hat{e}_2 & & \bigcirc \\ & \ddots & \hat{e}_{n-1} & \\ & & \hat{q}_{n-1} & \theta \\ \bigcirc & & & \hat{q}_n \end{bmatrix}$$

and θ is less than some agreed-on level of precision. Hence

$$\mathbf{B}_k = \begin{bmatrix} \hat{\mathbf{B}} & \mathbf{0} \\ \mathbf{0} & \hat{q}_n \end{bmatrix}. \tag{2}$$

(V) Step IV (and possibly II, III) is repeated on $\hat{\mathbf{B}}$ in (2). The process terminates in at most $n-1$ repetitions to give

$$\mathbf{UAV} = \begin{bmatrix} \mathbf{D} \\ \mathbf{0} \end{bmatrix} \text{ where } \mathbf{D} \text{ is diagonal,}$$

and $\mathbf{U}, \mathbf{V}$ are unitary.

(IV) Compute $\mathbf{A}^\dagger = \mathbf{U}^*[\mathbf{D}^\dagger \quad \mathbf{0}]\mathbf{V}^*$.

This completes our outline of the algorithm. We shall now discuss some of the steps in more detail.

Step I If $\mathbf{A} = [\mathbf{a}_1, \mathbf{a}_2, \ldots, \mathbf{a}_n]$, calculate $\mathbf{Q}_1$ by Proposition 1 so that $\mathbf{Q}_1\mathbf{a}_1$ is a multiple of $\mathbf{e}_1$. Let $\mathbf{Q}_1\mathbf{A} = \begin{bmatrix} \mathbf{r}_1 \\ \vdots \\ \mathbf{r}_m \end{bmatrix}$. Use Proposition 1 to find an $\tilde{\mathbf{H}}_2$ so that $[r_{12}, \ldots, r_{1n}]\tilde{\mathbf{H}}_2 = \alpha\mathbf{e}_1^*, \mathbf{e}_1 \in \mathbb{C}^{n-1}$. (That is,

$\tilde{\mathbf{H}}_2 \begin{bmatrix} r_{12} \\ \vdots \\ r_{1n} \end{bmatrix} = \alpha \mathbf{e}_1, \mathbf{e}_1 \in \mathbb{C}^{n-1}$). Let $\mathbf{H}_2 = \begin{bmatrix} 1 & \mathbf{0} \\ \mathbf{0} & \tilde{\mathbf{H}}_2 \end{bmatrix}$. Then

$$(\mathbf{Q}_1 \mathbf{A})\mathbf{H}_2 = \left[\begin{array}{c|cccc} x & x & 0 & \dots\dots & 0 \\ \hline 0 & x & & \dots\dots & x \\ \vdots & \vdots & & & \vdots \\ 0 & x & & \dots\dots & x \end{array}\right] \tag{2}$$

Now repeat this procedure on the lower right hand block of (3). Note that if $\mathbf{H}$ is an $n - l \times n - l$ Householder transformation, then $\begin{bmatrix} \mathbf{I}_l & \mathbf{0} \\ \mathbf{0} & \mathbf{H} \end{bmatrix}$ is an $n \times n$ Householder transformation. Continuing, we get (1) in at most $2n - 1$ Householder transformations.

Step II Clear.

Step III Suppose $e_k \neq 0$ but $q_k = 0$. The rotation $\mathbf{G}_j$ is defined as in Proposition 2 and places a zero in the (k, j)-position, the rotation effecting only the kth and jth rows of $\mathbf{B}$.

Step IV This is the most difficult part. It constitutes a modified implicit **QR**-procedure. Suppose $\mathbf{B}$ is in the form (1) with all $q_i, e_i \neq 0$. Set $\mathbf{B}_1 = \mathbf{B}$. Suppose that $\mathbf{B}_m$ has been calculated. We shall show how to get $\mathbf{B}_{m+1}, \mathbf{U}_m, \mathbf{V}_m$. Let $\mathbf{B}_m$ be in the form (1), $\mathbf{B}_m$ be $n \times n$. Set $f = (q_n^2 - q_{n-1}^2 + e_n^2 - e_{n-1}^2)/2e_n q_{n-1}$, and

$$t = \begin{cases} [f + (1 + f^2)^{1/2}] \text{ if } f \geq 0 \\ [f - (1 + f^2)^{1/2}] \text{ if } f < 0. \end{cases}$$

Set $\sigma = q_n^2 + e_n^2 - e_n q_{n-1}/t$.

Now $\mathbf{B}_m^{\mathrm{T}} \mathbf{B}_m - \sigma \mathbf{I}$ has $[x, x, 0, \dots, 0]^*$ for a first column. Calculate by Proposition 2 a rotation $\mathbf{R}_1$ to zero out the $(2, 1)$-entry of $\mathbf{B}_m^{\mathrm{T}} \mathbf{B}_m - \sigma \mathbf{I}$. This rotation would involve only the first two columns. Thus

$$\mathbf{B}_m \mathbf{R}_1 = \begin{bmatrix} x & x & & & \bigcirc \\ x & x & e_3 & & \\ & & q_3 & \ddots & \\ & & & \ddots & e_n \\ \bigcirc & & & & q_n \end{bmatrix}.$$

A series of Givens rotations are now performed on $\mathbf{B}_m \mathbf{R}_1$ to return it to bidiagonal form. Let $\mathbf{R}_i$ operate on columns i and $i + 1$; $\mathbf{T}_i$ operate on rows i and $i + 1$. Then

$$\mathbf{B}_{m+1} = \mathbf{T}_{n-1}(\dots \mathbf{T}_2((\mathbf{T}_1(\mathbf{B}_m \mathbf{R}_1))\mathbf{R}_2) \dots \mathbf{R}_{n-1}). \tag{4}$$

Each rotation takes a non-zero entry, zeros it, and sends the non-zero part to a new place. The pattern is as follows for a 5 × 5:

$$\begin{bmatrix} x & x & z_2 & & \bigcirc \\ z_1 & x & x & z_4 & \\ & z_3 & x & x & z_6 \\ & & z_5 & x & x \\ \bigcirc & & & z_7 & x \end{bmatrix}.$$

Here z_i is non-zero at stage i and zero at the other stages. Thus $\mathbf{T}_1$ zeros z_1 and creates a non-zero entry at z_2. $\mathbf{R}_2$ zeros z_2 and creates a non-zero entry at z_3. The $\mathbf{B}_{m+1}$ computed in this way will again be bidiagonal and in the form of (1). (This requires proof.)

Steps V and IV are reasonably self-explanatory.

4. (1)-inverses

When dealing with generalized inverses that are not uniquely defined, one cannot talk about error in the same sense as we did in Sections 1 and 2.

Probably the easiest way to compute a (1)-inverse is as follows. By reducing to echelon form, one can find permutation matrices $\mathbf{P}, \mathbf{Q}$ so that $\mathbf{PAQ} = \begin{bmatrix} \mathbf{U} & \mathbf{X} \\ \mathbf{Y} & \mathbf{Z} \end{bmatrix}$, $\mathbf{U}$ is square and rank $(\mathbf{U}) = $ rank $(\mathbf{A})$. Then $\mathbf{Q}\begin{bmatrix} \mathbf{U}^{-1} & \mathbf{0} \\ \mathbf{0} & \mathbf{0} \end{bmatrix}\mathbf{P}$ is a (1)-inverse of $\mathbf{A}$. Note that since there is some choice as to which columns go into $\mathbf{U}$ one might in some cases be able to control to some extent how well conditioned $\mathbf{U}$ is.

How to check a computed inverse will depend on how it is being used. For example, even if $\mathbf{A}$ is invertible it is sometimes possible to find $\mathbf{X}$ so that $\mathbf{AX} - \mathbf{I}$ is small but $\mathbf{XA} - \mathbf{I}$ is large.

Example 12.4.1 Let $\mathbf{A} = \begin{bmatrix} 1 & 0 \\ 0 & \varepsilon \end{bmatrix}$, $\mathbf{B} = \begin{bmatrix} 1 & 0 \\ 0 & 0 \end{bmatrix}$. Then $\mathbf{ABA} - \mathbf{A} = \begin{bmatrix} 0 & 0 \\ 0 & -\varepsilon \end{bmatrix}$ which is small. Thus $\mathbf{B}$ comes 'close' to being a (1)-inverse. However $\mathbf{B}$ gives $\mathbf{B}\begin{bmatrix} 1 \\ 1 \end{bmatrix} = \begin{bmatrix} 1 \\ 0 \end{bmatrix}$ as a solution of $\mathbf{Ax} = \begin{bmatrix} 1 \\ 1 \end{bmatrix}$ which has the actual solution $\mathbf{x} = \begin{bmatrix} 1 \\ \varepsilon^{-1} \end{bmatrix}$. If ε is small, then $\mathbf{B}\begin{bmatrix} 1 \\ 1 \end{bmatrix}$ is far from the correct solution even though $\mathbf{ABA} - \mathbf{B}$ is small and (1)-inverses solve consistent systems.

As Example 1 shows some care must be taken. If row operations are to be used to compute $\mathbf{A}^-$, then a well-conditioned matrix is desirable.

5. Computation of the Drazin inverse

Since

$$\mathbf{A}^{\mathrm{D}} = \mathbf{A}^l(\mathbf{A}^{2l+1})^\dagger \mathbf{A}^l \text{ for } l \geq \mathrm{Ind}(\mathbf{A}), \tag{1}$$

one might suppose that computation of the Drazin inverse would be, at worst, just a little more work than computation of the Moore–Penrose inverse. However, there are several problems. The first difficulty is that small perturbations in $\mathbf{A}$ can cause arbitrarily large perturbations in $\mathbf{A}^{\mathrm{D}}$. Unlike the Moore–Penrose, the singular value decomposition of $\mathbf{A}$ does not immediately help because $(\mathbf{UAW})^{\mathrm{D}}$ is not, in general, equal to $\mathbf{W}^*\mathbf{A}^{\mathrm{D}}\mathbf{U}^*$ when $\mathbf{U}, \mathbf{W}$ are unitary. The following weaker version of Theorem 2.1 shows that computation of $\mathbf{A}^{\mathrm{D}}$ does make some sense when using non-exact arithmetic. Of course, there is no theoretical difficulty if exact arithmetic is used and $\mathbf{A}$ is known exactly.

Theorem 12.5.1 *If* $\mathbf{A} \in \mathbb{C}^{n \times n}$ *and* $\{\mathbf{X}_r\} \subseteq \mathbb{C}^{n \times n}$ *is such that* $\{\mathbf{X}_r\mathbf{A}\mathbf{X}_r - \mathbf{X}_r\}$, $\{\mathbf{A}\mathbf{X}_r - \mathbf{X}_r\mathbf{A}\}$, $\{\mathbf{A}^{k+1}\mathbf{X}_r - \mathbf{A}^k\}$ *all converge to zero and if there is an* L *such that* $\|\mathbf{X}_r\| \leq \mathrm{L}$ *for all r, then* $\mathbf{X}_r \to \mathbf{A}^{\mathrm{D}}$.

Proof Suppose $\{\mathbf{X}_r\}$, $\mathbf{A}$ satisfy the assumptions of Theorem 1. Suppose $\mathbf{X}_r$ does not converge to $\mathbf{A}^{\mathrm{D}}$. Then there is a subsequence $\{\mathbf{X}_s\}$ of $\{\mathbf{X}_r\}$ such that $\|\mathbf{X}_s - \mathbf{A}^{\mathrm{D}}\| \geq \varepsilon > 0$ for all s and some $\varepsilon > 0$. Since $\{\mathbf{X}_s\}$ is bounded, it has a subsequence $\{\mathbf{X}_l\}$ which converges. Let $\mathbf{X}_0 = \lim_{l\to\infty} \mathbf{X}_l$. Then $\mathbf{X}_0 \neq \mathbf{A}^{\mathrm{D}}$ and $\mathbf{X}_0\mathbf{A} = \mathbf{A}\mathbf{X}_0$, $\mathbf{X}_0\mathbf{A}\mathbf{X}_0 = \mathbf{X}_0$, $\mathbf{A}^{k+1}\mathbf{X}_0 = \mathbf{A}$. Thus $\mathbf{X}_0 = \mathbf{A}^{\mathrm{D}}$ which is a contradiction. Hence $\mathbf{X}_r \to \mathbf{A}^{\mathrm{D}}$. ■

The next theorem is true with no assumption on $\mathrm{Ind}(\mathbf{A})$ but the proof came too late for inclusion.

Theorem 12.5.2 *Suppose that* $\mathbf{A} \in \mathbb{C}^{n \times n}$ *and* $\mathbf{A}$ *has index* 1. *If* $\{\mathbf{X}_r\}$ *is a sequence of matrices such that the sequences* $\{\mathbf{A}\mathbf{X}_r - \mathbf{X}_r\mathbf{A}\}$, $\{\mathbf{A}^2\mathbf{X}_r - \mathbf{A}\}$, *and* $\{\mathbf{X}_r\mathbf{A}\mathbf{X}_r - \mathbf{X}_r\}$ *all converge to zero, then* $\mathbf{X}_r \to \mathbf{A}^{\#}$.

Proof Suppose $\mathbf{A}$ has index one and $\{\mathbf{X}_r\}$ is a sequence which satisfies the assumptions of Theorem 2. $\mathbf{A}$ is similar by some non-singular $\mathbf{B}$ to a matrix of the form

$$\begin{bmatrix} \mathbf{C} & \mathbf{0} \\ \mathbf{0} & \mathbf{0} \end{bmatrix} \text{ where } \mathbf{C} \text{ is invertible.}$$

Let $\mathbf{B}\mathbf{X}_r\mathbf{B}^{-1} = \begin{bmatrix} \mathbf{X}_1(r) & \mathbf{X}_2(r) \\ \mathbf{X}_3(r) & \mathbf{X}_4(r) \end{bmatrix}$. Then by assumption,

$$\begin{bmatrix} \mathbf{C}\mathbf{X}_1(r) & \mathbf{C}\mathbf{X}_2(r) \\ \mathbf{0} & \mathbf{0} \end{bmatrix} - \begin{bmatrix} \mathbf{X}_1(r)\mathbf{C} & \mathbf{0} \\ \mathbf{X}_3(r)\mathbf{C} & \mathbf{0} \end{bmatrix} \to \mathbf{0}, \tag{2}$$

$$\begin{bmatrix} \mathbf{C}^2\mathbf{X}_1(r) & \mathbf{C}^2\mathbf{X}_2(r) \\ \mathbf{0} & \mathbf{0} \end{bmatrix} - \begin{bmatrix} \mathbf{C} & \mathbf{0} \\ \mathbf{0} & \mathbf{0} \end{bmatrix} \to \mathbf{0}, \tag{3}$$

and

$$\begin{bmatrix} \mathbf{X}_1(r)\mathbf{C}\mathbf{X}_1(r) & \mathbf{X}_1(r)\mathbf{C}\mathbf{X}_2(r) \\ \mathbf{X}_3(r)\mathbf{C}\mathbf{X}_1(r) & \mathbf{X}_3(r)\mathbf{C}\mathbf{X}_3(r) \end{bmatrix} - \begin{bmatrix} \mathbf{X}_1(r) & \mathbf{X}_2(r) \\ \mathbf{X}_3(r) & \mathbf{X}_4(r) \end{bmatrix} \to \mathbf{0}. \tag{4}$$

From (2), we get $\mathbf{X}_2(r) \to \mathbf{0}$, $\mathbf{X}_3(r) \to \mathbf{0}$, since $\mathbf{C}$ is invertible. But then $\mathbf{X}_4(r) \to \mathbf{0}$ from (4). Equation (3) yields $\mathbf{X}_1(r) \to \mathbf{C}^{-1}$. Hence $\mathbf{X}_r \to \mathbf{A}^{\#}$ as desired. ■

Another difficulty is that the index is not generally known. There are exceptions, such as was the case with the Markov chains in Chapter 8. Thus to use (1) one must either get an estimate on the index or use n instead of k. However, using n can introduce large errors for moderately sized matrices.

Example 12.5.1 Let $\mathbf{A} = \begin{bmatrix} \alpha\mathbf{I} & \mathbf{0} \\ \mathbf{0} & \mathbf{0} \end{bmatrix}$ where $\mathbf{A}$ is 20×20, $\alpha = 10^{-2}$, and $\mathbf{I}$ is an $r \times r$ identity for $1 \le r \le 19$. Then $\mathbf{A}^{41} = \begin{bmatrix} 10^{-82} & \mathbf{0} \\ \mathbf{0} & \mathbf{0} \end{bmatrix}$. If 10^{-75} is considered zero, then (1) would produce $\mathbf{A}^{\mathrm{D}} = \mathbf{0}$, whereas $\mathbf{A}^{\mathrm{D}}$ is actually $\begin{bmatrix} 100\mathbf{I} & \mathbf{0} \\ \mathbf{0} & \mathbf{0} \end{bmatrix}$.

Numerical experiments using (1) with reasonably conditioned matrices have shown that one can run into difficulty even with n less than 10.

The difficulty in Example 1 is the loss of 'small' eigenvalues.

Suppose $\mathbf{A} = \mathbf{P}\begin{bmatrix} \mathbf{J}_1 & & \\ & \ddots & \\ & & \mathbf{J}_r \end{bmatrix}\mathbf{P}^{-1}$ is the Jordan canonical form for $\mathbf{A}$.

Then

$$\mathbf{A}^m = \mathbf{P}\begin{bmatrix} \mathbf{J}_1^m & & & \\ & \ddots & & \\ & & \mathbf{J}_s^m & \\ & & & \mathbf{0} \end{bmatrix}\mathbf{P}^{-1}.$$

If any of the $t_i \times t_i$ $\mathbf{J}_i$ corresponds to an eigenvalue λ_i such that $\lambda_i^{2l+1-t_i}$ rounds to zero, then (1) will produce a commuting (2)-inverse and not $\mathbf{A}^{\mathrm{D}}$. Thus in checking a computed $\mathbf{A}^{\mathrm{D}}, \hat{\mathbf{A}}^{\mathrm{D}}$, it is important to check whether it reduces powers. However, if there is a $t_i \times t_i$ $\mathbf{J}_i$ such that $\lambda_i^{l-t_i}$ rounds to zero, then to machine accuracy we will have $\mathbf{A}^{l+1}\hat{\mathbf{A}}^{\mathrm{D}} = \mathbf{A}^l$ where as in fact $\hat{\mathbf{A}}^{\mathrm{D}}$ is not close to $\mathbf{A}^{\mathrm{D}}$. Thus if one has a 20×20 matrix of low index, using (1) with $n = 20$ could lead to erroneous answers that would not be detected. For these reasons we prefer other methods to (1).

Computation of the Jordan canonical form is also very sensitive so that one should not try and compute it in full detail unless it is needed. If one is going to calculate $\mathbf{A}^{\mathrm{D}}$ by (1), then rather than $\|\mathbf{A}\|(\|\mathbf{A}^{\mathrm{D}}\| + 1)$ or some such, a much better idea of the conditioning would be

$$\mathrm{C}(\mathbf{A}) = \|\mathbf{P}\|\,\|\mathbf{P}^{-1}\|\,(\|\mathbf{J}\|^l + \|\mathbf{J}^{\dagger}\|^l)$$

where $\mathbf{PJP}^{-1}$ is the Jordan form of $\mathbf{A}$. This $C(\mathbf{A})$ of course depends on $\mathbf{P}$. However, if it is not too large, the eigenvalues are not likely to be lost and eigenspaces are not nearly parallel.

The more one is willing to lump eigenvalues together and weaken the Jordan form, the less computational difficulty there should be.

Let $\mathbf{N}$ be an $n \times n$ nilpotent Jordan block. Then, $\mathbf{N}$ is nilpotent with index n and, of course, $\mathbf{N}^{\mathrm{D}} = \mathbf{0}$. Zero is the only eigenvalue of $\mathbf{N}$. Now let $\tilde{\mathbf{N}}$ be the same as $\mathbf{N}$ except for an $\varepsilon > 0$ in the $(n, 1)$ place. Then $\|\mathbf{N} - \tilde{\mathbf{N}}\| = \varepsilon$ but $\tilde{\mathbf{N}}$ has n eigenvalues of modulus $\varepsilon^{1/n}$. The singular values of $\tilde{\mathbf{N}}$, however, are $(n-1)$ ones and ε, while the singular values of $\mathbf{N}$ are $(n-1)$ ones and zero. If $\varepsilon = 10^{-20}$, $n = 20$ then $\tilde{\mathbf{N}}$ has an eigenvalue of 0.1 whereas $\|\mathbf{N} - \tilde{\mathbf{N}}\| = 10^{-20}$. It is because of this stability of the singular values, in particular the zero ones, and the instability of the zero eigenvalues that we suggest the following method for calculating $\mathbf{A}^{\mathrm{D}}$. This method is based on the orthogonal deflation method in [39]. Since there are established subroutines for calculating singular value decompositions, it should be fairly easy to implement.

Algorithm 12.5.1

(I) Given $\mathbf{A}$, calculate the singular value decomposition of $\mathbf{A} \in \mathbb{C}^{n \times n}$.

$$\mathbf{A} = \mathbf{U}\begin{bmatrix} \boldsymbol{\Sigma} & \mathbf{0} \\ \mathbf{0} & \mathbf{0} \end{bmatrix}\mathbf{V}.$$

If $\boldsymbol{\Sigma} \in \mathbb{C}^{n \times n}$, then $\mathbf{A}^{\mathrm{D}} = \mathbf{A}^{-1} = \mathbf{V}^*\boldsymbol{\Sigma}^{-1}\mathbf{u}^*$. If $\operatorname{Ind}(\mathbf{A}) > 0$, write

$$\mathbf{VAV}^* = \mathbf{VU}\begin{bmatrix} \boldsymbol{\Sigma} & \mathbf{0} \\ \mathbf{0} & \mathbf{0} \end{bmatrix} = \begin{bmatrix} \mathbf{A}_{11}^{(1)} & \mathbf{0} \\ \mathbf{A}_{21}^{(1)} & \mathbf{0} \end{bmatrix}.$$

(II) Now calculate the singular value decomposition of $\mathbf{A}_{11}^{(1)}$.

$$\mathbf{A}_{11}^{(1)} = \mathbf{U}_1\begin{bmatrix} \boldsymbol{\Sigma}_1 & \mathbf{0} \\ \mathbf{0} & \mathbf{0} \end{bmatrix}\mathbf{V}_1.$$

If $\mathbf{A}_{11}^{(1)}$ is invertible, go to Step IV. If not, then

$$\mathbf{V}_1\mathbf{A}_{11}^{(1)}\mathbf{V}_1^* = \mathbf{V}_1\mathbf{U}_1\begin{bmatrix} \boldsymbol{\Sigma}_1 & \mathbf{0} \\ \mathbf{0} & \mathbf{0} \end{bmatrix} = \begin{bmatrix} \mathbf{A}_{11}^{(2)} & \mathbf{0} \\ \mathbf{A}_{21}^{(2)} & \mathbf{0} \end{bmatrix}.$$

Thus

$$\begin{aligned}
\begin{bmatrix} \mathbf{V}_1 & \mathbf{0} \\ \mathbf{0} & \mathbf{I} \end{bmatrix}\mathbf{VAV}^*\begin{bmatrix} \mathbf{V}_1^* & \mathbf{0} \\ \mathbf{0} & \mathbf{I} \end{bmatrix} &= \begin{bmatrix} \mathbf{V}_1 & \mathbf{0} \\ \mathbf{0} & \mathbf{I} \end{bmatrix}\begin{bmatrix} \mathbf{A}_{11}^{(1)} & \mathbf{0} \\ \mathbf{A}_{21}^{(1)} & \mathbf{0} \end{bmatrix}\begin{bmatrix} \mathbf{V}_1^* & \mathbf{0} \\ \mathbf{0} & \mathbf{I} \end{bmatrix} \\
&= \left[\begin{array}{c|c} \mathbf{V}_1\mathbf{A}_{11}^{(1)}\mathbf{V}_1^* & \mathbf{0} \\ \hline \mathbf{A}_{21}^{(1)}\mathbf{C}_1^* & \mathbf{0} \end{array}\right] \\
&= \left[\begin{array}{cc|c} \mathbf{A}_{11}^{(2)} & \mathbf{0} & \mathbf{0} \\ \mathbf{A}_{21}^{(2)} & \mathbf{0} & \mathbf{0} \\ \hline \mathbf{A}_{31}^{(2)} & \mathbf{A}_{32}^{(2)} & \mathbf{0} \end{array}\right] \qquad (4)
\end{aligned}$$

(III) Continue in the manner of Step II, at each step calculating the singular value decomposition of $\mathbf{A}_{11}^{(m)}$ and performing the appropriate multiplication as in (4) to get $\mathbf{A}_{11}^{(m+1)}$. If some $\mathbf{A}_{11}^{(k)} = \mathbf{0}$, then $\mathbf{A}^{\mathrm{D}} = \mathbf{0}$. If some $\mathbf{A}_{11}^{(k)}$ is non-singular, go to Step IV.

(IV) We now have $k = \operatorname{Ind}(\mathbf{A})$ and

$$\mathbf{WAW}^* = \left[\begin{array}{c|cccc} \mathbf{A}_{11}^{(k)} & \mathbf{0} & \cdots & \cdots & \mathbf{0} \\ \hline \mathbf{A}_{21}^{(k)} & \mathbf{0} & \cdots & \cdots & \mathbf{0} \\ \mathbf{A}_{31}^{(k)} & \mathbf{A}_{32}^{(k)} & \ddots & & \vdots \\ \vdots & \vdots & & \ddots & \\ \mathbf{A}_{k+1,1}^{(k)} & \mathbf{A}_{k+1,2}^{(k)} & \cdots & \mathbf{A}_{k+1,k}^{(k)} & \mathbf{0} \end{array}\right] = \left[\begin{array}{c|c} \mathbf{B}_1 & \mathbf{0} \\ \hline \mathbf{B}_2 & \mathbf{N} \end{array}\right] \tag{5}$$

where $\mathbf{B}_1$ is invertible, $\mathbf{N}$ is nilpotent, and $\mathbf{W}$ is unitary. The Drazin inverse of (5) may be computed as

$$\begin{bmatrix} \mathbf{B}_1^{-1} & \mathbf{0} \\ \mathbf{XB}_1^{-1} & \mathbf{0} \end{bmatrix}$$

where $\mathbf{X}$ is the solution of $\mathbf{XB}_1 - \mathbf{NX} = \mathbf{B}_2$. The rows $\mathbf{x}_i$ of $\mathbf{X}$ are recursively solved as follows. If $\mathbf{N} = [n_{ij}]$, $\mathbf{r}_i$ the ith row of $\mathbf{B}_2$, then $\mathbf{x}_1 = \mathbf{r}_1\mathbf{B}_1$ and

$$\mathbf{x}_i = \left(\mathbf{r}_i + \sum_{j=1}^{i-1} n_{ij}\mathbf{x}_i\right)\mathbf{B}^{-1}$$

Note that the singular value decompositions are performed on successively smaller matrices. Also the unitary matrix $\mathbf{W}$ is the product of matrices of the form $\begin{bmatrix} \mathbf{V}_i & \mathbf{0} \\ \mathbf{0} & \mathbf{I} \end{bmatrix}$ and the size of $\mathbf{V}_i$ decreases with i. Thus the amount of computation decreases on each step.

If k is not large in comparison with the nullity of the core of $\mathbf{A}$, this method seems reasonable. If one suspects that k is comparable to the nullity of the core of $\mathbf{A}$, then it might be better to use a method like the double Francis QR algorithm and get

$$\mathbf{U}^*\mathbf{AU} = \mathbf{T}$$

where $\mathbf{T}$ is upper-triangle with the zero diagonal entries listed first. Thus

$$\mathbf{T} = \left[\begin{array}{c|c} \begin{matrix} 0 & x \\ 0 & 0 \end{matrix} & \mathbf{X} \\ \hline \bigcirc & \begin{matrix} x & \\ 0 & x \end{matrix} \end{array}\right] = \begin{bmatrix} \mathbf{N} & \mathbf{B}_1 \\ \mathbf{0} & \mathbf{B}_2 \end{bmatrix} \tag{6}$$

where $\mathbf{N}$ is nilpotent and $\mathbf{B}_2$ is invertible. Then Theorem 7.8.1 can be used on $\mathbf{T}$.

Since the singular value decomposition is considered as reliable a way as any of determining numerical rank, the deflation algorithm produces a reliable value of the index.

On the other hand, (6) provides information on the eigenvalues of **A**. As pointed out earlier if eigenvalues are small when taken to the kth power, then the algebraic definition cannot be used to numerically distinguish between $\mathbf{A}^{\mathrm{D}}$ and some commuting (2)-inverse. Thus the added information provided by (6) could be helpful.

If one does an operational count on finding the Drazin inverse of a 20×20 matrix with core-rank 8 and index 3, then the deflation method entails about the same work as the power method using $n = 20$, and avoids the risk of losing small eigenvalues.

If the deflation method reveals a small index, then (1) can be computed as a check using k instead of n.

Note that $R(\mathbf{I} - \mathbf{A}^{\mathrm{D}}\mathbf{A})$ can be immediately read off from (5) or (6) without any additional effort.

Another method that has some strong points is the one given in Theorem 7.8.2. There successive full-rank factorizations of successively smaller matrices are performed until an invertible matrix is reached. This method involves only elementary row operations, and matrix multiplications. It is fairly easy to program.

6. Previous algorithms

In other parts of this book we have presented several algorithms. This section will list the algorithms and those parts of the book that discuss computation. In general, the algorithms were only compared for operational counts. Error analyses are not given. The methods are all fairly easy to program and when tested by the authors worked well for small well-conditioned matrices (less than 10×10). All algorithms terminate in a finite number of steps.

1. Algorithm 1.3.1, (page 16). Computes $\mathbf{A}^{\dagger}$ from geometric definition.
2. Algorithm 1.3.2, (page 16). Computes $\mathbf{A}^{\dagger}$ by computing a full rank factorization.
3. Algorithm 3.2.1, (page 51). Computes $\mathbf{A}^{\dagger}$ when **A** is rank one modification of a matrix **B** for which $\mathbf{B}^{\dagger}$ has been computed.
4. Algorithm 3.3.1, (page 56). Computes $\mathbf{A}^{\dagger}$ by a sequence of rank one modifications.
5. Algorithm 3.3.2, (page 57). Computes $\mathbf{A}^{\dagger}$ when **A** is a hermitian matrix. Computation of any **A** may be reduced to computing the Moore–Penrose inverse of a hermitian matrix.
6. Algorithm 7.2.1, (page 125). Computes $\mathbf{A}^{\mathrm{D}}$ by computing core-nilpotent decomposition.
7. Theorem 7.5.2, (page 130). Computes $\mathbf{A}^{\mathrm{D}}$ from eigenvalues of **A** and their multiplicities.

8. Algorithm 7.5.1, (page 134). Computes $\mathbf{A}^{\mathrm{D}}$ by a finite number of recursively defined operations.
9. Section 8.5 Discusses computation of $\mathbf{A}^{\#}$ and $\mathbf{w}^*$ for an ergodic chain where $\mathbf{w}$ is the fixed probability vector.

Of course, throughout the text there are results, on partitioned matrices for example, that can be useful in special cases.

The results of Chapter 10 on continuity of generalized inversion may be useful in error analysis.

7. Exercises

1. (Alternate Proof of Theorem 12.2.1 for real matrices.)
Take $\mathbf{A}\in\mathbb{R}^{m\times n}$. Define $f:\mathbb{R}^{n\times m}\to\mathbb{R}^{m\times n}\times\mathbb{R}^{n\times m}\times\mathbb{R}^{m\times m}\times\mathbb{R}^{n\times n}$ by
$$f(\mathbf{X})=(\mathbf{AXA}-\mathbf{A},\mathbf{XAX}-\mathbf{X},\mathbf{AX}-\mathbf{X}^*\mathbf{A}^*,\mathbf{XA}-\mathbf{A}^*\mathbf{X}^*). \tag{1}$$
At any $\mathbf{X}_0\in\mathbb{R}^{n\times m}$, the derivative of f, denoted $f'(\mathbf{X}_0)$, is a linear transformation from $\mathbb{R}^{n\times m}$ into $\mathbb{R}^{m\times n}\times\mathbb{R}^{n\times m}\times\mathbb{R}^{m\times m}\times\mathbb{R}^{n\times n}$. Show that its value at $\mathbf{X}\in\mathbb{R}^{n\times m}$ is
$$f'(\mathbf{X}_0)\mathbf{X}=(\mathbf{AXA},\mathbf{XAX}-\mathbf{X},\mathbf{AX}-\mathbf{X}^*\mathbf{A}^*,\mathbf{XA}-\mathbf{A}^*\mathbf{X}^*). \tag{2}$$
Show that $f'(\mathbf{X}_0)$ is one-to-one for all $\mathbf{X}_0$, and that f is also one-to-one. Conclude that f has a continuous inverse from its range onto $\mathbb{R}^{n\times m}$, thus proving Theorem 12.2.1 for real matrices.
2. Suppose that $\mathbf{A}\in\mathbb{C}^{n\times n}$ has index k. Define $g:\mathbb{C}^{n\times n}\to\mathbb{C}^{n\times n}\times\mathbb{C}^{n\times n}\times\mathbb{C}^{n\times n}$ by
$$g(\mathbf{X})=(\mathbf{AX}-\mathbf{XA},\ \mathbf{A}^{k+1}\mathbf{X}-\mathbf{A}^k,\ \mathbf{XAX}-\mathbf{X}).$$
Show that
$$g'(\mathbf{X}_0)\mathbf{X}=(\mathbf{AX}-\mathbf{XA},\ \mathbf{A}^{k+1}\mathbf{X},\ \mathbf{X}_0\mathbf{AX}+\mathbf{XAX}_0-\mathbf{X}).$$
Show that $g'(\mathbf{X}_0)$ need not be one-to-one but that $g'(\mathbf{A}^{\mathrm{D}})$ is one-to-one.
3. Using Exercise 2 show that for any $\mathbf{A}\in\mathbb{C}^{n\times n}$, there exists a constant K such that if $g(\mathbf{X}_r)\to\mathbf{0}$ and $\|\mathbf{X}_r\|\le \mathrm{K}$ for a sequence $\{\mathbf{X}_r\}$, then $\mathbf{X}_r\to\mathbf{A}^{\mathrm{D}}$.

Exercises 4–9 are from [7], [8], and [13].

4. Suppose $\mathbf{X}_0\in\mathbb{C}^{n\times m}$ satisfies (i) $\mathbf{X}_0=\mathbf{A}^*\mathbf{B}_0$, $\mathbf{B}_0\in\mathbb{C}^{m\times m}$, $\mathbf{B}_0$ non-singular, (ii) $\mathbf{X}_0=\mathbf{C}_0\mathbf{A}^*$, $\mathbf{C}_0\in\mathbb{C}^{n\times n}$, $\mathbf{C}_0$ non-singular, (iii) $\|\mathbf{AX}_0-\mathbf{P}_{\mathrm{R}(\mathbf{A})}\|<1$, and (iv) $\|\mathbf{X}_0\mathbf{A}-\mathbf{P}_{\mathrm{R}(\mathbf{A}^*)}\|<1$, where $\|\cdot\|$ is a multiplicative matrix norm. Let $\mathbf{X}_{k+1}=\mathbf{X}_k(2\mathbf{P}_{\mathrm{R}(\mathbf{A})}-\mathbf{AX}_k)$. Then $\mathbf{X}_k\to\mathbf{A}^\dagger$ as $k\to\infty$.
5. Let λ_1 denote the largest eigenvalue of $\mathbf{AA}^*$. If $0<\alpha<2/\lambda_1$, set $\mathbf{X}_0=\alpha\mathbf{A}^*$ and define $\mathbf{X}_{k+1}=\mathbf{X}_k(2\mathbf{I}-\mathbf{AX}_k)$. Show $\mathbf{X}_k\to\mathbf{A}^\dagger$ as $k\to\infty$.
6. Define α as in Exercise 5. Let $\mathbf{X}_k=\alpha\sum_{p=0}^{k}\mathbf{A}^*(\mathbf{I}-\alpha\mathbf{AA}^*)^p$. Show $\mathbf{X}_k\to\mathbf{A}^\dagger$ as $k\to\infty$.

7. Show the convergence in Exercise 6 is of the first order, the convergence in Exercise 5 is of the second order.
8. Let α be as in Exercise 5. Let $\mathbf{Z}_o = \alpha\mathbf{A}\mathbf{A}^*$, $\mathbf{Z}_{k+1} = 2\mathbf{Z}_k - \mathbf{Z}_k^2$. Show $\mathbf{Z}_k \to \mathbf{A}\mathbf{A}^\dagger$ as $k \to \infty$ and $\|\mathbf{P}_{R(A)} - \mathbf{Z}_{k+1}\| \leq \|\mathbf{P}_{R(A)} - \mathbf{Z}_k\|^2$.
9. Let $\mathbf{Z}_k$ be as in Exercise 8. Show $\mathrm{Tr}(\mathbf{Z}_k)$ converges monotonically to rank $\mathbf{A}$.

Exercises 11–16 are from [14], [51].

10. Let $\mathbf{A} \in \mathbb{C}^{m \times n}$. Then $\mathbf{A} = \mathbf{M} - \mathbf{N}$ is a proper splitting if $R(\mathbf{A}) = R(\mathbf{M})$, $N(\mathbf{A}) = N(\mathbf{M})$. Let $\rho(\cdot)$ denote the spectral radius. In Exercises 11–16, $\mathbf{M}, \mathbf{N}$ is to be considered a proper splitting of $\mathbf{A}$.
11. Let $\mathbf{X}_{i+1} = \mathbf{M}^\dagger\mathbf{N}\mathbf{X}_i + \mathbf{M}^\dagger$ where $\mathbf{A} = \mathbf{M} - \mathbf{N}$ is a proper splitting of $\mathbf{A}$. Show $\mathbf{X}_{i+1} \to \mathbf{A}^\dagger$ if and only if $\rho(\mathbf{M}^\dagger\mathbf{N}) < 1$.
12. Show that if $\mathbf{M}_4^-$ is a (1, 4)-inverse for $\mathbf{M}$, then $(\mathbf{I} - \mathbf{M}_4^-\mathbf{N})^{-1}\mathbf{M}_4^-$ is well-defined and a (1, 4)-inverse for $\mathbf{A}$.
13. Show that if $\mathbf{M}_3^-$ is a (1, 3)-inverse for $\mathbf{M}$, then $(\mathbf{I} - \mathbf{M}_3^-\mathbf{N})^{-1}\mathbf{M}_3^-$ is well-defined and a (1, 3)-inverse for $\mathbf{A}$.
14. Show that $\mathbf{A}^\dagger = (\mathbf{I} - \mathbf{M}^\dagger\mathbf{N})^{-1}\mathbf{M}^\dagger$.
15. Show that if $\mathbf{M}^-$ is a (1, 3) or (1, 4)-inverse and $\mathbf{X}_{k+1} = \mathbf{M}^-\mathbf{N}\mathbf{X}_k + \mathbf{M}^-$, then $\mathbf{X}_{k+1}$ converges if and only if $\rho(\mathbf{M}^-\mathbf{N}) < 1$. If it converges it converges to a (1, 3) or (1, 4)-inverse, respectively,
16. Show that if $\mathbf{A}$ has full column rank, there exists a proper splitting of $\mathbf{A}$ such that $\mathbf{M}^\dagger = \mathbf{M}^*$.
17. Let $\mathbf{H} = \mathbf{I} - 2\mathbf{u}\mathbf{u}^*$ where $\mathbf{u} \in \mathbb{C}^n$ and $\|\mathbf{u}\|_2 = 1$. Show that $\mathbf{H} = \mathbf{H}^{-1}$.
18. Prove Proposition 12.3.1.
19. Prove Proposition 12.3.2.

*20. Prove Theorem 12.5.2 without assuming $\mathrm{Ind}(\mathbf{A}) \leq 1$. (See 'Continuity of the Drazin Inverse' by S. L. Campbell (to appear) for details.)

Bibliography

As mentioned in Chapter 0, reference [64] has an annotated 1775 item bibliography on generalized inverses. Accordingly, we have made no attempt to be complete. The references listed fall into three groups. Most are explicitly mentioned in the text. We have also referenced only that part of our work that appears in the book and was co-authored with others, principally N. J. Rose.

Finally, the idea of the Drazin inverse has recently proved useful in the study of singularly perturbed autonomous systems. This recent work, [19], [20], [22], [25], [26] and other related applications, [21], [58], [61] for example, were not included in the text due to page and time limitations, but are included in the references.

1 Athens, M. and Falb, P. L. *Optimal Control.* McGraw-Hill, New York, 1966.

2 Anderson, W. N. Jr. Shorted operators, *SIAM J. appl. Math.* **20**, 520–525, 1971.

3 Anderson, W. N. Jr. and Duffin, R. J. Series and parallel addition of matrices. *J. Math. Anal. Applic.* **26**, 576–594, 1969.

4 Anderson, W. N. Jr., Duffin, R. J. and Trapp, G. E. Matrix operations induced by network connections. *SIAM J. Control* **13**, 446–461, 1975.

5 Anderson, W. N. Jr. and Trapp, G. E. Shorted operators II. *SIAM J. appl. Math.* **28**, 60–71, 1975.

6 Bart, H., Kaashoek, M. A. and Lay, D. C. Relative inverses of meromorphic operator functions and associated holomorphic projection functions. *Math. Ann.* (to appear).

7 Ben-Israel, A. An iterative method for computing the generalized inverse of an arbitrary matrix, *Math. Comp.* **19**, 452–455, 1965.

8 Ben-Israel, A. A note on an iterative method for generalized inversion of matrices. *Math. Comp.* **20**, 439–440, 1966.

9 Ben-Israel, A. On error bounds for generalized inverses. *SIAM J. numer. Anal.* **3**, 585–592, 1966.

10 Ben-Israel, A. and Charnes, A. Generalized inverses and the Bott-Duffin network analysis, *J. math. Anal. Applic.* **7**, 428–435, 1963.
11 Ben-Israel, A. Linear equations and inequalities on finite dimensional, real or complex vector spaces: a unified theory. *J. math. Anal. Applic.* **27**, 367–389, 1969.
12 Ben-Israel, A. A note on partitioned matrices and equations. *SIAM Review* **11**, 247–250, 1969.
13 Ben-Israel, A. and Cohen, D. On iterative computation of generalized inverses and associated projections. *SIAM J. numer. Anal.* **3**, 410–419, 1966.
14 Berman, A. and Plemmons, R. J. Cones and iterative methods for best least-squares solutions of linear systems. *SIAM J. numer. Anal.* **11**, 145–154, 1974.
15 Businger, P. A. and Golub, G. H. Algorithm 358 singular value decomposition of a complex matrix. *Communications of ACM* **12**, 564–565, 1969.
16 Campbell, S. L. Differentiation of the Drazin inverse. *SIAM J. appl. Math.* **30**, 703–707, 1976.
17 Campbell, S. L. The Drazin inverse of an infinite matrix. *SIAM J. appl. Math.* **31**, 492–503, 1976.
18 Campbell, S. L. Linear systems of differential equations with singular coefficients. *SIAM J. math. Anal.* **8**, 1057–1066, 1977.
19 Campbell, S. L. On the limit of a product of matrix exponentials. *Linear multilinear Alg.* **6**, 55–59, 1978.
20 Campbell, S. L. Singular perturbation of autonomous linear systems II. *J. diff. Eqn.* **29**, 362–373, 1978.
21 Campbell, S. L. Limit behavior of solutions of singular difference equations. *Linear Alg. applic.* **23**, 167–178, 1979.
22 Campbell, S. L. Singular perturbation of autonomous linear systems IV (submitted).
23 Campbell, S. L. and Meyer, C. D. Jr. Recent applications of the Drazin inverse. In *Recent Applications of Generalized Inverses* M. Nashad, Ed. Pitman Pub. Co., London, 1979.
24 Campbell, S. L. Meyer, C. D. Jr. and Rose, N. J. Applications of the Drazin inverse to linear systems of differential equations. *SIAM J. appl. Math.* **31**, 411–425, 1976.
25 Campbell, S. L. and Rose, N. J. Singular perturbation of autonomous linear systems. *SIAM J. math. Anal.* **10**, 542–551, 1979.
26 Campbell, S. L. and Rose, N. J. Singular perturbation of autonomous linear systems III. *Houston J. Math.* **44**, 527–539, 1978.
27 Cederbaum, I. On equivalence of resistive n-port networks. *IEEE Trans. Circuit Theory*, Vol. CT-12, 338–344, 1965.
28 Cederbaum, I. and Lempel, A. Parallel connection of n-port networks. *IEEE Trans. Circuit Theory*, Vol. CT-14, 274–279, 1967.
29 Churchill, R. V. *Operational Mathematics.* McGraw-Hill, New York, 1958.

30 Cline, R. E. Representations for the generalized inverse of sums of matrices. *SIAM J. numer. Anal.* Series B, **2**, 99–114, 1965.

31 Cline, R. E. Representations for the generalized inverse of a partitioned matrix. *SIAM J. appl. Math.* **12**, 588–600, 1964.

32 Drazin, M. P. Pseudoinverses in associative rings and semigroups. *Amer. Math. Monthly* **65**, 506–514, 1968.

33 Faddeev, D. K. and Faddeeva, V. N. *Computational Methods of Linear Algebra*, (translated by Robert C. Williams). W. H. Freeman and Co., San Francisco, 1963.

34 Gantmacher, F. R. *The Theory of Matrices*, Volume II. Chelsea Publishing Company, New York, 1960.

35 Gallie, T. M. Calculation of the generalized inverse of a matrix, *Technical Report* CS-1975-7, Computer Science Department, Duke University.

36 Golub, G. and Kahan, W. Calculating the singular values and pseudo-inverse of a matrix. *SIAM J. numer. Anal.* Series B., **2**, 205–224, 1965.

37 Golub, G. H. and Pereyra, V. The differentiation of pseudo-inverses and non-linear least squares problems whose variables separate. *SIAM J. numer. Anal.* **10**, 413–432, 1973.

38 Golub, G. H. and Wilkinson, J. H. Ill conditioned eigensystems and the computation of the Jordan canonical form. *Technical Report*, STAN-CS-75-478.

39 Golub, G. H. and Reinsch, C. Singular value decomposition and least squares solutions. *Numer. Math.* **14**, 403–420, 1970.

40 Greville, T. N. E. Spectral generalized inverses of square matrices. *MRC Tech. Sum. Rep.* 823, Mathematics Research Center, University of Wisconsin, Madison, 1967.

41 Greville, T. N. E. The Souriau-Frame algorithm and the Drazin pseudoinverse. *Linear Alg. Applic.* **6**, 205–208, 1973.

42 Hakimi, S. L. and Manherz, R. K. The generalized inverse in network analysis and quadratic error-minimization problems. *IEEE Trans. Circuit Theory* Nov., 559–562, 1969.

43 Householder, A. S. *The Theory of Matrices in Numerical Analysis*. Blaisdell Publishing Co., New York, 1964.

44 Huelsman, L. P. *Circuits, Matrices, and Linear Vector Spaces*. McGraw-Hill, New York, 1963.

45 Jacobson, D. H. Totally singular quadratic minimization problems. *IEEE Trans. Automatic Cont.* **16**, 651–657, 1971.

46 Kemeny, J. G. and Snell, J. L. *Finite Markov Chains*. D. Van Nostrand Company, New York, 1960.

47 Kemeny, J. G. Snell, J. L. and Knapp, A. W. *Denumerable Markov Chains*. D., Van Nostrand Company, New York, 1966.

48 *Linear Inequalities and Related Systems* (Kuhn, H. W. and Tucker, A. W. Eds.), Princeton University Press, Princeton, N. J., 1956.

49 Lancaster, P. *Theory of Matrices*. Academic Press, New York, 1969.

50 Lay, D. C. Spectral properties of generalized inverses of linear

operators. *SIAM J. appl. Math.* **29**, 103–109, 1975.

51 Lawson, L. M. Computational methods for generalized inverse matrices arising from proper splittings. *Linear Alg. Applic.* **12**, 111–126, 1975.

52 Lawson, C. L. and Hanson, R. J. *Solving Least Squares Problems.* Prentice-Hall, New Jersey, 1974.

53 Meyer, C. D. Jr. Generalized inverses of triangular matrices. *SIAM J. appl. Math.* **18**, 401–406, 1970.

54 Meyer, C. D. Jr. Generalized inverses of block triangular matrices. *SIAM J. appl. Math.* **19**, 741–750, 1970.

55 Meyer, C. D. Jr. The Moore–Penrose inverse of a bordered matrix. *Linear Alg. Applic.* **5**, 375–381, 1972.

56 Meyer, C. D. Jr. Generalized inversion of modified matrices. *SIAM J. appl. Math.* **24**, 315–323, 1973.

57 Meyer, C. D. Jr. An alternative expression for the mean first passage matrix. *Linear Alg. Applic.* **22**, 41–47, 1978.

58 Meyer, C. D. Jr. and Plemmons, R. J. Convergent powers of a matrix with applications to iterative method for singular linear systems. *SIAM J. numer. Anal.* **14**, 699–705, 1977.

59 Meyer, C. D. Jr. and Rose, N. J. The index and the Drazin inverse of block triangular matrices. *SIAM J. appl. Math.* **33**, 1–7, 1976.

60 Meyer, C. D. Jr. and Shoaf, J. M. Updating finite Markov chains by using techniques of generalized matrix inversion. *J. Stat. Comp. and Simulation,* **11**, 163–181, 1980.

61 Meyer, C. D. Jr. and Stadelmaier, M. W. Singular M-matrices and inverse positivity. *Linear Alg. Applic.* **22**, 129–156, 1978.

62 Mihalyffy, L. An alternative representation of the generalized inverse of partitioned matrices, *Linear Alg. Applic.* **4**, 95–100, 1971.

63 Moore, R. H. and Nashed, M. Z. Approximations to generalized inverses. *University of Wis., MRC Tech. Summary Report* # 1294.

64 *Generalized Inverses and Applications* (Nashed, M. Z. Ed.), Academic Press, New York, 1976.

65 Noble, B. *Applied Linear Algebra.* Prentice-Hall, New Jersey, 1969.

66 Pearl, M. *Matrix Theory and Finite Mathematics.* McGraw-Hill, Inc., New York, 1973.

67 Penrose, R. A generalized inverse for matrices. *Proc. Cambridge Phil. Soc.* **51**, 406–413, 1955.

68 Penrose, R. On best approximate solutions of linear matrix equations. *Proc. Cambridge Phil. Soc.* **52**, 17–19, 1955.

69 Pyle, L. D. The generalized inverse in linear programming. Basic structure. *SIAM J. appl. Math.* **22**, 335–355, 1972.

70 Pyle, L. D. and Cline, R. E. The generalized inverse in linear programming—interior gradient projection methods. *SIAM J. appl. Math.* **24**, 511–534, 1973.

71 Rao, C. R. Some thoughts on regression and prediction, Part 1. *Sankhyā* **37**, Series C., 102–120, 1975.

72 Rao, J. V. V. Some more representations for the generalized inverse of a partitioned matrix. *SIAM J. appl. Math.* **24**, 272–276, 1973.

73 Rao, T. M. Subramanian, K. and Krishnamurthy, E. V. Residue arithmetic algorithms for exact computation of g-inverses of matrices. *SIAM J. numer. Anal.* **13**, 155–171, 1976.

74 Robert, P. On the group inverse of a linear transformation. *J. math. Anal. Applic.* **22**, 658–669, 1968.

75 Robertson, J. B. and Rosenberg, M. The decomposition of matrix valued measures. *Mich. math. J.* **15**, 353–368, 1968.

76 Rosenberg, M. Range decomposition and generalized inverse of non-negative hermitian matrices. *SIAM Review* **11**, 568–571, 1969.

77 Rose, N. J. A note on computing the Drazin inverse. *Linear Alg. Applic.* **15**, 95–98, 1976.

78 Rose, N. J. The Laurent expansion of a generalized resolvent with some applications. *SIAM J. appl. Math.* (to appear).

79 Schwerdtfeger, H. *Introduction to Linear Algebra and the Theory of Matrices*. P. Noordhoff, N. V., Groningen, Holland, 1961.

80 Scholnik, H. D. A new approach to linear programming, preliminary report.

81 Shinozaki, N., Sibuya, M. and Tanabe, K. Numerical algorithms for the Moore–Penrose inverse of a matrix: direct methods. *Ann. Inst. statist. Math.* **24**, 193–203, 1972.

82 Shinozaki, N., Sibuya, M. and Tanabe, K. Numerical algorithms for the Moore–Penrose inverse of a matrix: iterative methods. *Ann. Inst. statist. Math.* **24**, 621–629, 1972.

83 Shoaf, J. M. The Drazin inverse of a rank-one modification of a square matrix. Ph.D. Dissertation, North Carolina State University, 1975.

84 Simonnard, M. *Linear Programming*. Prentice-Hall, Inc., Englewood Cliffs, N. J., 1966.

85 Smythe, W. R. and Johnson, L. A. *Introduction to Linear Programming, with Applications*. Prentice-Hall, Inc., Englewood Cliffs, N. J., 1966.

86 Söderstrom, T. and Stewart, G. W. On the numerical properties of an iterative method for computing the Moore–Penrose generalized inverse. *SIAM J. numer. Anal.* **11**, 61–74, 1974.

87 Stallings, W. T. and Boullion, T. L. Computation of pseudo-inverse matrices using residue arithmetic. *SIAM Review* **14**, 152–537, 1972.

88 Stewart, G. W. On the continuity of the generalized inverse. *SIAM J. appl. Math.* **17**, 33–45, 1968.

89 Stewart, G. W. On the perturbation of pseudo-inverses, projections and linear least squares problems. *SIAM Review*, **19**, 634–662, 1977.

90 Szabo, S. and Tanaka, R. *Residue Arithmetic and Its Application to Computer Technology*. McGraw-Hill, New York, 1967.

91 Wedin, P.-A. Perturbation bounds in connection with singular value decomposition. *Nordisk Tidsskrift for Information Behandling* **12**, 99–111, 1972.

92 Young, D. M. and Gregory, R. T. *A Survey of Numerical Mathematics*, Vol. 2, Addision-Wesley, Reading, Mass. 1973.

Index

A CATALOG OF SELECTED

DOVER BOOKS

IN SCIENCE AND MATHEMATICS

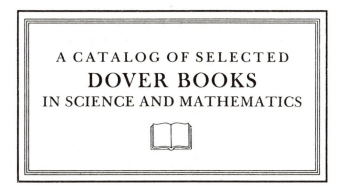